Plant Genomics and Biotechnology

Plant Genomics and Biotechnology

Edited by **Isabelle Nickel**

SYRAWOOD
PUBLISHING HOUSE
New York

Published by Syrawood Publishing House,
750 Third Avenue, 9th Floor,
New York, NY 10017, USA
www.syrawoodpublishinghouse.com

Plant Genomics and Biotechnology
Edited by Isabelle Nickel

International Standard Book Number: 978-1-68286-327-5 (Hardback)

Printed in China.

Contents

Preface

This book aims to highlight the current researches and provides a platform to further the scope of innovations in this area. This book is a product of the combined efforts of many researchers and scientists from different parts of the world. The objective of this book is to provide the readers with the latest information in the field.

There has been a revolution in the field of agriculture in the last few decades which has evolved the field of plant genomics and related techniques and processes of biotechnology as well. This book comprises advanced research in this field. Most of the topics introduced in this book cover new concepts, techniques and the applications of plant genomics such as DNA recombination, genetically modified crops, genetic sequencing and mapping, genomic structures, etc. The various sub-fields of plant genomics along with technological progress that have future implications are glanced at in this book. The book is appropriate for students seeking detailed information in this area as well as for experts.

I would like to express my sincere thanks to the authors for their dedicated efforts in the completion of this book. I acknowledge the efforts of the publisher for providing constant support. Lastly, I would like to thank my family for their support in all academic endeavors.

Editor

Potential functions of microRNAs in starch metabolism and development revealed by miRNA transcriptome profiling of cassava cultivars and their wild progenitor

Xin Chen[1,2], Jing Xia[3,4], Zhiqiang Xia[1,2], Hefang Zhang[1,2], Changying Zeng[1,2], Cheng Lu[1,2], Weixiong Zhang[3,4] and Wenquan Wang[1,2*]

Abstract

Background: MicroRNAs (miRNAs) are small (approximately 21 nucleotide) non-coding RNAs that are key post-transcriptional gene regulators in eukaryotic organisms. More than 100 cassava miRNAs have been identified in a conservation analysis and a repertoire of cassava miRNAs have also been characterised by next-generation sequencing (NGS) in recent studies. Here, using NGS, we profiled small non-coding RNAs and mRNA genes in two cassava cultivars and their wild progenitor to identify and characterise miRNAs that are potentially involved in plant growth and starch biosynthesis.

Results: Six small RNA and six mRNA libraries from leaves and roots of the two cultivars, KU50 and Arg7, and their wild progenitor, W14, were subjected to NGS. Analysis of the sequencing data revealed 29 conserved miRNA families and 33 new miRNA families. Together, these miRNAs potentially targeted a total of 360 putative target genes. Whereas 16 miRNA families were highly expressed in cultivar leaves, another 13 miRNA families were highly expressed in storage roots of cultivars. Co-expression analysis revealed that the expression level of some targets had negative relationship with their corresponding miRNAs in storage roots and leaves; these targets included *MYB33*, *ARF10*, *GRF1*, *RD19*, *APL2*, *NF-YA3* and *SPL2*, which are known to be involved in plant development, starch biosynthesis and response to environmental stimuli.

Conclusion: The identified miRNAs, target mRNAs and target gene ontology annotation all shed light on the possible functions of miRNAs in Manihot species. The differential expression of miRNAs between cultivars and their wild progenitor, together with our analysis of GO annotation and confirmation of miRNA: target pairs, might provide insight into know the differences between wild progenitor and cultivated cassava.

Keywords: MicroRNA, Target Gene, Wild Progenitor, Cassava (*Manihot esculenta* Crantz)

Background

MicroRNAs (miRNAs) are small (20–25 nucleotides) non-coding RNAs that have emerged as key players in post-transcriptional gene regulation in plants. They are generated from single-strand RNA precursors that are folded into stem-loop structures, and their abilities to bind to complementary sequences of target mRNAs results in cleavage or degradation of the target mRNAs or suppression of their translation [1-3]. Many studies have revealed important roles of miRNAs in development [4-7], adaptation to biotic and abiotic stresses and resistance to pathogen infection [8-10]. Many miRNAs are also specifically expressed during different stages of plant development and in specific plant organs or tissues [11-15]. For example, miR319 regulates transcription factors of the *TCP* family, which regulate multiple biological pathways, including hormone biosynthesis and signalling for cell proliferation and differentiation [4]; a set of miRNAs that affect plant hormone homeostasis

* Correspondence: wangwenquan@itbb.org.cn
[1]The Institute of Tropical Bioscience and Biotechnology (ITBB), Chinese Academy of Tropical Agricultural Sciences (CATAS), Haikou 571101, PR China
[2]Key Laboratory of Biology and Genetic Resources of Tropical Crops, Ministry of Agriculture, Haikou 571101, PR China
Full list of author information is available at the end of the article

and starch accumulation during grain filling in rice has also been identified [6].

Starch is an insoluble polymer of glucose residues produced by the majority of higher plant species, and is a major storage product of many of the seeds and storage organs produced agriculturally and used for human consumption [16]. Several studies reported that the targets of miRNA involved in the metabolism of carbon, sucrose, starch, lipid, etc. in switch grass [17], potato [18] and rice [6]. For example, miRn45-5p targeted *SPS* (sucrose-phosphate synthase) gene, miR1436 and miR1862 targeted *SS* (starch synthase) gene in rice. Cassava (*Manihot esculenta* Crantz), a woody shrub of the Euphorbiaceae, is one of the most important food crops in the world. It is remarkably productive in terms of its capacity to accumulate biomass and starch, and exhibits extraordinary environmental adaptability. Although about 169 miRNAs and 68 miRNA families have been predicted and characterised in cassava by both a computational approach [19] and small RNA sequencing [20], their potential roles in the production of biomass and starch have never been reported.

We are interested in identifying miRNAs and understanding miRNA functions in photosynthesis and starch accumulation in leaves and storage roots of cassava. Here, we addressed the issues of variation at the level of gene (mRNA) expression and regulation of expressed genes by miRNA in two plant organs (leaf and storage root). The two cassava cultivars (cv. KU50 and Arg7) and their wild progenitor (W14), which was deposited in Chinese Cassava Germplasm Garden and used in this study, have contrasting phenotypes in terms of photosynthetic capacity, starch accumulation and yield of storage roots (Table 1).

Results

Identification of miRNAs in cultivars and their wild progenitor species

Six small RNA samples from leaves and storage roots of wild progenitor W14 and cultivars KU50 and Arg7 were sequenced using an Illumina Hi-Seq2000 instrument. New miRNAs were identified in cassava in our recent studies. We identified miRNAs that did not have greater than 70% homology with any other miRNAs in any of the searchable databases, and designated these as new miRNAs. Briefly, we incorporated the sequencing data to conduct a genome-wide search for putative loci with miRNA signatures (see details in Methods), and the expression of the miRNAs detected was quantified based on the genome sequence of AM560 [21], as well as annotation in miRBase (www.mirbase.org). A total of 62 miRNA families were detected, including 19 previously reported ones [22] (Table 2). Among these 62 miRNA families, the presence/absence of members in 51 families was conserved across the three genotypes. Ten new families, i.e., new-1, −4, −7, −12 to −16, −27 and −28, were expressed in two cultivars but not in W14. The majority of the detected miRNAs were transcriptionally active in all of the three genotypes, as determined by a normalised number of reads (NNRs). For example, new-11 had more than one thousands of reads in all sequencing libraries. In total, 107 conserved miRNAs belonging to 29 annotated families and 39 new miRNAs from 33 families had detectable expression on the basis of the sequencing data (Additional file 1). Moreover, to confirm the above findings, the expression levels of 36 miRNA families (24 conserved and 12 new) were validated by reverse transcription polymerase chain reaction (Additional file 2).

Targets of cassava miRNAs

To get insight into or predict the functions of conserved and new miRNAs, a miRNA target search was performed to identify their putative targets (see Methods). A total of 360 loci on cassava unigenes were predicted to be targets of 26 conserved and 27 new miRNA families (Additional file 3), and the remaining 10 miRNA families had no predicted target. The number of targets of each miRNA family ranged from one to 25 (with an average of 9.4) for conserved miRNAs, and from one to seven (with an average of 2.9) for

Table 1 General characteristics of cultivar KU50, Arg7 and wild ancestor W14 used for mining and expression analysis of miRNAs and their target genes

Characteristic	KU50/Arg7	W14
Latin name	*M. esculenta* Crantz	*M. esculenta. ssp. flabellifolia*
Code in Chinese cassava germplasm garden	MS000168/MS000580	MS000581
Collection site and time	Thailand , 2002/CIAT, Columbia, 2007-01-20	CIAT, Columbia, 2004-07-30
Number of fruits	few	many
Propagation method	stems	seeds
Photosynthetic efficiency (μmol/m^2/s)	15.9–38.7	14.6–24.2
Storage root yield (kg/plant/yr)	3.0–10.0	0.5–1.0
Starch content of root (%)	28.0–32.0	3.0–5.0

Table 2 Discovery of miRNA families and members in wild progenitor W14 and cassava cultivars KU50 and Arg7

miRNA family	Members					
	W14		Arg7		KU50	
	Leaf	Root	Leaf	Root	Leaf	Root
156	11	11	11	11	9	9
159, *162*, 393, *403*, *408*, 2111	2	2	2	2	2	2
160	7	7	7	7	4	4
164	3	4	4	4	3	4
166	8	8	8	8	8	8
167	6	6	6	6	5	5
168, 391, *397*, *398*, 530, 827, 2950	1	1	1	1	1	1
169	5	5	5	5	4	4
171	9	9	9	9	9	9
172	0	2	0	2	0	2
319	5	6	6	6	6	6
390	3	3	3	3	2	2
394	3	3	3	3	3	3
395	4	4	4	4	4	4
396	4	4	4	4	3	3
399	7	7	8	7	4	3
477	5	5	5	5	4	4
535	3	3	3	3	1	1
New-1, −4, −7, −27, −28	0	0	1	1	1	1
New-3, −25, −34	1	1	1	1	0	1
New-5, −6, −26	2	2	2	2	2	2
New-8 to −11, −22 to −24, −30 to −32	1	1	1	1	1	1
New-12	0	0	1	0	1	1
New-13 to −16	0	0	1	1	1	1
New-17, −33	1	0	1	1	1	1
New-18	0	1	1	1	1	0
New-19	0	1	1	1	0	0
New-20	4	4	4	4	4	4
New-21, −29	0	1	1	1	1	1

Whereas "0" indicates that no miRNA is detected in the corresponding genotype/organ, a number from "1" to "11" indicates the number of detected members of the miRNA family; new miRNAs have names in the format of "new-#", for example, new-1.
Those miRNAs with names that are underlined were previously reported by Zeng et al. [22].

new miRNAs; the target number varied even within a miRNA family. Among conserved miRNAs, miR162 had one target, whereas six miRNA families (miR156, 164, 172, 319, 396 and 397) had more than 10 targets; among new miRNAs, five miRNA families (new-9, −10, −20, −22 and −26) had one target and four miRNA families (new-1, −7, −18 and −19) had more than five. Eight genes were targeted by at least two miRNAs; for example, MYB33 and MYB81 were targeted by miR159 and miR319, and SPL2 and SPL13b were targeted by miR156 and miR535. Many of these targets encode transcription factors; for example, they include the *SPL* (target of miR156 and miR535), *MYB* (miR159, miR319 and new-18), *HAM3* (miR171), *NAC* (miR164), *ARF* (miR160, miR167 and miR169), *TCP* (miR319), *GRF* (miR396 and miR477) and *SIGMA* (new-11) genes. Other miRNA targets that are functional genes included genes that encode the AGPase large subunit 2 (miR394), an F-box protein (miR394), RD19 (miR167), a member of the zinc finger family (miR172), *IRX*12 (miR397) and two disease resistance proteins (*NBS-LRR* class; miR396 and −30).

Most miRNAs were highly expressed in the leaves and storage roots of cultivars

The NNR for each miRNA should be no less than 50, and if the NNR of a miRNA is less than 50 then it is considered to be silent (not expressed at a detectable level). Whereas 34 miRNAs (24 families) displayed a two-fold difference in their expression level between the cultivars and W14 (Table 3), 29 miRNAs (18 families) were highly expressed in the leaves of cultivars, including miR169e, miR2950, miR319cd, miR391, miR393ab, miR394abc, miR395abcd, miR399d, miR535b, new-4, new-6ab, new-12, new-13, new-15, new-17, new-28, new-29 and new-32. On the other hand, six miRNA families were more highly expressed in leaves of W14 than in cultivars, including miR156ijk, miR396c, miR477e, new-11, new-30 and new-34.

Furthermore, a total of 44 miRNAs (23 families) were differentially expressed in storage roots between cultivars and the wild progenitor. Thirteen miRNA families were highly expressed in storage root of cultivars; these included miR156ijk, miR167acdef, miR169e, miR2111ab, miR397, miR399d, miR477cd, new-6ab, new-9, new-15, new-17, new-28 and new-29. Meanwhile, ten miRNA families were more highly expressed in the storage roots of wild progenitor W14; these included miR168, miR171abcdefghi, miR2950, miR393ab, miR396c, new-3, new-8, new-11 and new-34 (Table 3).

Confirmation of miRNA:target pairs by RLM-RACE

Binding of a miRNA to the complementary sequence in target mRNA leads to mRNA cleavage. The cleavage sites located at its complementary region (CR) are direct evidence of RNA-induced silencing complex (RISC)-mediated slicing of the target mRNA. Fifteen miRNA:target pairs were selected for slicing analysis using RNA ligase-mediated rapid amplification of the 5' cDNA ends technique (RLM-RACE). Of these, nine miRNAs (related to 12 pairs) exhibited differential expression between the cultivars

Table 3 miRNAs that are differentially expressed between cultivars and their wild progenitor

Organ	Highly expressed in cultivars	Highly expressed in the wild progenitor
Leaf	miR169e, miR2950, miR319cd, miR391, miR393ab, miR394abc, miR395abcd, miR399d, miR535b, new-4, new-6ab, new-12, new-13, new-15, new-17, new-28, new-29, new-32	miR156ijk, miR396c, miR477e, new-11, new-30, new-34
Root	miR156ijk, miR167acdef, miR169e, miR2111ab, miR397, miR399d, miR477cd, new-6ab, new-9, new-15, new-17, new-28, new-29	miR168, miR171abcdefghi, miR2950, miR393ab, miR396c, new-3, new-8, new-11, new-34

and W14. The corresponding primer sets are listed in Additional file 4. Cloning and sequencing of the PCR amplicons of remnant mRNAs enabled determination of the nucleotide position when a slicing event occurred. Finally, slicing events occurred in 14 of 15 miRNA:target pairs. The cleavage sites of eight pairs were at the 10th, 11th and 12th nucleotides in the CR (Table 4 and Figure 1). miR394 sliced two targets at different sites with different efficiencies: two slicing events at the 10th nucleotide in the CR and seven events downstream of the CR for *F-box* mRNA; there were also four slicing events at the 2nd and 14th nucleotides in CR and 14 slicing events upstream or downstream of CR occurred for ADP-glucose pyrophosphorylase large subunit 2 (*APL2*) mRNA.

Similarly, other miRNAs sliced their targets at the CR. For instance, miR160 sliced *ARF10*, with ten cleavage sites at the 11th nucleotide in the CR, and miR169 sliced *NF-YA3*, with six cleavage events at the CR. In addition, miR396 cleaved its target *GRF1* at the classic 11th site. For the remaining five pairs, cleavage sites could not be detected at the CR region, whereas they were detected up/downstream of this region. The cleavage sites of miR398:*Dir-like* and miR477:*CLB* were at the +8th and +37th nucleotides upstream of the CR, respectively. Finally, the cleavage sites of miR166:*GH17* and new-12: *EDR2* were all downstream of their CRs. Taken together, these findings indicated that many cleavage sites were not positioned at the CR or were even far away from it. Accordingly, these remnant mRNAs might have been randomly degraded mRNAs.

Co-expression analysis of the miRNAs and their targets in leaf and storage root of cassava cultivars and their wild progenitor

To further study the relationship between miRNAs and their targets, RNA-seq transcription profiling was performed to assay the expression of targets in leaf and storage root among these three *Manihot* genotypes. Co-expression analysis involved 32 miRNA families highly expressed in leaf or root (61 miRNAs) and their 87 corresponding targets. In general, the expression levels of 28 targets were negatively correlated with that of their corresponding miRNAs, which included 12 transcription factors, three plant hormone-related genes and one *SUT*. However, 22 targets showed a positive correlation with their corresponding miRNAs, including eight transcription factors, four plant hormone-related genes, one *APL2* and one *CTT* (Figure 2 and Additional file 5). In total, there were 27 transcription factors and plant hormone-related genes, and three starch biosynthesis- and sugar transport-related genes among these target genes, this might be explained by translation suppression or feedback regulation. Surprisingly, some of the miRNAs slicing their targets were confirmed, but being positively correlated with the targets; these miRNA:target pairs included miR156: *SPL13b*, miR169:*NF-YA3* and miR394:*APL2/F-box*.

On the basis of the miRNA:target pair confirmation and the co-expression analysis, 13 miRNA:target pairs were chosen for further assays of their regulatory relationship by quantitative real time PCR (qRT-PCR) in Arg7 storage roots at different growth stages; the corresponding primers are listed in Additional files 2 and 6. From the results, eight miRNAs and their corresponding targets had obvious negative correlations, except for miR160b; most of these miRNAs were expressed during the later stage of growth of the Arg7 root (Figure 3). For example, miR394 targets both cassava4.1_021267m (*APL2*) and cassava4.1_000867m (sucrose phosphate synthase 2F, *SPS2F*), which are involved in starch biosynthesis. *APL2* showed low expression from 120 DAP (day after planting) to 180 DAP, increased from 180 DAP and reached a peak at 210 DAP; it then fell close to its minimum at 240 DAP. In contrast, an increase in levels of miR394 at 180 DAP, followed by a sustained increase from 210 DAP to 240 DAP, suggested that *APL2* was negatively regulated by miR394 at a later stage. The second target, *SPS2F*, maintained a relatively stable expression level at the early stage, but then fell close to zero from 180 DAP to 240 DAP; this suggested that miR394 also negatively regulated *SPS2F* (Figure 3G). However, given that there was no negative/positive correlation between miR394 and *APL2* on the basis of the RNA-seq data in roots among the three *Manihot* genotypes, the regulation might only have existed in a specific growth stage of cassava storage root.

Discussion

The initial findings about miRNA in cassava were obtained with the aid of the castor bean genome information; specifically, 20 conserved miRNA families were reported [22], and then 17 conserved miRNA families were predicted in

Table 4 Confirmation of miRNA:target pairs by RLM-RACE

miRNA	Target	Annotation	Cleavage sites		
			Upstream	Complementary region	Downstream
miR156/miR535	cassava4.1_006419m	*SPL2*	/	10th/11th(8), 11th/12th(3)	−35th/−36th(1)
	cassava4.1_009657m	*SPL13b*	/	9th/10th(2), 10th /11th(4), 11th/12th(1)	−1st/−2nd (1)
miR159/miR319	cassava4.1_004606m	*MYB33*	/	11th/12th(10)	−52th/−53th(1),−115th/ −116th(1)
	cassava4.1_030321m	*MYB81*	/	11th/12th(6)	−59th /−60 th(1)
miR394	cassava4.1_021267m	*APL2*	+18th(1), +35th(3), +42th(1), +69th(2)	2nd(3), 14th(1)	−27th(1), −29th(1), −49th(1), −92th(1), −105th(1), −109th(1), −139th(1)
	cassava4.1_007038m	*F-box*	/	10th(2)	−31th(1), −52th(1), −55th(1), −59th(1), −68th(1), −75th(1), −102th(1)
miR160	cassava4.1_002668m	*ARF10*	/	11 th(10)	/
miR167	cassava4.1_009942m	*RD19*	+5th(1), +25th(1), +145th(1), +196th(1)		−41th(1)
miR169	cassava4.1_011576m	*NF-YA3*	+30th(1), +6th(1)	11th(5), 15th(1)	/
miR396	cassava4.1_003731m	*GRF1*	+69th(2)	6th(1), 11th(6)	−45th(1), −60th(1), −188th(1)
miR398	cassava4.1_024493m	*DIR-like*	+8th(10)	/	/
miR477	cassava4.1_008074m	*CLB*	+37th(8)	/	/
miR166	cassava4.1_009671m	*GH17*	/	/	−8th(3), −16th(6), −34th(1)
new-21	cassava4.1_006393m	*EDR2*	/	/	−112th(6), −160th(2), −176th(1), −177th(1), −178th(1), −182th(1)

Note: As an example, 10th/11th (8) indicate that there were eight cleavage sites at the 10th or 11th nucleotide in the complementary region of cassava4.1_006419m and miR156/535; alternatively, −35th/−36th (1) indicates that there was one cleavage site downstream of the complementary region of cassava4.1_006419m and miR156/535.

cassava by using *Arabidopsis* mature miRNAs as seed sequences [23]. Recently, 169 potential conserved miRNAs in cassava were identified by a computational approach, and 126 miRNAs (114 conserved and 12 new) were discovered by small RNA sequencing [19,20]. In this study, we discovered 107 conserved miRNAs (29 families) and 39 new miRNAs (33 families) by small RNA sequencing, and 36 of these families were detected again by RT-PCR. By miRNA transcriptome profiling, we identified 41 conserved miRNAs (17 families) and 19 new miRNAs

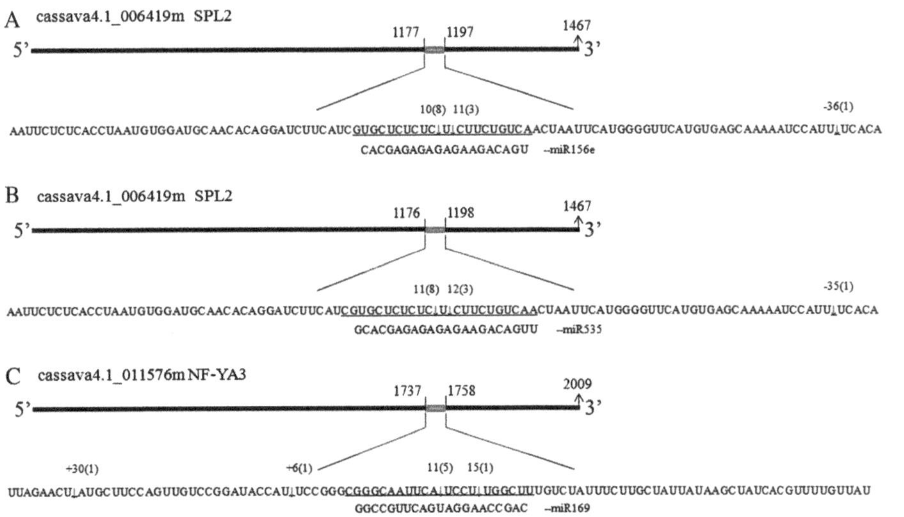

Figure 1 Identification of miRNA-guided cleavage products of target genes in cassava (partial results). The cleavage sites of selected targets as identified by 5′ RACE analysis. For each miRNA, the target sequence is shown on the top and the miRNA sequence on the bottom. The numbers indicate the fraction of cloned PCR products when PCR was terminated at different positions. **(A)** The site of cleavage of cassava4.1_006419m by miR156. **(B)** The site of cleavage of cassava4.1_006419m by miR535. **(C)** The site of cleavage of cassava4.1_011576m by miR169d.

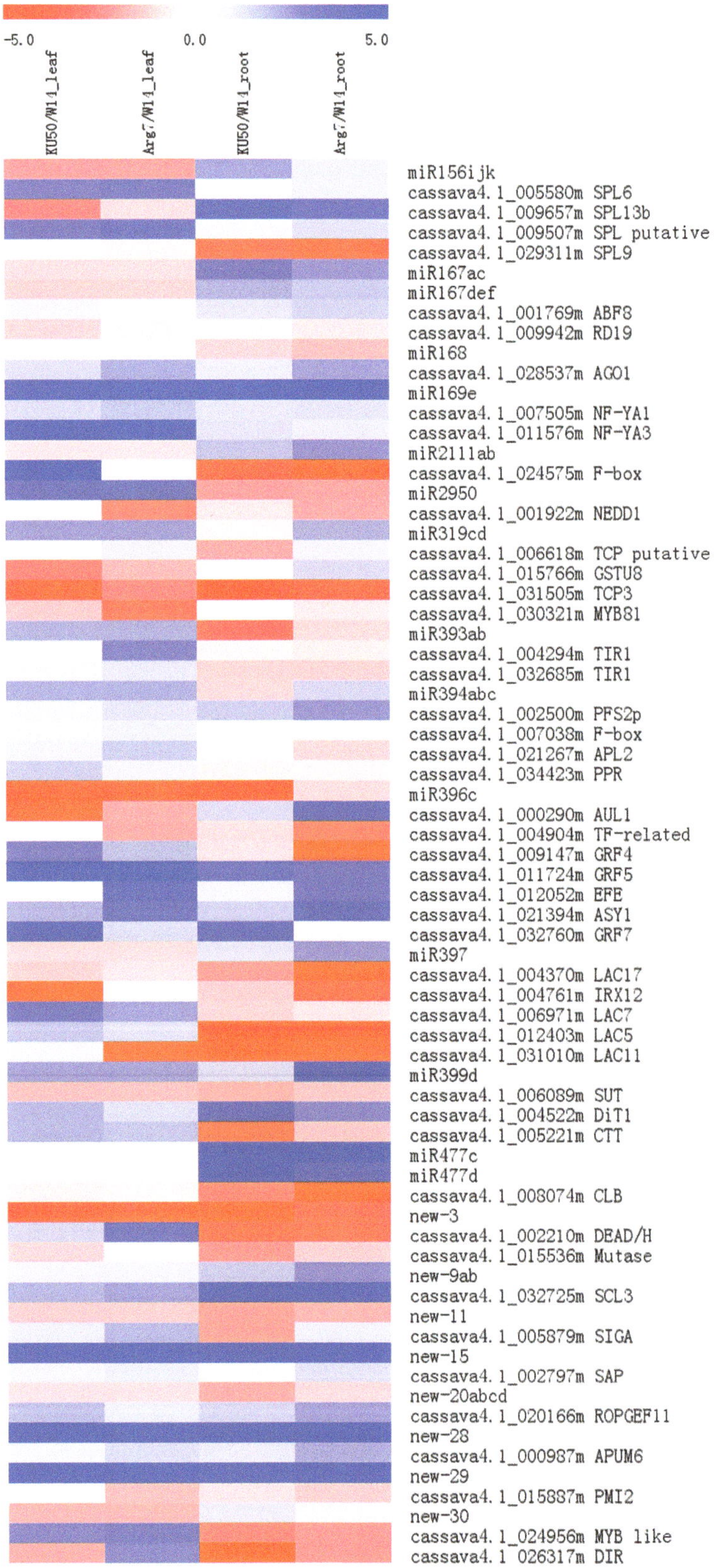

Figure 2 Correlations of the levels of expression of miRNAs and their corresponding targets in leaf and storage root between cultivars and their wild progenitor. Heat mapping was performed based on the log2- normalized expression ratio of cultivar/wild progenitor.

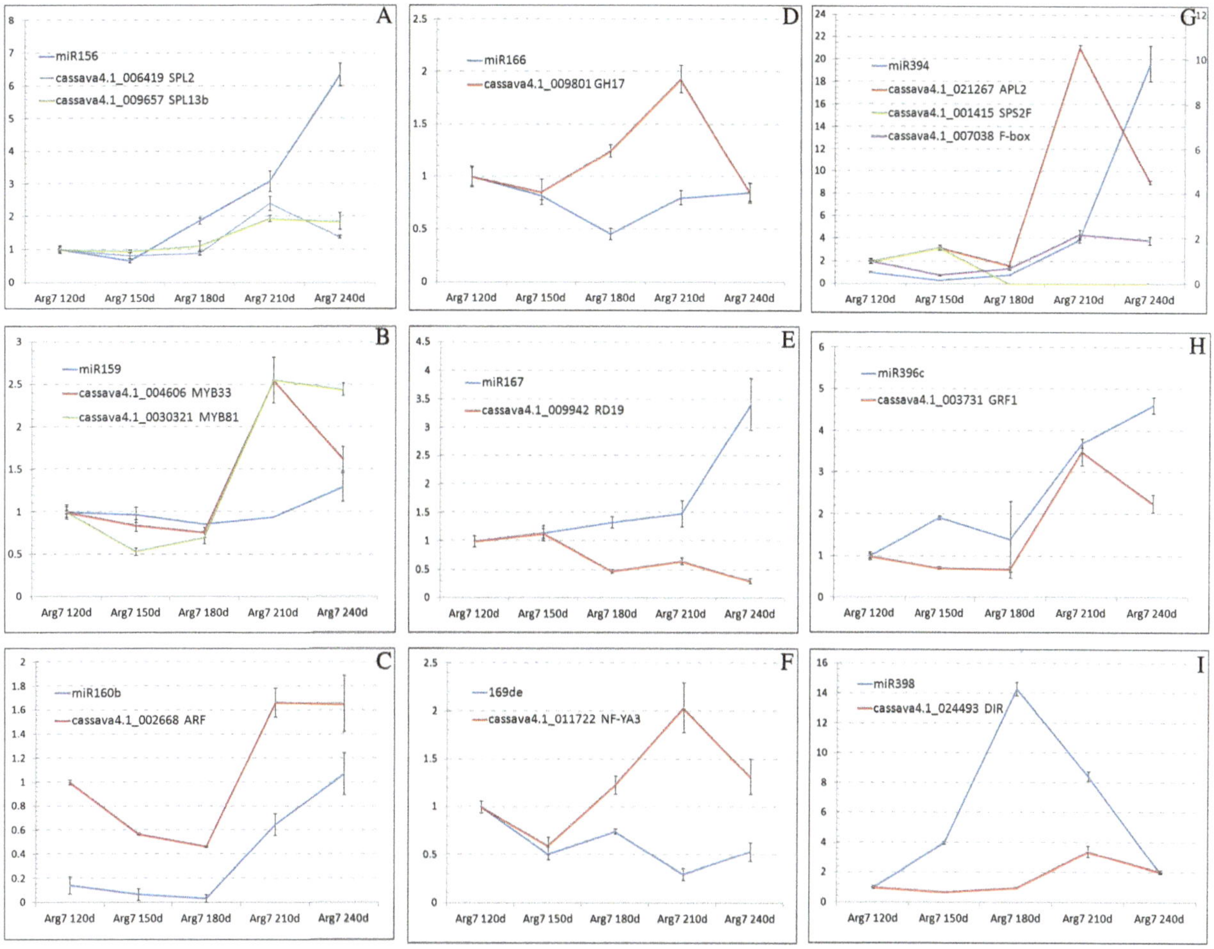

Figure 3 Expression correlations of miRNAs and their targets in storage root of cultivar Arg7. Quantification of the relative expression of miRNAs and their targets was carried out using the $^{\Delta\Delta}$CT method, with the U6 gene and *beta*-actin gene as references for miRNAs and targeted genes, respectively, and Arg7 120d as the control; Arg7 120d indicated that the storage root of Arg7 was sampled 120 days after planting, the other samples were labeled accordingly; the Y-axis means the times at which the levels of expression of miRNAs and their targets in other samples were comparable to those in Arg7 120d. **A-I**: means different miRNAs and their targets.

(15 families) that showed differential expression between the cultivars and their wild progenitor. Recently, Xia et al. also reported 22 cassava new miRNAs, and part of them responded to chilling stress [24].

We predicted that 360 cassava unigenes were targeted by 26 conserved and 27 new miRNA families, and that 57 of them (including *MYB, SPL, ARF, NAC* and *TCP*) encode transcription factors; several similar results have already been reported in cassava [19,20] and other species [5,25,26]. Several miRNAs are involved in starch accumulation in rice [6], and five of them (miR159, miR160, miR164, miR167 and miR319) also had the same targets in cassava. In addition, two other miRNAs (miR394 and miR399) targeted *APL2* and three sugar and carbohydrate metabolism-related genes (sugar transporter, invertase and carbohydrate transmembrane transporter), which have not been reported in other species. These miRNAs might be key regulators of starch biosynthesis in cassava.

A previous study reported that miR159 regulated *MYB* mRNAs, and miR319 predominantly acted on *TCP* mRNAs in *Arabidopsis* [27]. Both miR159 and miR319 shared sequence identity at 20 of 22 nucleotides, and the expression level of miR159 was far greater than that of miR319 in two cultivars and the wild progenitor. Therefore, similar to the case in *Arabidopsis*, MYB transcription factors were mainly regulated by miR159 in cassava. MYB could function as a transcriptional activator in ABA-inducible gene expression in *Arabidopsis* [28,29], and high concentrations of ABA could suppress the expression of starch synthesis genes in maize and rice [30,31]. It was inferred that miR159 could directly or indirectly affect starch biosynthesis in cassava.

miR396 targeted 20 cassava unigenes, including five *GRF* transcription factors. *MeGRF1* (cassava4.1_003731) was verified to be sliced by miR396 in cassava. Previous studies reported that high expression level of miR396 in root tips might result in reduced expression of six *MtGRF* genes [32], and that the miR396-*GRF1/GRF3* regulatory module acted as a developmental regulator in the reprogramming of root cells during cyst nematode infection in *Arabidopsis* [33]. These findings suggested that miR396 might regulate root development in cassava.

miR169 targeted *NF-YA* family genes, and over-expression of *NF-YA5* and *NF-YA3* or down-regulation of miR169 might enhance drought stress tolerance in *Arabidopsis* and soybean [34,35]. *MeNF-YA3* was negatively regulated by miR169, and the expression level of miR169 in the wild progenitor was lower than that in cultivars, and the evidence was that the wild progenitor had stronger drought tolerance than the cultivars usually.

miR398 targeted the mRNA of a disease resistance-responsive family protein (cassava4.1_024493, *Dir-like*) in cassava, but the cleavage sites of the miR398:*Dir-like* pair were not positioned within the CR: ten cleavage sites were all at the 8th nucleotide upstream of the CR. Given that the miRNA-guided cleavage occurred quite precisely at the 10th or 11th nucleotide from the 5' end of the miRNA in CR [24,36,37], this might be a surprising phenomenon that is difficult to explain.

Conclusion

Using next-generation sequencing technology, we carried out miRNA transcriptome and transcriptome profiling of two cultivated cassava and their wild progenitor. A total of 107 conserved miRNAs (29 families) and 39 new miRNAs (33 families) were identified, and most miRNAs were highly expressed in the cultivars. Of the 360 unigenes predicted to be the targets of 53 cassava miRNA families, 14 unigenes were confirmed. In addition, co-expression analysis between miRNAs and their targets was performed on the basis of the miRNA transcriptome and transcriptome profiling of leaves and storage roots; the expression levels of 28 targets were negatively correlated with that of their corresponding miRNAs. In conclusion, the differential expression of miRNAs between cultivars and their wild progenitor, together with our analysis of GO annotation and confirmation of miRNA:target pairs, might provide insight into how the wild progenitor was domesticated to cultivated cassava.

Methods

Plant materials

Two cultivars of the cultivated species *Manihot esculenta* Crantz (KU50 and Arg7) and W14, a subspecies of *Manihot esculenta spp. flabellifolia*, were used in this study. Both KU50 (a cultivar that is extensively planted in South East Asia) and Arg7 (a cultivar from Argentina) presented with higher photosynthesis and higher storage-root yield and starch content of storage roots. This distinguished them from W14, a native of Central Brazil, which had a lower rate of photosynthesis and very low storage root yield and starch content of the storage root. *Manihot esculenta spp. flabellifolia* was previously proposed to be the progenitor of cultivated cassava [38-40]. All three genotypes were grown in an experiment field in Haikou, China. Leaves and roots of these three genotypes were sampled at 150 DAP for sequencing of small RNAs and characterisation by RNA-seq. The roots of Arg7 were sampled on 120 DAP, 180 DAP, 210 DAP and 240 DAP for expression profile analysis of miRNA and targets by real-time PCR, and miRNA-target pair confirmation by RLM-RACE.

Small RNA extraction and Solexa sequencing

Small RNA samples of leaves and roots from the above three *Manihot* genotypes were extracted by using an miRNA isolation kit (Bioteke, Beijing, China) in accordance with the manufacturer's instructions. Small RNAs of fewer than 30 bases were isolated from these miRNA samples, and linked with a pair of Solexa adaptors to their 3' and 5' ends; then, the sample was reverse-transcribed into cDNA and amplified using the adaptor primers. The double-stranded miR-cDNAs were sequenced using Illumina's Solexa Sequencer in accordance with the manufacturer's instructions (BGI Company, Shenzhen, China).

RNA extraction and sequencing

Total RNA was extracted from leaves and storage roots using RNAplant reagent (Tiangen, Beijing, China) and purified using RNeasy Plant Mini Kit (Qiagen, Valencia, CA). The cDNA libraries for analysis using a Illumina Hiseq2000 instrument were prepared by following the protocol of Zhong et al. [41]. Six cDNA libraries of leaf and storage root were sequenced, and the sequenced reads were aligned to the cassava genome draft (http://www.phytozome.net/cassava) using TopHat and Cufflinks [42] and annotated using KEGG [43]. The fragments per kilobase per million reads (FPKMs) were used to normalise gene expression counts for each transcript. Transcripts with FPKM <3 were considered to be so rare as to not be expressed at all, as suggested for a study on white lupin [44].

Identification of conserved and new miRNAs in wild and cultivated species

1) Conserved miRNA
   Mature plant miRNA sequences from miRBase (http://www.mirbase.org/) were aligned to the AM560 genome (http://www.phytozome.net/cassava). We retrieved the flanking genomic

sequences around completely matched loci, with different upstream and downstream lengths, to form possible precursors of candidate miRNAs with the RNAfold program [45]. We chose those sequences with folding structures that have at least 18 bp in matched regions, one central loop and a folding energy ≤ −18 kcal/mol. The free tails in the secondary structures were then removed. Next, we applied the MiRcheck program [46] to select sequences that have ≤4 mismatches, ≤2 bulged or asymmetrically unpaired nucleotides and ≤2 continuous mismatches in the seed regions.

2) New miRNA.

We searched for new miRNAs in the three *Manihot* genotypes (namely, W14, Arg7 and KU50), using the corresponding small RNA-seq datasets. The newly identified miRNAs, combined with known miRNAs in *Manihot*, were then subjected to a homology search. We aligned mature and hairpin sequences of an miRNA to the cultivar-AM560 genome using the local alignment tool BLAST. We set the p-value obtained from BLAST to less than 1e-10 and manually examined the alignment to determine whether a BLAST hit was homologous to the input miRNA. We mapped the qualified reads from corresponding cultivar datasets to the identified homologous sequence using Bowtie [47] and counted the mappable reads. If no homologous sequences could be identified in a cultivar genome assembly, we then mapped reads from the sequencing datasets of the same cultivar to the input miRNA sequence, allowing two mismatches. We considered a miRNA to be not conserved in the cultivar genome assembly if both of the following criteria were met: 1) no homologous sequences identified and 2) insufficient mappable reads (fewer than 10 normalised reads) for analysis. If any of the reads mapped to the input miRNA sequences, we considered the miRNA to be conserved in the genome assembly even if we were not able to identify the homologous sequences.

Identification of differentially expressed miRNAs

Reads that aligned perfectly to the candidate miRNA-yielding transcripts were used to compute the digital expression levels of the miRNAs. Reads that mapped to multiple genomic loci were attributed to all derivative miRNAs. Read counts in each sample were normalised to adjust for sample variation. If N_{sample} is the number of qualified reads that aligned to the genome and cDNA sequences in that sample and C is the average value of N_{sample} of all samples, then the NNR for each miRNA in each sample is ($N_{miRNA}*C/N_{sample}$), where N_{miRNA} is the raw sequencing reads of the miRNA. Differentially expressed miRNAs were those that had at least two fold changes between the cultivars and their wild progenitor.

Target mRNA prediction and miRNA: target pair validation

We used the Hitsensor algorithm [48] to predict miRNA targets in AM560, which was downloaded from http://www.phytozome.net/cassava. Hitsensor searched for miRNA complementary sites in coding regions with a modified Smith-Waterman algorithm [49]. This algorithm scores these sites by giving rewards to key sequence-specific determinants, including the seed region (12–17-nucleotides long), local-AU content around the seed region and ≤3 mismatches.

RLM-RACE Gene Racer Kit (Invitrogen, CA, USA) was used to validate the predicted interaction. An RNA adaptor was ligated to the truncated mRNA, followed by reverse transcription with polyT. The next nested PCR with two gene-specific primers and two GeneRacer 5' forward primers was performed as described previously [22]. The amplified PCR products were gel-purified, cloned into the PMD-18T vector (Takara, Dalian, China) and sequenced.

Quantity detection of miRNA and targets

The amount of miRNA was quantified as described previously [50]. Firstly, the small RNA samples were converted to miR-cDNA by using an RT primer pool with reverse transcriptase. Then, a specific primer pair was designed for each miRNA, after which PCR amplification with SYBR Premix Ex TaqTM kit (Takara, Dalian, China) was carried out using a Rotor-Gene 6000 machine (Corbett Robotics, Australia), with the U6 gene as a control. Quantification of the relative expression of miRNAs was performed using the $^{\Delta\Delta}$CT method. Quantification of the target was also carried out using qRT-PCR, with the *beta*-actin gene as a control. The forward and reverse primers of the target were located at the two flanks of the binding region at which the miRNA interacted with its target mRNA; the primer pairs are listed in Additional file 6.

Supporting data

The small RNA sequencing data is deposited in Gene omnibus with accession number GSE52178. RNA-seq reads are deposited in GenBank/SRR sequence read archive under the accession codes SRR1299000, SRR1299003, SRR1299009, SRR1299006, SRR1298998 and SRR1298996.

Additional files

Additional file 1: Normalised digital read counts for each miRNA in leaf and storage root of three *Manihot* genotypes.

Additional file 2: Primer sets used to amplify microRNAs using qRT-PCR.

Additional file 3: Functional annotation of 360 predicted targeted genes.

Additional file 4: Gene-specific primers for target validation by RLM-RACE.

Additional file 5: Expression correlations of 21 miRNAs and their corresponding targets in leaf and root between cultivar and progenitor. "#" means that the slicing is confirmed by RLM-RACE between miRNA and target.

Additional file 6: Primer pairs used to amplify targets for analysis by quantitative real-time PCR.

Abbreviations
DAP: Day after planting; miRNA: MicroRNA; qRT-PCR: Quantitative reverse transcription polymerase chain reaction; NGS: Next-generation sequencing; NNR: Normalised number of reads; FPKM: Fragment per kilobase per million reads.

Competing interests
The authors declare that they have no competing interests.

Authors' contributions
XC prepared the manuscript. XC and CZ performed miRNA:target pair validation and qRT-PCR analysis. JX and ZX participated in miRNA prediction, the expression profiling of miRNA targets and the characterisation of miRNA targets. HZ and CL were involved in preparing the plant materials and deep sequencing, WW and WZ designed the project. All authors read and approved the final manuscript.

Acknowledgements
This work was supported financially by the National Basic Research and Development Program (2010CB126600), Natural Science Foundation of China (31101193) and the earmarked fund for China Agriculture Research System (CARS-12).

Author details
[1]The Institute of Tropical Bioscience and Biotechnology (ITBB), Chinese Academy of Tropical Agricultural Sciences (CATAS), Haikou 571101, PR China. [2]Key Laboratory of Biology and Genetic Resources of Tropical Crops, Ministry of Agriculture, Haikou 571101, PR China. [3]Institute for Systems Biology, Jianghan University, Wuhan 430056, China. [4]Department of Computer Science and Engineering, Washington University in St. Louis, St. Louis, Missouri, MO 63130, USA.

References

1. Lagos-Quintana M, Rauhut R, Lendeckel W, Tuschl T: **Identification of novel genes coding for small expressed RNAs.** *Science* 2001, **294**(5543):853–858.
2. Lau NC, Lim LP, Weinstein EG, Bartel DP: **An abundant class of tiny RNAs with probable regulatory roles in *Caenorhabditis elegans*.** *Science* 2001, **294**(5543):858–862.
3. Chen X: **MicroRNA biogenesis and function in plants.** *FEBS Lett* 2005, **579**(26):5923–5931.
4. Carla S, Edgardo GB, Silvana VS, Javier FP: *Role of MicroRNA miR319 in Plant Development, in R. Sunkar (ed.), MicroRNAs in Plant Development and Stress Responses, Signaling and Communication in Plants 15*, doi:10.1007/978-3-642-27384-1, Springer-Verlag Berlin Heidelberg 2012.
5. Ozhuner E, Eldem V, Ipek A, Okay S, Sakcali S, Zhang B, Boke H, Unver T: **Boron stress responsive miRNAs and their targets in barley.** *PLoS One* 2013, **8**(3):e59543.
6. Peng T, Sun H, Du Y, Zhang J, Li J, Liu Y, Zhao Y, Zhao Q: **Characterization and expression patterns of miRNAs involved in rice grain filling.** *PLoS One* 2013, **8**(1):e54148.
7. Chen XB, Zhang ZL, Liu DM, Zhang K, Li AL, Mao L: **SQUAMOSA promoter-binding protein-like transcription factors: star players for plant growth and development.** *J Integr Plant Biol* 2010, **52**(11):946–951.
8. Park W, Li J, Song R, Messing J, Chen X: **CARPEL FACTORY, a Dicer homolog, and HEN1, a novel protein, act in miRNA metabolism in *Arabidopsis thaliana*.** *Curr Biol* 2002, **12**:1484–1495.
9. Palatnik JF, Allen E, Wu X, Schommer C, Schwab R, Carrington JC, Weigel D: **Control of leaf morphogenesis by miRNAs.** *Nature* 2003, **425**(6955):257–263.
10. Aukerman MJ, Sakai H: **Regulation of flowering time and floral organ identity by a MiRNA and its APETALA2-like target genes.** *Plant Cell* 2003, **15**(11):2730–2741.
11. Sunkar R, Zhu JK: **Novel and stress-regulated miRNAs and other small RNAs from *Arabidopsis*.** *Plant Cell* 2004, **16**(8):2001–2019.
12. Sun G: **MicroRNAs and their diverse functions in plants.** *Plant Mol Biol* 2012, **80**(1):17–36.
13. Reinhart BJ, Weinstein EG, Jones-Rhoades MW, Bartel B, Bartel DP: **MiRNAs in plants.** *Genes Dev* 2002, **16**:1616–1626.
14. Sunkar R, Jagadeeswaran G: **In silico identification of conserved miRNAs in large number of diverse plant species.** *BMC Plant Biol* 2008, **8**:37.
15. Guo W, Wu G, Yan F, Lu Y, Zheng H, Lin L, Chen H, Chen J: **Identification of novel *Oryza sativa* miRNAs in deep sequencing-based small RNA libraries of rice infected with rice stripe virus.** *PLoS One* 2012, **7**(10):e46443.
16. Tetlow IJ, Morell MK, Emes MJ: **Recent developments in understanding the regulation of starch metabolism in higher plants.** *J Exp Bot* 2004, **55**(406):2131–2145.
17. Xie F, Frazier TP, Zhang B: **Identification and characterization of microRNAs and their targets in the bioenergy plant switchgrass (*Panicum virgatum*).** *Planta* 2010, **232**(2):417–434.
18. Xie F, Frazier TP, Zhang B: **Identification, characterization and expression analysis of MicroRNAs and their targets in the potato (*Solanum tuberosum*).** *Gene* 2011, **473**(1):8–22.
19. Patanun O, Lertpanyasampatha M, Sojikul P, Viboonjun U, Narangajavana J: **Computational identification of miRNAs and their targets in cassava (*Manihot esculenta* Crantz).** *Mol Biotechnol* 2013, **53**(3):257–269.
20. Pérez-Quintero ÁL, Quintero A, Urrego O, Vanegas P, López C: **Bioinformatic identification of cassava miRNAs differentially expressed in response to infection by *Xanthomonas axonopodis* pv. manihotis.** *BMC Plant Biol* 2012, **12**:29.
21. Prochnik S, Marri PR, Desany B, Rabinowicz PD, Kodira C, Mohiuddin M, Rodriguez F, Fauquet C, Tohme J, Harkins T, Rokhsar DS, Rounsley S: **The cassava genome: current progress, future directions.** *Trop Plant Biol* 2012, **5**(1):88–94.
22. Zeng CY, Wang WQ, Zheng Y, Chen X, Bo WP, Song S, Zhang WX, Peng M: **Conservation and divergence of miRNAs and their functions in *Euphorbiaceous* plants.** *Nucle Acid Res* 2010, **38**:981–995.
23. Amiteye S, Corra JM, Sharbe TF: **Overview of the potential of microRNAs and their target gene detection for cassava (*Manihot esculenta*) improvement.** *Afr J Biotech* 2011, **10**(14):2562–2573.
24. Xia J, Zeng CY, Chen Z, Zhang K, Chen X, Zhou YF, Song S, Lu C, Yang RJ, Yang Z, Zhou J, Peng H, Wang WQ, Peng M, Zhang WX: **Endogenous small-noncoding RNAs and their roles in chilling response and stress acclimation in Cassava.** *BMC Genomics* 2014, **15**:634.
25. Karlova R, van Haarst JC, Maliepaard C, van de Geest H, Bovy AG, Lammers M, Angenent GC, de Maagd RA: **Identification of miRNA targets in tomato fruit development using high-throughput sequencing and degradome analysis.** *J Exp Bot* 2013, **64**(7):1863–1878.
26. Carnavale Bottino M, Rosario S, Grativol C, Thiebaut F, Rojas CA, Farrineli L, Hemerly AS, Ferreira PC: **High-throughput sequencing of small RNA transcriptome reveals salt stress regulated miRNAs in sugarcane.** *PLoS One* 2013, **8**(3):e59423.
27. Palatnik JF, Wollmann H, Schommer C, Schwab R, Boisbouvier J, Rodriguez R, Warthmann N, Allen E, Dezulian T, Huson D, Carrington JC, Weigel D: **Sequence and expression differences underlie functional specialization of *Arabidopsis* microRNAs miR159 and miR319.** *Dev Cell* 2007, **13**(1):115–125.
28. Abe H, Urao T, Ito T, Seki M, Shinozaki K, Yamaguchi-Shinozaki K: ***Arabidopsis AtMYC2* (bHLH) and *AtMYB2* (MYB) function as transcriptional activators in abscisic acid signaling.** *Plant Cell* 2003, **15**(1):63–78.
29. Reyes JL, Chua NH: **ABA induction of miR159 controls transcript levels of two MYB factors during *Arabidopsis* seed germination.** *Plant J* 2007, **49**(4):592–606.
30. Chen J, Huang B, Li Y, Du H, Gu Y, Liu H, Zhang J, Huang Y: **Synergistic influence of sucrose and abscisic acid on the genes involved in starch synthesis in maize endosperm.** *Carbohydr Res* 2011, **346**(13):1684–1691.
31. Zhu GH, Ye NH, Yang JC, Peng XX, Zhang JH: **Regulation of expression of starch synthesis genes by ethylene and ABA in relation to the development of rice inferior and superior spikelets.** *J Exp Bot* 2011, **62**(11):3907–3916.
32. Bazin J, Khan GA, Combier JP, Bustos-Sanmamed P, Debernardi JM, Rodriguez R, Sorin C, Palatnik J, Hartmann C, Crespi M, Lelandais-Brière C: **MiR396 affects**

mycorrhization and root meristem activity in the legume *Medicago truncatula*. *Plant J* 2013, **74**(6):920–934.
33. Hewezi T, Maier TR, Nettleton D, Baum TJ: The *Arabidopsis* miRNA396-GRF1/GRF3 regulatory module acts as a developmental regulator in the reprogramming of root cells during cyst nematode infection. *Plant Physiol* 2012, **159**(1):321–335.
34. Li WX, Oono Y, Zhu J, He XJ, Wu JM, Iida K, Lu XY, Cui X, Jin H, Zhu JK: The *Arabidopsis* NFYA5 transcription factor is regulated transcriptionally and post-transcriptionally to promote drought resistance. *Plant Cell* 2008, **20**(8):2238–2251.
35. Ni Z, Hu Z, Jiang Q, Zhang H: GmNFYA3, a target gene of miR169, is a positive regulator of plant tolerance to drought stress. *Plant Mol Biol* 2013, **82**(1–2):113–129.
36. Carra A, Mica E, Gambino G, Pindo M, Moser C, Pè ME, Schubert A: Cloning and characterization of small non-coding RNAs from grape. *Plant J* 2009, **59**(5):750–763.
37. Beauclair L, Yu A, Bouché N: miRNA-directed cleavage and translational repression of the copper chaperone for superoxide dismutase mRNA in *Arabidopsis*. *Plant J* 2010, **62**(3):454–462.
38. Carvalho Luiz Joaquim CB, Schaal BA: Assessing genetic diversity in the cassava (*Manihot esculenta* Crantz) germplasm collection in Brazil using PCR-based markers. *Euphytica* 2001, **120**(1):133–142.
39. Olsen KM, Schaal BA: Microsatellite variation in cassava (*Manihot esculenta*, Euphorbiaceae) and its wild relatives: further evidence for a southern Amazonian origin of domestication. *Am J Bot* 2001, **88**(1):131–142.
40. Léa M, Doyle M, Gilda SM, Bruno C, Nick PR: The evolutionary fate of phenotypic plasticity and functional traits under domestication in manioc: changes in stem iomechanics and the appearance of stem brittleness. *PLoS One* 2013, **8**(9):1–20.
41. Zhong SL, Joung JG, Zheng Y, Chen YR, Liu B, Shao Y, Xiang JZ, Fei ZJ, Giovannoni JJ: High-throughput illumina strand-specific RNA sequencing library preparation. *Cold Spring Harb Protoc* 2011, **2011**(8):940–949.
42. Trapnell C, Roberts A, Goff L, Pertea G, Kim D, Kelley DR, Pimentel H, Salzberg SL, Rinn JL, Pachter L: Differential gene and transcript expression analysis of RNA-seq experiments with TopHat and Cufflinks. *Nat Protoc* 2012, **7**(3):562–578.
43. Kanehisa M, Goto S, Kawashima S, Okuno Y, Hattori M: The KEGG resource for deciphering the genome. *Nucle Acid Res* 2004, **32**:277–280.
44. O'Rourke JA, Yang SS, Miller SS, Bucciarelli B, Liu J, Rydeen A, Bozsoki Z, Uhde-Stone C, Tu ZJ, Allan D, Gronwald JW, Vance CP: An RNA-Seq transcriptome analysis of orthophosphate-deficient white lupin reveals novel insights into phosphorus acclimation in plants. *Plant Physiol* 2013, **161**(2):705–724.
45. Hofacker IL: Vienna RNA secondary structure server. *Nucle Acid Res* 2003, **31**:3429–3431.
46. Rhoades MW, Reinhart BJ, Lim LP, Burge CB, Bartel B, Bartel DP: Prediction of plant microRNA targets. *Cell* 2002, **110**(4):513–520.
47. Langmead B, Trapnell C, Pop M, Salzberg SL: Ultrafast and memory-efficient alignment of short DNA sequences to the human genome. *Genome Biol* 2009, **10**(3):R25.
48. Zheng Y, Zhang W: Animal microRNA target prediction using diverse sequence-specific determinants. *J Bioinform Comput Biol* 2010, **8**(4):763–788.
49. Smith TF, Waterman MS: Identification of common molecular subsequences. *J Mol Biol* 1981, **147**:195–197.
50. Wang XW: A PCR based platform for miRNA expression profiling studies. *RNA* 2009, **15**:716–723.

2

Genome-wide association study reveals a set of genes associated with resistance to the Mediterranean corn borer (*Sesamia nonagrioides* L.) in a maize diversity panel

Luis Fernando Samayoa[1*], Rosa Ana Malvar[1], Bode A Olukolu[2], James B Holland[2,3] and Ana Butrón[1]

Abstract

Background: Corn borers are the primary maize pest; their feeding on the pith results in stem damage and yield losses. In this study, we performed a genome-wide association study (GWAS) to identify SNPs associated with resistance to Mediterranean corn borer in a maize diversity panel using a set of more than 240,000 SNPs.

Results: Twenty five SNPs were significantly associated with three resistance traits: 10 were significantly associated with tunnel length, 4 with stem damage, and 11 with kernel resistance. Allelic variation at each significant SNP was associated with from 6 to 9% of the phenotypic variance. A set of genes containing or physically close to these SNPs are proposed as candidate genes for borer resistance, supported by their involvement in plant defense-related mechanisms in previously published evidence. The linkage disequilibrium decayed ($r^2 < 0.10$) rapidly within short distance, suggesting high resolution of GWAS associations.

Conclusions: Most of the candidate genes found in this study are part of signaling pathways, others act as regulator of expression under biotic stress condition, and a few genes are encoding enzymes with antibiotic effect against insects such as the *cystatin1* gene and the defensin proteins. These findings contribute to the understanding the complex relationship between plant-insect interactions.

Keywords: Candidate genes, Corn borer, Genome-wide association study, Insect resistance, Maize, Mixed linear models, *Sesamia nonagrioides*

Background

Corn borers are the primary maize pest in many environments [1,2]. Corn borers feeding on the pith of the stem results in yield losses because stem damage interferes with assimilate movement to developing kernels. They can also attack the ears, promoting secondary fungal infection, leading to contamination of grain with mycotoxins that may affect human and animal health [3,4].

There are different species of borers that attack maize in different parts of the world. The most economically important species are classified into two families: Crambidae and Noctuidae. Within the Crambidae family, the species with economic importance are: *Ostrinia nubilalis* Hubner in North America, Europe and North Africa; *Ostrinia furnacalis* Guenée in Asia; *Diatraea saccharalis* Fabricius from USA to Argentina; *Chilo partellus* Swinhoe in Southern USA, Central America and the Caribbean; *Diatraea lineolata* Walker in Central America, the Caribbean region and South America; and *Diatraea grandiosella* Dyar in North and Central America. Within the Noctuidae family, the main maize borers are: *Sesamia nonagrioides* Lefebvre in the Mediterranean region, *Busseola fusca* Fuller in sub-Saharan Africa, and *Sesamia calamistis* Hampson in West Africa. This study focuses on the noctuid Mediterranean corn borer (MCB) *Sesamia nonagroides* Lefebvre, the most important insect pest of maize in the Mediterranean region that includes Southern Europe [2,5,6].

* Correspondence: fsamayoa@mbg.csic.es
[1]Misión Biológica de Galicia, Spanish National Research Council (CSIC), P.O. Box 2836080 Pontevedra, Spain
Full list of author information is available at the end of the article

The use of transgenic corn which produces *Bacillus thutingiensis* (*Bt*) toxins is a good method for controlling these pests, but transgenic crops are not authorized in several European countries under any agricultural system [7] and are not allowed for organic production [8]. In addition, recent studies have reported a reduction of efficacy of *Bt* transgenes caused by evolved resistance of some important pests [9-11]. The stacking of several resistant genes has been proposed as one means to delay insect adaptation [12]. Natural sources of resistance to stem borers in maize could reveal promising genes for use in either breeding or transgenic approaches to resistance.

In Spain, there are three MCB generations per year and the second and subsequent generations are able of making significant damage on the stem and, secondarily, on the ear. Following artificial infestation, the level of maize host resistance to stem borer is measured by the tunnel length made by the larvae in the stem as well as by a visual scale for kernel damage. These traits have a complex genetic architecture because resistance depends on the plant-insect relationship, which is influenced by environmental conditions and the developmental stage of the host plant [13,14]. The line mean heritability estimates for tunnel length under corn borer infestation ranged among studies from moderate to relatively high ($h^2 = 0.50$ to 0.78) [15-18], depending on the genetic background.

At present, commercial materials with high levels of native resistance to these insects are not available even though breeding for increasing maize resistance to corn borers has been conducted during the last three decades in different regions around the world. Klenke et al. [19] reduced tunnel length by 4 and 6 cm after four cycles of recurrent selection for resistance to the first and second generations of ECB, respectively. Bosque-Pérez et al. [20] described successful results in the development of materials with resistance to *S. calamistis* and African sugarcane borer (*Eldana saccharina* Walker). Sandoya et al. [21] achieved a reduction of 1.8 cm per cycle of recurrent selection for tunnel length by MCB. In addition, a negative relationship between resistance to stem borers and yield has been found [22] when selection for improved yield under infestation was practiced [23]. In summary, classical breeding experiments have demonstrated some successful improvement in corn borer resistance, but natural levels of resistance in elite cultivars remain insufficient to manage the pest. Detection of stem borer resistance QTL could enhance breeding for this trait via marker-assisted breeding or genomic selection.

Genetic effects for resistance against borer attack fit an additive – dominant model, although additive effects appear to be the most important in determining resistance to tunnel length and kernel damage [14,24-28]. Therefore, the study of genetic factors involved in maize resistance to borers can be performed using highly inbred lines.

Several studies performed with segregating biparental populations have reported genomic regions containing minor and major quantitative traits loci (QTL) for resistance to stem and leaf attack by European corn borer (ECB; [15-17,29-31]. Fewer studies have reported QTLs for resistance to other borer species such as the Mediterranean corn borer (MCB) [18,32] or some tropical borers [33-35]. In maize, QTLs for resistance to tunneling by corn borers have been detected on all chromosomes, with the most commonly detected regions occurring on chromosomes 1, 2, 3, 5, and 9 [1]. Identification of causal genes underlying the QTL for resistance in these genomic regions could help breeders to transfer the allelic variants that confer resistance from the lines that carry them to elite breeding lines that lack these resistance alleles. The identification of the causal genetic variants or markers in high LD with causal variants in diverse materials would minimize the risk of dragging other genes with negative effect on the agronomic value during the transfer process.

Although conventional QTL mapping based on linkage maps of biparental populations has been an efficient approach to detect regions related to the resistance to corn borers, higher resolution is needed to detect the genes involved in the defense mechanism of the plant. Genome wide association study (GWAS) based on linked disequilibrium (LD) in diverse genetic samples is a relatively new approach which offers higher resolution mapping that under optimal conditions can pinpoint causal genes underlying quantitative trait variation. Exploiting advances in genotyping and sequencing technology, this approach has been successful in detecting genes associated with diseases in humans [36-38], animals [39-42] and different quantitative traits in plants [43-46]. In contrast to the conventional QTL mapping approach based on linkage in a biparental population, GWAS is based on LD among extant lines from different populations, such that a large number of markers covering the whole genome are required [45,47]. In diverse maize samples, LD is low, therefore, many more markers are needed than in autogamous species (with higher LD) to adequately explore the genetic architecture of complex traits [48]. The low LD offers the benefit of better resolution to delineate potential causal genes within small LD blocks.

Many single-nucleotide polymorphism (SNP) markers have been identified and scored on a maize diversity panel (composed of 302 inbred line) that represents the diversity available in public breeding sector around the world [49,50]. The population has been successfully used by the maize community to perform GWAS in economically important quantitative traits such as kernel composition [51], hypersensitive response [52] and *Fusarium* ear rot resistance [53].

To date, no GWAS for insect resistance in maize has been reported, although a few GWA studies that deal with plant defense mechanisms against insect attack have been reported in other plant species [54,55]. In this study, we performed GWAS to identify SNPs associated with resistance to MCB. GWAS was done in a subset (270 inbreds) of the maize diversity panel using the maize 50 k SNP genotyping array [56] plus a more recently developed set of 425 k SNPs found through genotyping by sequencing [57,58].

Results

Means, analysis of variance and heritabilities

Differences among inbred lines were highly significant ($P < 0.01$) for all resistance (TL, SD, and KR) and agronomic traits (PH, DTA, and DTS); while significant ($P < 0.05$) genotype by environment (G × E) interactions were observed for all traits as shown in Additional file 1: Table S1 and S2. The inbred means for TL ranged from 5.2 to 49.2 cm with an overall mean of 20.9 cm, and for SD from 4.1 to 22.6% with an overall mean of 11.9%, respectively. The inbred mean for KR ranged from 5.4 to 9 with an overall mean of 7.8 in the subjective scale. Higher values for TL were observed in 2012 (overall mean = 43.3 cm) compared with 2010 (overall mean = 17.5 cm) and 2011 (overall mean = 14.7 cm, Additional file 1: Table S3 and S4). Intermediate values for heritability on a line mean-basis were estimated for TL (h^2 = 0.60) and KR (h^2 = 0.52) across the three years, whereas a low heritability value was obtained for SD (h^2 = 0.25, Additional file 1: Table S1).

Correlation analysis

A significant and high ($r > 0.50$) phenotypic correlation was observed between TL and SD. Genetic correlation coefficients were also significant and high between TL and SD and between TL and PH (Table 1). KR showed significant phenotypic correlations with other resistance traits but the correlation coefficients were not higher than 0.5. Genetic correlations between KR and the agronomic traits (DTA, DTS and PH) were significant and high (Table 1).

Table 1 Genotypic[1] (above diagonal) and phenotypic[2] (below diagonal) correlation coefficient estimates for each pair of traits

	TL	SD	KR	DTA	DTS	PH
TL		0.97*	NS	NS	NS	0.51*
SD	0.74*		NS	NS	NS	NS
KR	−0.31*	−0.31*		0.75*	0.71*	0.51*
DTA	NS	NS	0.30*		0.99*	0.66*
DTS	NS	NS	0.28*	0.96*		0.65*
PH	0.24*	−0.06	0.24*	0.51*	0.48*	

TL, tunnel length; SD, stem damage; KR, kernel resistance; PH, plant height; DTA, days to anthesis; DTS, days to silking.
[1] Genotypic correlation coefficients were considered significant, *, when exceeded twice its standard error; NS, not significant.
[2] Phenotypic correlation coefficients were considered significant, *, at 0.05 probability level according to Steel and Torrie [59]; NS, not significant.

Association analysis of maize resistance to MCB and agronomic traits

The compressed mixed linear model computed for each trait in Tassel reduced the pairwise kinship matrix by clustering the 267 lines into 32 groups for TL, 201 groups for SD, 166 groups for KR, 192 for PH, and 240 for DTA, and DTS (Table 2). The proportion of the total phenotypic variation explained by background genetic effects was 39, 13, and 48% for TL, SD, and KR, respectively. By comparison, the background polygenic effects modeled by the K matrix accounted for 60, 94 and 90% of total variation for PH, DTA, and DTS, respectively.

Ten SNPs were identified as significantly associated with length of tunnels made by MCB (Figure 1, Table 3). Based on the additive effects, the major allele reduce TL for all significant SNPs except for the SNP located on chromosome 10 (Table 3). Four SNPs were significantly associated with SD made by MCB (Figure 1). The minor allele for those SNPs increases the SD from 1.1 to 2%. The total variance explained (R^2) by each SNP associated with TL and SD ranged from 7 to 9%. Eleven SNPs were significantly associated with KR (Figure 1) and the total

Table 2 Summary of the compressed mixed linear model analysis for three traits related to resistance to MCB attack and three agronomic traits in an inbred association panel evaluated in three years

Trait	n[a]	s[b]	Compression Level (c)[c]	$(\hat{\sigma}_g^2)$[d]	$(\hat{\sigma}^2)$[e]	$\left(\frac{\hat{\sigma}_g^2}{\hat{\sigma}_g^2+\hat{\sigma}^2}\right)$[f]
TL	267	32	8.34	32.80	50.61	0.39
SD	267	201	1.33	1.45	9.51	0.13
KR	265	166	1.60	0.15	0.16	0.48
PH	266	192	1.39	384.70	256.56	0.60
DTA	266	240	1.11	45.03	5.03	0.94
DTS	266	240	1.11	51.19	5.87	0.90

TL, tunnel length; SD, stem damage; KR, borer kernel resistance; PH, plant height; DTA, days to anthesis; DTS, days to silking.
[a] Total number of inbred lines included in the analysis.
[b] Number of groups obtained using a clustering approach based on K matrix with the optimum compression level option.
[c] Compression level is the average number of inbred lines per group estimated as n/s.
[d] Additive background genetic variance component estimated in Tassel by fitting the K matrix in the MLM without any SNP marker effects.
[e] Residual genotypic variance component estimated in Tassel.
[f] Proportion of phenotypic variance explained by the K matrix, estimated as background genetic variance divided by total phenotypic variance.

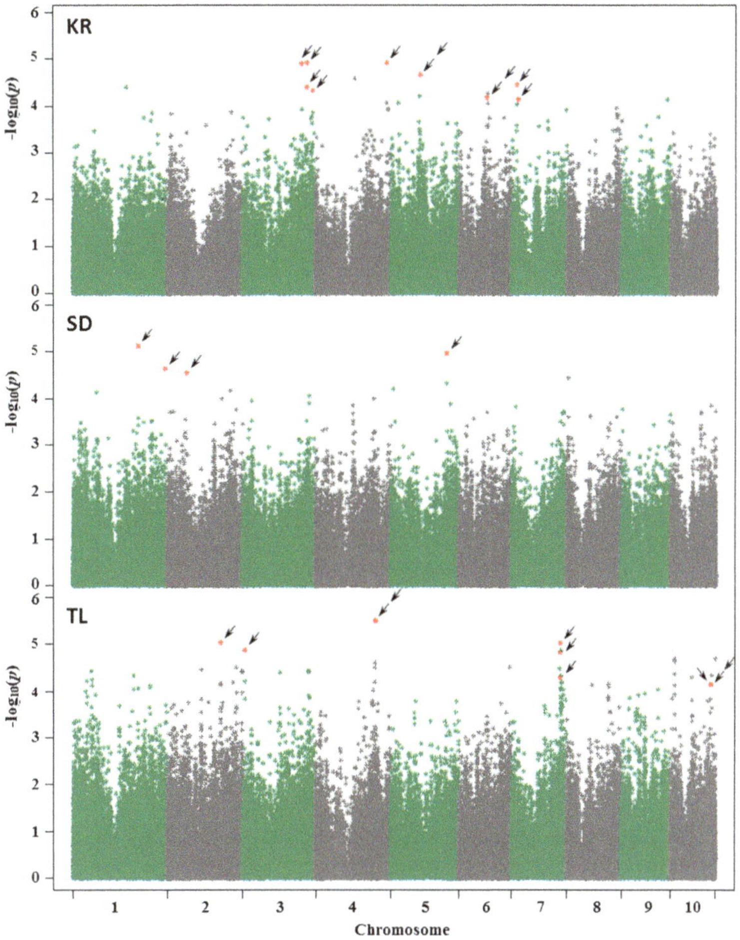

Figure 1 GWAS results for the three resistance traits to MCB attack in a maize association panel. Each graph (TL, SD, and KR) represent the *P*-values of the 246,477 SNPs tested for each resistance trait. Each row indicates the SNP significantly associated (RMIP ≥ 0.30) to each resistance trait analyzed.

variance (R^2) explained by each of them ranged from 6 to 8%. The minor allele for those SNP reduced from 0.15 to 0.40 points the ratio that accounts for kernel resistance.

Nineteen SNPs were significantly associated with PH and 48 and 43 SNPs were significantly associated with DTA and DTS, respectively (Additional file 1: Table S5, S6, and S7). But none of those SNPs coincide or are close to those detected for resistance traits.

Candidate genes selection

The filtered predicted gene set from the annotated B73 reference maize genome [60] was used to characterize the gene containing or nearby the SNP declared significant. Seven candidate genes containing or adjacent to the SNPs significantly associated with TL, four candidate genes containing or adjacent to the significant SNPs associated to SD, and ten candidate genes containing or adjacent to the significant SNPs associated to KR were proposed (Table 4).

In general, the LD (r^2) between significantly associated SNPs and SNPs around them decays ($r^2 \leq 0.10$) rapidly; but, in some cases, the LD spans as much as 0.5 Mb at an r^2 value greater than 0.20 (Figure 2). For each trait, LD estimates between significant SNPs were always below 0.1 except for SNPs significant for TL located on chromosome 7 as shown in Additional file 1: Table S8.

Discussion

Means, heritabilities, and correlations

Substantial differences among yearly means for TL made by MCB were observed across environments, with the highest values of TL observed in 2012. Artificial infestation is used to ensure the contact of the insect with the plant, but the severity of the attack is conditioned by environmental factors that in turn influence natural infestation. The prevalence of the pest can vary greatly between and within locations [5,6]. Data on the monthly samplings at different locations in Pontevedra (unpublished data) indicate a higher rate of natural infestation

Table 3 SNP identification (SNP ID), additive effect and allelic variants for the SNP, proportion of total variance explained by the SNPs significantly associated with resistance traits (TL, SD, and KR), and significance values for the association between the SNP and the phenotype (*P*-value and RMIP)

Trait[a]	SNP ID[b]	Alleles[c]	(N°)[d]	Additive effect[e]	*P*-value	(R^2)[f]	RMIP[g]
TL	S2_168004182	G/T	225/36	3.09	9.51×10^{-6}	0.08	0.51
TL	S3_7081859	T/A	207/55	2.52	1.38×10^{-5}	0.07	0.50
TL	S4_190444179	G/A	218/38	3.18	3.26×10^{-6}	0.09	0.37
TL	ss4_190679094[h]	G/A	226/41	3.09	3.37×10^{-6}	0.08	0.45
TL	S7_154739818	G/C	248/14	4.25	5.28×10^{-5}	0.07	0.37
TL	S7_154741622	C/T	217/47	2.84	9.84×10^{-6}	0.08	0.56
TL	S7_155702328	T/A	171/67	2.43	1.55×10^{-5}	0.08	0.38
TL	S10_133337924	G/T	133/99	−1.30	7.40×10^{-5}	0.07	0.30
TL	S10_133337925	G/C	133/99	−1.30	7.40×10^{-5}	0.07	0.30
TL	S10_133337950	T/C	133/99	−1.30	7.40×10^{-5}	0.07	0.30
SD	S1_208315891	C/A	126/106	1.09	7.86×10^{-6}	0.09	0.63
SD	S1_293163491	T/A	207/37	1.37	2.39×10^{-5}	0.08	0.38
SD	S2_59729532	A/T	235/17	1.90	2.93×10^{-5}	0.07	0.32
SD	ss5_176870721[h]	G/A	228/38	1.37	1.12×10^{-5}	0.08	0.41
KR	S3_187742562	C/A	244/12	−0.38	1.30×10^{-5}	0.08	0.48
KR	S3_204458505	T/A	186/65	−0.19	4.14×10^{-5}	0.07	0.40
KR	S3_204586960	C/G	251/13	−0.37	1.25×10^{-5}	0.08	0.44
KR	S3_222733400	C/T	142/110	−0.15	4.90×10^{-5}	0.07	0.34
KR	S4_227101950	C/T	242/17	−0.30	1.26×10^{-5}	0.08	0.43
KR	S4_227101985	A/T	242/17	−0.30	1.26×10^{-5}	0.08	0.43
KR	S5_93580059	C/T	230/28	−0.24	2.27×10^{-5}	0.07	0.31
KR	S6_88149024	A/G	187/55	−0.17	6.83×10^{-5}	0.07	0.48
KR	S6_88149036	G/C	187/55	−0.17	6.83×10^{-5}	0.07	0.48
KR	S7_15072370	G/A	230/24	−0.26	3.65×10^{-5}	0.07	0.46
KR	S7_19347596	A/G	250/11	−0.40	7.47×10^{-5}	0.06	0.36

[a] TL, tunnel length in cm; SD, stem damage in percentage; and KR, kernel resistance on a subjective visual scale of 1 to 9 in which 1 indicates completely damaged and 9 indicates no damage.
[b] The number before the underscore (_) indicates the chromosome number and the number after the underscore (_) indicates the physical position in bp within the chromosome.
[c] The letter before the diagonal (/) is the nucleotide more frequent ; and the letter after the diagonal the nucleotide less frequent.
[d] N° = number of inbred lines homozygous for a determined allelic variant. The number before the diagonal (/) represents the number of individuals with the mayor allele; and the number after the diagonal represents the number of individuals with the minor allele.
[e] The additive effect was calculated as half the difference between the mean of the homozygous for the minor and the mean of the homozygous for the major allele.
[f] R^2, proportion of the phenotypic variance explained by the SNP.
[g] RMIP, resample model inclusion probability.
[h] Based on SNPs from Illumina chip, the remaining locations without a superscript are based on SNPs obtained by GBS.

in 2012 compared with 2010 and 2011 in a plot adjacent to the GWAS trial. In addition, the experiment was harvested earlier in 2011 compared with 2010 and 2012, limiting the time that larvae could damage the plants. Therefore, differences on infestation levels and harvest time could be major causes of the observed differences in TL among years.

Significant genotype × environment interaction was observed for all traits, although no G × E interaction for TL was found in previous studies under infestation with MCB [18,61]. However, G × E interaction for resistance traits was not significant, except for KR, when we made the analysis discarding data from 2012. Therefore, the high values for TL in 2012 could be obscuring the genotype effect. In previous studies, typically no G × E interaction for KR was found [61-63], with some exceptions [18].

The heritabilities for resistance traits ranged from low to moderate while PH and flowering time were higher as expected. The heritability for TL estimated herein is within the range (h^2 = 0.40-0.77) of those obtained in other works with biparental crosses under infestation with MCB or ECB [16,18,64].

Table 4 Candidate genes for each significantly SNP associated with TL, SD, and KR and its respective encoding product

Chr[a]	Trait	SNP physical position (bp)	Gene ID	Encoding
2	TL	168,004,182	GRMZM2G504910	Tetratricopeptide repeat (TPR)
3	TL	7,081,859	GRMZM2G104081[b]	*hex1* (*hexokinase1*)
4	TL	190,444,179	GRMZM2G013128[b]	Double Clp-N motif-containing P-loop nucleoside triphosphate hydrolases superfamily protein
4	TL	190,679,094	GRMZM2G033820	Phospholipase A2
7	TL	154,739,818	GRMZM2G077008[bc]	Histidine kinase, hybrid-type, ethylene sensor
		154,741,622		
7	TL	155,702,328	GRMZM2G861541	Expressed protein
10	TL	133,337,924	GRMZM2G057084	Calcium/calmodulin-dependent protein kinase
		133,337,925		
		133,337,950		
1	SD	208,315,891	GRMZM2G101422	Expressed protein
1	SD	293,163,491	GRMZM2G060702[b]	Actin depolymerizing factor 4
2	SD	59,729,532	GRMZM2G389097	Leucine-rich receptor-like protein kinase family
5	SD	176,870,721	GRMZM2G325683[c]	Expressed protein
3	KR	187,742,562	GRMZM2G438551	*cystatin1*
3	KR	204,458,505	GRMZM2G111666	basic Helix-Loop-Helix (bHLH) transcription f.
3	KR	204,586,960	GRMZM2G091494[c]	Starch branching enzyme interacting protein-1
3	KR	222,733,400	GRMZM2G055578[c]	Glycine-rich protein
			GRMZM2G055629	Plant thionin family protein precursor/Defensin
4	KR	227,101,950	GRMZM2G116314	Ubiquitin thiolesterase
		227,101,985		
5	KR	93,580,059	GRMZM2G037308[c]	Phytosulfokine receptor
6	KR	88,149,024	GRMZM5G876960	Polyamine oxidase (propa-1,3-diamine-forming)
		88,149,036		
7	KR	15,072,370	GRMZM2G316256	Catalase//L-ascorbate peroxidase
7	KR	19,347,596	GRMZM2G042627	Kinase associated protein phosphatase

[a] Chr, chromosome; TL, tunnel length; SD, stem damage; and KR, kernel resistance.
[b] Gene containing the significant SNP within an exonic region.
[c] Gene containing the significant SNP within an intronic region.

A significant genetic relationship was observed between TL and PH ($r_g = 0.51$) in a panel of diverse origin. Nevertheless, we did not find a single SNP or linked SNPs associated with both traits, suggesting that it is primarily due to the polygenic background. This relationship has also been important in some previous experiments with MCB and ECB [16,32], but negligible in others [17,30].

A high genetic correlation coefficients between KR and flowering time ($r_g = 0.75$ for DTA and 0.71 for DTS) and between KR and PH were found, which suggest that late and taller genotypes will be the healthiest; but these results have to be taken with caution because infestation was made simultaneously for all genotypes placing the MCB eggs close to the ground and the higher distance between the eggs and the ear of the taller and later plants could have impeded the larval arrival to the ear. In addition, when stem tissue is more lignified at time of infestation (as would be the case for earlier flowering genotypes), the preference of MCB larvae for the stem tissue compared to ear tissue could be less evident.

Association analysis

There was a minimal variation in model fit of the compressed MLM among different traits because similar compression level values were observed for all traits ($c = 1.1$ to 1.6), except for TL, which had the highest value for compression level ($c = 8.3$). However, Zhang et al. [65] shown that this method controls the false positive rate well when the compression levels ranged from 1.5 to 10.

SNPs located on chromosomes 3 and 7 for TL co-localized with previously reported QTLs for TL by corn borers in genome bins 3.02 and 7.03 [30]. The proportion of the phenotypic variance explained ($R^2 = 7$ - 9%)

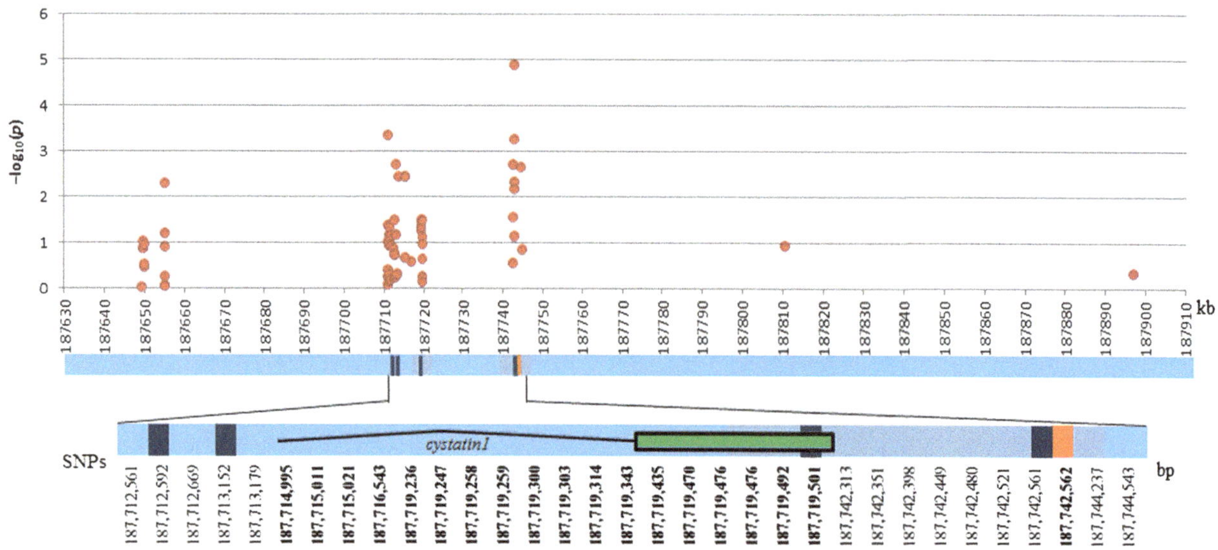

Figure 2 Linkage disequilibrium heat chart showing LD measure (r^2) between the SNP significantly associated with traits related to resistance to MCB attack and the closest 60 SNPs. Each bar represent a region (ranging from ~0.15 to ~1 Mbp) containing each significant SNP associated (black square) to resistance traits. The LD (value of r^2) with the 30 upstream SNPs were shown at the right side of the black square, and the LD with the 30 downstream SNPs were shown at the left side of the black square; on each bar, the extreme distances (in kbp) covered by the upstream and downstream SNPs are indicated.

by each SNP was comparable to that explained by QTLs for TL made by ECB and MCB (R^2 = 3.5 – 15.7%) in biparental crosses [16-18,64]. No QTLs for SD and KR have been previously reported in biparental crosses in the same regions where we located the significant SNPs, except one QTL for KR made by MCB at the bin 5.04 [66]. Therefore, association mapping uncovers additional genomic regions involved in maize resistance to corn borers that were not detected using biparental populations.

Candidate genes

We used the maize B73 genome v2 (RefGen_v2) available from the Maize GDB [67] (http://www.maizegdb.org/) to identify genes that either include or are close to the significantly associated SNPs. A region of approximately 0.2 Mb around the SNP was checked for annotated genes putatively involved in plant response to wounding and/or damage by microbes (insects or pathogens) based on bibliographic records.

Genes associated with TL made by MCB

The candidate gene adjacent to the significant SNP associated to TL on chromosome 2 encodes a Tetratricopeptide repeat (TPR) protein containing. The TPR is one of many repeat motifs that form structural domain mediating protein-protein interactions in several cellular process including translocation and degradations of proteins [68,69]. A recent study has proposed that the presence of those protein-protein interaction motifs could be acting as a modulator of the gene function and protein expression during the stress response caused by invading pathogens [70]. The *hex1* gene containing the significant SNP on chromosome 3 encodes *hexokinase*, a sugar sensor with numerous physiological functions within the cell including response to oxidative-stress and pathogen resistance [71-73]. A gene located on chromosome 4 that encodes a Double Clp-N motif-containing P-loop nucleoside triphosphate hydrolase superfamily protein contains a significant SNP. This gene shows a weak similarity to the AtHSP101 gen in *Arabidopsis*, that codifies for a heat shock protein required for acclimation to high temperatures, and probably could be involved in response to other stresses [74]. Furthermore the *ofp44* (OVATE-transcription factor 22) gene is relatively close to this SNP. It is known that some transcription factors from the Ovate family protein interact with other transcription factor families such as NAC domain protein1, MYB transcription factors, and KNOX homeodomain protein to regulate the synthesis of the three major components of secondary cell wall (lignin, cellulose and hemicellulose) in *A. thaliana* [75-78] and plants with a fortified cell wall are more resistance to corn borer attack [79].

The candidate gene for the other significant SNP on chromosome 4 encodes a Phospholipase A2 (PLA_2) which plays a very important role in signal transduction in plants since it is the precursor of oxylipins and jasmonated acid, two hormones which regulates defense genes against herbivores [80-82].

A 1 Mbp region on chromosome 7 contains three SNPs that were significantly associated with TL and were in significant LD with at least one of the other associated SNPs in the region ($r^2 > 0.2$; Additional file 1: Table S8). The SNPs at 154,739,818 and 154,741,622 bp are both located within a gene that putatively encodes for a Histidine kinase, hybrid-type, ethylene sensor; five other genes encoding Serine/Threonine kinase receptor and receptor-like ser/thre kinases family proteins (RLK) were also physically nearby. It is well known that both kinases and RLK proteins play a central role in signaling during pathogen recognition and the subsequent activation of plant defense mechanisms [83-85]. They are also involved in wound-mediated defense response [86], and maintenance of plant cell wall integrity [87]. Polymorphisms at the proposed kinase genes could also be responsible for the QTLs at bin 7.03 detected for TL in a biparental population [30]. The candidate gene encoding the maize Calcium/calmodulin-dependent protein kinase (CDPK) close to the three significant SNPs on chromosome 10, could be playing an important role in the activation of defense against the attack of MCB since it has been known that the CDPK is induced by mechanical wounding by herbivore attack inducing accumulation of jasmonic acid in maize and other species [88-91].

Genes associated with SD made by MCB

A gene encoding an Actin polymerizing factor 4 (APF4) contains the second SNP at chromosome 1 significantly associated to SD. One of the functions of APF4 protein is remodeling the actin of cytoskeleton under different stimulus, including wounding and pathogen attacks [92]. Some studies in *Arabidopsis* indicated that the APF4 mediated defense signal and it is also relates with actin dynamic of cytoskeleton during the innate immune signaling [93-95]. The candidate gene for the SNP significantly associated to SD on chromosome 2 encodes a Leucine-rich receptor-like protein kinase (LRR-RLK). LRR-RLK protein family plays an important role in cell-cell signaling and other signals involving peptide in ligands. They are involved in systemic activation of protease inhibitors in response to wounding by insect feeding [96,97]. On the other hand, the significant SNP located on chromosome 5 is within a gene encoding a protein with unknown function, and it is interesting that the SNP is close to the gene *nactf30* which encodes a NAC domain protein transcription factor. As already mentioned, this gene and other transcription factor family members regulate the synthesis of secondary cell wall [98,99]. In addition, other authors have associated the NAC domain proteins with response to stresses made by herbivore attack [100].

Genes associated with KR made by MCB

The SNP on chromosome 3 significantly associated to KR was located nearby the *cystatin1* gene, whose product is the corn kernel cysteine proteinase inhibitor//cysteine proteinase inhibitor I (*psei1*), an anti-metabolic protein synthetized and stored in the maize kernel [101]. The expression of some proteinase inhibitor genes are induced in response to mechanical wounding and insect damage [102], and it has been shown that cysteine proteinase inhibitor interferes with the digestive process of insects [103,104]. LD is low in this region, but it is interesting that the SNP significantly associated with KR was in LD with a SNP positioned in the exonic region of the gene (Figure 3). Therefore, if the association found between the SNP at position 187,742,562 and KR is due to the linkage between that SNP and a certain undetected polymorphism in the *cysteine1* gene, the effect of this polymorphism would be expected to be especially large because the linkage between the significant SNP and SNPs in the *cystatin1* gene is low, although significant.

Another significant SNP associated to KR was located close to a gene encoding a basic Helix-Loop-Helix (bHLH) transcription factor family member. Recent studies in other plant species have demonstrated that this family gene protein has an important role in regulating jasmonic acid response [105-109]. The third significant SNP associated with KR located on chromosome 3 was within a gene encoding a starch branching enzyme interacting protein, whose deficiency leads to decreased digestibility of maize kernel [110]. No previous evidence about its involvement in resistance to insect was found, however. A gene that encodes a Germin 1–2 protein is 106 kb downstream from this associated SNP; these proteins are implicated in the response to abiotic and biotic stresses including the response to mechanical wounding and insect damage [111-114]. Another significant SNP located on chromosome 3 at position 222,733,400 is within a gene that encodes a Glycine rich protein (GRP), a structural protein which could be playing an important role in the fortification of plant cell wall [115-117], also this protein is wound-inducible [118,119]. A set of plant Thionin family protein precursor genes were found nearby the SNP on chromosome 3 at position 222,733,400. This finding is particularly interesting since these proteins belong to the Defensin family protein which has antibacterial, antifungal, and insecticide activities [120-124].

The two significant SNPs associated to KR on chromosome 4 are adjacent to a gene that encodes for an Ubiquitin thiolesterase. These proteins form complex systems of selective protein degradation [125,126] and mediate the biosynthesis of plant hormone signaling such as salicylic, jasmonic, and abscisic acid and auxins [127,128].

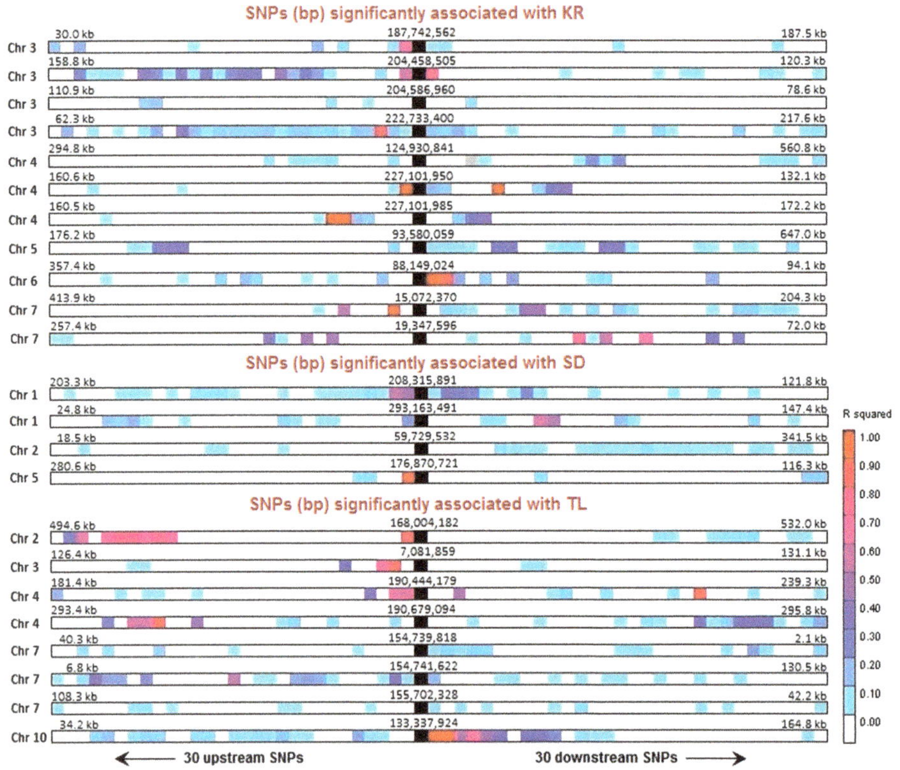

Figure 3 A region of approximately 280 kpb in chromosome 3 where a SNP significantly associated with borer kernel resistance (KR) has been found. The red points represent the *P*-values of the SNPs in the region mentioned above. The orange strip indicates the location of the SNP significantly associated with KR and the dark blue strips indicate genomic locations in LD ($r^2 > 0.2$) with the SNP significantly associated with KR. The solid black line and the green rectangle indicate the intronic and exonic region of the *cystatin1* gene, respectively. The SNP significantly associated with KR herein is in LD with the exonic region of the *cystatin1* gene, which encodes a maize cysteine proteinase inhibitor involved in plant resistance to insects.

The candidate gene containing the significant SNP associated to KR on chromosome 5 encodes a Phytosulfokine receptor (PSK), a recognition hormone whose level of expression has been increased by pathogens elicitors [129,130].

The candidate gene, adjacent to the significant SNP located on chromosome 6, encodes a Polyamine oxidase (propa-1,3-aimine-forming) (PAO). PAO plays an important role in stress tolerance by generating H_2O_2 which is a key component in signal transduction pathways leading to stress responses. In *Zea mays*, the function of PAO in wound-healing is likely due to increased lignin and suberine deposition as consequence of H_2O_2 release [131]. In addition, it has been described to play a role in cell wall stiffening and mediating abiotic and biotic stresses [132]. The candidate adjacent gene to the significant SNP associated to KR on chromosome 7 encodes a Catalase//L-ascorbate peroxidase, they are two major hydrogen peroxide-detoxifying enzymes whose activity is very important in the reduction of the oxidative stress caused by H_2O_2 [133,134], and the importance of these detoxifying enzymes in resistance to insect attack has been recently reported [135].

A candidate gene for the significant SNP on chromosome 7 at position 19,347,596 is adjacent to the SNP that encodes a Kinase associated protein phosphatase. This gene family has been proposed to regulate the response to different type of stresses including pathogens and herbivore attack [136]. It is interesting to note that another close gene to this SNP is the *ipt4* (*isopentenyl transferase4*) gene, which is involved in the regulation of cytokinin biosynthesis pathway [137]. The expression of an *itp* gen (fused with a wound-inducible promoter) in transgenic plants of *Nicotiana plumbaginifolia* decreased leaf consumption by *Manduca sexta* (lepidopterous) and reduced survival of *Myzus persicae* (aphid) [138]. Although there are several reports about the importance of cytokinins in the modulation of plant defense against pathogens and insect attack, their role is not clear [139].

Unlike markers linked to QTLs for resistance to corn borers, the SNPs associated with resistance in the present study could be incorporated in whole-genome predictor models in order to improve genomic selection [140]. Marker–assisted selection has proved an useful tool for improving resistance to the European corn borer

[141] of segregating bi-parental populations related to the mapping populations used for QTL detection, but could be inappropriate in non-structured populations. As additive effects are the most important genetic effects for resistance traits, crossing inbreds with improved resistance will render hybrids more resistant to attack by MCB larvae.

Conclusion

We conducted a genome wide association study for resistance traits to MCB with more than 245 kb SNP distributed through the whole genome. We found a set of significant SNPs associated to the three resistance traits to MCB attack. Of which 10 SNPs were significant associated to TL, 4 SNPs were associated to SD, and 11 SNPs were associated to KR. In general, each of these SNPs explain a considerable proportion of the phenotypic variance (R^2 = 6-9%). No co-localized SNPs were found for resistance and agronomic traits that could underlie the genetic correlations found between these traits.

Twenty one candidate genes were proposed for the three resistant traits, they were either containing or adjacent (within a window of ±130 kbp) to each significantly associated SNP.

Most of the candidate genes proposed herein are part of the signaling pathway, others act as regulator of expression under biotic stress condition, and a few genes are encoding enzymes with antibiotic effect against insects such as the *cystatin1* gen and the defensin proteins.

The identification of these polymorphisms associated to resistance traits to MCB attack can be useful to understand the molecular mechanisms that affect resistance and susceptibility of host plants to insect attack, in order to contribute to advance in the understanding of plant-insect interactions. Nevertheless further studies are necessary to validate the candidate genes identified herein.

Methods

Plant material and phenotypic data

The maize diversity panel (composed of 302 inbred lines) represents much of the diversity available in public breeding sector around the world. A subset of the maize diversity panel (henceforth we will refer to this population as "association panel") was evaluated for resistance to MCB attack at Pontevedra (42°24' N, 8°38' W, and 20 m above sea level), Spain, through three years (2010, 2011, and 2012). A subset of 270 inbred lines was assayed in an 18 × 15 α-lattice design with two replications in 2010 and 2011. In the third year a subset of 255 inbred lines was assayed (because we did not have enough seed for the remaining15 lines) in a 17 × 15 α-lattice design with two replications.

The trials were hand-planted and each experimental plot consisted of one row spaced 0.8 m apart from the other row with 29 two-kernel hills spaced 0.18 m apart. Plots were overplanted and thinned, obtaining a final density of ~70,000 plant ha^{-1}. The evaluations were performed under artificial infestation with eggs of MCB. The eggs for inoculation were obtained at the Misión Biológica de Galicia by rearing the insect as described by Eizaguirre and Albajes [142]. Five plants of each plot were infested with ~ 40 MCB eggs placed between the stem and the sheath of a basal leaf.

Data collected were: tunnel length (TL), the mean length of stem tunnels made by borers on the five infested plants, which were longitudinally split at the time of harvest; stem damage (SD) as the percentage of the stem damaged by MCB larvae on the five infested plants; kernel resistance to borer attack (KR) recorded at harvest as the damage on the main ear of the five infested plants according to a subjective visual resistance scale of 1 to 9 in which 1 indicates completely damaged and 9 indicates no damage; days to anthesis (DTA) and days to silking (DTS) as the days from planting to the date on which 50% of plants were shedding pollen or showing silks, respectively; and plant height (PH) on five representative plants as the distance from the ground to the top of the plant.

Genotypic data

We used a set of unique SNP markers derived from a Illumina maize 50 k array [56] and a genotyping-by-sequencing (GBS) strategy [48]. The two data sets were combined and filtered to exclude SNPs with more than 20% missing genotype data and minor allele frequency (MAF) less than 5%. Heterozygous genotypes were considered as missing data in the analysis. After filtering, a total of 246,477 SNPs (Additional file 2) distributed across the maize genome were used in this study.

A genetic kinship matrix (K) previously published by Olukolu et al. [52] was used for GWAS. The kinship matrix was estimated using a subset of 5000 SNPs without any missing genotypes and distributed approximately uniformly across the entire genome (at least 60 kbp between any two markers).

Statistical analyses

Best linear unbiased estimator (BLUE)

Each trial was analyzed separately with the SAS mixed model procedure (PROC MIXED) in SAS software version 9.3 [143] considering inbred lines as a fixed effect and replications and block within replication as random effects. Then, trials were combined using a mixed linear model across years and considering inbred lines as the only fixed effect. As large predicted values for stem damage and tunnel length were associated with larger residuals, a natural logarithmic transformation of TL and SD scores was used for obtaining BLUEs. The logarithmic transformation eliminated the relationship between residual variance and predicted values. Line BLUEs were back-transformed and then used to perform GWAS.

Heritabilities

Heritabilities ($\hat{h}^2$) for each year were estimated for each trait on a family-mean basis as described previously by Holland et al. [144]. The model for these analyses was similar to the model mentioned above with the exception that inbred lines were treated as random effects. The genetic (r_g) and phenotypic (r_p) correlations between traits were computed using REML estimation in SAS mixed model procedure following Holland [145].

Association analysis

Genome-wide association analysis based on mixed linear model (MLM) was performed in Tassel 4.1.26 [146]. The MLM used by Tassel was

$$y = X\beta + Zu + e$$

where **y** is the vector of phenotypes (BLUEs), **β** is a vector of fixed effects, including the SNP marker tested, **u** is a vector of random additive effects (inbred lines), **X** and **Z** represents matrices, and e is a vector of random residuals. The variance of random line effects was modeled as $\mathrm{Var}(u) = \mathbf{K}\sigma_a^2$, where **K** is the $n \times n$ matrix of pairwise kinship coefficient and σ_a^2 is the estimated additive genetic variance [147].

Restricted maximum likelihood estimates of variance components were obtained by using the optimum compression level (compressed MLM) and population parameters previously determined options (P3D) in Tassel [65]. The optimum compression level option reduces the computation demand by clustering the (n) total individuals into (s) groups based on their realized genomic relationships, allowing the original **K** matrix to be replaced by a smaller relationship matrix. The P3D option uses iteration to estimate population parameters such as genetic and residual variance only once in a model with no fixed marker effects, then uses those estimates without iteration in subsequent association tests for each marker. The combination of these two methods reduces computational time and improves model fit [65].

Threshold for GWAS

To identify SNPs with the most robust associations with traits, a subsampling or subagging procedure was employed in GWAS analysis [148,149]. Each of 100 subsampled datasets generated using the R software [150] comprised a random sample of 80% of inbred lines from the diversity population. Only SNP markers determined as significant at $P < 1 \times 10^{-4}$ and subsequently detected in ≥ 30 subsamples, i.e. resample model inclusion probability (RMIP) threshold of 0.30, were considered as significantly associated to the trait under study.

Linkage disequilibrium and candidate gene selection

We examined the linkage disequilibrium (LD) measure (r^2) with each SNP significantly associated with resistance traits in a region of 60 SNPs (ranging from 0.15 to 1 Mbp). For each trait, the linkage disequilibrium between significant SNPs was also calculated. The genes containing or adjacent to SNPs significantly associated with traits were identified and characterized by the use of MaizeGDB genome browser [67]. We examined a region around each significant SNP in order to identify candidate genes of interest. For most SNPs, more than two genes that could be involved in plant defense mechanism (based on previously published evidence) were preselected; the closest gene to the SNPs significantly associated with resistance was selected as candidate gene and the remaining genes were presented as other interesting genes that could be putatively involved in plant defense against insect herbivores.

Additional files

Additional file 1: Table S1. Mean squares and heritability estimates for MCB pest resistance traits evaluated in an association panel in three years. **Table S2** Mean squares and heritability estimates for agronomic traits evaluated in an association panel in three years. **Table S3** Phenotypic data of three resistance traits to MCB attack and three agronomic traits in a maize diversity panel. **Table S4** Annual and average means for traits related to resistance to MCB and three agronomic traits. **Table S5** Summary of the SNPs significantly associated to plant height. SNP identification (SNP ID), additive effect and allelic variants for the SNP, proportion of total variance explained by the SNPs significantly associated with plant height (PH) and significance values for the association between the SNP and the phenotype (*P*-value and RMIP). **Table S6** Summary of the SNPs significantly associated to days to anthesis. SNP identification (SNP ID), additive effect and allelic variants for the SNP, proportion of total variance explained by the SNPs significantly associated with days to anthesis (DTA) and significance values for the association between the SNP and the phenotype (*P*-value and RMIP). **Table S7** Summary of the SNPs significantly associated to days to silking. SNP identification (SNP ID), additive effect and allelic variants for the SNP, proportion of total variance explained by the SNPs significantly associated with days to silking (DTS) and significance values for the association between the SNP and the phenotype (*P*-value and RMIP). **Table S8** Linked disequilibrium (r^2) between the ten SNPs significantly associated to TL made by MCB.

Additional file 2: Genotypic data of a maize diversity panel with more than 245 kb SNP (GBS + Illumina). The data were filtered to exclude SNPs with more than 20% missing genotype data and minor allele frequency (MAF) less than 5%. Also the heterozygous are excluded. The array has 290 columns and 246,477 rows.

Abbreviations

APF4: Actin polymerizing factor 4; bHLH: Basic Helix-Loop-Helix; BLUE: Best linear unbiased estimator; Bt: Bacillus thuringiensis; CDPK: Calcium/calmodulin-dependent protein kinase; DTA: Days to anthesis; DTS: Days to silking; ECB: European corn borer; GBS: Genotyping by sequencing; GRP: Glycine rich protein; GWAS: Genome-wide association study; KR: Kernel resistance; LD: Linkage disequilibrium; LRR-RLK: Leucine-rich receptor-like protein kinase; MAF: Minor allele frequency; MCM: Mediterranean corn borer; PAO: Polyamine oxidase; PH: Plant height; PLA2: Phospholipase A2; PSK: Phytosulfokine receptor; QTL: Quantitative trait locus/loci; REML: Restricted maximum likelihood; RLK: Receptor like kinase; RMIP: Resample model inclusion probability; SD: Stem damage; SNP: Single nucleotide polymorphism; TL: Tunnel length; TPR: Tetratricopeptide repeat.

Competing interests

The authors declare that they have no competing interests.

Authors' contributions

LFS carried out field experiments, performed statistical and bioinformatics analysis of the data and made the draft of the manuscript. AB and RAM conceived the study, assisted LFS in field experiment design, and data collection and analysis. BAO and JBH prepared some arrays of data and assisted LFS in statistical and bioinformatics analysis. All authors read and approved the final manuscript.

Acknowledgements

This work was supported by the National Plan for Research and Development of Spain (projects AGL2012-33415). L.F. Samayoa acknowledges a contract JAE-Predoc from the Spanish Council for Scientific Research (CSIC). We also acknowledges to the Super Computation Center of Galicia (CESGA) for allowed to use their serves to do the preliminary analysis of this study. We thank Mike McMullen for inspiring this study and reviewing the manuscript.

Author details

[1]Misión Biológica de Galicia, Spanish National Research Council (CSIC), P.O. Box 2836080 Pontevedra, Spain. [2]Department of Crop Science, North Carolina State University, Raleigh, North Carolina 27695, USA. [3]U.S. Department of Agriculture-Agricultural Research Service, Plant Science Research Unit, Raleigh, North Carolina 27695, USA.

References

1. Meihls LN, Kaur H, Jander G. Natural variation in maize defense against insect herbivores. Cold Spring Harb Sym. 2012;77:269–83.
2. Malvar RA, Cartea ME, Revilla P, Ordás A, Alvarez A, Mansilla JP. Sources of resistance to pink stem borer and European corn borer in maize. Maydica. 1993;38:313–9.
3. Avantaggiato G, Quaranta F, Desiderio E, Visconti A. Fumonisin contamination of maize hybrids visibly damaged by *Sesamia*. J Sci Food Agric. 2002;83:13–8.
4. Visconti A, Marasas WFO, Miller JD, Riley R. Mycotoxins of growing interest: Fumonisins. In: Book Mycotoxins of growing interest: Fumonisins. 1999. p. 1–22.
5. Cordero A, Malvar RA, Butrón A, Revilla P, Velasco P, Ordás A. Population dynamics and life-cycle of corn borers in South Atlantic European Coast. Maydica. 1998;43:5–12.
6. Velasco P, Revilla P, Monetti L, Butrón A, Ordás A, Malvar RA. Corn borers (Lepidoptera: Noctuidae; Crambidae) northwestern Spain: population dinamics and distribution. Maydica. 2007;52:195–203.
7. Meissle M, Romeis J, Bigler F. *Bt* maize and integrated pest management - a European perspective. Pest Manag Sci. 2011;67:1049–58.
8. Speiser B, Tamm L. Regulation of plant protection in organic farming. In: Ehlers R-U, editor. Regulation of Biological Control Agents. Kiel, Germany: Springer; 2011. p. 113–25.
9. Campagne P, Kruger M, Pasquet R, Le Ru B, Van den Berg J. Dominant inheritance of field-evolved resistance to *Bt* corn in *Busseola fusca*. PLoS One. 2013;8:e69675.
10. Tabashnik BE, Brévault T, Carrière Y. Insect resistance to Bt crops: lessons from the first billion acres. Nat Biotechnol. 2013;31:510–21.
11. González-Cabrera J, García M, Hernández-Crespo P, Farinós GP, Ortego F, Castañera P. Resistance to Bt maize in *Mythimna unipuncta* (Lepidoptera: Noctuidae) is mediated by alteration in Cry1Ab protein activation. Insect Biochem Mol Biol. 2013;43:635–43.
12. Gould F. Sustainability of transgenic insecticidal cultivars: integrating pest genetics and ecology. Annu Rev Entomol. 1998;43:701–26.
13. Butrón A, Malvar R, Velasco P, Vales M, Ordás A. Combining abilities for maize stem antibiosis, yield loss, and yield under infestation and non infestation with pink stem borer. Crop Sci. 1999;39:691–6.
14. Cartea ME, Malvar RA, Butrón A, Vales MI, Ordás A. Inheritance of antibiosis to *Sesamia nonagrioides* (Lepidoptera: Noctuidae) in maize. J Econ Entomol. 1999;92:994–8.
15. Papst C, Bohn M, Utz HF, Melchinger AE, Klein D, Eder J. QTL mapping for European corn borer resistance (*Ostrinia nubilalis* Hb.), agronomic and forage quality traits of testcross progenies in early- maturing European maize (*Zea mays* L.) germplasm. Theor Appl Genet. 2004;108:1545–54.
16. Schön CC, Lee M, Melchinger AE, Guthrie WD, Woodman WL. Mapping and characterization of quantitative trait loci affecting resistancie against second-generation European corn borer in maize with the aid of RFLPs. Heredity. 1993;70:648–59.
17. Cardinal AJ, Lee M, Sharopova N, Woodman WL, Long MJ. Genetic mapping and analysis of quantitative trait loci for resistance to stalk tunneling by European corn borer in maize. Crop Sci. 2001;41:835–45.
18. Ordás B, Malvar RA, Santiago R, Sandoya G, Romay MC, Butrón A. Mapping of QTL for resistance to the Mediterranean corn borer attack using the intermated B73 × Mo17 (IBM) population in maize. Theor Appl Genet. 2009;119:1451–9.
19. Klenke J, Russel W, Guthrie W. Recurrent selection for resistance to European corn borer in a corn synthetic and correlated effects on agronomic traits. Crop Sci. 1986;26:864–8.
20. Bosque-Pérez N, Kling J, Odubiyi S. Recent advances in the development of sources of resistance to pink stalk borer and African sugarcane borer. In: Insect resistant maize: Recent advances and utilization Proc of an International Symposium. 1997. p. 234–40.
21. Sandoya G, Butrón A, Alvarez A, Ordás A, Malvar RA. Direct response of a maize synthetic to recurrent selection for resistance to stem borers. Crop Sci. 2008;48:113–8.
22. Butrón A, Romay MC, Peña-Asin J, Alvarez A, Malvar RA. Genetic relationship between maize resistance to corn borer attack and yield. Crop Sci. 2012;52:1176–80.
23. Ordás B, Butrón A, Alvarez A, Revilla P, Malvar R. Comparison of two methods of reciprocal recurrent selection in maize (*Zea mays* L.). Theor Appl Genet. 2012;124:1183–91.
24. Malvar RA, Butrón A, Revilla P, Ordás A. Resistance to the pink stem borer, *Sesamia nonagrioides*, in maize. Recent Res Devel Plant Sci. 2004;2:1–32.
25. Cartea ME, Malvar RA, Vales I, Butrón A, Ordás A. Inheritance of resistance to ear damage caused by *Sesamia nonagrioides* (Lepidoptera: Noctuidae) in maize. J Econ Entomol. 2001;94:277–83.
26. Hudon M, Chiang MS. Evaluation of resistance of maize germplasm to the univoltine European corn borer *Ostrinia nubilalis* (Hübner) and relationship with maize maturity in Quebec. Maydica. 1991;36:69–74.
27. Guthrie W, Russell W. Breeding methodologies and genetic basis of resistance in maize to the European corn borer. In: International Symposium on Methodologies for Developing Host Plant Resistance to Maize Insects Mexico, DF (Mexico) 9–14 Mar 1987. 1989.
28. Krakowsky MD, Brinkman MJ, Woodman WL, Lee M. Genetic components of resistance to stalk tunneling by the European corn borer in maize. Crop Sci. 2002;42:1309–15.
29. Krakowsky MD, Lee M, Holland JB. Genotypic correlation and multivariate QTL analyses for cell wall components and resistance to stalk tunneling by the European corn borer in maize. Crop Sci. 2007;47:485–8.
30. Krakowsky MD, Lee M, Woodman WL, Long MJ, Sharopova N. QTL mapping of resistance to stalk tunneling by the European corn borer in RILs of maize population B73 × De811. Crop Sci. 2004;44:274–82.
31. Orsini E, Krchov L, Uphaus J, Melchinger A. Mapping of QTL for resistance to first and second generation of European corn borer using an integrated SNP and SSR linkage map. Euphytica. 2012;183:197–206.
32. Ordás B, Malvar RA, Santiago R, Butrón A. QTL mapping for Mediterranean corn borer resistance in European flint germplasm using recombinant inbred lines. BMC Genomics. 2010;11:1–10.
33. Bohn M, Khairallah MM, González-de-León D, Hoisington DA, Utz HF, Deutsch JA, et al. QTL mapping in tropical maize: I. Genomic regions affecting leaf feeding resistance to sugarcane borer and other traits. Crop Sci. 1996;36:1352–61.
34. Groh S, González-de-León D, Khairallah MM, Jiang C, Bergvinson D, Bohn M, et al. QTL mapping in tropical maize III. Genomic regions for resistance to Diatraea spp and associated traits in two RIL populations. Crop Sci. 1998;1998:1062–72.
35. Bohn M, Khairallah M, Jiang C, González-de-León D, Hoisington D, Utz H, et al. QTL mapping in tropical maize: II. Comparison of genomic regions for resistance to *Diatraea spp*. Crop Sci. 1997;37:1892–902.

36. Weiss LA, Arking DE, Daly MJ, Chakravarti A, Brune CW, West K, et al. A genome-wide linkage and association scan reveals novel loci for autism. Nature. 2009;461:802–8.
37. Duerr RH, Taylor KD, Brant SR, Rioux JD, Silverberg MS, Daly MJ, et al. A genome-wide association study identifies IL23R as an inflammatory bowel disease gene. Science. 2006;314:1461–3.
38. Sladek R, Rocheleau G, Rung J, Dina C, Shen L, Serre D, et al. A genome-wide association study identifies novel risk loci for type 2 diabetes. Nature. 2007;445:881–5.
39. Barendse W, Reverter A, Bunch RJ, Harrison BE, Barris W, Thomas MB. A validated whole-genome association study of efficient food conversion in cattle. Genetics. 2007;176:1893–905.
40. Kijas JW, Townley D, Dalrymple BP, Heaton MP, Maddox JF, McGrath A, et al. A genome wide survey of SNP variation reveals the genetic structure of sheep breeds. PLoS One. 2009;4:e4668.
41. Bolormaa S, Hayes B, Savin K, Hawken R, Barendse W, Arthur P, et al. Genome-wide association studies for feedlot and growth traits in cattle. J Anim Sci. 2011;89:1684–97.
42. Fan B, Onteru SK, Du Z-Q, Garrick DJ, Stalder KJ, Rothschild MF. Genome-wide association study identifies loci for body composition and structural soundness traits in pigs. PLoS One. 2011;6:e14726.
43. Tian F, Bradbury PJ, Brown PJ, Hung H, Sun Q, Flint-Garcia S, et al. Genome-wide association study of leaf architecture in the maize nested association mapping population. Nat Genet. 2011;43:159–62.
44. Neumann K, Kobiljski B, Denčić S, Varshney R, Börner A. Genome-wide association mapping: a case study in bread wheat (*Triticum aestivum* L.). Mol Breed. 2011;27:37–58.
45. Zhu C, Gore M, Buckler ES, Yu J. Status and prospects of association mapping in plants. Plant Genome. 2008;1:5–20.
46. Peiffer JA, Flint-Garcia SA, De Leon N, McMullen MD, Kaeppler SM, Buckler ES. The genetic architecture of maize stalk strength. PLoS One. 2013;8:e67066.
47. Oraguzie NC, Rikkerink EH, Gardiner SE, Silva H. Association mapping in plants. New York: Springer-Verlag GmbH; 2007.
48. Romay MC, Millard MJ, Glaubitz JC, Peiffer J, Swarts K, Casstevens TM, et al. Comprehensive genotyping of the USA national maize inbred seed bank. Genome Biol. 2013;14:1–18.
49. Flint-Garcia SA, Thuillet A-C, Yu J, Pressoir G, Romero SM, Mitchell SE, et al. Maize association population: a high-resolution platform for quantitative trait locus dissection. Plant J. 2005;44:1054–64.
50. Liu K, Goodman M, Muse S, Smith JS, Buckler ES, Doebley J. Genetic structure and diversity among maize inbred lines as inferred from DNA microsatellites. Genetics. 2003;165:2117–28.
51. Cook JP, McMullen MD, Holland JB, Tian F, Bradbury P, Ross-Ibarra J, et al. Genetic architecture of maize kernel composition in the nested association mapping and inbred association panels. Plant Physiol. 2012;158:824–34.
52. Olukolu BA, Negeri A, Dhawan R, Venkata B, Sharma P, Garg A, et al. A connected set of genes associated with programmed cell death implicated in controlling the hypersensitive response in maize. Genetics. 2013;193:609–20.
53. Zila CT, Samayoa LF, Santiago R, Butrón A, Holland JB. A genome-wide association study reveals genes associated with fusarium ear rot resistance in a maize core diversity panel. G3-Genes Genomes Genet. 2013;3:2095–104.
54. Atwell S, Huang YS, Vilhjálmsson BJ, Willems G, Horton M, Li Y, et al. Genome-wide association study of 107 phenotypes in *Arabidopsis thaliana* inbred lines. Nature. 2010;465:627–31.
55. Chan EK, Rowe HC, Kliebenstein DJ. Understanding the evolution of defense metabolites in *Arabidopsis thaliana* using genome-wide association mapping. Genetics. 2010;185:991–1007.
56. Ganal MW, Durstewitz G, Polley A, Bérard A, Buckler ES, Charcosset A, et al. A large maize (*Zea mays* L.) SNP genotyping array: development and germplasm genotyping, and genetic mapping to compare with the B73 reference genome. PLoS One. 2011;6:e28334.
57. Elshire RJ, Glaubitz JC, Sun Q, Poland JA, Kawamoto K, Buckler ES, et al. A robust, simple genotyping-by-sequencing (GBS) approach for high diversity species. PLoS One. 2011;6:e19379.
58. Glaubitz JC, Casstevens TM, Lu F, Harriman J, Elshire RJ, Sun Q, et al. TASSEL-GBS: A High Capacity Genotyping by Sequencing Analysis Pipeline. PLoS One. 2014;9:e90346.
59. Steel RG, Torrie JH. Principles and procedures of statistics. New York, Toronto, London: McGraw-Hill Book Company; 1960.
60. Schnable PS, Ware D, Fulton RS, Stein JC, Wei F, Pasternak S, et al. The B73 maize genome: complexity, diversity, and dynamics. Science. 2009;326:1112–5.
61. Butrón A, Malvar RA, Velasco P, Cartea ME, Ordás A. Combining abilities and reciprocal effects for maize ear resistance to pink stem borer. Maydica. 1998;43:117–22.
62. Velasco P, Revilla P, Butrón A, Ordás B, Ordás A, Malvar RA. Ear damage of sweet corn inbreds and their hybrids under multiple corn borer infestation. Crop Sci. 2002;42:724–9.
63. Velasco P, Malvar RA, Butrón A, Revilla P, Ordás A. Ear feeding resistance of sweet corn inbreds to pink stem borer. J Am Soc Hortic Sci. 1999;124:268–72.
64. Bohn M, Schulz B, Kreps R, Klein D, Melchinger AE. QTL mapping for resistance against the European corn borer (*Ostrinia nubilalis* H.) in early maturing European dent germplasm. Theor Appl Genet. 2000;101:907–17.
65. Zhang Z, Ersoz E, Lai C-Q, Todhunter RJ, Tiwari HK, Gore MA, et al. Mixed linear model approach adapted for genome-wide association studies. Nat Genet. 2010;42:355–60.
66. Samayoa LF, Butrón A, Malvar RA. QTL mapping for maize resistance and yield under infestation with *Sesamia nonagrioides*. Mol Breeding. 2014;34:1331–44.
67. Harper LC, Schaeffer ML, Thistle J, Gardiner J, Andorf C, Campbell D, et al. The MaizeGDB Genome Browser tutorial: one example of database outreach to biologists via video. Database. 2011;2011:1–7.
68. Allan RK, Ratajczak T. Versatile TPR domains accommodate different modes of target protein recognition and function. Cell Stress Chaperones. 2011;16:353–67.
69. Blatch GL, Lässle M. The tetratricopeptide repeat: a structural motif mediating protein-protein interactions. Bioessays. 1999;21:932–9.
70. Pogorelko GV, Mokryakova M, Fursova OV, Abdeeva I, Piruzian ES, Bruskin SA. Characterization of three Arabidopsis thaliana immunophilin genes involved in the plant defense response against P. syringae. Gene. 2014;538:12–22.
71. Hellmann HA, Smeekens S. Sugar sensing and signaling in plants. Front Plant Sci. 2014;5:113.
72. Granot D, David-Schwartz R, Kelly G. Hexose kinases and their role in sugar-sensing and plant development. Front Plant Sci. 2013;4:1–17.
73. Morkunas I, Ratajczak L. The role of sugar signaling in plant defense responses against fungal pathogens. Acta Physiol Plant. 2014;8:409–14.
74. Lee U, Rioflorido I, Hong SW, Larkindale J, Waters ER, Vierling E. The Arabidopsis ClpB/Hsp100 family of proteins: chaperones for stress and chloroplast development. Plant J. 2007;49:115–27.
75. Zhong R, Richardson EA, Ye Z-H. The MYB46 transcription factor is a direct target of SND1 and regulates secondary wall biosynthesis in *Arabidopsis*. Plant Cell. 2007;19:2776–92.
76. Zhong R, Lee C, Zhou J, McCarthy RL, Ye Z-H. A battery of transcription factors involved in the regulation of secondary cell wall biosynthesis in *Arabidopsis*. Plant Cell. 2008;20:2763–82.
77. Liu Y. Investigation of a KNAT7-BLH-OFP transcription factor complex involved in regulation of secondary cell wall biosynthesis in *Arabidopsis thaliana*. In: MSc Thesis. Vancouver, Canada: University of British Columbia, Botany; 2010.
78. Mitsuda N, Seki M, Shinozaki K, Ohme-Takagi M. The NAC transcription factors NST1 and NST2 of *Arabidopsis* regulate secondary wall thickenings and are required for anther dehiscence. Plant Cell. 2005;17:2993–3006.
79. Barros-Rios J, Malvar RA, Jung H-JG, Santiago R. Cell wall composition as a maize defense mechanism against corn borers. Phytochemistry. 2011;72:365–71.
80. Ryan CA. The systemin signaling pathway: differential activation of plant defensive genes. BBA-Protein Struct M. 2000;1477:112–21.
81. Meijer HJ, Munnik T. Phospholipid-based signaling in plants. Annu Rev Plant Biol. 2003;54:265–306.
82. Zhu-Salman K, Bi J-L, Liu T-X. Molecular strategies of plant defense and insect counter-defense. Insect Sci. 2005;12:3–15.
83. Afzal AJ, Wood AJ, Lightfoot DA. Plant receptor-like serine threonine kinases: Roles in signaling and plant defense. Mol Plant-Microbe Interact. 2008;21:507–17.
84. Nürnberger T, Kemmerling B. Receptor protein kinases–pattern recognition receptors in plant immunity. Trends Plant Sci. 2006;11:519–22.
85. Duan Y, Ge C, Liu S, Wang J, Zhou M. A two-component histidine kinase *Shk1* controls stress response, sclerotial formation and fungicide resistance in *Sclerotinia sclerotiorum*. Mol Plant Pathol. 2013;14:708–18.
86. G-i A, Kost C, Boland W. Herbivore-induced, indirect plant defences. BBA-Mol Cell Biol L. 2005;1734:91–111.
87. Engelsdorf T, Hamann T. An update on receptor-like kinase involvement in the maintenance of plant cell wall integrity. Ann Bot. 2014;114:1339–47.

88. Szczegielniak J, Borkiewicz L, Szurmak B, Lewandowska-Gnatowska E, Statkiewicz M, Klimecka M, et al. Maize calcium-dependent protein kinase (ZmCPK11): local and systemic response to wounding, regulation by touch and components of jasmonate signaling. Physiol Plant. 2012;146:1–14.
89. Yang D-H, Hettenhausen C, Baldwin IT, Wu J. Silencing *Nicotiana attenuata* calcium-dependent protein kinases, *CDPK4* and *CDPK5*, strongly up-regulates wound-and herbivory-induced jasmonic acid accumulations. Plant Physiol. 2012;159:1591–607.
90. Szczegielniak J, Klimecka M, Liwosz A, Ciesielski A, Kaczanowski S, Dobrowolska G, et al. A wound-responsive and phospholipid-regulated maize calcium-dependent protein kinase. Plant Physiol. 2005;139:1970–83.
91. Ma F, Lu R, Liu H, Shi B, Zhang J, Tan M, et al. Nitric oxide-activated calcium/calmodulin-dependent protein kinase regulates the abscisic acid-induced antioxidant defence in maize. J Exp Bot. 2012;63:4835–47.
92. Staiger CJ, Gibbon BC, Kovar DR, Zonia LE. Profilin and actin-depolymerizing factor: modulators of actin organization in plants. Trends Plant Sci. 1997;2:275–81.
93. Tian M, Chaudhry F, Ruzicka DR, Meagher RB, Staiger CJ, Day B. Arabidopsis actin-depolymerizing factor AtADF4 mediates defense signal transduction triggered by the *Pseudomonas syringae* effector AvrPphB. Plant Physiol. 2009;150:815–24.
94. Henty-Ridilla JL, Li J, Day B, Staiger CJ. Actin depolymerizing factor4 regulates actin dynamics during innate immune signaling in *Arabidopsis*. Plant Cell. 2014;26:340–52.
95. Porter K, Shimono M, Tian M, Day B. Arabidopsis Actin-Depolymerizing Factor-4 links pathogen perception, defense activation and transcription to cytoskeletal dynamics. PLoS Pathog. 2012;8:e1003006.
96. Wu Y, Zhou JM. Receptor-Like Kinases in Plant Innate Immunity. J Integr Plant Biol. 2013;55:1271–86.
97. Jones DA, Jones J. The role of leucine-rich repeat proteins in plant defences. Adv Bot Res. 1997;24:90–168.
98. McCann MC, Carpita NC. Designing the deconstruction of plant cell walls. Curr Opin Plant Biol. 2008;11:314–20.
99. Zhong R, Demura T, Ye Z-H. SND1, a NAC domain transcription factor, is a key regulator of secondary wall synthesis in fibers of *Arabidopsis*. Plant Cell. 2006;18:3158–70.
100. Nogueira FT, Schlögl PS, Camargo SR, Fernandez JH. SsNAC23, a member of the NAC domain protein family, is associated with cold, herbivory and water stress in sugarcane. Plant Sci. 2005;169:93–106.
101. Abe M, Abe K, Kuroda M, Arai S. Corn kernel cysteine proteinase inhibitor as a novel cystatin superfamily member of plant origin. Eur J Biochem. 1992;209:933–7.
102. McMullen MD, Frey M, Degenhardt J. Genetics and biochemistry of insect resistance in maize. In: Bennetzen JL, Hake SC, editors. *Handbook of maize: Its biology*. NY, USA: Springer; 2009. p. 271–89.
103. Ussuf K, Laxmi N, Mitra R. Proteinase inhibitors: plant-derived genes of insecticidal protein for developing insect-resistant transgenic plants. Curr Sci India. 2001;80:847–53.
104. Fabrick J, Behnke C, Czapla T, Bala K, Rao A, Kramer K, et al. Effects of a potato cysteine proteinase inhibitor on midgut proteolytic enzyme activity and growth of the southern corn rootworm, *Diabrotica undecimpunctata howardi* (Coleoptera: Chrysomelidae). Insect Biochem Mol Biol. 2002;32:405–15.
105. Zhang H, Hedhili S, Montiel G, Zhang Y, Chatel G, Pré M, et al. The basic helix-loop-helix transcription factor CrMYC2 controls the jasmonate-responsive expression of the *ORCA* genes that regulate alkaloid biosynthesis in *Catharanthus roseus*. Plant J. 2011;67:61–71.
106. Niu Y, Figueroa P. Characterization of JAZ-interacting bHLH transcription factors that regulate jasmonate responses in *Arabidopsis*. J Exp Bot. 2011;62:2143–54.
107. Sasaki-Sekimoto Y, Jikumaru Y, Obayashi T, Saito H, Masuda S, Kamiya Y, et al. Basic Helix-Loop-Helix transcription factors JASMONATE-ASSOCIATED MYC2-LIKE1 (JAM1), JAM2, and JAM3 are negative regulators of jasmonate responses in Arabidopsis. Plant Physiol. 2013;163:291–304.
108. Qi T, Huang H, Wu D, Yan J, Qi Y, Song S, et al. Arabidopsis DELLA and JAZ Proteins Bind the WD-Repeat/bHLH/MYB Complex to Modulate Gibberellin and Jasmonate Signaling Synergy. Plant Cell. 2014;26:1118–33.
109. Song S, Qi T, Fan M, Zhang X, Gao H, Huang H, et al. The bHLH subgroup IIId factors negatively regulate jasmonate-mediated plant defense and development. PLoS Genet. 2013;9:e1003653.
110. Xia H, Yandeau-Nelson M, Thompson DB, Guiltinan MJ. Deficiency of maize starch-branching enzyme i results in altered starch fine structure, decreased digestibility and reduced coleoptile growth during germination. BMC Plant Biol. 2011;11:1–13.
111. Dunwell JM, Gibbings JG, Mahmood T, Saqlan Naqvi S. Germin and germin-like proteins: evolution, structure, and function. Crit Rev Plant Sci. 2008;27:342–75.
112. Davidson RM, Reeves PA, Manosalva PM, Leach JE. Germins: A diverse protein family important for crop improvement. Plant Sci. 2009;177:499–510.
113. Breen J, Bellgard M. Germin-like proteins (GLPs) in cereal genomes: gene clustering and dynamic roles in plant defence. Funct Integr Genomics. 2010;10:463–76.
114. Armstrong JS, Abdel-Mageed H, Fokar M, Allen R, Adamczyk Jr JJ. Dietary effects of cotton tissue expressing germin like protein on beet armyworm (Lepidoptera: Noctuidae) growth, survival and pupation. Fla Entomol. 2013;96:693–700.
115. Ryser U, Schorderet M, Guyot R, Keller B. A new structural element containing glycine-rich proteins and rhamnogalacturonan I in the protoxylem of seed plants. J Cell Sci. 2004;117:1179–90.
116. Ringli C, Keller B, Ryser U. Glycine-rich proteins as structural components of plant cell walls. Cell Mol Life Sci. 2001;58:1430–41.
117. Mangeon A, Junqueira RM, Sachetto-Martins G. Functional diversity of the plant glycine-rich proteins superfamily. Plant Signal Behav. 2010;5:99–104.
118. Showalter AM. Structure and function of plant cell wall proteins. Plant Cell. 1993;5:9.
119. Sturm A. A wound-inducible glycine-rich protein from Daucus carota with homology to single-stranded nucleic acid-binding proteins. Plant Physiol. 1992;99:1689–92.
120. Bohlmann H, Broekaert W. The role of thionins in plant protection. Crit Rev Plant Sci. 1994;13:1–16.
121. De Coninck B, Cammue B, Thevissen K. Modes of antifungal action and *in planta* functions of plant defensins and defensin-like peptides. Fungal Biol Rev. 2013;26:109–20.
122. Shiau Y-S, Horng S-B, Chen C-S, Huang P-T, Lin C, Hsueh Y-C, et al. Structural analysis of the unique insecticidal activity of novel mungbean defensin VrD1 reveals possibility of homoplasy evolution between plant defensins and scorpion neurotoxins. J Mol Recognit. 2006;19:441–50.
123. Bohlmann H, Apel K. Thionins. Annu Rev Plant Biol. 1991;42:227–40.
124. Kessler A, Baldwin IT. Plant responses to insect herbivory: The emerging molecular analysis. Annu Rev Plant Biol. 2002;53:299–328.
125. Hershko A, Ciechanover A. The ubiquitin system for protein degradation. Annu Rev Biochem. 1992;61:761–807.
126. Zeng L-R, Vega-Sánchez ME, Zhu T, Wang G-L. Ubiquitination-mediated protein degradation and modification: an emerging theme in plant-microbe interactions. Cell Res. 2006;16:413–26.
127. Kelley DR, Estelle M. Ubiquitin-mediated control of plant hormone signaling. Plant Physiol. 2012;160:47–55.
128. Santner A, Estelle M. The ubiquitin-proteasome system regulates plant hormone signaling. Plant J. 2010;61:1029–40.
129. Igarashi D, Tsuda K, Katagiri F. The peptide growth factor, phytosulfokine, attenuates pattern-triggered immunity. Plant J. 2012;71:194–204.
130. Rodriguez PA. Analysis of *AtPSKR1*, an LRR receptor protein kinase, and other PSK-signalling components in plant defence responses. In: PhD Thesis. Tübingen, Germany: University of Tübingen, Mathematics and Natural Sciences; 2011.
131. Angelini R, Tisi A, Rea G, Chen MM, Botta M, Federico R, et al. Involvement of polyamine oxidase in wound healing. Plant Physiol. 2008;146:162–77.
132. Wimalasekera R, Tebartz F, Scherer GF. Polyamines, polyamine oxidases and nitric oxide in development, abiotic and biotic stresses. Plant Sci. 2011;181:593–603.
133. Rizhsky L, Hallak-Herr E, Van Breusegem F, Rachmilevitch S, Barr JE, Rodermel S, et al. Double antisense plants lacking ascorbate peroxidase and catalase are less sensitive to oxidative stress than single antisense plants lacking ascorbate peroxidase or catalase. Plant J. 2002;32:329–42.
134. Willekens H, Chamnongpol S, Davey M, Schraudner M, Langebartels C, Van Montagu M, et al. Catalase is a sink for H_2O_2 and is indispensable for stress defence in C_3 plants. EMBO J. 1997;16:4806–16.
135. Kaur R, Gupta AK, Taggar GK. Role of catalase, H_2O_2 and phenolics in resistance of pigeonpea towards *Helicoverpa armigera* (Hubner). Acta Physiol Plant. 2014;36:1513–27.
136. Bajsa J, Pan Z, Duke SO. Serine/threonine protein phosphatases: Multi-purpose enzymes in control of defense mechanisms. Plant Signal Behav. 2011;6:1921.
137. Vyroubalová Š, Václavíková K, Turečková V, Novák O, Šmehilová M, Hluska T, et al. Characterization of new maize genes putatively involved in cytokinin metabolism and their expression during osmotic stress in relation to cytokinin levels. Plant Physiol. 2009;151:433–47.

138. Smigocki A, Neal Jr J, McCanna I, Douglass L. Cytokinin-mediated insect resistance in *Nicotiana* plants transformed with the *ipt* gene. Plant Mol Biol. 1993;23:325–35.
139. Giron D, Frago E, Glevarec G, Pieterse CM, Dicke M. Cytokinins as key regulators in plant-microbe-insect interactions: connecting plant growth and defence. Funct Ecol. 2013;27:599–609.
140. De Los CG, Gianola D, Allison DB. Predicting genetic predisposition in humans: the promise of whole-genome markers. Nat Rev Genet. 2010;11:880–6.
141. Flint-Garcia SA, Darrah LL, McMullen MD, Hibbard BE. Phenotypic versus marker-assisted selection for stalk strength and second-generation European corn borer resistance in maize. Theor Appl Genet. 2003;107:1331–6.
142. Eizaguirre M, Albajes R. Diapause induction in the stem corn borer *Sesamia nonagrioides* (Lepidoptera, Noctuidae). Entomol Gen. 1992;17:277–83.
143. SAS Institute Inc. SAS 9.3 Guide to software updates. Cary, NC, USA, SAS Institute Inc. 2011; 314 p
144. Holland JB, Nyquist WE, Cervantes-Martínez CT. Estimated an interpreting heritability for plant breeding: An update. In: Janick J, editor. Plant Breeding Reviews, vol. Volume 22. Hoboken, New Jersey, USA: Jonh Wiley & Sons; 2003. p. 9–112.
145. Holland JB. Estimating genotypic correlations and their standard errors using multivariate restricted maximum likelihood estimation with SAS Proc MIXED. Crop Sci. 2006;46:642–56.
146. Bradbury PJ, Zhang Z, Kroon DE, Casstevens TM, Ramdoss Y, Buckler ES. TASSEL: software for association mapping of complex traits in diverse samples. Bioinformatics. 2007;23:2633–5.
147. Yu J, Pressoir G, Briggs WH, Vroh I, Yamasaki I, Doebley JF, et al. A unified mixed-model method for association mapping that accounts for multiple levels of relatedness. Nat Genet. 2006;38:203–8.
148. Valdar W, Holmes CC, Mott R, Flint J. Mapping in structured populations by resample model averaging. Genetics. 2009;182:1263–77.
149. Panagiotou OA, Ioannidis JP. What should the genome-wide significance threshold be? Empirical replication of borderline genetic associations. Int J Epidemiol. 2012;41:273–86.
150. R Core Team. R: A languange and environment for statistical computing. In: Book R: A languange and environment for statistical computing (Editor ed. ^eds.). Vienna, Austria: City: R Foundation for Statistical Computing; 2013.

3

Identification, cloning and characterization of an ultrapetala transcription factor *CsULT1* from *Crocus*: a novel regulator of apocarotenoid biosynthesis

Nasheeman Ashraf[1*], Deepti Jain[2] and Ram A Vishwakarma[3]

Abstract

Background: *Crocus sativus* is a triploid sterile plant with long red stigmas which form commercial saffron. Saffron is the site for synthesis and accumulation of apocarotenoids like crocin, picrocrin and safranal which are responsible for its color, flavour and aroma making it world's most expensive spice. These compounds are formed by oxidative cleavage of zeaxanthin by carotenoid cleavage dioxygenases. Although the biosynthetic pathway of apocarotenoids is known to a considerable extent, the mechanism that regulates its tissue and developmental stage specific expression is not known.

Results: In the present work, we identified, cloned and characterized ultrapetala transcription factor called *CsULT1* from *Crocus*. The gene contains an 80 amino acid long conserved SAND domain. The *CsULT1* transcript was more abundant in stigma and showed increase in expression from pre anthesis stage till anthesis and decreased in post anthesis stage which corroborated with the accumulation pattern of crocin indicating its possible role in regulation of apocarotenoid biosynthesis. *CsULT1* was found to be transcriptionally active and localized in nucleus. Its expression is induced in response to phytohormones like auxin, methyljasmonate and salicylic acid. Overexpression of *CsULT1* in *Crocus* calli resulted in enhanced expression of key pathway genes like *phytoene synthase* (*PSY*), *phytoene desaturase* (*PDS*), *beta carotene hydroxylase* (*BCH*) and *carotenoid cleavage dioxygenases* (*CCDs*) indicating its role in regulation of apocarotenoid biosynthesis.

Conclusion: This work presents first report on isolation and characterization of ultrapetala gene from *Crocus*. Our results suggest that *CsULT1* is a novel regulator of *Crocus* apocarotenoid biosynthesis. We show for the first time involvement of plant SAND domain proteins in regulating secondary metabolic pathways.

Keywords: Ultrapetala, Crocus, Stigma, Saffron, Carotenoids, Apocarotenoids, SAND domain

Background

Crocus sativus L. (Iridaceae) is a sterile triploid plant propagated vegetatively through corms [1]. The desiccated stigma of *C. sativus* forms saffron and is source of various carotenoids and unique compounds called apocarotenoids which are produced by oxidative tailoring of carotenoids [2]. Apocarotenoids are synthesized in a number of plants including maize, tomato, *Arabidopsis*, *Crocus* etc. but *Crocus* finds a special place because it is the only plant which produces crocin, picrocrocin and saffranal in significant quantities [3]. The saffron apocarotenoids are formed by zeaxanthin cleavage [4] followed by specific glycosylation steps [5]. Because of the presence of these unique apocarotenoids *Crocus* stands apart from other related crops and is considered as one of the world's costliest spices [6]. Besides, saffron apocarotenoids also have tremendous pharmacological properties and have been used for the treatment of a wide range of cancers [7,8].

* Correspondence: nashraf@iiim.ac.in
[1]Plant Biotechnology Division, CSIR- Indian Institute of Integrative Medicine, Sanat Nagar, Srinagar J&K-190005, India
Full list of author information is available at the end of the article

Carotenoids and their cleavage products are synthesized by plastid localized methylerythritol phosphate (MEP) pathway. Biosynthesis of these compounds is regulated throughout the life cycle of a plant and dynamic changes occur in their composition to match the prevailing developmental requirements and response to external environmental stimuli [9]. Although the carotenoid biosynthetic pathway has been studied to a considerable extent in many plants including *Crocus* but the fundamental knowledge regarding the regulation of carotenogenesis in plant cells is still in its infancy [10]. In *Crocus*, apocarotenoids are synthesized only in stigma part of the flower and that too in developmental stage specific manner, but nothing is known about the mechanism that regulates its synthesis. Therefore, it will be quite interesting to take a step towards unravelling the regulatory pathway of carotenoid/apocarotenoid biosynthesis in *Crocus*.

The major goal of the present study was to identify transcription factors that regulate apocarotenoid biosynthesis in saffron. It is a well established fact that stigma part of the *Crocus* flower is the actual site for synthesis of many important apocarotenoids [11,12], however, we still attempted to study pattern of crocin accumulation (crocin being an important metabolite) in different parts of the flower and at different stages of stigma development. We also selected five transcription factors belonging to five different gene families from saffron gene database [1] and investigated their temporal and spatial expression profile. The results demonstrated that *ULTRAPETALA* (*ULT*) gene shows higher expression in stigma tissue and the expression increased till the day of anthesis and subsequently decreased. This expression profile matched with the accumulation pattern of crocin in saffron thereby indicating a possible role of this gene in regulating biosynthesis of apocarotenoids. The *ULT* encodes a small cysteïne rich protein containing a B-box like motif and a SAND domain, a DNA binding motif previously reported only in animal transcription factors [13]. This transcription factor has been proposed to act as regulator of developmental gene expression. In *Arabidopsis*, it functions in floral stem cell termination pathway [14]. *ULT* has been shown to act as trithorax group chromatin remodelling factor which regulates function of Agamous locus during stem cell termination [15]. Recently it has been demonstrated that *ULT1* acts as an antirepressor that promotes transcriptional activation by antagonizing PcG-induced histone methylation and, via physical interaction with *ATX1* that deposits H3K4me3 activating marks, promotes an open chromatin conformation to recruit proteins involved in transcriptional initiation and elongation [16]. More recently *ULT* was found to be involved in gynoecium formation [17]. *ULT1* function thus represents a novel chromatin-mediated mechanism that activates genes controlling stem cell fate in plants. This observation expands the repertoire of plant epigenetic regulators involved in developmental pathways and suggests involvement of chromatin mediated pathways in controlling dynamics of transcription during such pathways.

In this report we describe identification, isolation and characterization of *ULT* gene, *CsULT1*, from *Crocus sativus*. *CsULT1* is preferentially expressed in stigma and induced by phytohormones such as MJ, SA, 2,4-D. Further, *CsULT1* is localized in nucleus and is transcriptionally active. *Crocus* transformation has not yet been established. Here we studied transient overexpression of *CsULT1* in *Crocus* calli and observed that its overexpression upregulates some key carotenoid/apocarotenoid pathway genes. This work represents, to our knowledge, the first functional characterization of a *C. sativus ULT* gene and also first report on a transcriptional regulator of apocarotenoid biosynthetic pathway.

Methods

Plant material

Crocus sativus was grown in the experimental farm at Indian Institute of Integrative medicine (IIIM), Srinagar, India (longitude: 34°5′24′′N; latitude: 74°47′24′′ and altitude 1585 m above sea level). It was used as source plant material for the present study. The voucher specimen was deposited at Janaki Ammal Herbarium (RRLH), IIIM, Jammu. The details of the specimen are: (Accession number: 22893; Accession date: 12/01/2015; name of collector: Nasheeman Ashraf; Place of collection: IIIM, Srinagar Farm; Date of collection: 01/01/2015). For tissue specific expression profiling, on the day of flower opening, tepals, anthers and stigma were collected from flowers separately, frozen in liquid nitrogen and stored in −80°C till further use. For developmental stage specific expression, stigma was collected at three different stages viz three days before anthesis, on the day of anthesis and two day after anthesis. For hormone treatments, flowers were grown in pots and were mist sprayed with 100 μM methyljasmonate, 1 mM salicylic acid, 50 μM 2, 4-D and 100 μM ABA. Tissue samples were collected after 12 and 24 h of hormone treatment. For overexpression studies, calli overexpressing *CsULT1* and vector control calli were taken for RNA isolation. For each experiment, tissue from three biological replicates was pooled in.

Sample preparation and HPLC analysis

Crocin analysis was done as described by Moraga *et al.* [12]. For extract preparation, 0.5 mg tissue from tepals, anthers and stigma (collected at three different stages) was crushed with a micropestle in 700 μl Tris–HCl (50 mM, pH 7.5 containing 1 M NaCl), and incubated for 10 minutes on ice. This was followed by addition of

700 μl of chloroform. The extract was then incubated on ice for an additional 10 min. Centrifugation at 3000 g for 5 min at 4°C was done to separate the phases. The lower chloroform phase was evaporated and the dried residues were stored together with the upper aqueous phases at −80°C until high-performance liquid chromatography (HPLC) analysis. The LCMS apparatus of Nexera UHPLC (130 MPa) equipped with MS-8030 (Shimadzu) was used for the Study and data was generated using lab solutions software. Enable RP-C18 column (250 mm × 4.6 mm, 5 μm) was used. The injection volume was 5 μl and flow rate 0.3 ml/min. Mobile Phase A (Water and Acetonitrile ratio 1:1) and mobile phase B (0.1% Acetic acid in water) were used in a linear gradient flow and column temperature was set at 75°C initially.

Gene expression analysis using quantitative real time PCR

Total RNA was extracted from pooled tissue using TRIzol reagent and used for cDNA synthesis by Reverse Transcription kit (Fermentas) following manufacturer's instructions. qRT-PCR was performed in triplicates in ABI StepOne Real time (Applied biosystems). The reaction was carried out in a total volume of 20 μl, consisting of 10 μL of 2X SYBR Green Master Mix, 0.2 μM (each) gene specific primers for all the genes studied and 100 ng of template cDNA. The cycling parameters were 95°C for 20 s, followed by 40 cycles of 95°C for 15 s and 60°C for 1 min. The sequence of all the primers used in this study is given in Additional file 1. The specificity of each primer pair was validated by a dissociation curve (a single peak was observed for each primer pair) (Additional file 2). The relative quantification method (ΔΔ-CT) was used to evaluate quantitative variation between the replicates examined. The amplification of actin cDNA was used as an endogenous control to normalize all data.

Cloning of full length *CsULT1* gene

The partial clone of *CsULT1* was obtained using cDNA synthesized from *Crocus* flower RNA and primers (ULT-F and ULT-R) designed from EST sequence (cr.saCl000502:1) present in NCBI (www.ncbi.nlm.nih.gov/nucest). Sequence analysis of the partial clone revealed that it has the 3'end and only 5'end needs to be amplified in order to obtain the full length clone. Thus the full length cDNA clone of *CsULT1* was obtained by performing 5′RACE using gene specific primer (ULT-5') and UAP primer provided with the 5′RACE kit (Clontech) following manufacturer's instructions. The amplified product was run on 1% agarose gel and purified with gel extraction kit (Qiagen). The purified product was then cloned in the pGEM-T Easy vector and sequenced. For the amplification of full length clone, gene specific primers were designed from the full length nucleotide sequence as obtained from alignment of partial clone and the 5'RACE product. The full length cDNA clone was amplified by PCR using cDNA as template and the gene specific primer pair (CsULT-F and CsULT-R). The PCR product was run on 1% agarose gel, purified by gel extraction kit (Qiagen) and subsequently cloned into the pGEM-T Easy vector. The cycling conditions used were 3 min at 94°C, 30 cycles (30s at 94°C, 30 s at 60°C and 1 min at 72°C) and final extension for 10 min at 72°C. The nucleotide sequence of *CsULT1* was submitted to GenBank and the accession number is KM670459.

Sequence analyses

The full length nucleotide sequence of *CsULT1* was translated using Translate tool (http://web.expasy.org/translate/) and the properties of deduced amino acid sequence were estimated using ProtParam (http://web.expasy.org/protparam/). Multiple sequence alignment and phylogenetic analysis was performed using the ClustalW with the default parameters through the service of the European Bioinformatics Institute (http://www.ebi.ac.uk/Tools/msa/clustalw2).

Subcellular localization

The subcellular localization of *CsULT1* was studied by performing transient expression assay in onion epidermal cells. For this, *CsULT1* with restriction sites for *Nco*I and *Spe*I was amplified using ULTCam-F and ULTCam-R1 primer pair and fused in frame with 5' terminus of GFP reporter gene in pCAMBIA-1302. The cycling parameters were same as described above. The fusion construct of *CsULT1*-GFP was bombarded on to the onion peels using biolistic gene delivery device PDS-1000/He (Bio-Rad, USA). The onion peels were then incubated for 24 hours before visualizing in confocal microscope.

Transactivation assay

Full length protein coding sequence of *CsULT1* was cloned in yeast (*Saccharomyces cerevisiae*) expression vector pGBKT7 (Clontech) at *NdeI-EcoRI* site to express CsULT1 protein fused to GAL4 DNA-binding domain (GAL4-BD). The primers used were ULTGBKT-F and ULTGBKT-R and cycling parameters are same as described above. The resulting construct was transformed into Y187 yeast strain. The positive transformants were selected onto synthetic medium lacking tryptophan and leucine. Cells from two independent transformants were collected and assayed for β-galactosidase activity by using ortho-nitrophenyl-β-D-galactoside (ONPG) as substrate as described in clontech manual (PT3024-1).

Plant expression vector and transformation of *Crocus* calli

The *CsULT1* gene with *NcoI* and *SpeI* restriction sites was PCR amplified using ULTCam-F and ULTCam-R2 primer pair and cloned into pCAMBIA1302 vector, containing

CaMV 35S promoter. For the transient expression, the *Crocus* calli were arranged at the center of petri dish and the biolistic gene delivery device PDS-1000/He (Bio-Rad, USA) was used for transgene delivery via microprojectile bombardment. Plasmid DNA (at the concentration of 1 μg/μL) was coated on the surface of gold particles and bombarded on to the calli. The particle delivery system was adjusted to 1100 psi of helium pressure and 27 mmHg of vacuum pressure inside the chamber. After bombardment, the calli were transferred to fresh media and after five days they were again transferred to media containing hygromycin for selection of transgenic structures. After 10 days, the calli which did not harbour the *CsULT1*-pCAMBIA construct turned blackish whereas the ones with the gene construct looked fresh and were used for further experimental studies. The transgenic calli were screened by genomic PCR. For this, genomic DNA was isolated from independent *CsULT1* overexpression and empty control calli using the DNeasy Plant mini kit (Qiagen). The presence of *CsULT1* was confirmed by genomic PCR using gene specific primer and reverse primer corresponding to GFP. Transgenic and control calli were used for measuring the transcript levels of few carotenoid pathway genes including *PSY* (GenBank accession: AJ888514), *PDS* (GenBank accession: AY183118), *BCH* (GenBank accession: AJ937791). *CCD4b* (GenBank accession: EU523663.1) and *CCD2* (GenBank accession: KJ541749) using quantitative real time PCR as described above.

Results

Analysis of Crocin in different tissues and developmental stages

Since crocin is the most important metabolite in saffron and responsible for its coloring property, we measured its quantity in different parts of *Crocus* flower and at various stages of stigma development (pre anthesis, anthesis and post anthesis) using HPLC. Results indicated that crocin was present only in stigma part of flower. We were not able to detect crocin in other parts of the flower like tepals and anther. Further, its content showed increasing trend from pre anthesis to anthesis stage and later again decreased after anthesis (Figure 1). This was in confirmation with earlier reports [12] where they have shown that the major apocarotenoids like crocin and picrocrocin are detected in orange stage and increased rapidly during the following stages of stigma development till they reached maximum in scarlet stage at anthesis. This confirms that stigma is the site for synthesis and accumulation of major *Crocus* apocarotenoids and their synthesis is congruent with development of stigma reaching highest at anthesis stage.

Isolation and expression profiling of *CsULT1*

We aimed at identification of transcription factors which regulate biosynthesis of *Crocus* apocarotenoids. Towards this, five transcription factors belonging to different gene families were selected from saffron ESTs (www.ncbi.nlm.nih.gov/nucest). The selected genes were Myb (cr.saCl000348:1), MADS box (cr.saCl001329:1), WRKY (cr.saCl000652:1), Zinc finger (cr.saCl000359) and ULT (cr.saCl000502:1). The expression pattern for all these genes was investigated in various tissue types and at different developmental stages using quantitative real time PCR. Our results demonstrated that a *ULT* transcription factor showed higher induction in stigma part of the flower and its expression increased till the day of anthesis and then subsequently decreased (Figure 2a and b). This expression pattern corroborated with the accumulation pattern of apocarotenoids suggesting involvement of this gene in regulating biosynthesis of these compounds. Among other genes studied, only *Myb* showed higher expression in stigma as compared to other flower parts, however, its expression at different developmental stages of stigma did not match with the pattern of apocarotenoid accumulation (Additional file 3).

Full length *ULT* was cloned by RT- PCR and 5′ RACE and was named as *CsULT1* [GenBank accession number: KM670459]. The gene contains 708 bp open reading frame coding for 235 amino acids long protein (Additional file 4) with a predicted molecular mass of 26.5kD and pI 8.32. Domain search revealed presence of conserved SAND domain in *CsULT1* which normally consists of evolutionarily conserved 80 to 100 amino acid long DNA binding motif [18]. The sequence alignment of ULTs from various organisms has revealed two conserved cores in SAND domains viz TPxxFE and KDWK. While TPxxFE is perfectly conserved among all ULT proteins in plants, KDWK shows variability at primary level however, the secondary structure is conserved [19]. Alignment of *CsULT1* with other plant ortholgs (Figure 3) showed high sequence homology with ULT from *Phoenix* (79.57%), *Vitis vinifera* (78%), *Populus* (76%) and *Medicago* (75%). Sequence alignment showed that ULT proteins show significant homology along the entire length of the protein except at the extreme N terminus. Moreover, TPxxFE motif was present in *CsULT1* and was conserved among all the proteins used for alignment. Phylogenetic analysis of selected sequences placed *CsULT1* close to its homolog from *Phoenix* (Figure 4).

Subcellular localization of *CsULT1*

In order to have a preliminary understanding about the mechanism underlying the regulatory activity of *CsULT1*, its subcellular localization was investigated. Programs like Prosite and PSORT revealed absence of any sorting signal and predicted *CsULT1* to be localized in cytosol. For confirming the localization experimentally, *CsULT1* was cloned in frame with GFP reporter gene. The expression of the fusion gene construct *CsULT1*-GFP was driven by the 35S promoter of cauliflower mosaic virus (*CaMV-*

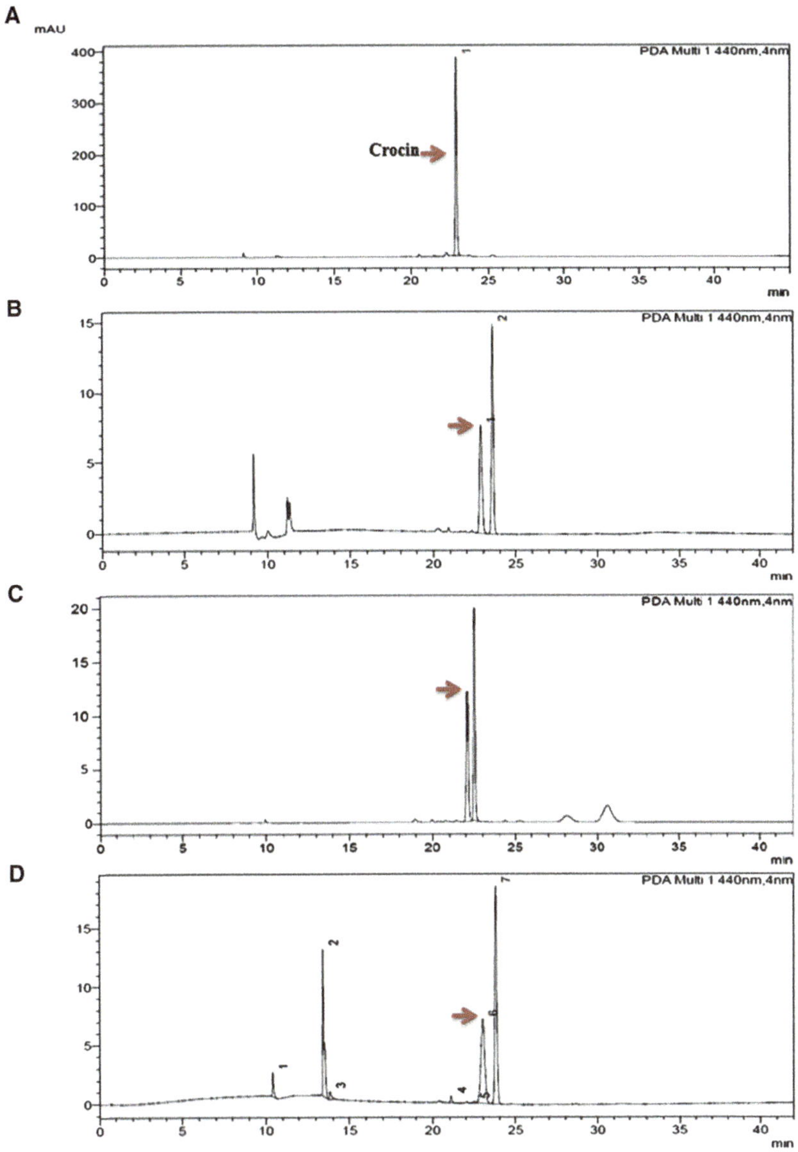

Figure 1 HPLC chromatograms for crocin. (A) represents chromatogram for crocin standard, **(B-D)** chromatogramas for *Crocus* stigma collected at pre anthesis, anthesis and post anthesis stages.

35S). The fusion gene was introduced into onion (*Allium cepa)* epidermal cells by particle bombardment. While the control GFP accumulated throughout the cell, *CsULT1*-GFP was localized in the nucleus (Figure 5). This might be because of the fact that ULT proteins are small enough and can diffuse passively into the nucleus through the nuclear pores [15].

Transactivation assay

To investigate the ability of *CsULT1* to activate transcription, a transient expression assay was performed using a GAL4-responsive reporter system in yeast cells. For this, the full-length coding region of *CsULT1* was fused to the GAL4 DNA-binding domain (BD) to generate pGBKT7-*CsULT1-BD* construct which was then transformed into yeast strain Y187. The transformants were assayed for their ability to activate transcription from the GAL4 upstream activation sequence. The transformed yeast cells harboring pGBKT7-*CsULT1-BD* construct grew well in SD medium lacking tryptophan and leucine, and showed β-galactosidase activity, whereas cells containing pGBKT7 (negative control) showed no β-galactosidase activity (Figure 6). This data confirmed transcriptional activity of *CsULT1*.

Induction of *CsULT1* by phytohormones

To investigate the effect of various phytohormones on expression of *CsULT1*, *Crocus* flowers were treated with

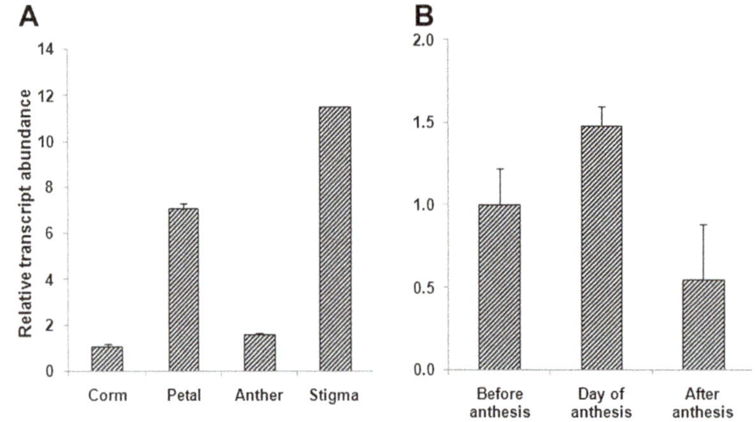

Figure 2 qRT-PCR analysis of *CsULT1* expression. (A) in different tissues of *C. sativus* **(B)** at different developmental stages. Transcript levels were normalized by actin. Data are means and SD from three biological replicates.

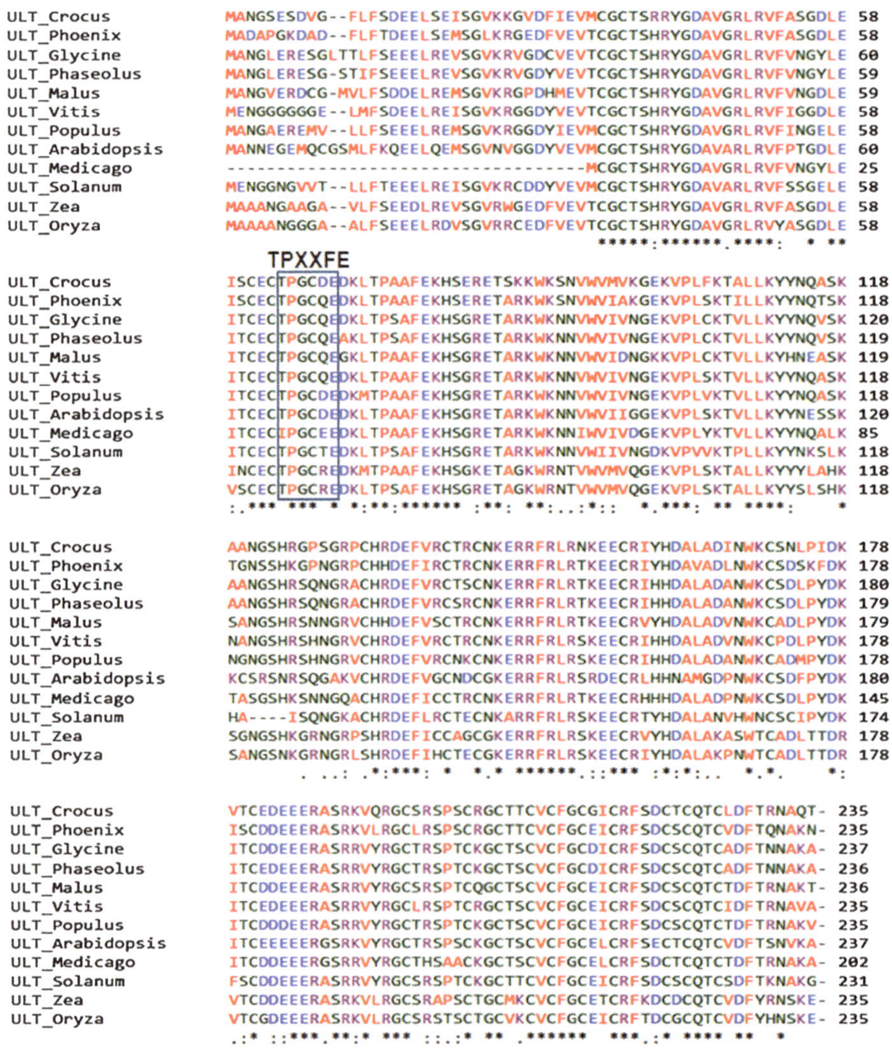

Figure 3 Multiple sequence alignment of CsULT1. The deduced amino acid sequence of CsULT1 is aligned with homologs from other plants.

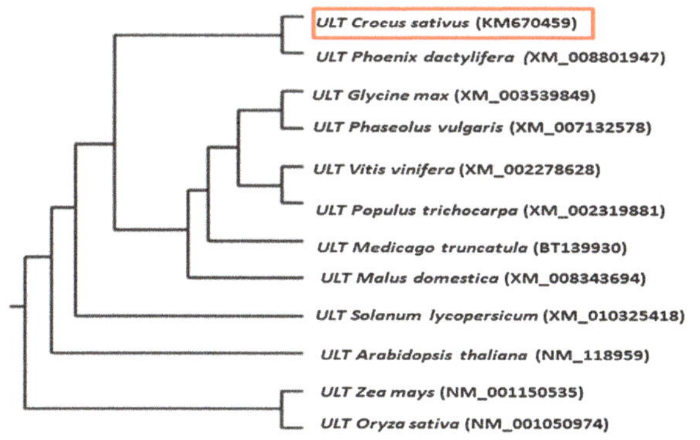

Figure 4 Phylogenetic analysis of CsULT1. A neighbor-joining phylogenetic tree of CsULT1 and selected ULT proteins from other plant species. The statistical reliability of individual nodes of the tree is assessed by bootstrap analyses with 1,000 replications.

salicylic acid (SA), methyljasmonate (JA), 2,4-D and abscisic acid (ABA) in a time course study. The expression of *CsULT1* was measured by qPCR using RNA isolated from treated tissue samples. Compared with the uninduced control, *CsULT1* expression increased in response to all these hormones used (Figure 7). In response to SA treatment, expression of *CsULT1* increased upto 122 fold (log 7 fold) at 24 hr post treatment while in response to JA, the expression enhanced approximately upto 150 fold (log 7 fold) at 12 h post treatment. *CsULT1* showed maximum change in expression in response to 2,4-D where it showed 175 fold (log 7.4) induction at 12 hr of treatment. However, there was not much significant change in expression in response to ABA.

Transient over-expression of *CsULT1* in *C. sativus* calli increases MEP pathway gene expression

For gaining an understanding on the role of *CsULT1* in *Crocus* apocarotenoid biosynthesis, the gene was transiently expressed in *Crocus* calli under the control of CaMV-35S promoter. The presence of transgene in transiently transformed calli was confirmed by genomic PCR. In the *CsULT1*-overexpressing calli, the *CsULT1* gene was expressed 2.5 fold higher than the empty vector control. Further, we checked expression of few of the MEP pathway genes and observed that *PSY, PDS, BCH, CsCCD4b* and *CsCCD2* genes showed upregulation in *CsULT1* overexpressing calli (Figure 8a). *PSY* and *PDS* catalyze the initial rate limiting steps in carotenoid biosynthetic pathway. Further, *BCH* is involved in the formation of zeaxanthin from beta carotene [11] and this zeaxanthin acts as the substrate for the formation of *Crocus* apocarotenoids by *CsCCD2* enzyme [20]. *CsCCD4b* (another member of CCD gene family) is also involved in the formation of apocarotenoids from carotenoid substrates. Therefore, enhanced expression of the

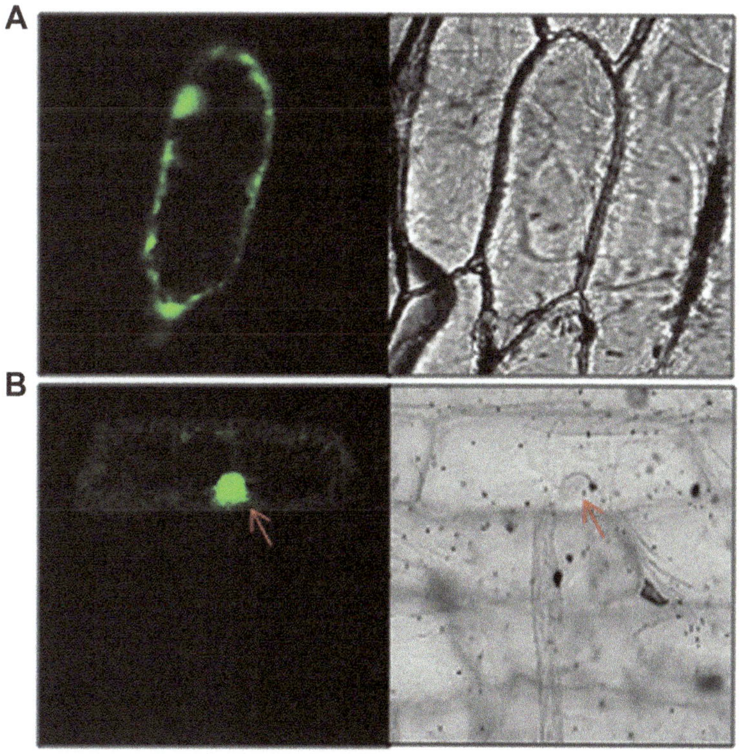

Figure 5 Subcellular localization of *CsULT1*. (A) GFP is accumulated throughout the cell **(B)** *CsULT1-GFP* is localized to the nucleus.

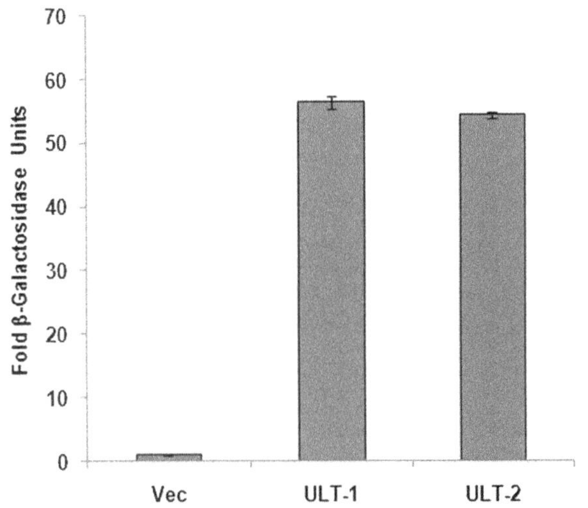

Figure 6 Transactivation analysis of *CsULT1* by β-galactosidase assay. Vec represents empty vector control and ULT-1 and ULT-2 represent two independent colonies used for the assay. Values are taken as average of three independent experiments of the transformants and presented as fold increase in activity.

above mentioned genes may result in increased zeaxanthin pool which may subsequently be tailored to form apocarotenoids.

Various phytohormones were shown to induce expression of *CsULT1* which in turn induced expression of key pathway genes of carotenoid metabolism. Therefore we were keen to investigate change in expression of pathway genes in response to phytohormone application. It was observed that SA, JA and 2,4-D induced expression of carotenoid pathway genes (Figure 8b) therefore indicating their possible role in mediating the function of *CsULT1* in regulating carotenoid/apocarotenoid biosynthesis. Taken together, these results suggest that *CsULT1* has a role in regulating metabolic flux towards the biosynthesis of apocarotenoids in *Crocus*.

Discussion

Crocus is known to accumulate large amounts of the apocarotenoids like crocetin (and its glycosylated forms, crocins), picrocrocins and saffranal in stigma part of the flowers. The proposed biosynthetic pathway is initiated through the symmetric cleavage of zeaxanthin at the 7,8/7,8 positions by a CCD2 enzyme [20]. The two cleavage products formed are 3-OH-β-cyclocitral and crocetin

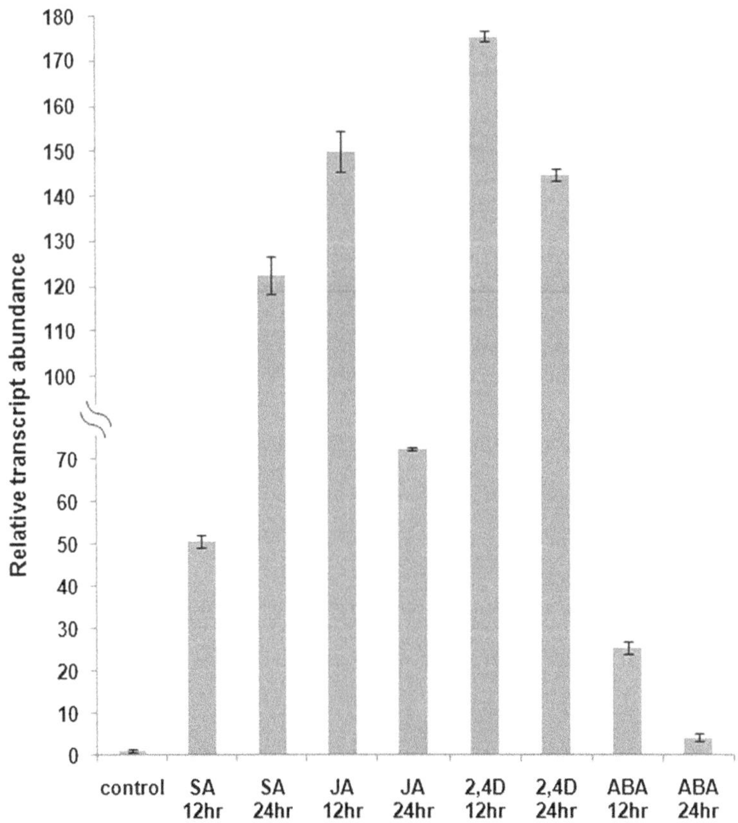

Figure 7 qRT PCR showing relative transcript level of *CsULT1* in response to various hormones. Transcript levels were normalized by actin transcript level. Error bars indicate SD of three replicates.

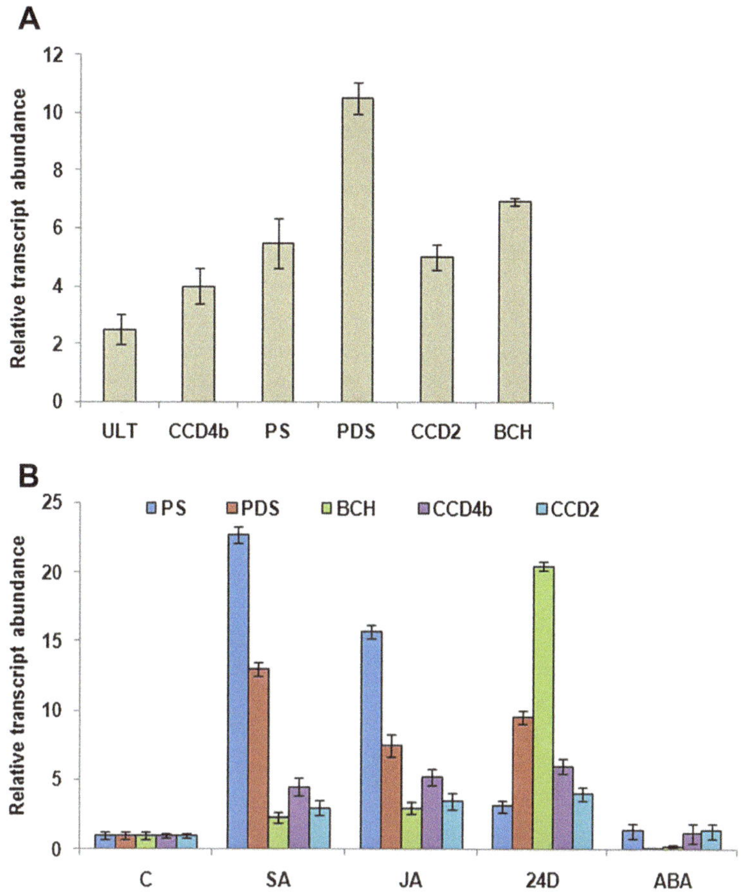

Figure 8 Relative expression levels of selected carotenoid pathway genes (A) in *CsULT1* overexpressing calli (B) in response to various phytohormones. The actin gene was used as an internal control. Each relative gene expression represents the average of three replicates with error bars representing SD.

dialdehyde which are further dehydrogenated and glycosylated to yield picrocrocin and crocins respectively. Aim of our study was to identify transcription factors regulating synthesis of these *Crocus* apocarotenoids. For this, we used combined approach of transcript and metabolite profiling. Since crocetin (which is subsequently converted into crocin) and picrocrocin are products of same cleavage step and crocin is more stable, we investigated crocin levels in different parts of *Crocus* flower and in stigma collected at three different developmental stages (pre anthesis, anthesis and post anthesis). We could detect crocin only in stigma while in other parts it was below detection levels. Further, crocin content increased from pre anthesis stage to anthesis and later decreased post anthesis (Figure 1). In earlier reports also same trend has been described for crocin accumulation [12]. Next, expression profile of *CsULT1* was investigated which indicated that it follows the same trend (Figure 2) and corroborates with accumulation pattern of crocin suggesting its possible role in regulating the crocin biosynthetic pathway. The apocarotenoid accumulation varies with developmental stages and thus in order to fit into this narrow window of developmental changes, chromatin needs to be flexibly regulated so as to confer stable expression states that can be reset owing to changes in the progression of development. Since recently *ULT* has been shown to act in chromatin mediated pathways, its involvement in regulating plant secondary metabolic pathways would be a new dimension to its functional domain.

Domain analysis showed that *CsULT1* contains a SAND domain which represents conserved 80-residue amino acid sequence and is found in a number of nuclear proteins, many of which function in chromatin-dependent transcriptional control [13]. These include proteins linked to various human diseases, such as the Sp100 (Speckled protein 100 kDa), NUDR (Nuclear DEAF-1 related), GMEB (Glucocorticoid Modulatory Element Binding) proteins and AIRE-1 (Autoimmune regulator 1) proteins [18]. Many of these proteins have been shown to bind DNA, but no clear sequence or structural relationship to known DNA binding motifs has been established. Based on the conservation of positively charged residues, including a characteristic

KDWK sequence motif, the SAND domain has been suggested to mediate the DNA binding of these proteins.

In animals SAND domain-containing proteins are found in nucleus or cytoplasm, or have dual localization being present in both nucleus and cytoplasm [21-23]. In confirmation with this data pertaining to animal proteins, plant ultrapetala proteins with SAND domains are also demonstrated to localize to both the nucleus and the cytosol [15]. However, our study showed that *CsULT1* is localized in nucleus (Figure 5). SAND domain containing proteins have been found to be transcriptionally active and are involved in regulation of gene expression [17]. We also investigated transcriptional activity of *CsULT1* and our results demonstrated that it activated beta galactosidase enzyme proving that it is transcriptionally active (Figure 6).

Plant developmental, metabolic and stress pathways have been shown to be influenced and controlled by various phytohormones. In order to gain an insight about effect of different phytohormones on *CsULT1* expression, we investigated effect of SA, JA, 2,4D and ABA on expression of *CsULT1*. Our results indicated that *CsULT1* is induced in response to all the hormones studied. However, effect of SA, JA and 2,4-D was much more profound than ABA (Figure 7). Jasmonates are known elicitors of plant secondary metabolism and trigger extensive transcriptional reprogramming which ultimately leads to activation of whole metabolic pathway [24]. Induction of *CsULT1* in response to JA might be part of this transcriptional activation scenario which as a final outcome leads to activation of carotenoid metabolic pathway. SA has been reported to induce expression of many carotenogenesis related genes [25]. Since *CsULT1* is a probable regulator of carotenogenesis, its induction in respone to SA is thus in confirmation with earlier reports. Auxin has been demonstrated to have a profound effect on stigma development [26]. Therefore enhanced expression of *CsULT1* in response to 2,4-D treatment might suggest a parallel role of auxin in regulating stigma development vis a vis apocarotenoid biosynthesis in *Crocus*. Although ABA treatment also enhanced expression of *CsULT1*, it was much less as compared to other hormones.

Several attempts have been made to establish *Crocus* transformation but no success has been achieved so far. Lack of transformation protocol is a limitation for functional characterization of genes in *Crocus*. The site of apocarotenoid biosynthesis is *Crocus* stigma, however, many of the pathway genes are expressed in callus also. Considering the limitation of transformation system, we transiently overexpressed *CsULT1* in *Crocus* calli by particle bombardment in order to confirm its role in regulating carotenoid/apocarotenoid pathway. The expression of *CsULT1* in transgenic calli was 2.5 fold as compared to vector control. This value is not good enough but since transformation in *Crocus* has not been established and callus is not the actual site of gene expression, the reported increase in expression can be considered as significant. Further, the expression analysis of a few carotenoid pathway genes was carried out and the results showed increase in expression level of *PSY* and *PDS* genes which catalyze initial rate limit steps of this pathway. This suggests role of *CsULT1* in regulating carotenoid biosynthesis in *Crocus*. Till so far there is only one report on regulation of *PSY* gene expression by phytochrome interacting factor (*PIF1*) which binds directly to *PSY* promoter and thereby regulates carotenoid accumulation during daily cycles of light and dark in mature plants [27]. Another member of AP2 gene family (*RAP2.2*) binds to of *PSY* promoter and is shown to modestly regulate the transcript levels of *PSY* and *PDS* in Arabidopsis [28]. Also *BCH* which is involved in conversion of beta carotene into zeaxanthin showed enhanced expression in transgenic calli (Figure 8a). In earlier reports *CsCCD4b* was considered responsible for cleaving zeaxanthin to produce apocarotenoids. However, later it was shown to cleave beta carotene at the 9,10 and/or the 9,10 positions, yielding beta-ionone. Recently a new isoform of CCDs (*CsCCD2*) was identified and isolated from *Crocus* and was shown to cleave zeaxanthin sequentially at 7,8 and 7,8 double bonds suggesting that *CsCCD2* catalyzes the first dedicated step in crocin biosynthesis [20]. We investigated change in expression of *CsCCD4b* as well as *CsCCD2* in transgenic calli and observed that their expression was enhanced around 4 and 5 fold respectively (Figure 8a). This suggests that apart from regulating synthesis of crocin and picrocrocin from zeaxanthin, *CsULT1* also plays role in regulating biosynthesis of other apocarotenoids including beta ionone. Thus *CsULT1* might regulate expression of more than one members of *CCD* gene family. Except for *PIF1* and *RAP2.2* no other transcription factors have been identified till so far which regulate expression of genes involved in carotenogenesis in plants. Therefore the present work will form a platform for enhancing our knowledege on regulation of this important pathway.

Carotenoids are involved in many biological functions including plastid biogenesis, photosynthesis, photomorphogenesis etc. Carotenoid metabolic pathway is also linked with many other pathways like ABA and GA biosynthesis. Therefore, carotenoid metabolism might be regulated at multifaceted levels in plants. Further, because of this close coordination of many pathways, content and composition of carotenoids is important. Thus biosynthesis of carotenoids and their turn-over to produce apocarotenoids needs to be tightly regulated in order to maintain their steady levels in plants. Hormones are known to play key roles in regulating various metabolic pathways. They also help in coordinating interplay between various pathways. Our results also demonstrated effect of phytohormones on *CsULT1* expression. In order to further our understanding on mechanism of regulation of carotenoid biosynthesis by *CsULT1*, we also investigated effect of various phytohormones on carotenoid pathway genes particularly those which showed enhanced expression in *CsULT1* overexpressing calli. Our results indicated that expression of all the genes studied was enhanced in response to SA, JA and 2, 4-D, however, there was no significant increase in expression of these genes in response to ABA (Figure 8b). Earlier also, SA and JA have been shown to affect expression of many carotenogenic pathway genes [29]. If considered individually, *PSY* was induced more in response to SA and JA while *PDS* expression increased more under the influence of SA and 2,4-D. *BCH* showed much higher response to 2,4-D than other hormones while *CCD4b* and *CCD2* showed more or less similar trend in expression in response to all the hormones except ABA. ABA did not cause any significant effect on expression of carotenoid pathway genes. Earlier, exogenous application of ABA was shown to repress transcript levels of many chloroplast genes [30]. It was also found to reduce chlorophyll levels and also endogenous ABA levels [31]. This might be the reason that we did not observe any significant increase in expression levels of carotenoid pathway genes in response to ABA. Thus in general, change in the expression of these genes followed the same trend as that of *CsULT1*. This suggests that these phytohormones might play a role in mediating effect of *CsULT1* on regulation of carotenoid metabolism. Further, different genes showed varied induction levels in response to different hormones which might be because of the fact that control of plant metabolic pathways is a complex phenomenon and network of cross-communicating hormone signaling pathways are involved so as to maintain overall metabolite homeostasis within the plant system.

Taken together, these results suggest that *CsULT1* plays role in diverting metabolic flux towards enhanced production of apocarotenoids. Recently *ULT* in *Arabidopsis* has been shown to be involved in gynoecium development and patterning [17]. In *Crocus*, apocarotenoid biosynthesis is congruent with stigma development. This indicates that *CsULT1* might have a parallel role in regulating *Crocus* gynoecium (stigma) development vis a vis apocarotenoid biosynthesis. However, this needs to be further verified by experimentation in *Crocus* itself or in other alternative plant system.

Conclusions

In this study an ultrapetala transcription factor from *Crocus*, *CsULT1*, was identified, cloned and characterized. The *CsULT1* transcript was expressed more in stigma till flower anthesis. Application of phytohormones like 2,4D, JA and SA enhanced *CsULT1* expression. The gene is nuclear localized and is transcriptionally active. Moreover, ectopic expression of this gene altered expression of some important genes of carotenoid/apocarotenoid pathway confirming that *CsULT1* plays a regulatory role in *Crocus* apocarotenoid biosynthesis. Furthermore, the hormones which affected expression of *CsULT1* were also shown to enhance expression of pathway genes indicating their role in mediating regulatory role of *CsULT1*.

Additional files

Additional file 1: Table S1. List of primer sequences.

Additional file 2: Figure S1. Melt curve depicting single peak in qRT-PCR.

Additional file 3: Figure S2. qRT-PCR analysis of different *Crocus* transcription factors.

Additional file 4: Figure S3. Nucleotide and deduced amino acid sequence of *CsULT1*.

Abbreviations

ULT: Ultrapetala; PSY: Phytoene synthase; PDS: Phytoene desaturase; CCD: Carotenoid cleavage dioxygenase; BCH: Beta cyclohydroxylase; MEP: Methylerythritol phosphate.

Competing interests

The authors declare that they have no competing interests.

Authors' contributions

NA: designed and performed experiments and wrote the manuscript; DJ: performed experiments; RAV: conceptualized experiments and did critical revision of manuscript. All authors read and approved the final manuscript.

Acknowledgements

This work was supported by grants SIMPLE (BSC-0109) from CSIR and GAP-1177 from DST, Government of India.

Author details

[1]Plant Biotechnology Division, CSIR- Indian Institute of Integrative Medicine, Sanat Nagar, Srinagar J&K-190005, India. [2]National Institute of Plant Genome Research, Aruna Asaf Ali Marg, New Delhi 110067, India. [3]Medicinal Chemisrty Division, CSIR- Indian Institute of Integrative Medicine, Canal Road, Jammu J&K 180001, India.

References

1. D'Agostino N, Pizzichini D, Chiusano ML, Giuliano G. An EST database from saffron stigmas. BMC Plant Biol. 2007;7:53.
2. Auldridge ME, McCarty DR, Klee HJ. Plant carotenoid cleavage oxygenases and their apocarotenoid products. Curr Opin Plant Biol. 2006;9:315–21.
3. Liao YH, Houghton PJ, Hoult JR. Novel and known constituents from Buddleja species and their activity against leukocyte eicosanoid generation. J Nat Prod. 1999;62:1241–5.
4. Pfander H, Schurtenberger H. Biosynthesis of C20-carotenoids in *Crocus sativus*. Phytochemistry. 1982;21:1039–42.
5. Rubio A, Rambla JI, Santaella M, Gomez MD, Orzaez D, Granell A, et al. Cytosolic and plastaglobule targeted Carotenoid dioxygenases from Crocus sativus are both involved in β-ionone-release. J Biol Chem. 2008;283:24816–25.
6. Winterhalter P, Rouseff RL. Carotenoid-derived aroma compounds. Washington, DC: American Chemical Society; 2002.
7. Abdullaev FI, Espinosa-Aguirre JJ. Biomedical properties of saffron and its potential use in cancer therapy and chemoprevention trials. Cancer Detect Prev. 2004;28:426–32.

8. Zhang Z, Wang CZ, Wen XD, Shoyama Y, Yuan CS. Role of saffron and its constituents on cancer chemoprevention. Pharm Biol. 2013;51:920–4.
9. Cunningham FX, Gantt E. Genes and enzymes of carotenoid biosynthesis in plants. Annu Rev Plant Physiol Plant Mol Biol. 1998;49:557–83.
10. Lu S, Li L. Carotenoid metabolism: biosynthesis, regulation and beyond. J Integr Plant Biol. 2008;50:778–85.
11. Castillo R, Fernandez J, Gomez-Gomez L. Implications of carotenoid biosynthetic genes in apocarotenoid formation during the stigma development of crocus sativus and its closer relatives. Plant Physiol. 2005;139:674–89.
12. Moraga AR, Rambla JL, Ahrazem O, Granell A, Gómez-Gómez L. Metabolite and target transcript analyses during *Crocus sativus* stigma development. Phytochemistry. 2009;70:1009–16.
13. Bottomley MJ, Collard MW, Huggenvik JI, Liu Z, Gibson TJ, Sattler M. The SAND domain structure defines a novel DNA-binding fold in transcriptional regulation. Nat Struct Biol. 2001;8:626–33.
14. Fletcher JC. The ULTRAPETALA gene controls shoot and floral meristem size in Arabidopsis. Development. 2001;128:1323–33.
15. Carles CC, Lertpiriyapong K, Reville K, Fletcher JC. The ULTRAPETALA1 gene functions early in Arabidopsis development to restrict shoot apical meristem activity, and acts through WUSCHEL to regulate floral meristem determinacy. Genetics. 2004;167:1893–903.
16. Carles CC, Fletcher JC. The SAND domain protein ULTRAPETALA1 acts as a trithorax group factor to regulate cell fate in plants. Genes Dev. 2009;23:2723–8.
17. Monfared MM, Carles CC, Rossignol P, Pires HR, Fletcher JC. The ULT1 and ULT2 trxG genes play overlapping roles in Arabidopsis development and gene regulation. Mol Plant. 2013;6:1564–79.
18. Gibson TJ, Ramu C, Gemünd C, Aasland R. The APECED polyglandular autoimmune syndrome protein, AIRE-1, contains the SAND domain and is probably a transcription factor. Trends Biochem Sci. 1998;23:242–4.
19. Carles CC, Fletcher JC. Missing links between histones and RNA Pol II arising from SAND? Epigenetics. 2010;2010(5):381–5.
20. Frusciante S, Diretto G, Bruno M, Ferrante P, Pietrella M, Prado-Cabrero A, et al. Novel carotenoid cleavage dioxygenase catalyzes the first dedicated step in saffron crocin biosynthesis. PNAS. 2014;111:12246–51.
21. Gross CT, McGinnis W. DEAF-1, a novel protein that binds an essential region in Deformed response element. EMBO J. 1996;15:1961–70.
22. Jimenez-Lara AM, Heine MJ, Gronemeyer H. Cloning of a mouse glucocorticoid modulatory element binding protein, a new member of the KDWK family. FEBS Lett. 2000;468:203–10.
23. Peterson P, Pitkänen J, Sillanpaa N, Krohn K. Autoimmune polyendocrinopathy candidiasis ectodermal dystrophy (APECED): a model disease to study molecular aspects of endocrine autoimmunity. Clin Exp Immunol. 2004;135:348–57.
24. De Geyter N, Gholami A, Goormachtig S, Goossens A. Transcriptional machineries in jasmonate-elicited plant secondary metabolism. Trends Plant Sci. 2012;176:349–59.
25. Gao Z, Meng C, Zhang X, Xu D, Miao X, Wang Y, et al. Induction of salicylic acid (SA) on transcriptional expression of eight carotenoid genes and astaxanthin accumulation in *Haematococcus pluvialis*. Enzyme Microb Technol. 2012;51:225–30.
26. Sundberg E, Østergaard L. Distinct and dynamic auxin activities during reproductive development. 2012. Published by Cold Spring Harbor Laboratory Press.
27. Toledo-Ortiz G, Huq E, Rodríguez-Concepción M. Direct regulation of phytoene synthase gene expression and carotenoid biosynthesis by phytochrome-interacting factors. PNAS. 2010;107:11626–31.
28. Welsch R, Maass D, Voegel T, DellaPenna D, Beyer P. Transcription factor RAP2.2 and its interacting partner SINAT2: stable elements in the carotenogenesis of Arabidopsis leaves. Plant Physiol. 2007;145:1073–85.
29. Gao Z, Meng C, Zhang X, Xu D, Zhao Y, Wang Y, et al. Differential expression of carotenogenic genes, associated changes on astaxanthin production and photosynthesis features induced by JA in H. pluvialis. PloS One. 2012;7:e42243.
30. Yamburenko MV, Zubo YO, Vanková R, Kusnetsov VV, Kulaeva ON, Börner T. Abscisic acid represses the transcription of chloroplast genes. J Exp Bot. 2013;64:4491–502.
31. Meier S, Tzfadia O, Vallabhaneni R, Gehring C, Wurtzel ET. A transcriptional analysis of carotenoid, chlorophyll and plastidial isoprenoid biosynthesis genes during development and osmotic stress responses in *Arabidopsis thaliana*. BMC Syst Biol. 2011;5:77.

4

Identification of the relationship between Chinese *Adiantum reniforme* var. *sinense* and Canary *Adiantum reniforme*

Ai-Hua Wang[1,2], Ye Sun[1], Harald Schneider[3], Jun-Wen Zhai[4], Dong-Ming Liu[1], Jin-Song Zhou[5], Fu-Wu Xing[1], Hong-Feng Chen[1*] and Fa-Guo Wang[1*]

Abstract

Background: There are different opinions about the relationship of two disjunctively distributed varieties *Adiantum reniforme* L. var. *sinense* Y.X.Lin and *Adiantum reniforme* L. *Adiantum reniforme* var. *sinense* is an endangered fern only distributed in a narrowed region of Chongqing city in China, while *Adiantum reniforme* var. *reniforme* just distributed in Canary Islands and Madeira off the north-western African coast. To verify the relationship of these two taxa, relative phylogenetic analyses, karyotype analyses, microscopic spore observations and morphological studies were performed in this study. Besides, divergence time between *A. reniforme* var. *sinense* and *A. reniforme* var. *reniforme* was estimated using GTR model according to a phylogeny tree constructed with the three cpDNA markers *atp*A, *atp*B, and *rbc*L.

Results: Phylogenetic results and divergence time analyses–all individuals of *A. reniforme* var. *sinense* from 4 different populations (representing all biogeographic distributions) were clustered into one clade and all individuals of *A. reniforme* var. *reniforme* from 7 different populations (all biogeographic distributions are included) were clustered into another clade. The divergence between *A. reniforme* var. *reniforme* and *A. reniforme* var. *sinense* was estimated to be 4.94 (2.26-8.66) Myr. Based on karyotype analyses, *A. reniforme* var. *reniforme* was deduced to be hexaploidy with $2n = 180$, $X = 30$, while *A. reniforme* var. *sinense* was known as tetraploidy. Microscopic spore observations suggested that surface ornamentation of *A. reniforme* var. *reniforme* is psilate, but that of *A. reniforme* var. *sinense* is rugate. Leaf blades of *A. reniforme* var. *sinense* are membranous and reniform and with several obvious concentric rings, and leaves of *A. reniforme* var. *reniforme* are pachyphyllous and coriaceous and are much rounder and similar to palm.

Conclusion: *Adiantum reniforme* var. *sinense* is an independent species rather than the variety of *Adiantum reniforme* var. *reniforme*. As a result, we approve *Adiantum nelumboides* X. C. Zhang, nom. & stat. nov. as a legal name instead of the former *Adiantum reniforme* var. *sinense*. China was determined to be the most probable evolution centre based on the results of phylogenetic analyses, divergence estimation, relative palaeogeography and palaeoclimate materials.

Keywords: Chromosome numbers, cpDNA, Flow cytometry, Molecular clock dating, Morphological characters, Phylogenetic position, Relationship identification, SEM observation

* Correspondence: h.f.chen@scbg.ac.cn; wangfg@scib.ac.cn
[1]Key Laboratory of Plant Resources Conservation and Sustainable Utilization, South China Botanical Garden, Chinese Academy of Sciences, Guangzhou 510650, China
Full list of author information is available at the end of the article

Background

Adiantum reniforme L. var. *sinense* Y.X.Lin (Chinese name "He ye jin qian cao") was first discovered in Chongqing city in China in 1978 [1]. It was published in *Acta Phytotaxonomica Sinica* as a variety of *Adiantum reniforme* L. because of their similar morphological characters in 1980. It is only distributed along the Yangtze River from Shizhu County to the Wanzhou District of Chongqing, which stretches for almost 100 kilometres through Xi-tuo, Xinxiang, Wu-ling, Chang-ping and other places [2-4]. It has a narrow distribution zone and an endangered status. *A. reniforme* var. *sinense* was listed as a class II protected fern in China [2]. The plant is known to have medicinal uses including heat-clearing and detoxifying, promoting diuresis and relieving stranguria, curing icteric hepatitis and stones [5]. As a result, the plant has been over-collected by local people. Additionally, the construction of the Three Gorges Dam from 1993 to 2009 caused destruction of habitats and reduced its population size, which reduced gene flow among populations [6]. Many studies have been conducted to protect *A. reniforme* var. *sinense* from extinction. These studies included field habitat investigations [2], the use of spore propagation technology [7] and increases in population gene diversity [6,8,9]. *A. reniforme* var. *sinense* was previously shown to be tetraploid (2n = 120, X = 30) in Lin YX [10]. Scanning electron microscopy (SEM) analysis of *A. reniforme* var. *sinense* suggested that its spores are actinomorphic and trilete with polar surface triangles. Additionally, the equatorial surface is semicircular or super-semicircular, and the surface ornamentation is psilate [11]. *Adiantum* belongs to the family Pteridaceae, although different opinions exist regarding whether *Adiantum* is monophyletic or paraphyletic with vittarioid ferns [12-17]. A phylogenetic tree of Chinese *Adiantum* was constructed using five cpDNA primers for the following genes: *atp*A, *atp*B, *rbc*L, *trn*L-F and *trn*S. This analysis indicated that *Adiantum* was monophyletic and *A. reniforme* var. *sinense* was closely related to *Adiantum* Ser. *Venusta,* which was established by Ching Renchang in *Flora Republicae Popularis Sinicae, Tomus* 3(1) [18].

There are a limited number of reports of *A. reniforme* var. *reniforme*. The first specimens were collected in Madeira, and it was first published in *Species Plantarum* by Linnaeus in 1753. The plant is found in the Canary Islands and Madeira off the north-western African coast. Manton [19] considered *A. reniforme* var. *reniforme* as decaploid (2n = 300, X = 30) after her study on the specimens kept in Kew garden but collected in Madeira and Tenerife. In 1985, Mary Gibby restudied ploidy and the chromosomes of materials collected in the Canary Island and suggested that it was tetraploid (2n = 120, X = 30). However, there is no photographic record of this result. Subsequent studies have demonstrated that ploidy levels of all ferns in the Canary Islands are no more than hexaploid [20]. Consequently, the ploidy of *A. reniforme* var. *reniforme* is controversial, and the differences in chromosome number between the Canary population and the Madeira population are unclear.

There are similar morphological characters between *A. reniforme* var. *sinense* and *A. reniforme* var. *reniforme*. So, it seems reasonable that they are varieties. However, the China-Canary distribution disjunction of these two taxa makes their relationships doubtful. Zhang XC [21] treated *A. reniforme* var. *reniforme* as an independent species in the book "*Lycophytes and ferns of China*" but without explanation. As described above, the spore morphology, karyotype analysis and phylogenetic analysis of *A. reniforme* var. *reniforme* are currently unknown. Because of the limited morphological characters of these two taxa, for example, only one single leaf blade with one petiole, it is not convictive for the treatment that they were varieties between each other just based on their limited morphological characters (see Figure 1). Additional studies are required to determine whether *A. reniforme* var. *sinense* is a variety or an independent species. To make the taxonomy relationship between *A. reniforme* var. *sinense* and *A. reniforme* var. *reniforme* clear and deduce mechanisms of the intercontinental disjunction, we have analysed 7 populations consisting of almost 96 individuals of *A. reniforme* var. *sinense* from China and 8 populations consisting of almost 164 individuals of *A. reniforme* var. *reniforme* from Canary and Madeira.

Methods

Materials

In this study, 24 individuals from 11 populations of both the *Adiantum reniforme* var. *reniforme* and *A. reniforme* var. *sinense* representing all biogeographic distributions were sampled and sequenced. The 31 species of *Adiantum* and *Vittaria flexuosa* (outgroup) were downloaded from GenBank to construct a phylogeny tree of *Adiantum* with the combined cpDNA markers *atp*A, *atp*B, *trn*L-F and *trn*S. Furthermore, three plastid genes (*rbc*L, *atp*A, and *atp*B) from 24 outgroup species were downloaded to test the divergence time of *Adiantum reniforme* var. *reniforme* and *A. reniforme* var. *sinense.* All taxa included in this study, voucher information and collection sites are listed in Additional file 1 and Addition file 2.

DNA extraction, amplification and sequencing

Total DNA was extracted from 20 mg silica-gel-dried leaf material using a modified CTAB DNA extraction protocol [22]. The *atp*A gene was amplified with primers "ESATPF412F"and"ESTRNR46F" [23]. "ESATB172F" and "ESATPE45R" were used for amplifying and sequencing the *atp*B gene [14]. "1 F" and "1379R" were used to

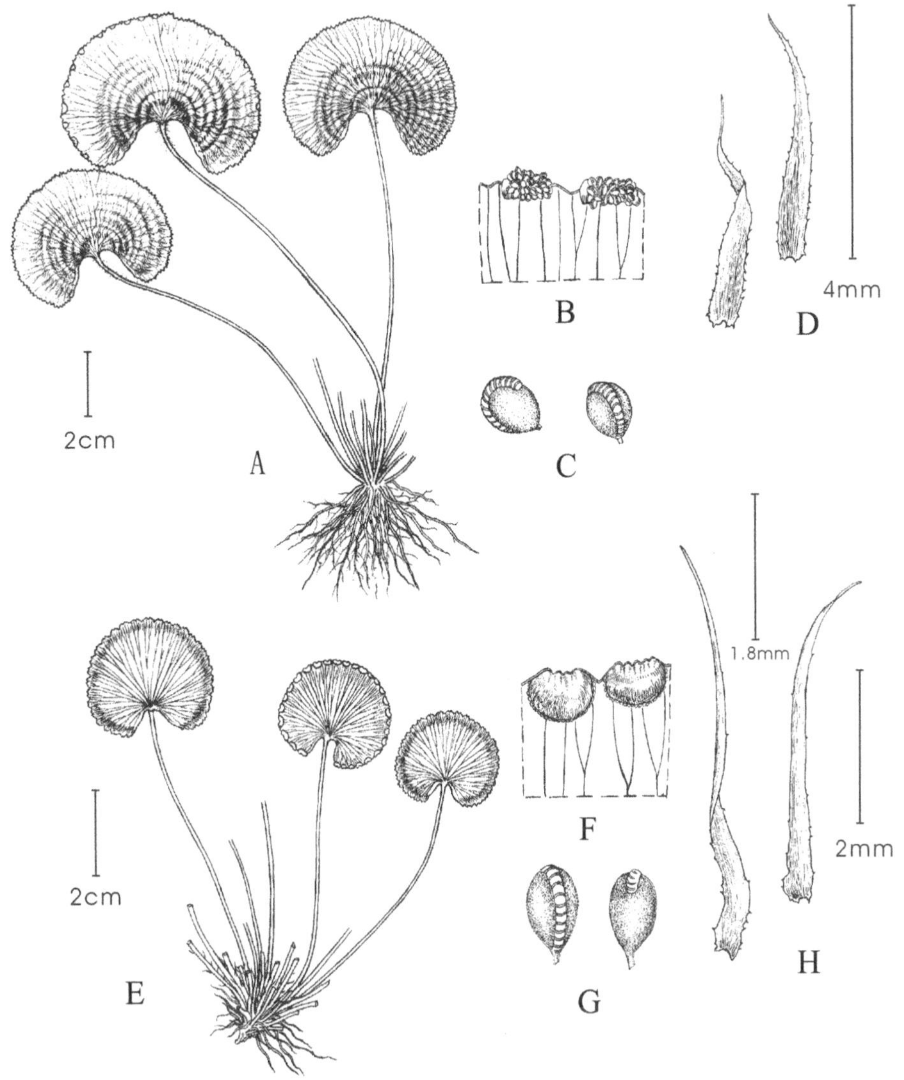

Figure 1 Morphological characters of *A. reniforme* var. *sinense* and *A. reniforme* var. *reniforme*. **A**, **B**, **C**, and **D** represent the leaf, sporangiorus, sporangium and scales of A. *reniforme* var. sinense, respectively. **E**, **F**, **G**, and **H** represent the related leaf, sporangiorus, sporangium and scales of A. *reniforme* var. sinense, respectively.

amplify and sequence the *rbc*L gene [24]. The *trn*L-F region was amplified and sequenced with primers "p1" and "f" [25,26]. Primers "*trn*S" [27] and "*rps*4.5" [28] were used to amplify and sequence the *rps*4-*trn*S region. All amplifications were performed in a 30-μL reaction mixture. The PCR reactions contained the following reagents: 1.0-2.4 μL of each primer (5p), 17-60 ng sample DNA, 1.5 U of Taq DNA polymerase, 10 × buffer (including Mg^{2+}), 0.25 mmol · L^{-1}dNTP, and ultrapure water (ddH_2O). The *atp*A and *atp*B 30-μL reaction mixtures were incubated at 95°C for 10 min, cycled 35 times (95°C for 1 min, 50°C for 1 min, and 72°C for 100 s), followed by a final extension for 10 min at 72°C. The *rbc*L and *trn*L-F PCR reactions were incubated at 95°C for 3 min, cycled 35 times (95°C for 1 min, 51°C for 1 min, and 72°C for 80 s), followed by a final extension for 10 min at 72°C. The *rps*4-*trn*S PCR reactions were incubated at 95°C for 3 min, cycled 35 times (94°C for 30 s, 58°C for 45 s, and 72°C for 80 s), followed by a final extension for 10 min at 72°C. The PCR products were purified and sequenced with an ABI 3730XL by Majorbio Company.

Phylogenetic analyses

The sequences were assembled with Sequencher 4.14 and then adjusted manually through Bioedit v.7.1.3 [29] and aligned using the program Clustal X version 2.0 [30]. Phylogenetic trees of each individual and the combined markers (*atp*A, *atp*B, *rbc*L, *trn*L-F, and *rps*4-*trn*S) were constructed using maximum parsimony (MP) and Bayesian Markov chain Monte Carlo (MCMC) inference.

The maximum parsimony analyses were performed with PAUP* 4.0b10 [31], treating gaps as missing data and using the heuristic search options with 1000 random replicates and tree-bisection-reconnection (TBR) branch swapping. All characteristics were unordered and equally weighted. For Bayesian analyses, MrModeltest2 (v2.3; [32]) based on the Akaike information criterion (AIC) was used to identify the best-fit molecular evolution model for each of the DNA markers. We constructed Bayesian trees using MrBayes 3.1 [33] with the best-fit model GTR + I + G. Trees were generated for 1,000,000 generations, sampling every 100 generations. Four chains were used with a random initial tree. For each of the individual data partitions and the combined dataset, the first 2500 sample trees were discarded as burn-in to ensure that the chains reached stationarity. Nodes receiving bootstrap support (BS) of < 70% in the MP analyses or PP of < 0.95 in the BI analyses were not considered to be well supported.

Molecular clock dating

Bayesian molecular dating studies were performed with the combined dataset of *rbc*L, *atp*A and *atp*B. Sequences of 24 outgroup species were downloaded from NCBI. The divergence time estimation of each clade in *Adiantum* and their credibility intervals were implemented in BEAUTI / BEAST 1.7.4 [34]. The BEAST analyses were performed with the GTR model, the uncorrelated relaxed lognormal clock model and the coalescent exponential growth tree. We used the 65.5 ± 0.3 Myr, which was the crown of the ceratopteridoids clade [35], as the calibration point. Posterior distributions of parameters were approximated using three independent MCMC analyses of 20,000,000 generations with 10% burn-in. Convergence was examined using Tracer 1.5 [36].

Karyotype analysis

To deduce the ploidy levels of *A. reniforme* var. *reniforme*, *A. reniforme* var. *sinense* was used as an internal standard because of its clear sporophytic chromosomes (2n = 120, X = 30), as displayed in Lin YX [10]. There were 32 sporophytic materials from different populations of both taxa examined by flow cytometric analyses to confirm the accuracy of ploidy levels for *A. reniforme* var. *reniforme* (Table 1). The leaves have membranous and hard leaf blades, so young and fresh blades spreading from circinate leaves were used. Small pieces of plant leaves were chopped with a double-edged razor in a Petri dish containing 0.4 mL mixed buffer (including ice-cold Otto buffer combined with DAPI fluorochrome, as patented by Partec Comneruim). Then, an additional 1.6 mL of buffer was mixed with the cells in the Petri dish and the cells were filtered through a 30-μm-mesh filter into a 5-mL cytometry tube. The tube was incubated in the dark at room temperature for 5-10 min. Each sample was analysed on a flow cytometer (Cyflow Space, Partec) equipped with a high-pressure mercury arc lamp for UV excitation. For each sample, a minimum of 2,000 nuclei were analysed. The fluorescence peaks and relative fluorescence intensity were analysed by the software Flomax.

SEM observation

For SEM analysis, mature spores from different populations were dispersed on stubs directly after being collected. The spores were gold-coated in a JFC-1600 Auto Fine Coater and observed using a JEOL JSM-6360LV Scanning Electron Microscope at 25 kV at the South China Botanical Garden, Chinese Academy of Sciences. The spore mean sizes of 7 populations of *A. reniforme* var. *sinense* and 7 populations of *A. reniforme* var. *reniforme* were measured by Smile View software (20 spores per population), and a scatter diagram was made with SPSS. The descriptive terminology in *Spores of Polypodiales (Filicales) from China* [11] and *Plant identification terminology: An illustrated glossary* [37] was followed.

Results

Phylogenetic and molecular divergence time analyses

The topologies derived from analyses of the individual datasets were similar to those obtained from the combined data. Therefore, we emphasised the results of the combined data. The sequences of 23 Chinese species and 8 foreign species of *Adiantum* and *Vittaria flexuosa* (outgroup) were downloaded from GenBank. The combined 4-marker (*atp*A, *atp*B, *trn*L-F and *rps*4-*trn*S) dataset included 56 taxa and consisted of 5,210 nucleotides, of which 1961 were variable (37.6%) and 1,468 were phylogenetically informative (28.2%). Rooted with the specified outgroup *Vittaria flexuosa*, the MP analysis on the combined 4-marker dataset yielded one maximally parsimonious tree of 3,911 steps, a consistency index (CI) of 0.6423, and a retention index (RI) of 0.8944. The tree obtained from the BI analyses had similar topology as the MP strict consensus tree (Figure 2).

All individuals of *A. reniforme* var. *sinense* from different populations were clustered into one clade, and all individuals of *A. reniforme* var. *reniforme* from different populations were clustered into another clade (Figure 2). Our analysis strongly supported that Canary Islands and Madeira *A. reniforme* var. *reniforme* was sister to Chinese *A. reniforme* var. *sinense* (1.0/100). The genetic distance (GD) between *A. reniforme* var. *reniforme* and *A. reniforme* var. *sinense* was calculated by constructing NJ trees using Mega5.0 based on the combined 4-marker data. Compared with the GD between *A. caudatum* and *A. malesianum* (GD = 0.004 ± 0.001) and the distance between *A. flabellulatum* and *A. induratum* (GD = 0.008 ± 0.002), the

Table 1 Relative fluorescence intensity (DAPI measurements) for the *A. reniforme* var. *sinense* and *A. reniforme* var. *reniforme*, summarised by the phytogeographic regions

Taxon	Ploidy level	Accession number	Region	Relative fluorescence intensity	Relative fluorescence intensity (mean ± s.d.)	Overall mean (±s.d.)
A.reniforme var. *sinense*	4X	WAH009	xi-tuo, shi zhu, China	62.06	65.44 ± 3.59	65.44 ± 3.59
		WAH007	xi-tuo, shi zhu, China	65.06		
		WAH003	xi-tuo, shi zhu, China	69.2		
A.reniforme var. *reniforme*	?	LPCG002	Cubo de la Galga, La Palma	103.09	97.78 ± 4.06	
		LPCG009	Cubo de la Galga, La Palma	100.88		
		LPCG011	Cubo de la Galga, La Palma	99.42		
		LPCG003	Cubo de la Galga, La Palma	90.45		
		LPCG004	Cubo de la Galga, La Palma	96.77		
		LPCG001	Cubo de la Galga, La Palma	95.92		
		LPCGO14	Cubo de la Galga, La Palma	97.94		
		LPB023	Bermúdec, La Palma	82.18	84.11 ± 2.96	
		LPB006	Bermúdec, La Palma	80.99		
		LPB007	Bermúdec, La Palma	86.83		
		LPB010	Bermúdec, La Palma	86.43		
		TBI001	Barranco del Infierno, Tenerife	95.74	92.75 ± 6.85	
		TBI011	Barranco del Infierno, Tenerife	97.49		
		TBI014	Barranco del Infierno, Tenerife	99		
		TBI017	Barranco del Infierno, Tenerife	88.89		
		TBIO05	Barranco del Infierno, Tenerife	82.64		
		TPH008	Punta del Hidalgo, Tenerife	97.32		
		TPH021	Punta del Hidalgo, Tenerife	93.34		
		TPH010	Punta del Hidalgo, Tenerife	104.57		
		TPH003	Punta del Hidalgo, Tenerife	84.62		
		TPH0017	Punta del Hidalgo, Tenerife	86.86	93.34 ± 8.06	92.92 ± 7.24

value between *A. reniforme* var. *reniforme* and *A. reniforme* var. *sinense* (GD = 0.011 ± 0.003) was much longer.

The divergence between *A. reniforme* var. *reniforme* and *A. reniforme* var. *sinense* was estimated to be 4.94 (2.26-8.66) Myr, while *A. flabellulatum* and *A. induratum* were dated to diverge 4.06 (1.25-7.80) Myr ago (see Figure 3).

Chromosome analysis

The ploidy level of *A.reniforme* var. *reniforme* was estimated by comparison with the known tetraploidy *A. reniforme* var. *sinense*. Based on DAPI staining, 21 accessions of *A. reniforme* var. *reniforme* showed relative fluorescence intensities of 92.92 ± 7.24, and 3 accessions of the internal standard *A. reniforme* var. *sinense* showed relative fluorescence intensities of 65.44 ± 3.59 (Table 1). We deduced that *A. reniforme* var. *reniforme* was hexaploidy with 2n = 180, X = 30 because the relative fluorescence intensity of the *A. reniforme* var. *reniforme* accessions was approximately 1.5-fold higher than the *A. reniforme* var. *sinense* accessions. The chromosome number of *A. reniforme* var. *sinense* was determined to be 2n = 120, X = 30 [10]. The flow cytometry histograms of both plants are shown in Figure 4 (left).

SEM observation and morphological character differences

The spore shapes of both taxa are tetrahedric and are similar in polar and equatorial views. However, the spores are clearly different with respect to surface ornamentation. The spores are actinomorphic and trilete with polar surface triangles, and the equatorial surface is semicircular or super-semicircular. The surface ornamentation of *A. reniforme* var. *reniforme* is psilate, while that of *A. reniforme* var. *sinense* is rugate (see Figure 4). The mean sizes of 7 populations of *A. reniforme* var. *sinense* were 37.1 ± 3.7 μm, which is shorter than the 7 populations of *A. reniforme* var. *reniforme* (47.8 ± 3.9 μm). The spore equatorial axis sizes of *Adiantum* vary from 32 to 55 μm [11], and our findings are consistent with these data.

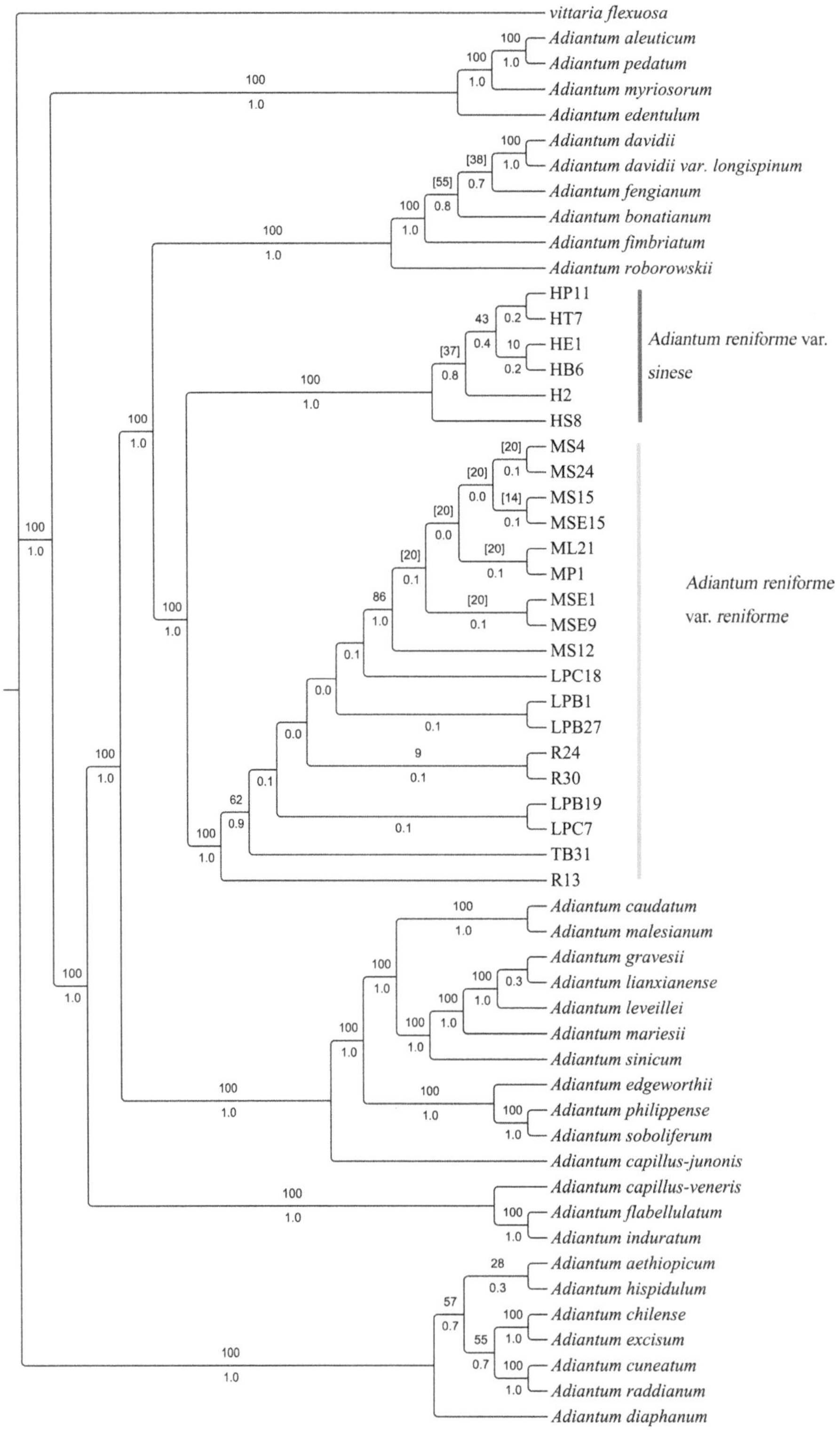

Figure 2 Strict consensus tree of two maximally parsimonious trees derived from the analysis of the plastid atpA, atpB, trnL-F, and rps4-trnS sequences (tree length = 3,911 steps, CI = 0.6423, and RI = 0.8944). The bootstrap values for 1,000 replicates are shown above the lines, and the Bayesian posterior probabilities are shown below the lines. Front alphabets of HP11, HT7, R13 are the short names of different populations of these two taxa, and the latter numbers represent single individuals.

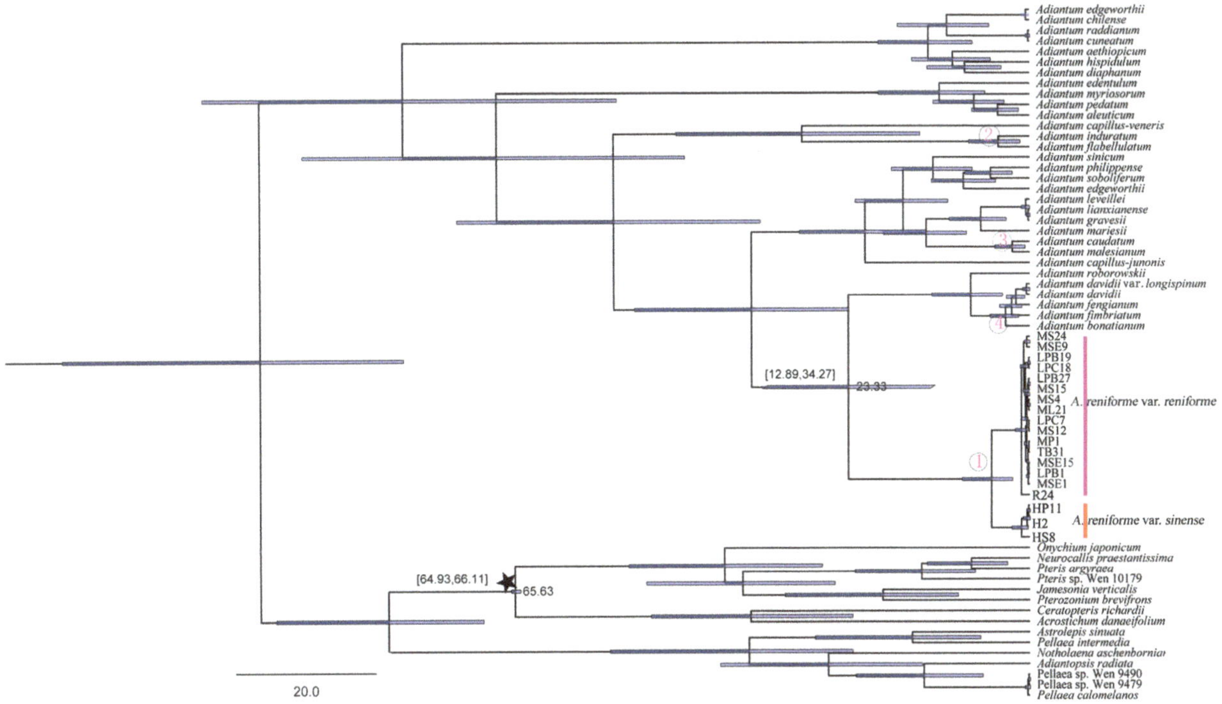

Figure 3 **Chronogram of *Adiantum* inferred from BEAST with combined sequences (*atp*A, *atp*B and *rbc*L).** The calibration scheme is indicated with black asterisks. Node 1: *A. reniforme* var. *reniforme* and *A. reniforme* var. *sinense*; Node 2: *A. flabellulatum* and *A. induratum*.

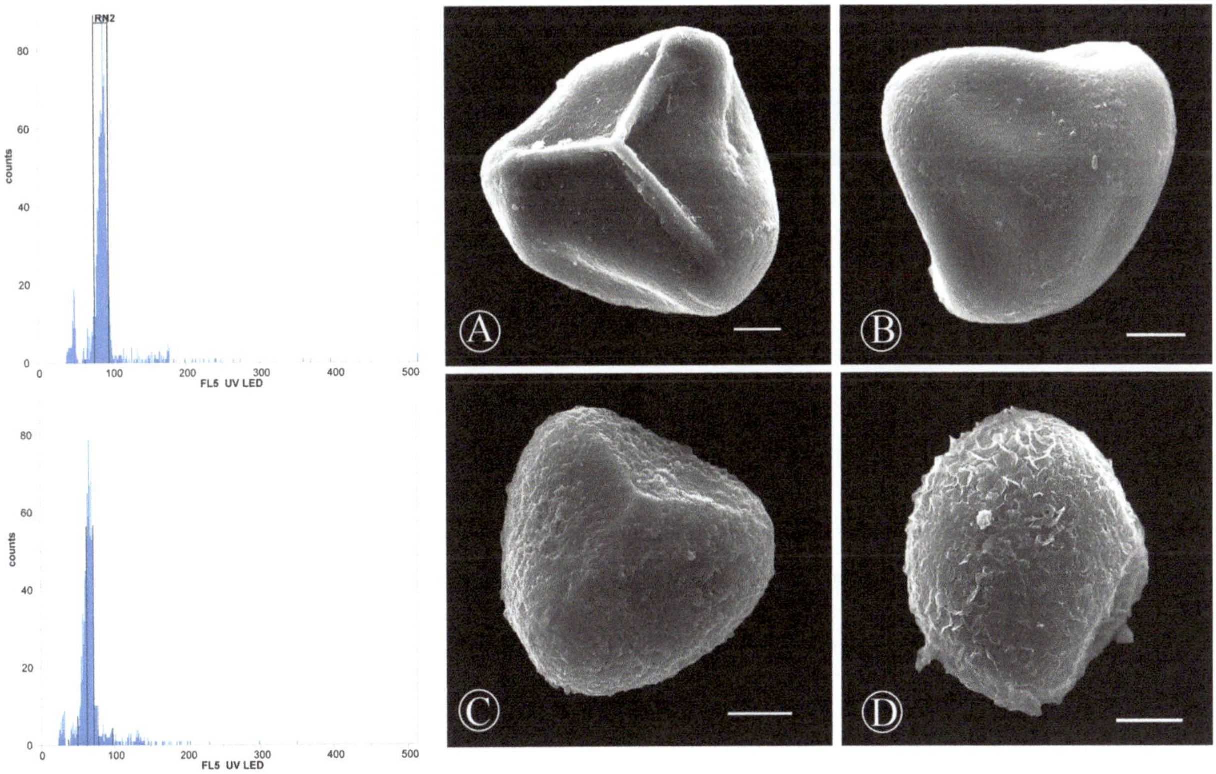

Figure 4 **Flow Cytometric Histogram and SEM Observation of *A. reniforme* var. *reniforme* and *A. reniforme* var. *sinense*.** **A** and **C**: proximal surface; **B** and **D**: distal surface.

The morphological characters of these two taxa are obviously different. The leaf blades of *A. reniforme* var. *sinense* are membranous and reniform. Each blade has several concentric rings and yellowish-brown scales. The leaves of *A. reniforme* var. *reniforme* are pachyphyllous and coriaceous and are much rounder and similar to palm. The leaves lack any concentric rings and have deep brown scales (see Figure 1).

Discussion

Relationship between *A. reniforme* var. *reniform* and *A. reniforme* var. *asariforme*

The Canary Islands *A. reniforme* var. *reniforme* was determined to be hexaploid in this study based on flow cytometric analyses of sporophytic material. An additional experiment was performed to determine chromosome numbers with conventional squashes of root tip cells but failed because of the huge numbers and crowded chromosomes. Thus, the chromosomes could not be counted using light microscopy.

The ploidy level of *A. reniforme* var. *reniforme* is the same as *A. reniforme* var. *asariforme* if the description in *Flora Republicae Popularis Sinicae*, 3(1) [5] is correct. According to *Flora Republicae Popularis Sinicae*, 3(1), *A. reniforme* var. *asariforme* is another variety of *A. reniforme* var. *reniforme* and is only distributed in South Africa, Madagascar, and Mauritius. Its pachyphyllous and coriaceous leaves have deep brown scales that contain tight and slender white hairs on both surfaces of leaves. The taller and stronger plant size and its hexaploidy are considered the major differences from *A. reniforme* var. *sinense*. However, taller and stronger plants of *A. reniforme* var. *reniforme* are found in fields in La Palma. Its leaves are also pachyphyllous and coriaceous and have deep brown scales. The leaf shape is very similar to the leaf of *A. reniforme* var. *asariforme* based on comparisons of their respective specimens. Therefore, it is reasonable that researchers have treated *A. reniforme* var. *asariforme* as a variety of *A. reniforme* var. *reniforme* [38]. Tardieu-Blot claimed that *A. reniforme* var. *asariforme* was conspecific with *A. reniforme* var. *reniforme* [20]. Further evidence is required to clearly define the relationship between these two varieties.

Evolution of intercontinental disjunctions between Chinese *A. reniforme* var. *sinense* and Canary *A. reniforme* var. *reniforme*

Three issues have to be discussed to explain the evolution of China-Madagascar-Canary intercontinental disjunctions. The first issue is the original centre of these three taxa. Second, how did the spores spread between each location? Finally, what is the genesis evolution and phylogenetic status of ser. *Reniformia* in *Adiantum* and Pteridaceae?

There are three probable original centres: China; Madagascar or South Africa; the Canary Islands or the western Mediterranean. According to our phylogenetic analysis and molecular divergence estimation results, China is speculated to be the most probable centre. There is strong evidence showing that Chinese *A. reniforme* var. *sinense* is sister to Canary *A. reniforme* var. *reniforme* (BP100; PP1.0; Figure 3). Clades of these two species together form ser. *Reniformia* [5], which has morphological synapomorphies of simple and kidney-shaped blades and clustered short-creeping rhizomes. Ser. *Reniformia* is suggested to be monophyletic and is sister to Ser. *Venusta* (Figure 3), which consists of 10 species and 4 varieties only distributed in Chinese temperate regions. The divergence between *A. reniforme* var. *reniforme* and *A. reniforme* var. *sinense* was estimated to be 4.94 (2.26-8.66) Myr in the Pliocene, and ser. *Reniformia* and Ser. *Venusta* was estimated to diverge in 23.33 (12.89-34.27) Myr in the Miocene. These results indicated that Ser. *Reniformia* and Ser. *Venusta* had a common ancestor at least 23.33 Myr ago but diverged later. The divergence may be related to the intense uplift of the Qinghai-Tibet plateau in the Neocene [39]. The average altitude of the Qinghai-Tibet plateau may have reached 2000 m at 22 Myr [40], during which the landform diversity of the Qinghai-Tibet plateau and climate aridification may have led to the divergence of ser. *Reniformia* from Ser. *Venusta* in China. The Himalayas uplifted rapidly 5.4-2.7 Myr [41], and *A. reniforme* var. *reniforme* diverged from *A. reniforme* var. *sinense* 4.94 (2.26-8.66) Myr. These results indicate that the divergence of the two species may be closely related to the rapid uplift of the Himalayas. Paleomonsoon had existed in China in the Eogene and intensified with the uplift of the Qinghai-Tibet plateau in the Neocene [42]. Northwestern Eurasia high pressure centres have passed through Southeast Asian nations such as China and India to the Indian Ocean since the Miocene [40,42]. The long distance dispersal of ferns is more common than seed plants because ferns are dispersed by small, windblown spores that are produced in very large numbers and are capable of travelling thousands of kilometres [43-45]. Thus, it was very possible for spores of Chinese *A. reniforme* var. *sinense* to reach the Indian Ocean and Madagascar through winter monsoons and other general atmosphere circulation in winter. Spores of *A. reniforme* var. *sinense* in Madagascar also can get back to China through summer southwest monsoons from the Southern Indian Ocean. However, gene flow was hindered by the high altitude caused by the rapid uplift of the Himalayas in the Pliocene, which caused speciation over time. If China was the origin centre of *A. reniforme*, the dispersal sequence would be as follows: China to Madagascar and then to Canary.

The Canary Islands consist of seven volcanic islands, namely El Hierro, La Palma, La Gomera, Tenerife, Gran Canaria, Fuerteventura, and Lanzarote (from west to east, respectively), located off the north-western African coast. They formed by multiple volcanic episodes [46-48] but showed different evolutionary histories [49]. The western islands of La Palma, El Hierro, and Tenerife are the younger archipelago and are still in their shield stage, which began at most 7.5 Myr ago. The oldest island Fuerteventura began its shield stage 20.6 Myr ago [50]. A fossil of *A. reniforme* var. *reniforme* was discovered in Meximieux near Lyons in the Rhone Valley in Europe [20]. Thus, the Canary Islands may be glacial refugia of *A. reniforme* var. *reniforme* in Quaternary.

Conclusions

Adiantum reniforme var. *sinense* is an independent species rather than a variety of *A. reniforme* var. *reniforme* based on morphological differences, spore observations, chromosome analyses, phylogeny research of the genus *Adiantum* and molecular divergence estimations. Our data are different from Lin YX [1] but in accordance with treatment of Zhang XC [21]. The name *Adiantum nelumboides* X. C. Zhang should be applied to the Chinese taxon as a legal name and the commonly used name for *A. reniforme* var. *sinense* will be treated as a synonym. China is deduced to be the most probable evolution centre of ser. *Reniformia*, and the divergence between *A. reniforme* var. *sinense* and *A. reniforme* var. *reniforme* may be related to the intense uplift of the Qinghai-Tibet plateau in the Neocene. The Canary Islands and Madeira were probably glacial refugia of *A. reniforme* var. *reniforme* in the Quaternary, based on the fossil evidence found in Meximieux near Lyons in the Pliocene.

Availability of supporting data

The data sets supporting the results of the article are available in GenBank under accession numbers KJ742731-KJ742799 and KJ779969-KJ780019. All of the phylogenetic sequence data in this study are deposited in GenBank (National Center for Biotechnology Information) with the link http://www.ncbi.nlm.nih.gov/nuccore/.

Additional files

Additional file 1: Table S1. Voucher information and GenBank accession numbers for taxa used in the phylogenetic study on *Adiantum*.

Additional file 2: Table S2. Samples examined in the study to estimate divergence times.

Competing interests

The authors declare that they have no competing interests.

Authors' contributions

AHW carried out the molecular phylogeny study and microscopic spore observations and flow cytometry, participated in data analysis and drafted the manuscript; YS conducted the data analysis, and contributed to the supervision and discussion of the research; HS and JWZ revised the manuscript; DML, JSZ contributed to collect part materials; HFC and FWX provided plant samples and contributed to the supervision of the research; FGW provided plant samples, performed morphological studies, conducted interpretation for the data and results and discussions, and contributed to the supervision of the research. All authors read and approved the final manuscript.

Acknowledgements

The authors thank Senior Engineer Xiao-Ying HU (South China Botanical Garden, Chinese Academy of Sciences, Guangzhou, China) for her help with SEM studies, Qing-Wen ZENG, Hui YU (South China Botanical Garden, Chinese Academy of Sciences, Guangzhou, China) and Jin-Song ZHOU (College of Chinese Traditional Medicine, Guangzhou University of Chinese Medicine, Guangzhou, China) for their help collecting plant samples, and Yun-Xiao LIU (South China Botanical Garden, Chinese Academy of Sciences, Guangzhou, China) for their help with the morphology figures. This work was funded by the Main Direction Program of Knowledge Innovation of the Chinese Academy of Sciences (Grant Nos. KSCX2-EW-Q-8), and the Key Laboratory of Plant Resource Conservation and Sustainable Utilization, South China Botanical Garden, Chinese Academy of Sciences (201214ZS), and Guangdong Provincial Key Laboratory of Applied Botany, South China Botanical Garden, Chinese Academy of Sciences.

Author details

[1]Key Laboratory of Plant Resources Conservation and Sustainable Utilization, South China Botanical Garden, Chinese Academy of Sciences, Guangzhou 510650, China. [2]University of Chinese Academy of Sciences, Beijing 100049, China. [3]Department of Life Sciences, Natural History Museum, London SW75BD, UK. [4]College of Landscape Architecture, Fujian Agriculture and Forestry University, Fuzhou 350002, China. [5]College of Chinese Traditional Medicine, Guangzhou University of Chinese Medicine, Guangzhou 510006, China.

References

1. Lin YX: **New taxa of *Adiantum* L. in China.** *Acta phytotax Sin* 1980, **18**:101–105 [in Chinese with English summary].
2. Xu TQ, Zhen Z, Jin YX: **On the distribution characteristic of the variety Adiantun reniforme var. sinense.** *Wuhan Bot Res* 1987, **5**:247–251 [in Chinese with English summary].
3. Liu YC: **Flora geography of national wild conservative plants in Chongqing.** *Southwest China Normal Univ (Nature)* 2000, **25**:439–446.
4. Peng J, Long Y, Liu YL, Li XG: **The rare and endangered species in Chongqing.** *Wuhan Bot Res* 2000, **18**:42–48 [in Chinese with English summary].
5. Shing KH, Wu SK: **Adiantaceae.** In *Flora Republicae Popularis Sinicae, 3(1).* Edited by Ching RC, Shing KH. Beijing: Science Press; 1990:173–216.
6. Kang M, Huang H, Jiang M, Lowe AJ: **Understanding population structure and historical demography in a conservation context: population genetics of an endangered fern.** *Diversity and Distributions* 2008, **14**(5):799–807.
7. Pan L: *Study on Growth Charaeteristics and Propagation of Adiantum reniforme var. sinense.* Wuhan Botanical Garden, Chinese Academy of Science.: Wuhan; 2007.
8. Liu XQ, Wahiti GR, Chen LQ: **Genetic variation in the endangered fern *Adiantum reniforme* var. *sinense* (Adiantaceae) in China.** *Annales Botanici Fennici* 2007, **44**(1):25–32.
9. Pan LQ, Ji H, Chen LQ: **Genetic diversity of the natural populations of *Adiantum reniforme* var. *sinense*.** *Biodivers Sci* 2005, **13**(2):122–129 [in Chinese with English summary].
10. Lin YX: **The sexual propagation and chromosome number of *Adiantum reniforme* L. var. *sinense* Y. X. Lin.** *Cathaya* 1989, **1**:143–148.
11. Wang QX, Dai XL: *Spores of Polypodiales (Filicales) from China.* Beijing: Science Press; 2010:10–170.

12. Smith AR, Pryer KM, Schuettpelz E, Korall P, Schneider H, Wolf PG: **A classification for extant ferns.** *Taxon* 2006, **55:**705–731.
13. Schneider H, Schuettpelz E, Pryer KM, Cranfill R, Magallón S, Lupia R: **Ferns diversified in the shadow of angiosperms.** *Nature* 2004, **428**(6982):553–557.
14. Schuettpelz E, Pryer KM: **Fern phylogeny inferred from 400 leptosporangiat species and three plastid genes.** *Taxon* 2007, **56:**1037–1050.
15. Schuettpelz E, Schneider H, Huiet L, Windham MD, Pryer KM: **A molecular phylogeny of the fern family Pteridaceae: assessing over all relationships and the affinities of previously unsampled genera.** *Mol Phylogenet Evol* 2007, **44:**1172–1185.
16. Ruhfel B, Lindsay S, Davis CC: **Phylogenetic placement of Rheopteris and the polyphyly of Monogramma (Pteridaceae s.l.): evidence from rbcL.** *Syst Bot* 2008, **33:**37–43.
17. Bouma WLM, Ritchie P, Perrie LR: **Phylogeny and generic taxonomy of the New Zealand Pteridaceae ferns from chloroplast rbcL DNA sequences.** *Australian systematic botany* 2010, **23**(3):143–151.
18. Lu JM, Wen J, Lutz S, Wang YP, Li DZ: **Phylogenetic relationships of Chinese Adiantum based on five plastid markers.** *J Plant Res* 2012, **125**(2):237–249.
19. Manton I: *Problems of cytology and evolutions in the Pteridophyta.* London: Cambridge University Press; 1950.
20. Manton I, Lovis JD, Vida G, Gibby M: **Cytology of the fern flora of Madeira.** *Bulletin of the British Museum Natural History Botany* 1986, **15**(2):123–161.
21. Zhang XC: *Lycophytes and Ferns of China.* Beijing: Peking University Press; 2012:258.
22. Doyle JJ, Doyle JL: **A rapid DNA isolation procedure for small quantities of fresh leaf tissue.** *Phytochem Bull* 1987, **19:**11–15.
23. Schuettpelz E, Korall P, Pryer KM: **Plastid atpA data provide improved support for deep relationships among ferns.** *Taxon* 2006, **55:**897–906.
24. Little DP, Barrington DS: **Major evolutionary events in the origin and diversification of the fern genus Polystichum (Dryopteridaceae).** *Am J Bot* 2003, **90:**508–514.
25. Taberlet P, Gielly L, Pautou G, Bouvet J: **Universal primers for amplification of three non-coding regions of chloroplast DNA.** *Plant Mol Biol* 1991, **17:**1105–1109.
26. Lu JM, Li DZ, Gao LM, Cheng X, Wu D: **Paraphyly of Cyrtomium (Dryopteridaceae): Evidence from rbcLand trnL-F sequence data.** *J Plant Res* 2005, **118:**129–135.
27. Shaw J, Lickey EB, Beck JT, Farmer SB, Liu W, Miller J, Siripun KC, Winder CT, Schilling EE, Small RL: **The tortoise and the hare II: Relative utility of 21 noncoding chloroplast DNA sequences for phylogenetic analysis.** *Am J Bot* 2005, **92:**142–166.
28. Souza-Chies TT, Bittar G, Nadot S, Carter L, Besin E, Lejeune B: **Phylogenetic analysis of Iridaceae with parsimony and distance methods using the plastid gene rps4.** *Plant Systematics and Evolution* 1997, **204:**109–123.
29. Hall TA: *BioEdit: a user-friendly biological sequence alignment editor and analysis*; 1999.
30. Larkin MA, Blackshields G, Brown NP, Chenna R, McGettigan PA, McWilliam H, Valentin F, Wallace LM, Wilm A, Lopez R, Thompson JD, Gibson TJ, Higgins DG: **Clustal W and Clustal X version 2.0.** *Bioinformatics* 2007, **23**(21):2947–2948.
31. Swofford DL: **PAUP*: phylogenetic analysis using parsimony (* and other methods). version 4.0.b10.** Sunderland, MA: Sinauer Associates 2002.
32. Nylander JAA: *Mrmodeltest (version 2): Program Distributed by the Author.* Uppsala: Evolutionary Biology Centre, Uppsala University; 2004.
33. Ronquist F, Huelsenbeck JP: **MrBayes 3: Bayesian phylogenetic inference under mixed models.** *Bioinformatics* 2003, **19:**1572–1574.
34. Drummond AJ, Suchard MA, Xie D, Rambaut A: **Bayesian phylogenetics with BEAUti and the BEAST 1.7.** *Mol Biol Evol* 2012, **29**(8):1969–1973.
35. Lu JM, Li DZ, Lutz S, Soejima A, Yi T, Wen J: **Biogeographic disjunction between eastern Asia and North America in the Adiantum pedatum complex (Pteridaceae).** *Am J Bot* 2011, **98**(10):1680–1693.
36. Rambaut A, Drummond AJ: Tracer Version 1.5. 2007. Available online: http://beast.bio.ed.ac.uk/Tracer (accessed on 21 December 2011).
37. Harris JG, Harris MW: *Plant identification terminology: An illustrated glossary.* Beijing: Science Press; 2001:1–302.
38. Sim TR: **Ferns of South Africa.** CUP Archive. 1915. Available online: http://www.clarkes.co.za/book/ferns-of-south-africa.
39. Sun H, Li ZM: **Evolution and development of the Tethys flora in China after uplift of Tibet Plateau.** *Adv Earth Sci* 2003, **18**(6):852–862.
40. Li JJ: **Landform evolution and Asian monsoon of the Qing hai -Xi zang Plateau.** *Marine Geology and Quaternary Geology* 1999, **19**(1):1–9.
41. Zhu DG, Meng XG, Shao ZG, Yang CB, Han JE, Yu J, Meng QW, Lü RP: **The Formation and Evolution of Zhada Basin in Tibet and the Uplift of the Himalayas.** *Acta geoscientica sinica* 2006, **27**(3):193–200.
42. Peng H: **Discussion about the impact of the Qinghai-Tibet Plateau's uplift on China climate.** *Geographical Research* 1989, **8**(3):85–92.
43. Barrington DS: **Ecological and historical factors in fern biogeography.** *J Biogeogr* 1993, **20:**275–280.
44. Smith AR: **Phytogeographic principles and their use in understanding fern relationships.** *J Biogeogr* 1993, **20:**255–264.
45. Wolf PG, Schneider H, Ranker TA: **Geographic distributions of homosporous ferns: Does dispersal obscure evidence of vicariance?** *J Biogeogr* 2001, **28:**263–270.
46. Schmincke HU: **Volcanic and chemical evolution of the Canary Islands.** In *Geology of the Northwest African continental margin.* Edited by von Rad U, Hinz K, Sarnthein M, Seibold E. Berlin Heidelberg New York: Springer; 1982:273–306.
47. Carracedo JC: **The Canary Islands: an example of structural control on the growth of large oceanic-island volcanoes.** *Journal of Volcanology and Geothermal Reaearch* 1994, **60:**225–241.
48. Carracedo JC: **A simple model for the génesis of large gravitational landslide hazards in the Canary Islands.** In *Volcano Instability on the Earth and Other Planets*, Geol. Soc. Lond. Spec. Publ, Volume 110. Edited by McGuire WJ, Jones AP, Neuberg J. ; 1996:125–135.
49. Abratis M, Schmincke HU, Hansteen TH: **Composition and evolution of submarine volcanic rocks from the central and western Canary Islands.** *Int J Earth Sci (Geol Rundsch)* 2002, **91:**562–582.
50. Stillman CJ: **Giant Miocene landslides and the evolution of Fuerteventura, Canary Islands.** *J Volcanol Geotherm Res* 1999, **94:**89–104.

5

Non-targeted metabolite profiling of citrus juices as a tool for variety discrimination and metabolite flow analysis

Vicent Arbona[1*], Domingo J Iglesias[2] and Aurelio Gómez-Cadenas[1]

Abstract

Background: Genetic diversity of citrus includes intrageneric hybrids, cultivars arising from cross-pollination and/or somatic mutations with particular biochemical compounds such as sugar, acids and secondary metabolite composition.

Results: Secondary metabolite profiles of juices from 12 commercial varieties grouped into blonde and navel types, mandarins, lemons and grapefruits were analyzed by LC/ESI-QTOF-MS. HCA on metabolite profiling data revealed the existence of natural groups demarcating fruit types and varieties associated to specific composition patterns. The unbiased classification provided by HCA was used for PLS-DA to find the potential variables (mass chromatographic features) responsible for the classification. Abscisic acid and derivatives, several flavonoids and limonoids were identified by analysis of mass spectra. To facilitate interpretation, metabolites were represented as flow charts depicting biosynthetic pathways. Mandarins 'Fortune' and 'Hernandina' along with oranges showed higher ABA contents and ABA degradation products were present as glycosylated forms in oranges and certain mandarins. All orange and grapefruit varieties showed high limonin contents and its glycosylated form, that was only absent in lemons. The rest of identified limonoids were highly abundant in oranges. Particularly, Sucrenya cultivar showed a specific accumulation of obacunone and limonoate A-ring lactone. Polymethoxylated flavanones (tangeritin and isomers) were absolutely absent from lemons and grapefruits whereas kaempferol deoxyhexose hexose isomer #2, naringin and neohesperidin were only present in these cultivars.

Conclusions: Analysis of relative metabolite build-up in closely-related genotypes allowed the efficient demarcation of cultivars and suggested the existence of genotype-specific regulatory mechanisms underlying the differential metabolite accumulation.

Keywords: Fruit quality, Liquid chromatography, Mass spectrometry, Orange, Phenotyping, Secondary metabolites

Background

In the Rutaceae family, citrus constitutes a highly heterogeneous taxonomic group including several species such as sweet oranges (*Citrus sinensis* L. Osbeck), mandarins (*C. clementina* hort. Ex Tan. and *C. reticulata* Blanco), lemons (*Citrus × limon* L. Burm.f.) and grapefruits (*C. paradisi* Macf.). Besides these species, there are other related species with agronomic uses as rootstocks or for ornamental purposes (e.g. *Poncirus trifoliata* L. Raf.). Usually, the different cultivars within a species show low genetic variability but do have particular desirable phenotypic characteristics such as precocity or delayed harvesting, seedless fruits, sugar and acid accumulation, easiness to peel, etc. However, alteration of the harvesting period is one of the most desirable traits, either when precocity or delayed harvesting is achieved. This alteration has additional impacts on fruit quality, as environmental variables change over the year and irradiation, temperature and humidity influence fruit growth, accumulation of sugars and acids and other non-palatable chemical constituents [1-3]. It is difficult to have a control on the buildup of these compounds in fruits over the maturation process. This fraction of citrus juice is constituted, among others, by carotenoids, triterpenoids, flavonoids and other secondary metabolites known to have an impact on health [4,5]. It

* Correspondence: vicente.arbona@camn.uji.es
[1]Laboratori d'Ecofisiologia i Biotecnologia, Departament de Ciències Agràries i del Medi Natural, Universitat Jaume I, E-12071 Castelló de la Plana, Spain
Full list of author information is available at the end of the article

has been previously shown that different citrus juices have different carotenoid profiles depending on genotype and growth conditions [6] that could have an impact on citrus nutritional properties. To this respect, within a particular growth area, the genotype is expected to be the major contributing factor determining fruit compositional properties, and therefore genetic mutations that give rise to new varieties would also affect fruit chemical composition [7]. Nevertheless, despite the enormous amount of information available it has been so far impossible to establish a reliable model of metabolite flow in citrus fruits. A possible utilization pathway for citric acid was proposed linking it to acetyl-CoA through ATP-citrate lyase after isomerization to isocitrate catalyzed by aconitase [8]. This acetyl-CoA could be in turn channeled to the biosynthesis of secondary metabolites such as limonoids, carotenoids and xanthophylls through the methyl-eriothritiol phosphate pathway. Moreover, biosynthesis of flavonoids and other phenylpropanoids is fueled by intermediates generated during glycolysis and pentose phosphate pathway. To add more complexity to the model, levels of these compounds are determined by the activity of different enzymes that are, in turn, responsible for their biosynthesis, their degradation/biotransformation and/or the conjugation to different chemical moieties. In this sense, as the enzyme activity is generally associated to gene expression, metabolites could be considered the end-products of gene expression [9]. Therefore, to better understand the regulation of secondary metabolism in citrus fruits a comprehensive and unbiased analysis of this class of compounds is required. To this regard, non-targeted LC/MS metabolite profiling has proved to be a valuable tool for phenotyping environmentally- or genetically-induced variations in secondary metabolite composition [10] as well as to evaluate the impact of stress on plant biochemistry [11]. This technique has been previously used to assess adulteration of citrus juice with grape or apple ones [12] and, more recently, to phenotype wild type and mutant orange varieties [7].

The aim of this work was to investigate the differences in secondary metabolite composition within and between five important commercial citrus fruit groups: oranges (blonde and navel), mandarins, grapefruits and lemons (see Table 1). A detailed identification of selected metabolite features was considered in this work to further investigate secondary metabolite flows in every variety, linking diversification to particular metabolite profiles.

Methods

Fruit harvesting, sample collection and preparation for analyses

Citrus fruits from different genotypes and varieties (see Table 1) were harvested at commercial maturity from trees at the germplasm bank (Institut Valencià d'Investigacions Agràries, IVIA, Moncada, València). Commercial maturity refers to the timing of harvest to meet specific market and consumer requirements. In citrus, this is assessed by means of the maturity index (°Brix/acidity, see Table 1 for usual maturity index values). Genotypes were characterized according to an enlarged modification of the "Descriptor for Citrus" from the International Plant Genetic Resources Institute (IPGRI) [13]. At least four fruits, one from each direction on the tree, were collected from three replicate trees (n = 3) grafted onto the same rootstock. Juice extraction was performed by

Table 1 List of genotypes included in this study

#	Name	Species*	Type*	Harvesting period**	°Brix/acidity**
1	Eureka	*Citrus × limon* L. Burm.f.	Lemon	October to February	3-10
2	Fino	*Citrus × limon* L. Burm.f.	Lemon	October to April	
3	Marsh	*Citrus paradisi* Macf.	Grapefruit	November to March	4-8
4	Star Ruby	*Citrus paradisi* Macf.	Grapefruit	October to March	
5	Fortune	*Citrus reticulata* Blanco (Clementine mandarin × Dancy mandarin)	Mandarin	February to April	8-11
6	Nadorcott	*Citrus reticulata* Blanco (open pollination of Murcott mandarin)	Mandarin	January to March	8-13
7	Pixie	*Citrus reticulata Blanco (open pollination of Kincy mandarin)*	Mandarin	December to February	10-28
8	Hernandina	*Citrus clementina hort. ex Tanaka*	Mandarin	January to February	16-28
9	Sucrenya	*Citrus sinensis L. Osbeck*	Blonde orange	December to March	>40
10	Midknight	*Citrus sinensis L. Osbeck*	Blonde orange	March to June	8-14
11	Washington	*Citrus sinensis L. Osbeck*	Navel orange	December to February	8-13
12	Lane late	*Citrus sinensis L. Osbeck*	Navel orange	January to April	10-16

(*)Information retrieved from University of California, Riverside Citrus variety collection website (http://citrusvariety.ucr.edu). (**) information retrieved from http://www.ivia.es.

manual squeezing and juice of fruits from the same tree was pooled. Juice aliquots were immediately stored at −80°C until analyses with no further processing. Right before chromatographic analyses, frozen fruit juices were thawed at room temperature, centrifuged and the supernatants filtered through PTFE syringe filters (0.2 μm pore size) directly to vials.

Chromatographic and QTOF-MS conditions

Fruit juices were separated by reversed phase HPLC using acetonitrile (B) and water (A), both supplemented with formic acid to a concentration of 0.1% (v/v), as solvents and a C18 column (5-μm particle size, 100 9 2.1 mm, XTerra™, Waters). The separation module, a Waters Alliance 2965 was operated in gradient mode at a flow rate of 300 μl min^{-1} for 30 min as follows: 0–2 min 95:5 (A:B) followed by an increase in B from 5 to 95 in the following 26 min (2.01-28.00 min), thereafter returning to initial conditions (29.01-30.00 min) that were maintained for 5 min for column reconditioning. Column eluates were introduced into a QTOF-MS (Micromass Ltd., Manchester, UK) through an ESI source operated in positive and negative mode. Nitrogen was used as the nebulization as well as the desolvation gas and working flows were set at 100 and 800 L h^{-1}, respectively. Source block temperature was kept at 120°C and desolvation gas at 350°C. Capillary, cone, and extractor voltages were set at 4 kV, 25 eV, and 3 eV, respectively. Before analyses, the QTOF-MS was calibrated by infusing a mixture of NaOH and HCOOH at a flow rate of 25 μl min^{-1}. After calibration, the average error was less than 5 ppm. During acquisition, a one-ppm solution of Leu-enkephalin ($[M+H]^+ = 556.2771$) was continuously post column infused as a lockmass reference. Data were acquired under continuous mode in the 50–1000 amu range, scan duration was set at 1.0 s, and interscan delay was set at 0.1 s.

Data processing

Data processing was achieved using Masslynx v.4.1 and raw data files were analyzed using xcms following conversion to netCDF with the databridge software provided by Masslynx. Chromatographic peak detection was performed using the matchedFilter algorithm, applying the following parameter settings: snr = 3, fwhm = 15 s, step = 0.01 D, mzdiff = 0.1 Da, and profmethod = bin. Retention time correction was achieved in three iterations applying the parameter settings minfrac = 1, bw = 30 s, mzwid = 0.05 Da, span = 1, and missing = extra = 1 for the first iteration; minfrac = 1, bw = 10 s, mzwid = 0.05 Da, span = 0.6, and missing = extra = 0 for the second iteration; and minfrac = 1, bw = 5 s, mzwid = 0.05 Da, span = 0.5, and missing = extra = 0 for the third iteration. After final peak grouping (minfrac = 1, bw = 5 s) and filling in of missing features using the fillPeaks command of the xcms package, a data matrix consisting of mass features (including accurate mass values and retention time) and peak area values per sample was obtained.

Statistical analyses

Hierarchical Cluster Analysis (HCA) was performed with pvclust package running under R 3.2 and PLS-DA was performed using SIMCA-P+ 11.0 (Umetrics, Umea, Sweden). HCA, followed by bootstrap resampling (n = 1000) to validate grouping, was performed on raw data without any variable selection to observe natural grouping of samples. The classification provided by unsupervised HCA confirmed homogeneity of sample groups (Figure 1) and allowed using genotype denomination as parameter to feed PLS-DA. This strategy was further used to select potential variables contributing to the provided classification. Prior to analyses, data were normalized to total ion intensity. The potential variables contributing to the classification were selected based on variable importance in the projection (VIP > 2.0) values. Relevant variables were then confirmed after integration of chromatographic peaks and analysis of variance (ANOVA) of peak areas throughout the 12 sample groups. The metabolites were tentatively identified by elucidation of structures with MS fragments, comparison of accurate *m/z* value and MS fragmentation pattern with literature and co-injection with pure standards when available. All standards were purchased from Sigma-Aldrich (Madrid, Spain) except for ABA and derivatives that were obtained from the Plant Biotechnology Institute of the National Research Council (Canada).

Results and discussion

Non-targeted analysis of secondary metabolite features in citrus fruit juices

The analyses, carried out by means of reversed-phase liquid chromatography coupled to a QTOF-MS operated in positive and negative ionization modes, rendered a number of chromatograms that were extracted with XCMS [14]. The resulting datasets were subjected to HCA using the R package pvclust and presented as dendrograms in Figure 1. The results showed grouping of sample replicates in tight clusters according to the juice source fruit (see Table 1). In addition, relationship between clusters was in agreement with the expected phylogenetic relationships among varieties showing a perfect separation of the represented groups: grapefruits, lemons, oranges (blonde and navel types) and mandarins (see Additional file 1: Figure S1). All varieties could be resolved using different component combinations after PLS-DA. In addition, loadings plots indicated that some

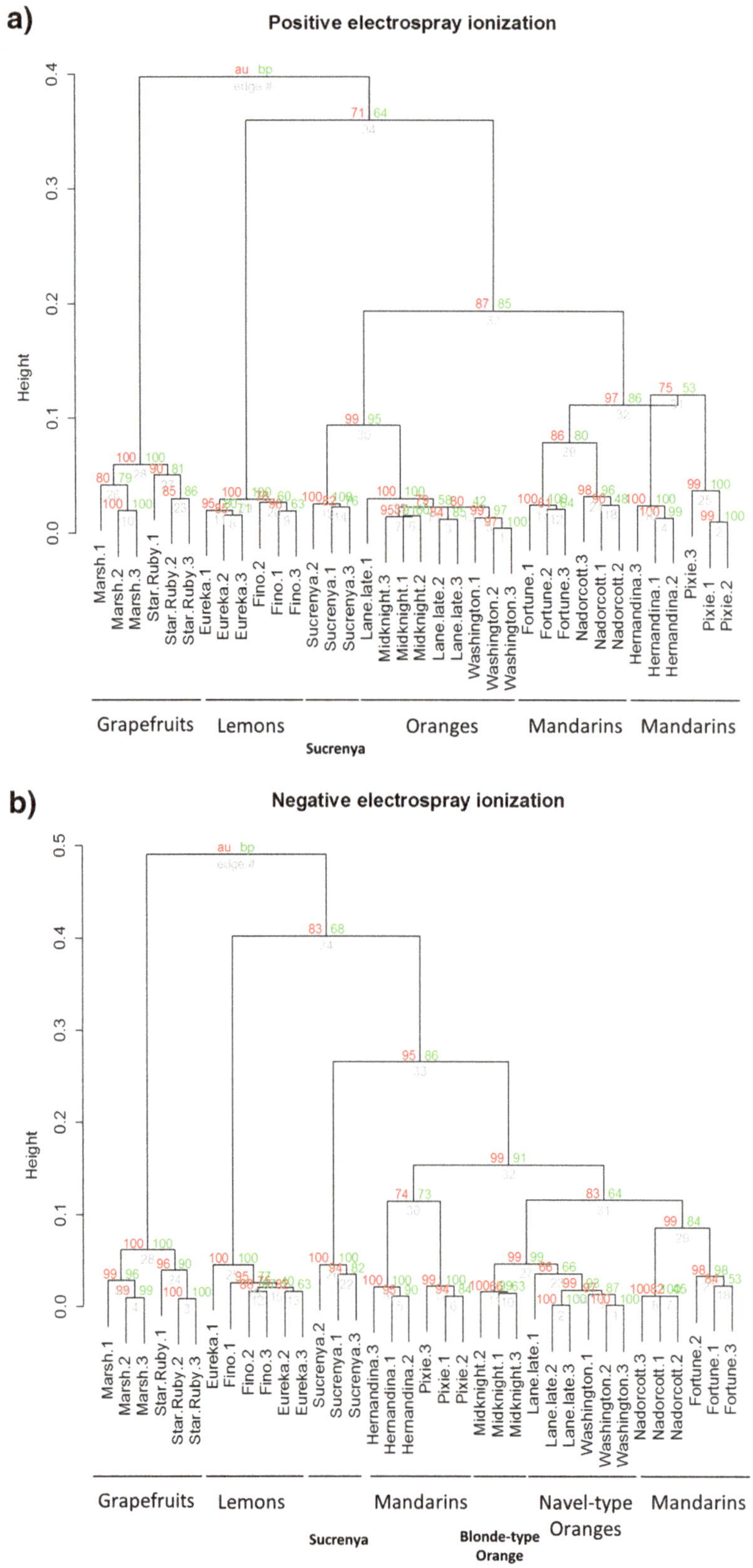

Figure 1 Hierarchical clustering dendrograms obtained from (a) positive and (b) negative electrospray metabolite profiles of citrus juices. On every node, approximate unbiased (red, au) and bootstrap values (green, bp) are presented.

variables were important in defining the different sample groups (Additional file 2: Figure S2a through f). Component 2 resolved well 'Washington' navel from the rest whereas 'Sucrenya' resolved along component 3. A combination of components 5 and 6 allowed the resolution of grapefruits and the two varieties included in this group. Component 5 alone allowed the discrimination of 'Hernandina' from the rest of varieties. A better resolution for grapefruits was obtained along component 8. Meanwhile, component 7 resolved well 'Nadorcott' and 'Midknight' varieties. Lemons resolved along component 10 whereas component 9 discriminated 'Pixie' from the rest. A combination of components 9 and 10, allowed demarcation of 'Fortune' and 'Lane late' although these two varieties were better resolved along component 11 in combination with component 1 (Additional file 2: Figure S2f). The two grapefruit varieties were the utmost distant species included in the study followed by lemons, both constituting highly tight clusters in the HCA (Figure 1). This is probably due to their clear phylogenetic origin, grapefruits are crosses between sweet orange *Citrus sinensis* and *Citrus maxima* (pummelo), whereas lemons arise from the cross of sour orange *Citrus aurantium* and *Citrus medica* (citron, see Additional file 1: Figure S1 for more details). Two major clusters originated from grouping oranges ('Sucrenya', 'Lane late', 'Midknight' and 'Washington') and mandarins ('Hernandina', 'Pixie', 'Fortune' and 'Nadorcott') that are also phylogenetically related. To this respect, although 'Lane late', 'Midknight' and 'Washington' always occurred together, 'Sucrenya' appeared as a separate cluster probably due to its acidless juice characteristics. Moreover, in both ionization modes the methodology efficiently demarcated mandarins in two groups: 'Fortune'/'Nadorcott', arising from clementine × mandarin cross-pollination and an open pollination of 'Murcott' mandarin (see Table 1) respectively,and 'Hernandina'/'Pixie', resulting from a bud mutation from 'Fina' clementine and an open pollination of 'Kincy' mandarin, correspondingly. Mandarins are self-incompatible citrus species that usually produce seedless fruits unless flowers are cross-pollinated with compatible species. These cross-pollination has been extensively used to generate new commercial cultivars with particular fruit traits that differ from those of each parental. Examples of this are 'Fortune' and Nadorcott', often classified as mandarin hybrids, which share several fruit morphology, color and aroma characteristics. On the other hand, 'Hernandina' and 'Pixie', although are classified as two different species, they show more similar phenotypic traits, including morphology, flavor and period of maturation [15]. It is likely that despite of differences in their genetic origin the respective overcrosses yielded varieties with similar metabolite phenotypes quite different from the rest of varieties included in this study. Profiling of citrus juices in negative electrospray also gave the required resolution to discriminate genotypes included in the navel and blonde groups: 'Lane late'/'Washington' and 'Sucrenya'/'Midknight', respectively. In this sense, it is worthwhile to note that 'Sucrenya' always occurred as a separate group from oranges. This is likely a result of its particular juice traits. This variety usually shows very low titratable acid contents, compared to the rest of blonde or navel-type varieties [8]. Nevertheless, although all related varieties (within the same group) clustered together, it was still possible to clearly differentiate each of them (Figure 1).

Variable selection and annotation of compounds

In order to identify those variables contributing to the observed classification (Figure 2), a PLS-DA was carried out using the entire XCMS output using sample classification provided by HCA. PLS-DA calculates a regression model between the multivariate dataset (each variable consisting of a *m/z* and retention time value) and a response variable that only contains class information

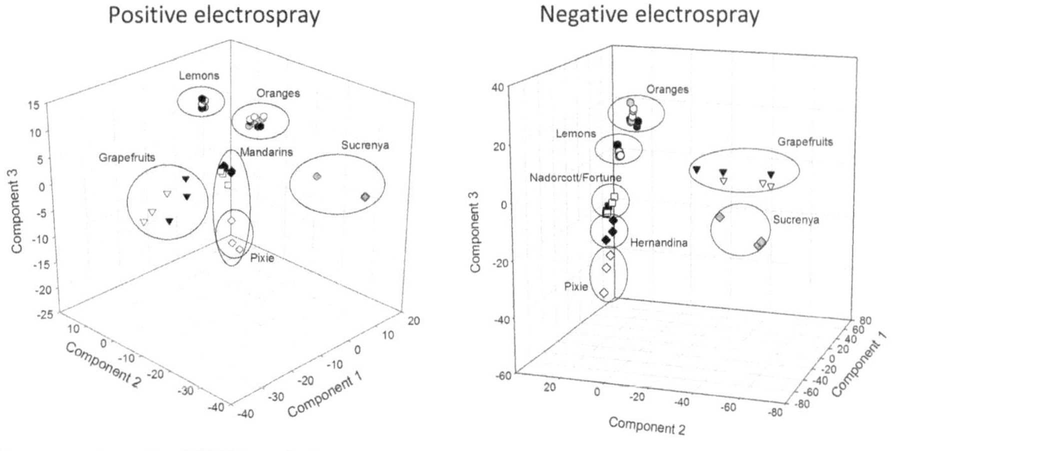

Figure 2 Scores 3D scatter plots after PLS-DA analysis.

(e.g. the variety classification provided by HCA). This analysis yielded a number of variables (chromatographic peaks, each represented by *m/z* and retention time values) ranked from very important (VIP > 2 to 1.5) to irrelevant (VIP values lower than 1). Scores 3D scatter plots from PLS-DA results indicated an optimal performance of the model to differentiate big groups of fruits: lemons, grapefruits, oranges and mandarins (Figure 2) and, in addition, some varieties were clearly differentiated within their respective groups such as both grapefruit cultivars, 'Pixie' mandarin and 'Sucrenya' blonde orange. Nevertheless, by representing other combinations of components the model is also able to clearly differentiate closely-related varieties within a group (data not shown). In general, varieties were grouped according to genotype and not harvesting period (Table 1). It seems clear that environmental growth conditions have an influence on fruit secondary metabolite composition as shown in [6]. In that case, carotenoid composition of orange and mandarin varieties grown in Mediterranean, subtropical and tropical conditions was evaluated showing clear differences. However, when the same parameter was evaluated in varieties grown in the same climatic conditions, little changes could be observed throughout the year. Therefore, the differences in secondary metabolite composition observed in the present work are likely to arise as particular genotype traits rather than being induced by environmental changes. Biochemical evolution of fruits throughout the ripening process is also an important aspect. In this work, all fruits were harvested at optimum commercial maturity. It is likely that fruit metabolite composition changes during fruit growth and maturation and also during the postharvest period. However, it is expected that they keep their characteristic traits. Recently, it was shown that even after industrial orange juice processing it was possible to identify adulteration with other juice sources, such as apple or grapefruit [12]. This demonstrates that industrial juice processing is not sufficient to remove or mask the discriminant metabolite of orange juice. Moreover, metabolomic analysis of pulp extracts of an orange bud mutant variety and its parental at different harvesting dates revealed higher differences between varieties than among sampling dates. Therefore, it could be hypothesized that differences among varieties could be minimized throughout the ripening process; however, the discriminant metabolite traits allowing demarcation of genotypes would still remain present.

Chromatographic mass features showing a VIP value higher than 1.5 were located and further inspected using Masslynx 4.1. software to attain structure elucidation and annotation of compounds. A number of potential metabolites were identified and annotated based on structural elucidation, literature search and comparison with commercial standards, when available (Table 2). According to their putative annotation, all compounds were grouped into metabolite classes and their relative accumulation represented as metabolite flow charts (Figures 3, 4 and 5). ABA and its derivatives were identified based on mass spectra and/or comparison with commercial standards. It has been previously shown that variations in the expression of *NCED2* and *3* are correlated with endogenous ABA levels. To this respect, juice sacs of satsuma mandarin had higher ABA levels than those of lemons or sweet oranges along with higher NCED expression [16]. This could be somehow associated to differences found in carotenoid content among citrus varieties [6]. Besides changes in expression and activity of NCEDs, carotenoid precursor availability could influence ABA content. Citrus fruits are also important sources of flavonoids, including several kaempferol, hesperetin, naringenin and isorhamnetin derivatives that were putatively identified based on the literature and the comparison with commercial standards. In addition, three metabolites showing a *m/z* compatible with their annotation as tangeretin ($[M+H]^+$ 373.1397, ΔDa –0.011) were detected under positive electrospray ionization (Table 2). This would indicate the presence of different tangeretin isomers with identical composition but methoxylated in different positions. Moreover, some limonoids were annotated in citrus samples. These compounds are triterpenoids derived from squalene by formation of a polycyclic molecule containing a furanolactone core structure [17] and some of them are known to provide bitter taste to citrus juices namely limonin, nomilin, obacunone and nomilinic acid. Limonoids can also be released from their respective glycosylated forms upon cleavage after freeze damage or other environmental stress conditions [18]. These compounds have been associated to fruit quality and reported to have important health benefits [17,19,20]. Besides, some bitter limonoids can be present as tasteless A-ring lactones that were also tentatively annotated in this work. In addition, some compounds involved in other mixed pathways, such as the aminoacids Phe and Trp, involved in aromatic and indolic compound biosynthesis [21,22] and a ferulic acid hexoside, derived from the phenylpropanoid pathway, were also annotated.

For an easier interpretation of data, flow charts depicting biosynthetic pathways (constructed according to the current information available on Kegg, http://www.genome.jp/kegg/) are presented in this work. This allows classifying most metabolites as part of specific biosynthetic pathways, the relative concentration of each metabolite throughout all analyzed genotypes is represented as a color scale (Figures 2, 3 and 4) following the same sample order as in Table 1. The validity of each metabolite marker was assessed by ANOVA comparing peak areas

Table 2 Identification of compounds

Compound	ESI +	annotation positive	ESI -	annotation negative	Rt (min)	Rt (s)	annotation level	ChEBI code
Abscisic acid and derivatives								
Dihydrophaseic acid glycosil ester (DPAGE)	**247.1445**	$[M+H-H_2O]^+$	**443.1941**	$[M-H]^-$	4.15	249.3	2, 3 [23]	
	265.1483	$[M+H-Glucose]^+$	479.1836	$[M+Cl]^-$				
	467.2023	$[M+Na]^+$	489.1993	$[M+HCOOH]^-$				
	483.1752	$[M+K]^+$						
Phaseic acid glycosyl ester (PAGE)	**247.1447**	$[M+H-Glucose-H_2O]^+$	**442.1841**	$[M-H]^-$	5.83	350.0	1, 3	
	467.2059	$[M+Na]^+$						
	483.1753	$[M+K]^+$						
Dihydrophaseic acid (DPA)	**265.1454**	$[M+H-H_2O]^+$	**281.1455**	$[M-H]^-$	8.46	507.6	2, 3 [23]	CHEBI:23757
	247.1357	$[M+H-2 \times H_2O]^+$						
	305.1456	$[M+Na]^+$						
Abscisic acid glycosyl ester (ABAGE)	229.1329	$[M+H-H_2O]^+$	**425.1833**	$[M-H]^-$	10.26	615.6	1	CHEBI:62436
	247.1379	$[M+H-Glucose]^+$	471.1915	$[M-H+HCOOH]^-$				
	265.1528	$[M+H-Hexose]^+$	263.1404	$[M-Hexose]^-$				
	449.1775	$[M+Na]^+$						
	465.1740	$[M+K]^+$						
Phaseic acid (PA)	**247.1357**	$[M+H-2 \times H_2O]^+$	**279.1400**	$[M-H]^-$	11.56	693.6	1	CHEBI:28205
	265.1490	$[M+H-H_2O]^+$						
	229.1490	$[M+H-3 \times H_2O]^+$						
Abscisic acid (ABA)	**247.1385**	$[M+H-H_2O]^+$	**263.1374**	$[M-H]^-$	12.52	751.2	1	CHEBI:2365
	303.1071	$[M+K]^+$						
	265.1495	$[M+H]^+$						
	328.1577	$[M+Na+CH_3CN]^+$						
Limonoids and glycosides								
Limonin glycoside	**471.2049**	$[M+H-Glucose]^+$	**649.2438**	$[M-H]^-$	10.20	612.0	2, 3 [24]	CHEBI:16063
	673.2702	$[M+Na]^+$						
	689.2392	$[M+K]^+$						
	489.2241	$[M+H-Hexose]^+$						
Deacetyl Nomilinic acid glycoside	nd		**669.2733**	$[M-H]^-$	10.55	633.0	2, 3 [24]	
Limonin A-ring lactone*	**471.2007**	$[M+H-H_2O]^+$	**487.1945**	$[M-H]^-$	10.93	655.8	2, 3 [24]	CHEBI:16226
Deacetyl Nomilin glycoside	nd		**651.2624**	$[M-H]^-$	11.08		2, 3 [24]	
Nomilinic acid glucoside	nd		**711.2627**	$[M-H]^-$	11.82	709.2	2, 3 [24]	
Nomilin glycoside	**515.2332**	$[M+H-Glucose]^+$	**693.2737**	$[M-H-H_2O]^-$	11.87	712.2	2, 3 [24]	
	533.2402	$[M+H-Hexose]^+$	711.2837	$[M-H]^-$				
	455.2494	$[M+H-CH4O]^+$						
	695.2495	$[M+H]^+$						
	487.2391	$[M+H-CO]^+$						
	419.2000	$[M+H-2xH_2O]^+$						
Obacunone glycoside	**531.3160**	$[M+H-C_4H_8O_3]^+$	**633.2513**	$[M-H]^-$	12.25	735.0	2, 3 [24]	
	455.2069	$[M+H-Glucose]^+$						

Table 2 Identification of compounds *(Continued)*

Nomilin A-ring lactone*	**533.2700**	$[M+H-H_2O]^+$	**531.2330**	$[M-H]^-$	14.75	885.0	2, 3 [24]	
Limonin	**471.2031**	$[M+H]^+$	515.1922	$[M+HCOOH]^-$	16.60	996.0	1	
	427.2233	$[M-CO_2]^+$	**469.1904**	$[M-H]^-$				
	512.2452	$[M+CH_3CN]^+$	505.1670	$[M+Cl]^-$				
Nomilin	**515.2448**	$[M+H]^+$	**513.2211**	$[M-H]^-$	17.50	1050.0	2, 3 [24]	
	455.2251	$[M+H-C_2H_4O_2]^+$						
Obacunone	nd		**453.2200**	$[M-H]^-$	21.38	1282.8	2, 3 [24]	
Flavonoids								
Eriodictyol 7-O rutinoside	**595.1725**	$[M+H]^+$	**593.1387**	$[M-H]^-$	8.92	535.2	2, 3 [25]	
	451.0975	$[M+H-Deoxyhexose]^+$	449.1096	$[M-H-Deoxyhexose]^-$				
	289.0905	$[M+H-Hexose-Deoxyhexose]^+$						
Rutin	**611.1700**	$[M+H]^+$	**609.1772**	$[M-H]^-$	10.25	615.0	1	CHEBI:28527
	449.1563	$[M+H-Hexose]^+$						
	303.0947	$[M+H-Hexose-Deoxyhexose]^+$						
Narirutin	**581.1946**	$[M+H]^+$	**579.1660**	$[M-H]^-$	10.82	649.2	1	CHEBI:28705
	419.1390	$[M+H-Hexose]^+$	615.1440	$[M+Cl]^-$				
	273.0783	$[M+H-Hexose-Deoxyhexose]^+$	271.0668	$[M-H-Hexose-Deoxyhexose]^-$				
	435.1369	$[M+H-Deoxyhexose]^+$						
	401.1318	$[M+H-Glucose]^+$						
	603.1908	$[M+Na]^+$						
Isorhamnetin-3-O-rutinoside	**625.1985**	$[M+H]^+$	**623.1828**	$[M-H]^-$	10.86	651.6	2, 3 [25]	
	317.0667	$[M+H-Rutinose]^+$						
	479.1347	$[M+H-Deoxyhexose]^+$						
Naringin	**581.1829**	$[M+H]^+$	**579.1614**	$[M-H]^-$	11.14	668.4	1	CHEBI:28819
	435.1303	$[M-Hexose]^+$						
	419.1327	$[M-Hexose-H_2O]^+$						
	273.0775	$[M+H]^+$						
Hesperidin	**611.1993**	$[M+H]^+$	**609.1722**	$[M-H]^-$	11.27	676.2	1	CHEBI:28775
	449.1449	$[M-Hexose]^+$	301.0767	$[M-Hexose-Deoxyhexose]^-$				
	303.0947	$[M-Hexose-Deoxyhexose]^+$	279.1298					
	495.1524	$[M+H-C_5H_8O_3]^+$						
Neohesperidin	**611.2115**	$[M+H]^+$	**609.1772**	$[M-H]^-$	11.53	691.8	2, 3 [25]	CHEBI:59016
	449.1539	$[M-Hexose]^+$						
	303.0948	$[M-Hexose-Deoxyhexose]^+$						
Kaempferol Deoxyhexoside Hexoside #1	**595.1475**	$[M+H]^+$	**593.1871**	$[M-H]^-$	12.93	775.8	2, 3 [25]	
	433.1568	$[M+H-Hexose]^+$	639.1884	$[M+HCOOH]^-$				
	287.1010	$[M+H-Hexose-Deoxyhexose]^+$						
Kaempferol Deoxyhexoside Hexoside #2	**287.0959**	$[M+H-Deoxyhexose-Hexose]^+$	**593.1887**	$[M-H]^-$	13.20	792.0	2, 3 [25]	

Table 2 Identification of compounds *(Continued)*

	595.2134	$[M+H]^+$	639.1846	$[M+HCOOH]^-$				
	433.1555	$[M\text{-}Hexose]^+$						
	449.1511	$[M\text{-}Deoxyhexose]^+$						
Tangeretin #1	**373.1397**	$[M+H]^+$	nd		15.04	902.4	3	CHEBI:9400
	436.1960	$[M+NaCH_2CN]^+$						
Tangeretin #2	**373.1397**	$[M+H]^+$	nd		15.93	955.8	3	CHEBI:9400
	436.1960	$[M+NaCH_2CN]^+$						
	411.1044	$[M+K]^+$						
	395.1297	$[M+Na]^+$						
Tangeritin #3	**373.1397**	$[M+H]^+$	nd		18.14	1088.4	3	CHEBI:9400
	436.1960	$[M+NaCH_2CN]^+$						
	411.1044	$[M+K]^+$						
	395.1297	$[M+Na]^+$						
Miscellaneous compounds								
Phenylalanine	**166.0693**	$[M+H]^+$	nd		2.10	126.0	1	CHEBI:17295
	120.0588	$[M+H\text{-}NH_3]^+$						
Tryptophan	205.1006	$[M+H]^+$	nd		4.56	273.6	1	CHEBI:16828
	188.0719	$[M+H\text{-}NH_3]^+$						
	144.0951	$[M+H\text{-}NH_3\text{-}CO_2]^+$						
Ferulic acid hexoside	**379.1035**	$[M+Na]^+$	**355.0981**	$[M\text{-}H]^-$	8.10	486.0	1, 3	
	177.0488	$[M+H\text{-}Hexose\text{-}H_2O]^+$						
	195.0595	$[M+H\text{-}Hexose]^+$						
	395.0840	$[M+K]^+$						

Annotation level: 1) co-injected with pure standards, 2) annotated matching published data and mass spectral results and 3) annotation made based on mass spectral data, *) tentatively annotated and nd) not determined. *m/z* values in bold are quantifier ions.

throughout sample groups (Additional file 3: Table S1). This was achieved using the quantifier ion (an ion with the highest intensity within the spectrum of a given metabolite, marked in bold in Table 2) to extract metabolite peaks with Masslynx 4.1. software.

ABA and derivatives

The pathway, starting from ABA, has two major branches: the catabolic and the conjugating branch. The first one starts with the conversion of ABA into 8′-hydroxy ABA (catalyzed by ABA 8′-hydroxylase), which spontaneously isomerizes to PA. This metabolite is further catabolized to DPA by a soluble reductase [26]. The conjugating branch involves the temporary storage of ABA into a glycosylated form catalyzed by an UDP-ABA glycosyl transferase (Figure 3). The most widespread form is ABAGE which is the result of esterification at the C1 position of the carboxyl group [26,27]. In turn, active ABA can be released from ABAGE by a glycosidase (BGLU18, [28]).

ABA levels in fruits of the 'Sucrenya' orange were the highest. Whereas, high contents of this hormone were also found in 'Hernandina', 'Midknight', 'Washington', and 'Lane late'; and 'Fortune', ABA levels were much lower in lemons, grapefruits and Pixie and Nadorcott mandarin cultivars. In general, varieties showing low ABA content had also low concentrations of ABA catabolites, including ABAGE (Figure 3). Conversely, 'Sucrenya' that showed the highest ABA levels had also the highest PA and ABAGE levels among all varieties. These results suggested a different ABA metabolic fingerprinting for each variety. ABA levels seem to be regulated by degradation to DPA followed by conjugation in 'Hernandina'. On the other hand, ABA metabolism in 'Nadorcott' and 'Pixie' as well as in 'Lane late' and 'Washington' oranges appeared to be channeled to the production of glycosylated forms of PA and DPA, respectively, showing scarce accumulation of their free forms. Surprisingly, the other blonde-type variety, 'Midknight', did not accumulate any catabolite or ABA derivative, suggesting that the control of ABA levels took place by regulating its biosynthesis (NCED activity). On the contrary, in Fortune ABA levels appeared to be regulated in by diverting metabolic flow to PA and PAGE synthesis. The rest of cultivars accumulating low ABA contents such as lemons a general downregulation of the pathway was found whereas in

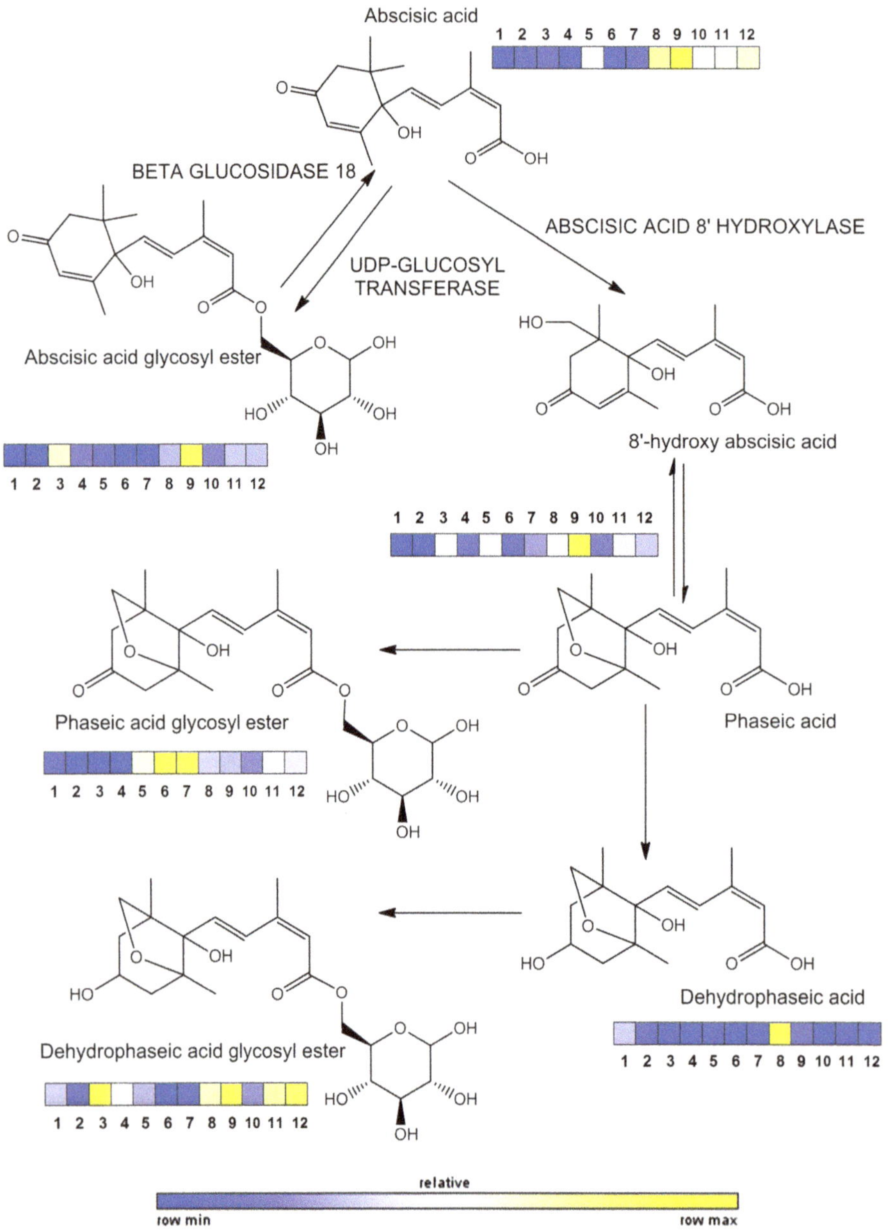

Figure 3 Scheme of ABA metabolism, including chemical structure of free and conjugated forms and products of degradation. On every compound a color scale indicates relative amounts in juices of each variety studied. Sample ID followed the same order as in Table 1.

grapefruits metabolite flow was directed to DPAGE synthesis (with a particular behavior of Marsh genotype that accumulated significant amounts of PA and ABAGE). Noteworthy, only 'Sucrenya' orange and 'Marsh' grapefruit showed significantly higher ABAGE levels than the rest of varieties. Overall, this indicates that citrus fruits and especially juice sacs preferentially induce the degradation pathway to reduce ABA levels (being conjugation of ABA a less relevant mechanism). In previous reports, higher ABA levels were found in juice sacs of satsuma mandarin (*Citrus unshiu*) compared to 'Lisbon' lemon or 'Valencia' orange [16]. This could be explained in part by a higher ability of satsuma mandarin for carotenoid and xanthophyll biosynthesis in juice sacs together with a higher metabolite flow towards xanthoxin and ABA [29]. On the contrary, although carotenoid availability in mandarins is higher than in oranges [15], it is likely that availability of xanthophyll substrates needed for NCED activity is much lower probably channeling these precursors to other metabolic pathways, thus contributing to lower ABA levels in this group (Figure 3). Nevertheless, in 'Nadorcott' and 'Pixie' cultivars, increased degradation to PA along with its conjugation to hexoses rendering PAGE could also contribute to decreased

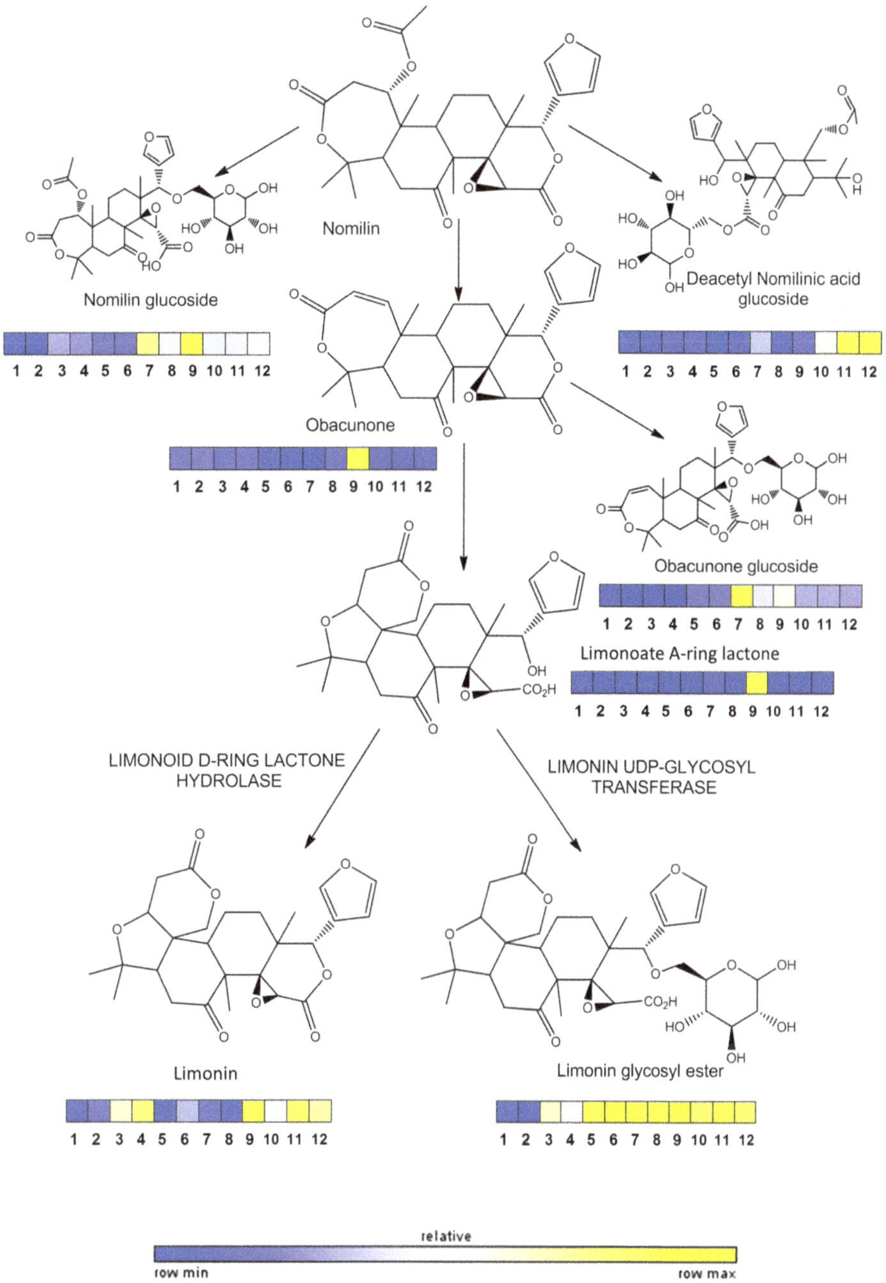

Figure 4 Scheme of limonoid metabolic pathway arising from nomilin. On every compound a color scale indicates relative amounts in juices of each variety studied. Sample ID followed the same order as in Table 1.

ABA levels. In contrast, the lemon and grapefruit varieties showed lower levels of ABA and catabolites and observation coincident with the reported low carotenoid levels in juice sacs [30].

Limonoids

Limonoids are highly oxygenated triterpenes present in Rutaceae and Meliaceae. These compounds are derived from squalene, although the first true limonoid precursor is nomilin that can be directly glucosylated by a limonoid UDP-glucosyl transferase or also deacetylated (Figure 4) rendering obacunone. Cleavage of C-O bond at the D-ring and reorganization of the D-ring renders the tasteless limonoate A-ring lactone that can be alternatively glycosylated (as occurs during normal maturation) or converted into bitter limonin [17]. All identified limonoids were present in all varieties at different levels but especially in lemons showed very low values (Figure 3). Particularly, limonin glycosyl ester was present at similar levels in all

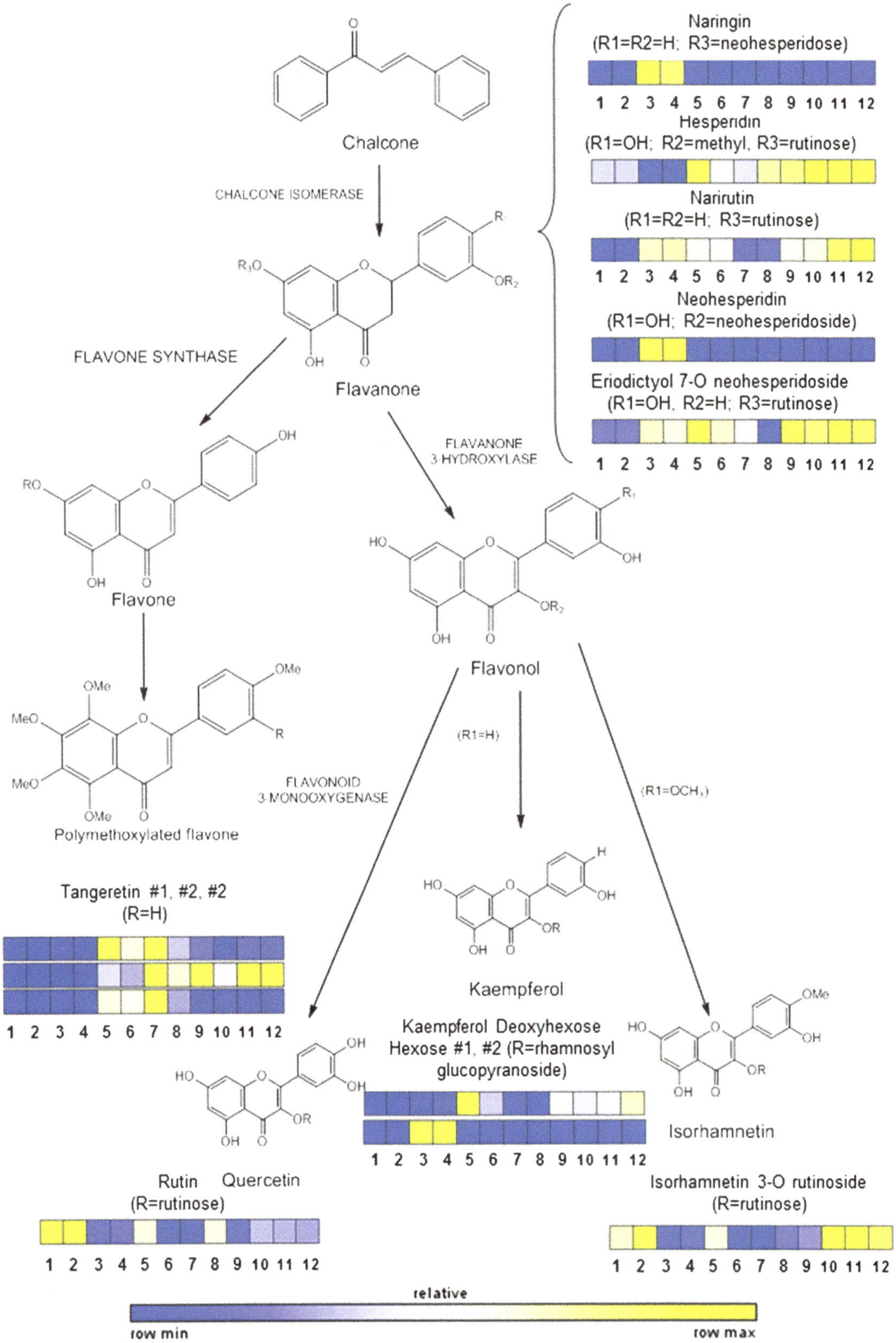

Figure 5 Scheme of flavonoid metabolic pathway arising from chalcone (not analyzed). On every compound a color scale indicates relative amounts in juices of each variety studied. Sample ID followed the same order as in Table 1.

orange and mandarin varieties but showed slightly lower levels in the two grapefruit varieties. These two genotypes, together with orange cultivars, also contained high concentrations of the bitter limonin whereas mandarins showed very low values. In addition, 'Sucrenya' was the only variety that had significant amounts of limonoate A-ring lactone and obacunone, suggesting a highly active biosynthesis in this variety. Nomilin could not be detected in this study but its glycosylated and deacetylated derivatives (Figure 4 and Table 2). Nomilin glucoside levels were much higher in 'Sucrenya' and 'Pixie' cultivars than in the rest of citrus varieties that showed very low levels. Levels of Deacetylated nomilin glucoside in the navel-type oranges ('Washington' and 'Lane late') were the highest whereas they were slightly lower in 'Midknight' and in trace amounts in 'Pixie'. Obacunone glucoside levels were high in this variety and lower levels, by decreasing order, were detected in 'Sucrenya', 'Hernandina' and the rest of oranges. This indicated that all limonoid pool was diverted into production of glycosides, as expected in normal maturation, but some varieties also presented significant amounts of bitter limonin [31], including lemons, 'Sucrenya', 'Midknight' and navel-type oranges. Nomilin glucoside, obacunone, obacunone glucoside, limonoate A-ring lactone and its glucoside and limonin are over-accumulated in 'Sucrenya' compared to the other blonde-type variety 'Midknight'. This suggests a particularly active limonoid biosythetic pathway in 'Sucrenya' whereas in 'Midknight' all intermediates are readily channeled to the production of deacetyl nomilic acid glucoside, limonin and limonin glucoside. Indeed, limonin glucoside was highly abundant in almost all studied citrus juices (0.035% of juice weight in mexican lime, as described in [19]) that could also be cleaved to render free limonin upon induction of a glucosidase [32]. The concentration of limonoid metabolites highly increased in citrus affected by bacterial Greening Disease or Huanglongbing (HLB) [33] suggesting a role in defense against bacterial infection. Moreover, limonoids have exhibited significant antioxidant and antitumorigenic activity [19,20]. However, their specific physiological role in citrus is still unknown. The results presented here also point out differences in palatable fruit quality among varieties at optimum commercial maturation stage, likely associated to genetic differences.

Flavonoids

This class of compounds has been involved in the antioxidant and beneficial health properties of citrus. Indeed, the high radical scavenging activity of citrus juices has been almost exclusively associated to flavonoids and other phenolic constituents [19]. In citrus, the most abundant flavonoids are flavanones, flavones and flavonols being the methoxylation and glycosylation the main reactions rendering derivatives [34]. In this study, from the same flavanone core several derivatives were identified by substitution with methyl groups or hexose moieties: naringin, hesperidin, narirutin, neohesperidin, and eriodictyol (Figure 5 and Table 2). From this group, the most widespread compounds were hesperidin, narirutin and eriodictyol 7-O-neohesperidoside, whereas naringin and neohesperidin were exclusively present in grapefruits. Flavonoid synthesis starts from the flavanone naringenin by successive transfer of glycosyl groups (a first step by which glucose is transferred to oxygen in position 7 generating a 7-O-glucoside). In turn, a 1,6 rhamnosyl transferase renders the hesperidosides (or rutinosides) hesperidin and narirutin. Conversely, action of 1,2 rhamnosyl transferase on flavanone 7-O-hexosides generates neohesperidosides: neohesperidin and naringin. A very low expression of 1,6 rhamnosyl transferases in citron, pummelo and grapefruit and absence of 1,2 rhamnosyl transferases in mandarins and oranges have been recently reported [35]. These results would explain the exclusive occurrence of neohesperidosides in grapefruit cultivars in the present work (Figure 4). Apparently, this low expression is enough to grant occurrence of rutinosides such as narirutin and eriodictyol 7-O rutinoside in grapefruits. Another group of flavonoids, flavonols, synthesized from the same flavanones by hydroxylation include isorhamnetin, kaempferol and quercetin. Rutin showed the highest accumulation in lemons, although it was present in most citrus cultivars included in this study, showing significantly lower levels in 'Fortune' and 'Hernandina'. This could point out a higher flavonoid 3-monooxygenase activity in these genotypes. Isorhamnetin 3-O rutinoside derived from addition of an hexose moiety on oxygen in position 3 catalyzed by 3-glycosyl transferase followed by 1,6 rhamnosyl transferase [35]. It is now clear that 1,6 rhamnosyl transferase is present and active to different levels in most cultivated citrus species. Therefore, the selectivity relies on the previous action of 7- or 3-glycosyl transferases. To this respect, the results obtained suggest that 3-glycosyl transferases are likely to be rather active in grapefruit cultivars, therefore rendering flavonol 7-O rutinosides. Isorhamnetin 3-O-rutinoside, product of methoxylation and subsequent glycosylation of a flavonol moiety was found to be highly abundant in both navel oranges and 'Midknight' and in the two lemon cultivars which could point out at flavonol A-ring methoxylation being highly active in these genotypes. On the other hand, polymethoxylated flavones, derived from flavanone in a reaction catalyzed in turn by flavone synthase and flavone A-ring methyl transferases were completely absent in lemon and grapefruit cultivars, suggesting a lower enzyme activity or expression.

Miscellaneous compounds

Precursor compounds such as phenylalanine, ferulic acid hexoside and tryptophan were grouped under this epigraph (Figure 6). Phenylalanine, along with Tyr, is the precursor of all aromatic compounds (among which flavonoids and phenolic acids are found) through reaction catalyzed by PAL and CHS. Results indicated that this precursor compounds are not limiting for all derived compounds and therefore differences in flavonoid composition are due to variations in the expression of genes encoding for metabolic enzymes acting downstream CHS. To this respect, ferulic acid hexoside was scarce in juices of 'Sucrenya' but, conversely, this variety did not show any limitation in flavonoid biosynthesis (Figure 5), suggesting that biosynthetic restrains did not affect steps upstream ferulic acid. In the rest of genotypes, this metabolite was moderately abundant with the exception of 'Fortune' and 'Midknight' which juices had significantly higher levels of this compound. Content of tryptophan, an aminoacid precursor of indolic compounds and the auxin indole-3-acetic acid, was found to be extremely scarce in the vast majority of citrus varieties assayed with the exception of the two grapefruit cultivars studied, with values four-fold higher than the average levels.

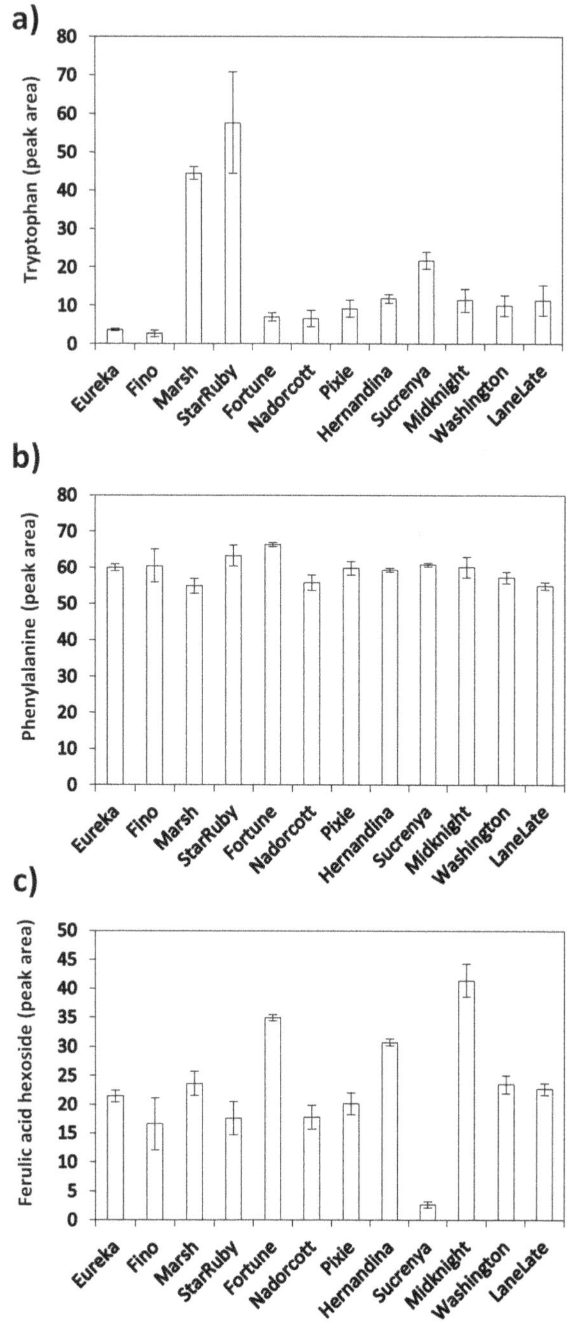

Figure 6 Relative levels of miscellaneous metabolites identified in citrus juices: tryptophan (a), phenylalanine (b) and a ferulic acid hexoside (c). Sample ID is given in x-axis.

Conclusions

Organoleptic quality is associated not only to primary attributes such as soluble solids (sugars) and acids (mainly citric acid) but other minor compounds such as triterpenoids, flavonoids, coumarins and anthocyanidins. Recently, these compounds have gained scientific and commercial attention due to their beneficial effects on human health and also as important phylogenetic markers. It is well known that metabolites are the downstream products of gene expression and, as such, subjected to a thorough selection process. Therefore, secondary metabolites can be used either as quality traits or as markers for the selection and/or certification of different fruit sources [12] or for the physiological evaluation of plant genotypes [7]. To this regard, in this work we have focused only in commercial subspecies arising from reciprocal crosses between different citrus ancestor lines: *Citrus maxima* (pummelo), *Citrus reticulata* (mandarin), *Citrus medica* (citron) and *Citrus aurantifolia* (mexican lime). To further investigate the inheritance of specific metabolite traits, an exhaustive analysis including these ancestor species should be performed. Nevertheless, the results presented in this manuscript indicate that LC/ESI-QTOF-MS non-targeted metabolite profiling is an efficient technique to profile secondary metabolites in citrus juices with little sample processing (squeezing, centrifuging and filtering). In addition, this technique could be coupled to multivariate analysis as data mining technique to allow separation of different fruit sources: lemons, grapefruits, mandarins, navel and blonde oranges and, more importantly, the differentiation of varieties within a particular group.

Additional files

Additional file 1: Figure S1. Phylogenetic tree depicting relationships between all known parental ancestor lines and commercial genotypes.

Additional file 2: Figure S2. 2D scores (left) and loadings (right) plots depicting different projections: a) component 1 vs. component 2, b) component 2 vs. component 3, c) component 5 vs. component 6, d) component 7 vs. component 8, e) component 9 vs. component 10 and f) component 1 vs. component 11. In scores plots, clearly demarcated variety sample groups are indicated; in loadings plots variables potentially contributing to variety demarcation are indicated in red.

Additional file 3: Table S1. Confirmation of Metabolite Candidates for Classification of Citrus Varieties by ANOVA followed by Fisher Least Significant Difference test.

Abbreviations

LC/MS: Liquid chromatography coupled to mass spectrometry; PTFE: Polytetrafluoroethylene, teflon; QTOF-MS: Quadrupole time-of-flight mass spectrometry; HPLC: High performance liquid chromatography; ESI: Electrospray ionization; HCA: Hierarchical cluster analysis; PLS-DA: Partial least squares discriminant analysis; VIP: Variable importance for the projection (related to PLS-DA); ABA: Abscisic acid; NCED: Nine cis-epoxycarotenoid dioxygenase; ANOVA: Analysis of variance; PA: Phaseic acid; DPA: Dehydrophaseic acid; ABAGE: Abscisic acid glycosyl ester; PAL: Phenylalanine ammonia lyase; CHS: Chalcone synthase.

Competing interests

The authors declare that they have no competing interests.

Authors' contributions

VA, DJI and AGC designed and planned experiments. DJI harvested and authenticated citrus samples. VA performed sample processing, instrument measurements and analysis of data. VA and AGC wrote the manuscript and elaborated results. All authors have read and approved the final version of the manuscript.

Acknowledgements

This project was funded by the Spanish Ministerio de Economia y Competitividad (MINECO) and Universitat Jaume I through grants AGL2013-42038-R and P1 1B2013-23, respectively, to Aurelio Gómez-Cadenas. LC/ESI-QTOF-MS measurements were carried out at Central Instrument Facilities (SCIC) of Universitat Jaume I, assistance of Dr. Cristian Vicent in mass spectrometric measurements is greatly acknowledged.

Author details

[1]Laboratori d'Ecofisiologia i Biotecnologia, Departament de Ciències Agràries i del Medi Natural, Universitat Jaume I, E-12071 Castelló de la Plana, Spain. [2]Institut Valencià d'Investigacions Agràries (IVIA), Moncada, Spain.

References

1. Distefano G, Las Casas G, Caruso M, Todaro A, Rapisarda P, La Malfa S, et al. Physiological and molecular analysis of the maturation process in fruits of Clementine Mandarin and one of its late-ripening mutants. J Agric Food Chem. 2009;57:7974–82.
2. Koca U, Berhow M a, Febres VJ, Champ KI, Carrillo-Mendoza O, Moore G a. Decreasing unpalatable flavonoid components in Citrus: the effect of transformation construct. Physiol Plant. 2009;137:101–14.
3. González-Molina E, Moreno D a, García-Viguera C. Genotype and harvest time influence the phytochemical quality of Fino lemon juice (*Citrus limon* (L.) Burm. F.) for industrial use. J Agric Food Chem. 2008;56:1669–75.
4. Xu G, Liu D, Chen J, Ye X, Shi J. Composition of major flavanone glycosides and antioxidant capacity of three citrus varieties. J Food Biochem. 2009;33:453–69.
5. Giménez-Bastida J a, Martínez-Florensa M, Espín J-C, Tomás-Barberán F a, García-Conesa M-T. A citrus extract containing flavanones represses plasminogen activator inhibitor-1 (PAI-1) expression and regulates multiple inflammatory, tissue repair, and fibrosis genes in human colon fibroblasts. J Agric Food Chem. 2009;57:9305–15.
6. Dhuique-Mayer C, Fanciullino A-L, Dubois C, Ollitrault P. Effect of genotype and environment on citrus juice carotenoid content. J Agric Food Chem. 2009;57:9160–8.
7. Pan Z, Li Y, Deng X, Xiao S. Non-targeted metabolomic analysis of orange (Citrus sinensis [L.] Osbeck) wild type and bud mutant fruits by direct analysis in real-time and HPLC-electrospray mass spectrometry. Metabolomics. 2013;10:508–23.
8. Terol J, Soler G, Talon M, Cercos M. The aconitate hydratase family from Citrus. BMC Plant Biol. 2010;10:222.
9. Arbona V, Manzi M, de Ollas C, Gómez-Cadenas A. Metabolomics as a tool to investigate abiotic stress tolerance in plants. Int J Mol Sci. 2013;14:4885–11.
10. Arbona V, López-climent MF, Pérez-Clemente RM, Gómez-cadenas A. Maintenance of a high photosynthetic performance is linked to flooding tolerance in citrus. Env Exp Bot. 2009;66:135–42.
11. Böttcher C, Westphal L, Schmotz C, Prade E, Scheel D, Glawischnig E. The multifunctional enzyme CYP71B15 (PHYTOALEXIN DEFICIENT3) converts cysteine-indole-3-acetonitrile to camalexin in the indole-3-acetonitrile metabolic network of *Arabidopsis thaliana*. Plant Cell. 2009;21:1830–45.
12. Vaclavik L, Schreiber A, Lacina O, Cajka T, Hajslova J. Liquid chromatography–mass spectrometry-based metabolomics for authenticity assessment of fruit juices. Metabolomics. 2011;8:793–803.
13. IPGRI. Descriptors for Citrus. Rome (Italy): International Plant Genetic Resources Institute (IPGRI); 1999. p. 75.
14. Arbona V, Iglesias DJ, Talón M, Gómez-Cadenas A. Plant phenotype demarcation using nontargeted LC-MS and GC-MS metabolite profiling. J Agric Food Chem. 2009;57:7338–47.
15. Goodner KL, Rouseff RL, Hofsommer HJ. Orange, mandarin, and hybrid classification using multivariate statistics based on carotenoid profiles. J Agric Food Chem. 2001;49:1146–50.
16. Kato M, Matsumoto H, Ikoma Y, Okuda H, Yano M. The role of carotenoid cleavage dioxygenases in the regulation of carotenoid profiles during maturation in citrus fruit. J Exp Bot. 2006;57:2153–64.
17. Manners GD. Citrus limonoids: analysis, bioactivity, and biomedical prospects. J Agric Food Chem. 2007;55:8285–94.
18. Schoch TK, Manners GD, Hasegawa S. Analysis of limonoid glucosides from citrus by electrospray ionization liquid chromatography-mass spectrometry. J Agric Food Chem. 2001;49:1102–8.
19. Patil JR, Chidambara Murthy KN, Jayaprakasha GK, Chetti MB, Patil BS. Bioactive compounds from Mexican lime (*Citrus aurantifolia*) juice induce apoptosis in human pancreatic cells. J Agric Food Chem. 2009;57:10933–42.
20. Perez JL, Jayaprakasha GK, Valdivia V, Munoz D, Dandekar DV, Ahmad H, et al. Limonin methoxylation influences the induction of glutathione S-transferase and quinone reductase. J Agric Food Chem. 2009;57:5279–86.
21. Tovar MJ, Romero MP, Girona J, Motilva MJ. L-Phenylalanine ammonia-lyase activity and concentration of phenolics in developing olive (*Olea europaea* L cv Arbequina) fruit grown under different irrigation regimes. J Sci Food Agric. 2002;82:892–8.
22. Sanchez-Vallet A, Ramos B, Bednarek P, López G, Piślewska-Bednarek M, Schulze-Lefert P, et al. Tryptophan-derived secondary metabolites in *Arabidopsis thaliana* confer non-host resistance to necrotrophic *Plectosphaerella cucumerina* fungi. Plant J. 2010;63:115–27.
23. Turecková V, Novák O, Strnad M. Profiling ABA metabolites in Nicotiana tabacum L. leaves by ultra-performance liquid chromatography-electrospray tandem mass spectrometry. Talanta. 2009;80:390–9.
24. Tian Q, Schwartz SJ. Mass spectrometry and tandem mass spectrometry of *Citrus* limonoids. Anal Chem. 2003;75:5451–60.
25. Djoukeng JD, Arbona V, Argamasilla R, Gomez-cadenas A. Flavonoid profiling in leaves of citrus genotypes under different environmental situations. J Agric Food Chem. 2008;56:11087–97.
26. Nambara E, Marion-Poll A. Abscisic acid biosynthesis and catabolism. Ann Rev Plant Biol. 2005;56:165–85.
27. Priest DM, Ambrose SJ, Vaistij FE, Elias L, Higgins GS, Ross ARS, et al. Use of the glucosyltransferase UGT71B6 to disturb abscisic acid homeostasis in *Arabidopsis thaliana*. Plant J. 2006;46:492–502.
28. Schroeder JI, Nambara E. A quick release mechanism for abscisic acid. Cell. 2006;126:1023–5.
29. Zhang L, Ma G, Kato M, Yamawaki K, Takagi T, Kiriiwa Y, et al. Regulation of carotenoid accumulation and the expression of carotenoid metabolic genes in citrus juice sacs in vitro. J Exp Bot. 2012;63:871–86.

30. Matsumoto H, Ikoma Y, Kato M, Kuniga T, Nakajima N, Yoshida T. Quantification of carotenoids in citrus fruit by LC-MS and comparison of patterns of seasonal changes for carotenoids among citrus varieties. J Agric Food Chem. 2007;55:2356–68.
31. Breksa AP, Ibarra P. Colorimetric method for the estimation of total limonoid aglycones and glucoside contents in citrus juices. J Agric Food Chem. 2007;55:5013–7.
32. Breksa AP, Hidalgo MB, Wong RY. Stability of limonin glucoside in beverage matrices. J Sci Food Agric. 2008;88:2194–200.
33. Manthey JA. Differences in secondary metabolites in leaves from orange (*Citrus sinensis* L.) trees affected with Greening Disease (Huanglongbing) (HLB). Proc Florida State Hortic Soc. 2008;121:285–8.
34. Gattuso G, Barreca D, Gargiulli C, Leuzzi U, Caristi C. Flavonoid composition of *Citrus* juices. Molecules. 2007;12:1641–73.
35. Frydman A, Liberman R, Huhman DV, Carmeli-Weissberg M, Sapir-Mir M, Ophir R, et al. The molecular and enzymatic basis of bitter/non-bitter flavor of citrus fruit: evolution of branch-forming rhamnosyltransferases under domestication. Plant J. 2012;73:166–78.

Transcriptome profiling of root microRNAs reveals novel insights into taproot thickening in radish (*Raphanus sativus* L.)

Rugang Yu[1,2], Yan Wang[1], Liang Xu[1], Xianwen Zhu[3], Wei Zhang[1], Ronghua Wang[1], Yiqin Gong[1], Cecilia Limera[1] and Liwang Liu[1*]

Abstract

Background: Radish (*Raphanus sativus* L.) is an economically important root vegetable crop, and the taproot-thickening process is the most critical period for the final productivity and quality formation. MicroRNAs (miRNAs) are a family of non-coding small RNAs that play an important regulatory function in plant growth and development. However, the characterization of miRNAs and their roles in regulating radish taproot growth and thickening remain largely unexplored. A Solexa high-throughput sequencing technology was used to identify key miRNAs involved in taproot thickening in radish.

Results: Three small RNA libraries from 'NAU-YH' taproot collected at pre-cortex splitting stage, cortex splitting stage and expanding stage were constructed. In all, 175 known and 107 potential novel miRNAs were discovered, from which 85 known and 13 novel miRNAs were found to be significantly differentially expressed during taproot thickening. Furthermore, totally 191 target genes were identified for the differentially expressed miRNAs. These target genes were annotated as transcription factors and other functional proteins, which were involved in various biological functions including plant growth and development, metabolism, cell organization and biogenesis, signal sensing and transduction, and plant defense response. RT-qPCR analysis validated miRNA expression patterns for five miRNAs and their corresponding target genes.

Conclusions: The small RNA populations of radish taproot at different thickening stages were firstly identified by Solexa sequencing. Totally 98 differentially expressed miRNAs identified from three taproot libraries might play important regulatory roles in taproot thickening. Their targets encoding transcription factors and other functional proteins including *NF-YA2*, *ILR1*, *bHLH74*, *XTH16*, *CEL41* and *EXPA9* were involved in radish taproot thickening. These results could provide new insights into the regulatory roles of miRNAs during the taproot thickening and facilitate genetic improvement of taproot in radish.

Keywords: *Raphanus sativus*, Taproot, Thickening, microRNA, Solexa sequencing

Background

Radish (*Raphanus sativus* L., 2n = 2x = 18) is an economically important root vegetable crop belonging to the Brassicaceae family [1]. The fleshy taproot comprises the main edible portion of the plant. Therefore, the taproot thickening phase is a critical period of root development that mainly determines yield and quality in radish. During taproot thickening process, an abundance of storage compounds and secondary metabolites are synthesized, which mainly determine the economic value of radish taproot and provide nutrients and medicinal function for human beings [2]. It is therefore of significance to clarify the molecular genetic mechanism underlying taproot thickening in radish.

The fleshy taproot thickening of radish is a complex biological process involving morphogenesis and dry matter accumulation [1]. Previous studies of the taproots

* Correspondence: nauliulw@njau.edu.cn
[1]National Key Laboratory of Crop Genetics and Germplasm Enhancement; Engineering Research Center of Horticultural Crop Germplasm Enhancement and Utilization, Ministry of Education of P.R.China; College of Horticulture, Nanjing Agricultural University, Nanjing 210095, P.R. China
Full list of author information is available at the end of the article

have been focused mainly on the morphological and physio-biochemical levels. For example, the taproot axis of radish is composed of the hypocotyl and true root tissue [3], and the thickening of taproot was mainly due to the activity of a vascular cambium and the differentiation of secondary xylem and phloem [3,4]. Additionally, some studies have demonstrated that taproot development in radish was controlled by complex interactions among genetic, environmental and physiological factors [1,5]. However, root development and response to the environment are thought to be controlled by gene regulatory networks [6]. To date, great advances about gene regulation in root development have been made in several plant species [7], such as *Arabidopsis thaliana* [6,8], *Zea mays* [9], and *Oryza sativa* [10]. Unlike other roots, the taproot of radish is a storage root, the knowledge about gene regulation and the molecular mechanism is little known in storage root development, including radish. Recently, radish genome sequencing and the radish root transcriptomics studies have facilitated the investigation of the molecular mechanisms in radish taproot development [11,12]. Nevertheless, the key gene isolation and molecular mechanism underlying radish taproot thickening remain elusive.

MicroRNAs (miRNAs) are class of important non-protein-coding regulatory small RNAs (20 to 24 nt) that mediate gene expression at transcriptional and post-transcriptional level by repressing gene translation or degrading target mRNAs [13-16]. During the last decades, miRNAs have been discovered as regulators of numerous physiological and developmental processes during the life cycle of plants, including root development. For example, in *Arabidopsis*, miR164 targets NAC domain containing protein 1 (*NAC1*) to regulate lateral root development [14]; miR169 isoform targets nuclear transcription factor Y subunit A (*NF-YA*) to regulate primary root growth [15]; miR160 is involved in adventitious rooting and root cap development through the regulation of auxin response factors (*ARFs*) [16].

Recently, high-throughput sequencing technology has become a valuable tool to discover a large set of diverse plant miRNAs. Up to now, a large number of miRNAs in different plant species have been registered in miRBase 21.0 database (http://www.mirbase.org/cgi-bin/browse.pl). Additionally, several studies using this approach have identified some miRNAs and explored the roles of miRNAs in root development in *Medicago truncatula* [17], maize [18,19], rice [20] and potato [21]. In maize, 246 conserved, 32 novel and some dramatically differentially expressed miRNAs were identified in different maize roots [18]. Additionally, 137 known and 159 novel miRNAs, and 30 differentially expressed miRNAs, as well as 15 target genes, were identified during the early development of the maize brace root [19]. As one of the most important root vegetable crop, the regulatory roles of microRNAs in radish have been extensively studied in recent years. Some conserved miRNAs and novel miRNAs were identified from radish roots based on the *R. sativus* EST and GSS sequences [22,23]. Although a significant fraction of miRNAs associated with some important agronomic traits including cadmium (Cd) accumulation and embryogenesis have been successfully identified in radish [24,25], there is as yet no report on the characterization of miRNAs and their roles in regulating taproot growth and thickening in radish. To investigate the miRNA-mediated regulatory mechanism during this process, Solexa sequencing of three small RNA libraries from 'NAU-YH' taproots collected at pre-cortex splitting stage (Stage1, 10 DAS), cortex splitting stage (Stage2, 20 DAS) and expanding stage (Stage3, 40 DAS) were performed, respectively. As a result, some known and new miRNA families were isolated from these three taproot libraries, from which the differentially-expressed miRNAs involved in taproot thickening were identified. Subsequently, the targets of differentially expressed miRNAs were predicted and their potential functions were discussed. In addition, expression profiling of several miRNAs and their targets were further validated by RT-qPCR technology. These results would firstly reveal the miRNA-mediated regulatory network during radish taproot thickening, and provide novel insights into the molecular genetic mechanisms underlying storage root development in radish.

Methods

Plant growth and sample collection

The radish (*Raphanus sativus* L.) advanced inbred line 'NAU-YH' was used in this study. Seeds were germinated on moist filter paper in darkness for 3 d, and then transplanted into plastic pots with mixture of soil and peat substrate (1:1, V/V), and cultured in the greenhouse. Samples of taproots were collected at three different development stages: pre-cortex splitting stage (Stage1, 10 DAS), cortex splitting stage (Stage2, 20 DAS) and expanding stage (Stage3, 40 DAS). Taproot developmental stages of 'NAU-YH' were identified using the established morphological traits (Figure 1). The subsamples of taproots were collected from five developmental stages: 10, 15, 20, 40, and 50 DAS, respectively, for RT-qPCR verification. All samples were snap-frozen in liquid nitrogen and stored at −80°C for further use.

Transcriptome and small RNA sequencing

Total RNA was extracted from the taproot of 'NAU-YH' at pre-cortex splitting stage (stage1), cortex splitting stage (stage2), and expanding stage (stage3) using Trizol regent (Invitrogen, USA) following the manufacturer's protocol. Equal amounts of total RNA from the three samples were mixed to construct a transcriptome library

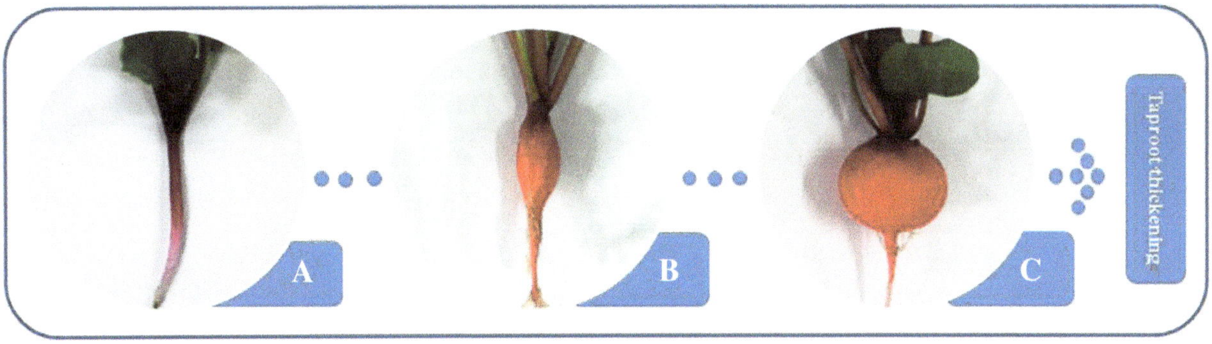

Figure 1 The morphology of 'NAU-YH' taproot in three different thickening stages. (A) Morphology of the pre-cortex splitting stage, 10 DAS. **(B)** Morphology of the cortex splitting stage, 20 DAS. **(C)** Morphology of the expanding stage, 40 DAS.

using an Illumina TruSeq RNA Sample PrepKit following the manufacturer's instructions. After removing sequence reads containing low-quality sequences (reads with ambiguous bases 'N'), adapter sequences, and reads with more than 10% Q<20 bases, mRNA transcriptome *de novo* assembly was performed using the Trinity program [26].

The extracted RNA from the taproot samples of three thickening stages were respectively used for three small RNA libraries construction including stage1, stage2 and stage3. Small RNAs of 18–30 nt in length were separated and purified by denaturing polyacrylamide gel electrophoresis. After dephosphorylation and ligation of a pair of Solexa adaptors to their 5′ and 3′ends, the products were reverse-transcribed and amplified by RT-PCR. Both the paired-end transcriptome and sRNA sequencing were performed at the Beijing Genomics Institute (BGI)-Shenzhen, China.

Data analysis

After Solexa sequencing, the clean reads were obtained from raw reads by getting rid of the contaminated reads including sequences with 5′-primer contaminants, and poly(A) tails, without 3′-primer and the inserted tag, either shorter than 18 nt or longer than 30 nt. Then the unique RNAs were aligned with the radish reference sequences including the mRNA transcriptome sequences, EST sequences (http://www.ncbi.nlm.nih.gov/nucest/?term=radish) and genomic survey sequences (GSS, ftp://ftp.ncbi.nlm.nih.gov/repository/UniGene/Raphanus_sativus/Rsa.seq.uniq.gz) using SOAP2 program [27]. Sequences ranging from 18 to 30 nt (reads with no "N", no more than 4 bases with quality score <10 and no more than 6 bases with quality score <13) were collected for further analysis. Firstly, the sequences matching non-coding RNAs [tRNAs, rRNAs, small nucleolar RNAs (snoRNAs) and small nuclear RNAs (snRNAs)] deposited in the Rfam 10.1 (http://www.sanger.ac.uk/Software/ac.uk/Software/Rfam) and NCBI GeneBank databases (http://www.ncbi.nlm.nih.gov/GenBank/) were eliminated. Then, using a BLASTn search, the remaining sequences with a maximum of two mismatches mapped onto known plant mature miRNAs in miRBase 21.0 (http://www.mirbase.org/index.shtml) were considered as known miRNAs.

The remaining unannotated sRNAs were used to predict novel miRNA using Mireap software (https://sourceforge.net/projects/mireap/), and the stem-loop structure of miRNA precursor was constructed by M-fold program [28]. Basic criteria by Meyers et al. (2008) and Kong et al. (2014) were used for identifying the potential novel miRNA candidates [18,29].

Differential expression analysis of miRNAs in three libraries

To identify the differentially expressed miRNAs among three different taproot thickening stages, the miRNA expression profiles among three sRNA libraries (stage1 versus stage2; stage1 versus stage3; stage2 versus stage3) were comprehensively compared. The clean read of the tag for each miRNA was normalized to one million [25]. After normalization, if the expression level was less than one between two libraries, differential expression analysis was not performed owing to their too low expression level; if the normalized read count of a given miRNA is zero, the expression value is set to 0.01 for further analysis.

The differentially expressed miRNAs were screened with a threshold of fold change ≥ 1.0 or ≤ -1.0 (the $\log_2$ treatment/control) and with P-value < 0.05 at stage2 and stage3 versus stage1, and stage3 versus stage2, where stage1 and stage2 served as the control, respectively. The P-value was calculated according to previously described by Li et al. [27]. Candidate targets of differentially expressed miRNAs were predicted by aligning the miRNA sequences with the available radish reference sequences (GSS, EST and our mRNA transcriptome sequences) using the plant small RNA target analysis server (psRNATarget) with default parameters [30]. The KOBAS 2.0 program (http://kobas.cbi.pku.edu.cn/home.do) and Blast2GO program (http://www.blast2go.com/) were used to annotate the functions of the potential target sequences [25].

RT-qPCR validation of miRNAs and their potential targets

Total RNA were isolated from the five taproot samples (10, 15, 20, 40 and 50 DAS, respectively) using Trizol reagent (Invitrogen, USA) and then treated with PrimeScript® RT reagent Kit (Takara, Dalian, China) to reverse transcribe into cDNA. MicroRNA was extracted from five radish taproot samples using RNAiso for small RNA kit (Takara, Dalian, China) and reverse transcribed into cDNA using a One Step PrimeScript® miRNA cDNA Synthesis Kit (Takara, Dalian, China). The cDNA was quantified by an iCycler IQ real-time PCR detection system (BIO-RAD) using a 20 μl reaction mixture, which consisted of 2 μl of diluted cDNA, 0.2 μM forward and reverse primer, and 10 μl of 2× SYBR Green PCR Master Mix (Takara, Dalian, China). The amplification reaction for miRNAs and their targets was performed, respectively, according to the previous reports [24,25,31]. The equation ratio $2^{-\Delta\Delta C_T}$ was applied to calculate the relative expression level of miRNAs and targets using 5.8S rRNA and *Actin* gene as the reference gene, respectively. The primers for real-time RT-qPCR were designed using Beacon Designer 7.0 software (Additional file 1A and B). In addition, the statistical analysis with SAS Version 9.0 software (SAS Institute, Cary, North Carolina, USA) was performed using Duncan's multiple range test at the $P < 0.05$ level of significance.

Results

Root transcriptome and small RNA sequencing

A total of 51.2 million clean reads were generated in the transcriptome sequencing. By trinity assembly, totally 130,953 contigs with a mean length of 352 nt and 70,168 unigenes with an average length of 717 nt were obtained, which were then combined with the available GSS and EST sequence records in NCBI database to perfect the radish reference sequences for isolating miRNAs associated with radish taproot thickening and development.

To identify miRNAs involved in radish taproot thickening and development, three small RNA libraries, stage1 (10 DAS), stage2 (20 DAS) and stage3 (40 DAS), were constructed, and then sequenced by the Illumina Solexa system. As a result, 17,160,426 (stage1), 19,055,129 (stage2) and 17,263,334 (stage3) raw reads were generated, respectively (Table 1). After removing low-quality reads and trimming adaptor sequences, 16,819,905 clean reads (4,318,929 unique) for stage1, 18,853,348 clean reads (6,575,007 unique) for stage2 and 17,082,616 clean reads (4,542,390 unique) for stage3 were obtained for further analysis (Additional file 2). Among these clean reads, comparative analysis of the common and specific reads of sRNAs between random two libraries, more than 60% of the total sRNAs were common to two different libraries, while the unique sequence reads were common only accounted for small fraction (10%–13%), indicating that there was a less abundant but variety pool of stage-specific small RNAs (Additional file 3A-F). The length of most of sRNA reads (18 to 30 nt) were 21 to 24 nt in these three stages (Figure 2). In stage1 and stage2 library, the highest proportion (>31.53%) of sRNAs was 24 nt in length, followed by 21 nt (>19.03%), which was consistent with previous studies in other species such as *M. truncatula* [17], maize [18] and potato [21]. However, the highest proportion (36.53%) of sRNAs was 21 nt in stage 3 library, followed by 24 nt (24.16%). This result was also observed in grape, in which the number of 21 nt sequence reads were more than five times of 24 nt reads [32]. Overall, these results suggest the existence of complex and diverse sRNA populations in radish.

A total of 725,181 (stage1), 858,371 (stage2) and 906,459 (stage3) unique sequences were successfully mapped to the radish reference sequences, respectively (Additional file 2). Subsequently, for annotation, the acquired sRNA sequences were matched with NCBI GenBank, Rfam, and

Table 1 The result of sRNA sequences from three libraries

Category	Stage1		Stage2		Stage3	
	Count	Percentage	Count	Percentage	Count	Percentage
Raw reads	17160426		19055129		17263334	
High-quality	17087884	100%	18975347	100%	17190771	100%
3′ adapter null	2578	0.02%	2945	0.02%	2380	0.01%
Insert null	2292	0.01%	1415	0.01%	1190	0.01%
5′ adapter contaminants	152282	0.89%	83280	0.44%	56475	0.33%
Smaller than 18 nt	104474	0.61%	31266	0.16%	46772	0.27%
Poly (A)	6353	0.04%	3093	0.02%	1338	0.01%
Clean reads	16819905	98.43%	18853348	99.36%	17082616	99.37%

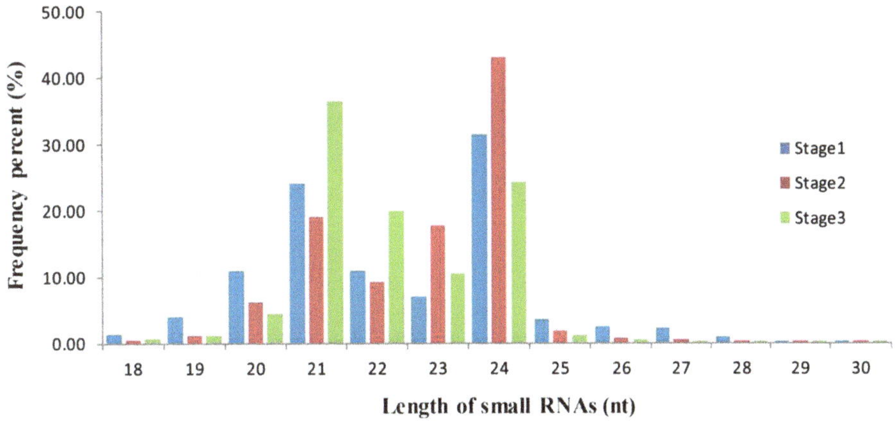

Figure 2 The length distribution of small RNAs in three libraries.

miRbase 21.0 database. The non-coding sRNAs were classified into six categories including miRNA, rRNAs, snRNAs, snoRNAs, tRNAs and those detected but without annotation (Additional file 2). Of all the sRNA categories, un-annotated sRNAs accounted for an average of 69.96% in total acquired sRNAs (Additional file 2). There were large variations about the number of matching unique miRNAs in these three different stages of taproot thickening, 18,078 (stage1), 36,239 (stage2) and 23,604 (stage3) unique miRNAs reads were matched to known miRNAs, respectively, implying that miRNA-mediated gene silencing is involved in the regulation of radish taproot thickening.

Identification of known miRNAs during radish taproot thickening

To identify the known miRNAs, the small RNA sequences were mapped with known mature miRNAs from plants in miRBase 21.0 with a maximum of two mismatches. A total of 175 known miRNAs (148, 150 and 141 in the stage1, stage2 and stage3 libraries, respectively) from 57 families were detected during the radish taproot thickening process (Additional file 4A, B, C and D). Among these miRNAs, 120 (68.57%) known miRNAs were detected in all three libraries, while 145 miRNAs were shared in at least two of three small RNA libraries, and only 30 miRNAs (16, 7 and 7 in the stage1, stage2 and stage3 libraries, respectively) were stage-specifically expressed, implying that the component of miRNAs during taproot thickening was relatively stable (Additional file 3G). In this study, 144 know miRNA sequences belonging to 31 conserved miRNA families were confirmed (Additional file 4D). For example, miR156, miR158, miR159, miR160, miR167, miR394 and miR398 are conserved in a variety of plant species (Table 2 and Additional file 4D). Of these, several miRNAs, such as miR156, miR158, miR159, miR160, miR166, miR168 and miR2118, were expressed at relatively high levels, suggesting that they are highly expressed in root and possibly important regulators for radish root development. In addition, it could be found that 31 known miRNA sequences representing 26 non-conserved miRNA families, such as miR400, miR774, miR812, miR825, miR831, miR1510, miR3630 and miR8005, were previously identified only from one or few plant species.

Furthermore, the members of known miRNA families were also analyzed in this study. Among conserved miRNA families (Figure 3A), the miR156 was the largest family with 19 members, followed by miR166, miR159, miR169 and miR396, with 14, 11, 10 and 10 members, respectively. Of remaining 26 miRNA families, 22 comprised two to seven members, and others had only one member (Table 2). In addition to the conserved miRNA families, the other 26 non-conserved miRNA families comprised only one or two members (Table 2). The expression levels of known miRNA families were also analyzed. Among the 31 conserved miRNA families, the expression levels of several miRNA families including miR158, miR160 and miR166 showed high abundance, each with total read >100,000 (Figure 3B). In contrast, very low level of expression was found in some miRNA families including miR161, miR395 and miR828. Meanwhile, 31 conserved miRNA families were also found to be more abundant than non-conserved miRNAs (Table 2, Figure 3B and C). In addition, various members within the same family showed considerably variable in expression levels, for example, the number of miR156 family member reads ranged from one to 719,515 in three libraries (Additional file 4D). Moreover, the same member within different stages also indicated different read numbers, for instance, the abundance of miR156a in stage1, stage2 and stage3 were 505,759, 50,613 and 16,238 reads, respectively, implying that there were various functional

Table 2 Summary information of known miRNA families and their transcript abundance identified in all libraries

miRNA family	No. of members	miRNA reads			Normalized read count			Fold change		
		Stage1	Stage2	Stage3	Stage1	Stage2	Stage3	Log_2 (stage2/stage1)	Log_2 (stage3/stage1)	Log_2 (stage3/stage2)
Conserved miRNA										
miR156	19	1255950	197473	60169	30069.08	2684.56	950.56	−3.49	−4.98	−1.50
miR158	2	557972	982854	368856	20520.33	27460.11	11365.77	0.42	−0.85	−1.27
miR159	11	57202	53482	8715	16.29	0.01	6.67	−10.67	−1.29	9.38
miR160	5	1871124	1233815	437740	16.83	12.20	5.09	−0.46	−1.72	−1.26
miR161	1	113	15	0	6.72	0.80	0.01	−3.08	−9.39	−6.31
miR162	4	2343	1994	1126	0.01	0.27	0.06	4.73	2.55	−2.18
miR164	6	31856	9637	7293	1631.16	422.95	349.19	−1.95	−2.22	−0.28
miR166	14	560689	400549	158105	1040.43	643.12	276.19	−0.69	−1.91	−1.22
miR167	7	11164	13684	9386	96.43	151.64	120.71	0.65	0.32	−0.33
miR168	4	182143	183619	99034	10707.73	9669.95	5744.26	−0.15	−0.90	−0.75
miR169	10	30809	21665	18662	1.66	0.95	0.70	−0.80	−1.24	−0.44
miR171	5	158	470	51	0.01	12.46	2.87	10.28	8.16	−2.12
miR172	6	1234	6039	4154	19.03	26.31	53.21	0.47	1.48	1.02
miR390	5	22804	7150	2234	1301.37	358.88	118.37	−1.86	−3.46	−1.60
miR391	2	4090	5165	275	242.45	272.10	16.10	0.17	−3.91	−4.08
miR393	2	35	36	59	0.06	0.16	0.01	1.42	−2.57	−3.99
miR394	2	702	426	654	0.01	2.12	1.52	7.73	7.25	−0.48
miR395	3	39	50	18	1.90	0.01	0.01	−7.57	−7.57	0.00
miR396	10	2795	1792	1427	109.75	32.46	22.95	−1.76	−2.26	−0.50
miR397	1	2316	172	30	137.69	9.12	1.76	−3.92	−6.29	−2.38
miR398	3	804	1638	28150	47.50	86.46	0.01	0.86	−12.21	−13.08
miR399	4	65	87	26	0.01	0.05	0.18	2.41	4.13	1.73
miR403	1	2175	1604	1053	129.31	85.08	61.64	−0.60	−1.07	−0.46
miR408	3	11998	3335	400	712.19	176.31	23.42	−2.01	−4.93	−2.91
miR482	3	68	2640	2324	0.01	140.03	0.01	13.77	0.00	−13.77
miR535	2	1402	2121	709	0.65	0.80	0.41	0.28	−0.67	−0.96
miR824	2	4469	330	344	219.80	13.05	16.16	−4.07	−3.77	0.31
miR827	1	353	122	27	20.99	6.47	1.58	−1.70	−3.73	−2.03
miR828	2	56	18	33	3.33	0.95	1.05	−1.80	−1.66	0.14
miR2111	2	10	229	47	0.59	4.93	0.94	3.05	0.66	−2.40
miR2118	2	92829	15700	12723	2834.86	0.01	0.01	−18.11	−18.11	0.00
Total	144	4709767	3147905	1223824	280011.51	166968.59	71641.49	−0.75	−1.97	−1.22
Non-conserved miRNA										
miR400	1	147	72	41	8.74	4.28	2.40	−1.03	−1.86	−0.83
miR774	1	307	10640	5198	18.25	632.58	304.29	5.12	4.06	−1.06
miR812	1	0	3145	1503	0.00	186.98	87.98	18.26	17.18	−1.09
miR825	1	1675	0	0	99.58	0.00	0.00	−17.35	−17.38	−0.02
miR831	1	267	295	760	15.87	17.54	44.49	0.14	1.49	1.34
miR845	2	11735	7945	1722	697.69	472.36	100.80	−0.56	−2.79	−2.23
miR858	2	22	28	153	1.31	1.66	3.40	0.35	1.38	1.03
miR859	1	444	956	347	26.40	56.84	20.31	1.11	−0.38	−1.48
miR862	1	76	224	58	4.52	13.32	3.40	1.56	−0.41	−1.97

Table 2 Summary information of known miRNA families and their transcript abundance identified in all libraries *(Continued)*

miR1310	1	1376	273	296	81.81	16.23	17.33	−2.33	−2.24	0.09
miR1510	2	1419	0	1905	84.36	0.00	111.52	−17.11	0.40	17.52
miR1511	2	14510	22556	8500	862.67	1341.03	497.58	0.64	−0.79	−1.43
miR1513	1	484	1001	0	28.78	59.51	0.00	1.05	−15.59	−16.63
miR1520	1	36042	7457	6131	2142.82	443.34	358.90	−2.27	−2.58	−0.30
miR1885	1	3846	1383	1444	228.66	82.22	84.53	−1.48	−1.44	0.04
miR3630	2	266	52	0	15.81	3.09	0.00	−2.35	−14.72	−12.37
miR4378	1	0	9787	1082	0.00	581.87	63.34	19.90	16.70	−3.20
miR5654	1	8098	2411	738	481.45	143.34	43.20	−1.75	−3.48	−1.73
miR5763	1	0	370	669	0.00	22.00	39.16	15.18	16.01	0.83
miR5774	1	0	0	1234	0.00	0.00	72.24	0.00	16.89	16.89
miR6164	1	892	971	615	53.03	57.73	36.00	0.12	−0.56	−0.68
miR7504	1	45	113	5348	2.68	6.72	313.07	1.33	6.87	5.54
miR7510	1	218	383	171	12.96	22.77	10.01	0.81	−0.37	−1.19
miR7532	1	8125	2752	2304	483.06	163.62	134.87	−1.56	−1.84	−0.28
miR8005	1	0	1118	0	0.01	59.2998	0.01	12.53	0.00	−12.53
miR8041	1	18	62	27	1.07	3.69	1.58	1.78	0.56	−1.22
Total	31	90012	73994	40246	5351.52	3865.41	3469.08	−0.47	−0.63	−0.16
Total	175	4799779	3221899	1264070	285363.03	170834.01	75110.57	−0.74	−1.93	−1.19

divergences within miRNA family during the radish taproot thickening.

Identification of novel miRNA candidates during radish taproot thickening

Based on the key characteristics of novel miRNA [18,29], the formation of stem loop structure of precursor is prerequisite for a new miRNA. In total, 107 potential novel miRNAs (90 miRNA families) were predicted from three libraries (Additional file 5A). The stem loop structures of these predicted miRNA precursors were shown in Additional file 6. In addition to stem-loop structure prediction, detection of complementary sequences is another way to increase the authenticity of predicted novel miRNAs [29]. Among these potential novel miRNAs, five potential novel miRNA with complementary sequences were detected as the novel miRNA candidates (Additional file 5B).

In this study, the predicted hairpin length of these 107 potential novel miRNA precursors ranged from 47 to 354 nt. The folding of minimum free energy (MFE) value of these miRNA precursors ranged from −18.3 to −95.2 kcal/mol with an average of −40.1 kcal/mol (Additional file 5A). In addition, only seven out of 107 predicted miRNAs candidates were shared by all three libraries, while 53, 73 and 39 miRNAs were detected in stage1, stage2 and stage3 libraries, respectively (Additional file 5A).

The 107 potential novel miRNAs exhibited lower expression levels with the abundance ranging from five to 3,318 reads, as compared with known miRNAs. In addition, the numbers of all novel complementary miRNAs reads ranging from five to 114 were clearly less than those for their corresponding mature miRNAs, which was consistent with the idea that miRNA* strands were degraded rapidly during the biogenesis of mature miRNAs [33]. Interestingly, rsa-nmiR2-5p (read count of 17 vs. 14 in stage2 library) and rsa-miR18-5p (read count of 20 vs. 11 in stage2 library) showed similar abundance between novel miRNA and complementary miRNA (Additional file 5B), indicating that both the miRNA and their complementary miRNA might be functional in regulating gene expression during the taproot thickening process in radish.

Differentially expressed miRNAs during radish taproot thickening

Differential expression analysis was performed to identify differentially expressed miRNAs during the taproot thickening process. Based on the selected criteria (At least one comparison has a fold change $\log_2$ scale value ≥ 1.0 or ≤ -1.0 with P-value < 0.05), in all, 85 known miRNAs and 13 novel miRNAs were identified as differentially expressed miRNAs (Additional file 7). It was shown here that two important transitions (Stage1 to Stage2/Stage2 to Stage3) were analyzed during taproot thickening (Additional file 8). The differentially expressed miRNAs were divided into seven clusters according to

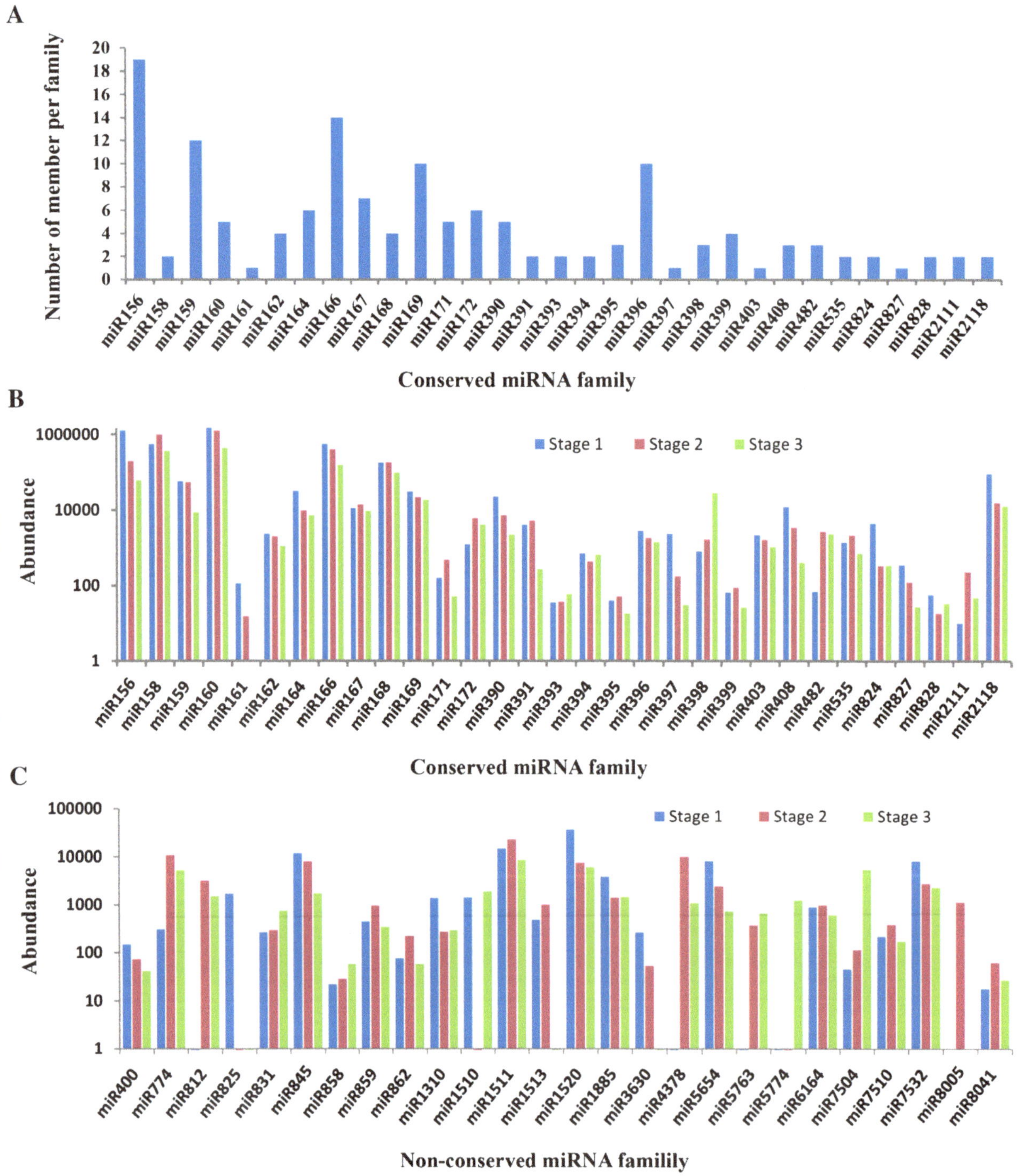

Figure 3 Sizes and abundance of identified known miRNA families in radish. The distribution of conserved miRNA family size **(A)** and the abundance of conserved **(B)** and non-conserved **(C)** miRNA family.

their highly similar expression patterns at the different stages of taproot thickening (Additional file 8 and Figure 4). The results indicated that 34 miRNAs had a down-regulated pattern during taproot thickening (Cluster 1 in Additional file 8). As from stage1 to stage2, the expression of 42 miRNAs including miR156a, miR157a, miR160b, miR169m, miR390a and miR397a, declined obviously (Clusters 1, 2 and 3 in Additional file 8), whereas 13 miRNAs in Cluster 1 exhibited a gradually decline. As from stage2 to stage3, 64 miRNAs exhibited down-regulated pattern (Clusters 1 and 5 in Additional file 8). In contrast, six miRNAs had an up-regulated

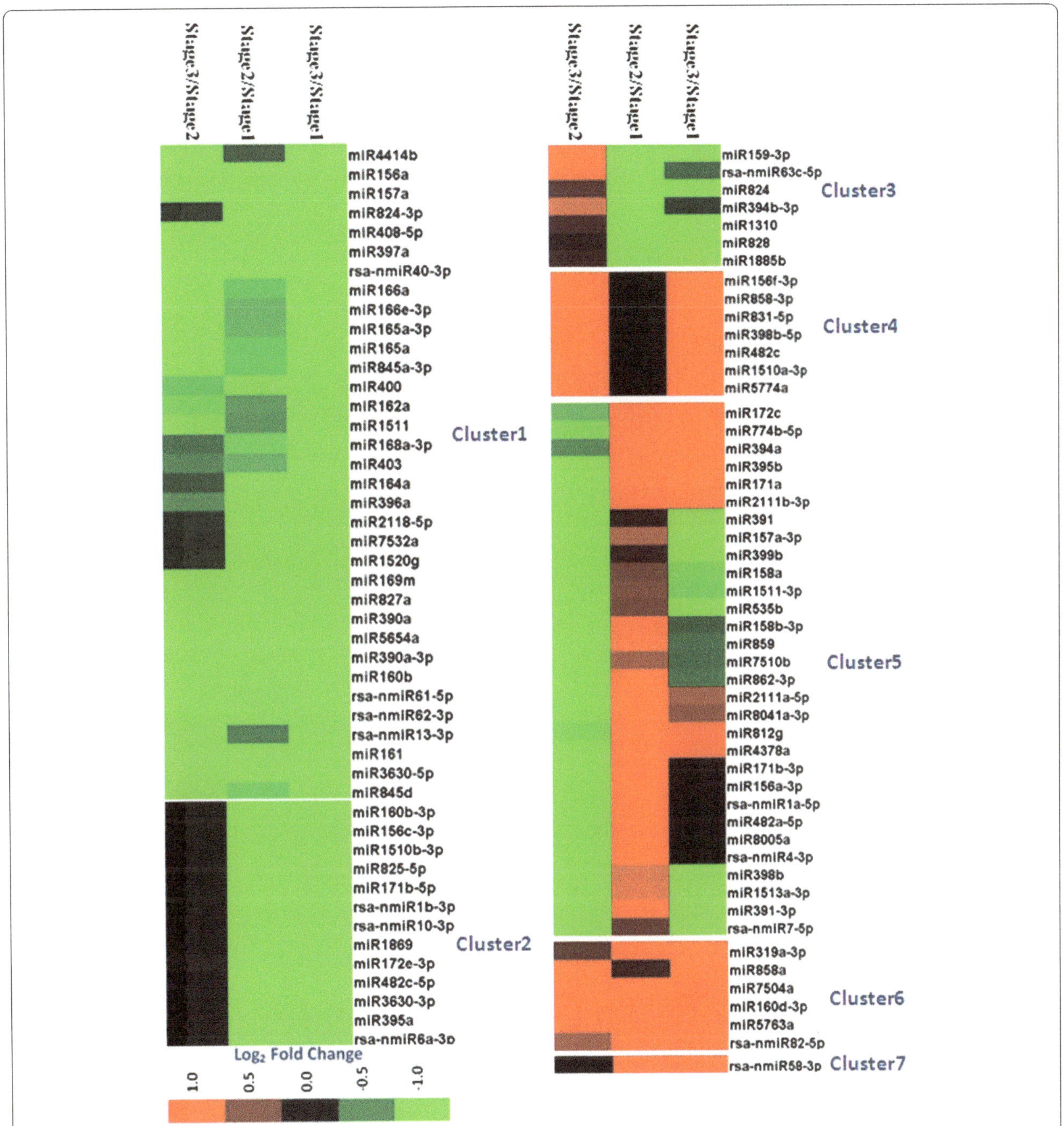

Figure 4 Clustering of differentially expressed miRNAs in three libraries. The bar represents the scale of relative miRNA expression (Log_2 Fold change).

pattern during taproot thickening (Cluster 6 in Additional file 8). The expressions of 37 miRNAs increased from stage1 to stage2 (Clusters 5, 6 and 7 in Additional file 8), and 20 miRNAs increased from stage2 to stage3 (Clusters 3, 4 and 6 in Additional file 8). Moreover, some miRNAs were preferentially expressed in only one taproot thickening stage. For example, rsa-nmiR6a-3p and rsa-nmiR4-3p were enriched at stage1 and stage2, respectively (Clusters 2, 4 and 5 in Additional file 8). Additionally, the miRNAs in Cluster 3 decreased obviously from stage1 to stage2, but increased from stage2 to stage3, whereas the miRNAs in Cluster 5 increased from stage1 to stage2, and decreased from stage2 to stage3.

Among the 31 conserved miRNA families, 13 and 18 miRNA families were up and down-regulated at stage2 as compared with stage1, respectively (Figure 5A, Table 2). Meanwhile, five and 24 miRNA families were up and down-regulated in stage3 compared with stage2,

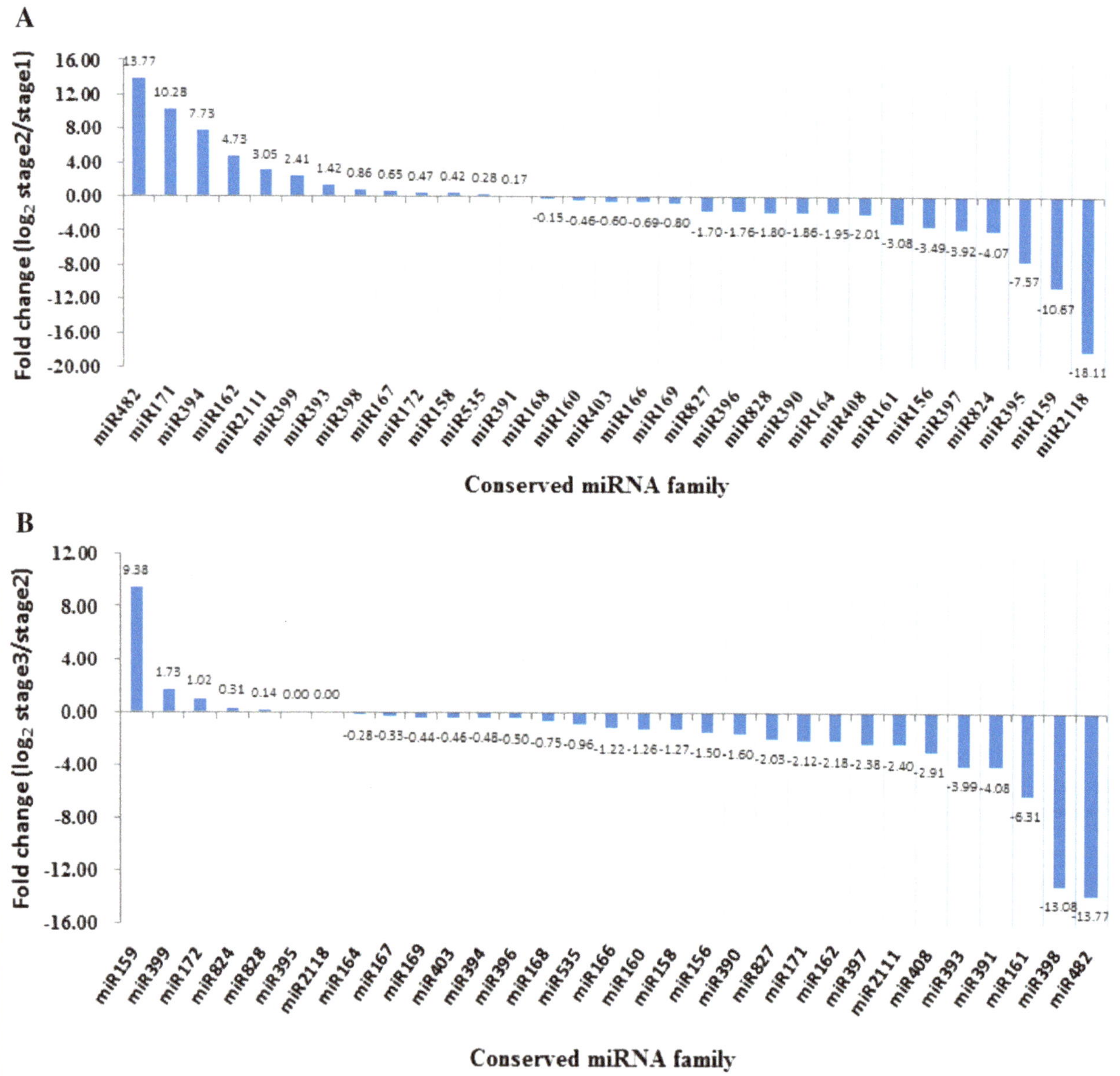

Figure 5 Comparatively relative expression of differentially expressed conserved miRNA family in radish. Comparison of stage 1 and stage 2 **(A)**, and comparison of stage 2 and stage 3 **(B)**.

respectively (Figure 5B, Table 2). Of these, five miRNA families were differentially expressed at a ratio greater than 10-fold (Figure 5). These results implied that these miRNA sequences and miRNA families might play essential regulatory roles during radish taproot thickening.

Prediction of potential target genes of differentially expressed miRNAs

To further clarify biological functions of the differentially expressed miRNAs during taproot thickening process, a total of 482 target sequences for 85 differentially expressed miRNAs were predicted (Additional file 9). Among these sequences, 191 potential target genes for 78 differentially expressed miRNAs were further annotated by BLAST search against *Arabidopsis* sequences using KOBAS 2.0 program (Additional file 9). Among them, 176 and 20 target genes were predicted for 67 known and 11 novel miRNAs, respectively (Additional file 9). It could be found that there are many single miRNAs targeted multiple genes and multiple miRNAs regulated a single gene. As a result, lots of these target genes were annotated as transcription factors (TFs). For instance, miR156, miR159 and miR774 family members were identified to target the squamosa promoter-binding-like protein genes (*SPLs*). miR160 family members were identified to target the auxin response factor genes (*ARFs*) and vascular plant one zinc finger protein genes (*VOZ1*). miR172 family members were identified to target the floral homeotic protein APETALA 2 gene (*AP2*), SC35-like splicing factor 33 gene (*SCL33*) and transcription factor IIIA gene (*TFIIIA*). The targets of miR164, miR169 and miR396 family members belonged to NAC-domain containing protein genes (*NACs*), nuclear transcription factor Y subunit A-2 protein gene (*NF-YA2*) and basic helix-loop-helix transcription factor bHLH74 gene (*bHLH74*), respectively. On the other hand, some target genes were annotated as other functional proteins. For instance, glutamine synthetase gene (*GS2*), IAA-amino acid hydrolase ILR1 gene (*ILR1*), laccases gene (*LACs*), xyloglucan endotransglucosylase/hydrolase protein 16 gene (*XTH16*), alkaline/neutral invertase gene (*INV-E*), protein CLAVATA3/ESR-related 41 gene (*CLE41*), expansin A9 gene (*EXPA9*), calmodulin 7 gene (*CAM7*) and protein phosphatase 2A regulatory B subunit gene (*PP2A-B*) were identified as the targets of miR156, miR172, miR397, miR858, miR5654, miR7532, miR8005, rsa-nmiR4 and rsa-nmiR6, respectively. The sulfate adenylyltransferase gene (*APS4*) and ATP sulfurylase 1 genes (*APS1*) were targeted by miR395. Additionally, some target genes were annotated as uncharacterized and hypothetical protein.

The majority of these identified target genes were annotated as transcription factors and other functional proteins which involved in plant growth and development, metabolism, cell organization and biogenesis, signal sensing and transduction, and plant defense response, such as auxin signaling, nutrition metabolism, sucrose metabolism, cell growth and cell wall expansion. All these results suggested that the differentially expressed miRNAs might play crucial regulatory roles during the taproot thickening process. The putative roles of miRNAs involved in taproot thickening of radish are summarized in Figure 6.

GO term annotation of the differentially expressed miRNAs targets

To further investigate the function of the differentially expressed miRNAs, the predicted 191 target genes were collected to be performed a Gene Ontology (GO) term annotation using the Blast2GO program (http://www.blast2go.com) (Additional file 10). GO term annotation results indicated that 16 different biological processes, nine different molecular functions and nine different cellular components were predicted (Figure 7). Among biological process, cellular process, metabolic process,

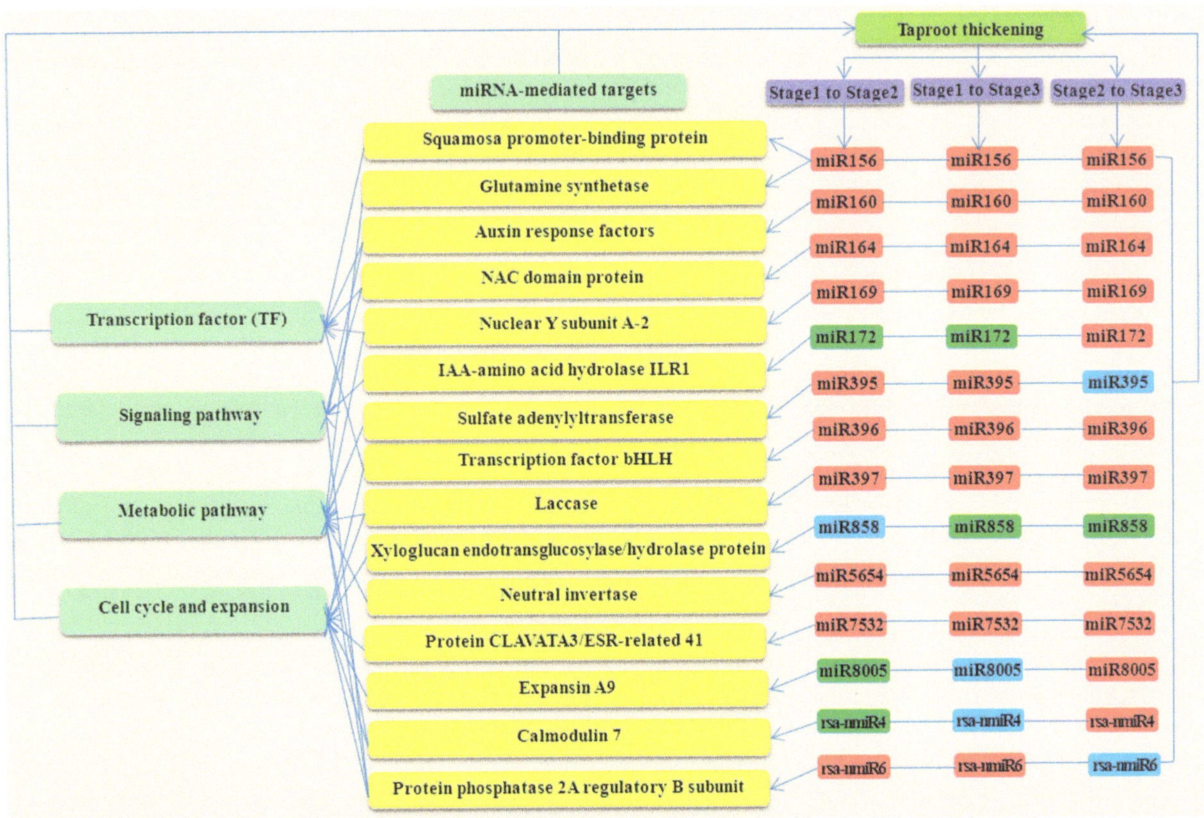

Figure 6 The putative role of miRNAs during radish taproot thickening. Green boxes: up regulated; rose-bengal boxes: down regulated; blue boxes: unchanged.

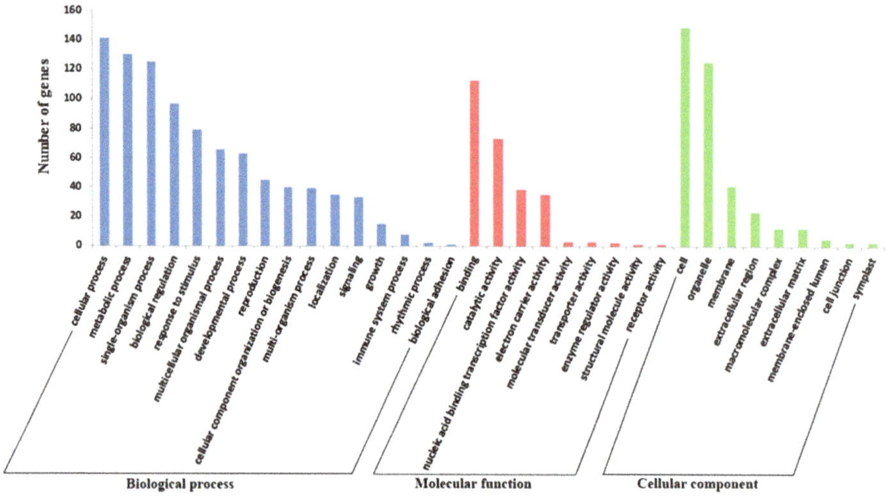

Figure 7 Gene ontology of the predicted targets for differentially expressed miRNAs.

single-organism process, biological regulation, response to stimulus, multicellular organismal process and developmental process are the most significantly enriched GO terms (Figure 7). Interestingly, the enriched GO terms also showed to be involved in various development and metabolic processes, such as carbohydrate metabolic process (GO: 0005975), sucrose metabolic process (GO: 0005985), root system development (GO: 0022622), root morphogenesis (GO: 0010015) and root development (GO: 0048364) (Additional file 10). These results suggested that the taproot thickening process in three stages is potentially regulated by the miRNAs and their corresponding target genes.

Validation of the expression patterns of differentially expressed miRNAs and their targets by RT-qPCR

To further confirm the expression pattern of the differentially expressed miRNAs, and their putative potential targets during radish taproot thickening process, five differentially expressed miRNAs and eight corresponding targets were randomly selected and validated with RT-qPCR. As a result, all five differentially expressed miRNAs and their targets were obviously differentially expressed among various stages of taproot development (10, 15, 20, 40 and 50 DAS) (Figure 8). Among them, miR156a, miR164a and miR169m were almost down-regulated during the taproot development (Figure 8A, C and D), while miR172c was up-regulated and peaked at 20 DAS compared with 10 DAS and 15 DAS, and then down-regulated compared with 40 DAS and 50 DAS (Figure 8E). The results showed well consistency with the expression pattern analyzed by small RNA high-throughput sequencing. However, miR4414b showed dynamic change (Figure 8B), which was not well consistent with the results of the sequencing.

Furthermore, it could be found that miRNAs and their target genes had anti-correlated expression tendencies at various taproot development stages in radish. miR156a, miR164a and miR172c as well as their corresponding target transcripts (*TOC1*, miR156a target gene, gi|167492752|gb|FD975103.1|FD975103; *NAC080*, miR164a target gene, gi|158663918|gb|EX773809.1|EX773809; *ILR1*, CL5916.Contig2_NAU-YH and *AP2*, gi|166139427|gb|FD572123.1|FD572123, miR172c target genes) had contrasting expression tendencies during various taproot thickening stages (Figure 8A-1, C-1, E-2 and E-3), suggesting that these miRNAs may regulate their potential target expressions, and the target genes of miRNAs may be involved in radish taproot thickening. However, no correlation was also observed between the expression of some miRNAs and their targets during the taproot thickening process. For instance, miR169m, miR4414b and their targets had similar expression tendency over the various taproot development stages (Figure 8B-1 and D-1), which indicating that these two predicted genes may not be targets of miR4414b and miR169m. Meanwhile, miRNA172 and its target transcript (*TFIIIA*, gi|332778718|gb|FY435119.1|FY435119) showed unique no correlation expression patterns (Figure 8E-4), suggesting that the putative target may implement a particular function during radish taproot-thickening.

Discussion

Radish is an important root vegetable crop and its edible part is the taproot, which directly determines the yield and quality [1]. It is therefore of significance to understand the mechanism of radish taproot formation. miRNAs regulate multiple developmental events in plants. To date, much effort has been put in studying miRNA mechanisms underlying different plant development processes, such as flower development [34], seed and seedling

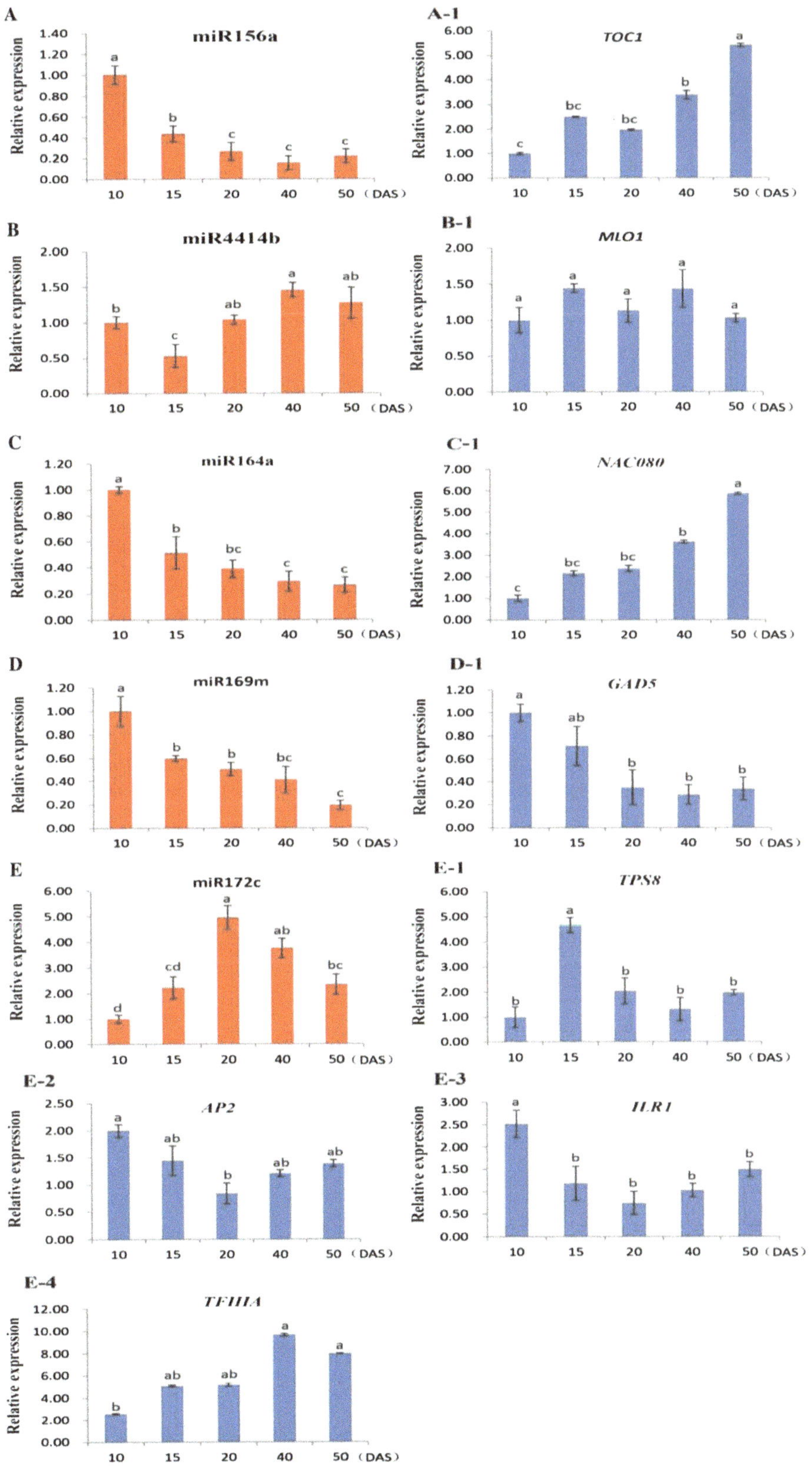

Figure 8 Quantitative expression analyses of five differentially expressed miRNAs (A ~ E) and their target genes (A-1 ~ E-4). Each bar shows the mean ± SE of triplicate assays. The values with different letters indicate significant differences at $P < 0.05$ according to Duncan's multiple range tests.

development [35,36], and root development [19-21]. Up to now, several studies on expression profiles of miRNA associated with plant root development were conducted in many important plant species, such as maize [19], rice [20], *M. truncatula* [17] and potato [21], while an overall expression profiles of miRNA during the phase of thickening in radish taproot is still unexplored. Although some potential conserved and novel miRNAs have been predicted from radish root based on the *R. sativus* EST and GSS sequences [22,23], all of these miRNAs were obtained from taproot collected at one period, which greatly constrained the investigation of the miRNAs regulation mechanism underlying the taproot thickening. In this study, a population of both known and novel miRNAs from different thickening stages of fleshy taproot were firstly identified and characterized. Furthermore, the differentially expressed miRNAs and their potentially target genes associated with taproot thickening were also investigated in radish.

The development of cortex splitting of radish taproot marks the entry into a growth stage that mainly involves root thickening [1]. In this study, using a Solexa sequencing technology, small RNAs in developing radish taproot from three different developmental stages: pre-cortex splitting stage (Stage1, 10 DAS), cortex splitting stage (Stage2, 20 DAS) and expanding stage (Stage3, 40 DAS), which cover the key morphological changes that occur during the taproot thickening process, were firstly isolated.

Identification of taproot thickening-related miRNAs by Solexa sequencing in radish

Since the miRNAs in Arabidopsis was firstly discovered, thousands of mature sequences in plants have been registered in miRbase, and some miRNAs were identified to be indispensable for the development and formation of plant roots [37]. Recently, by the high-throughput sequencing, 137 known and 159 novel miRNAs were obtained during the early development of the maize brace root [19], and 83 known and 24 novel miRNAs were identified in root tissues and root callus tissues in *M. truncatula* [17]. Previously, although some conserved miRNAs have been reported in radish [22-25], the miRNAs involved in the process of taproot thickening have not been discovered. In this study, 175 known miRNAs (57 miRNA families) and 107 potential novel miRNA candidates (90 miRNA families) were identified during the taproot thickening process. Among the 57 identified known miRNA families, over a half (31) are conserved with other species predicted previously [38]. Moreover, some miRNA families with high expression levels in this study are also expressed in the roots of other plant species, such as miR156, miR159, miR164, miR166, miR167, miR168 and miR172, which were abundantly found in maize brace roots [19]. These results suggested that these miRNAs may play crucial and conserved roles in plant root development. In addition, with the high-throughput sequencing technology, 107 potential novel miRNAs were identified during radish taproot thickening in this study (Additional file 5A). Among these novel miRNAs, some of them may only be detected during specific developmental stages. Compared to the conserved miRNAs, most of the novel miRNAs exhibited lower abundance levels, as previously reports indicating that the novel miRNAs were often expressed at relatively lower levels than conserved miRNAs [25]. However, although expressing at low level, these new miRNAs might play developmental-specific or species-specific roles during taproot thickening in radish.

Dynamic expression patterns of miRNAs associated with radish taproot-thickening

Most of differentially expressed miRNAs have been identified as being involved in the regulation of plant growth and development in diverse plants [13,19,34-36]. In the present study, 85 known miRNAs belonging to 54 miRNA families, and 13 novel miRNAs belonging to 12 miRNA families were identified to be differentially expressed during radish taproot thickening (Additional file 7). Among these, more miRNAs were observed to be down-regulated than up-regulated during the taproot thickening process. For instance, miR156, miR159, miR160, miR166, miR390, miR397, miR408, miR5654, rsa-nmiR40 and rsa-nmiR62 were down-regulated significantly during the period stage1 to stage2 and stage2 to stage3. Additionally, most of these down-regulated miRNAs were also highly expressed in three libraries (Cluster 1 in Additional file 8), implying that miRNAs play vital roles during taproot thickening in radish. Meanwhile, a few miRNAs, such as miR5763 and miR7504 were increasingly up-regulated during the period stage1 to stage2 and stage2 to stage3 (Cluster 6 in Additional file 8). Some miRNAs, such as miR171, miR172, miR774, miR812, miR2111, rsa-nmiR1, rsa-nmiR4 and rsa-nmiR7, were increasingly up-regulated from stage1 to stage2, and then down-regulated from stage2 to stage3 (Cluster 5 in Additional file 8). It was reported that in plant development process, the up- or down-regulation of miRNAs might play a more important role in the regulation of network [39-41]. Therefore, it was possible that 98 miRNAs showing differential expression patterns may also play crucial roles in regulating the thickening of radish taproot, although more investigations are needed to further clarify regulation mechanism associated with various miRNAs during taproot thickening.

In addition, previous study has shown that several known miRNAs were differentially regulated during the early development of maize brace root, including six miRNA families (miR164, miR167, miR171, miR390,

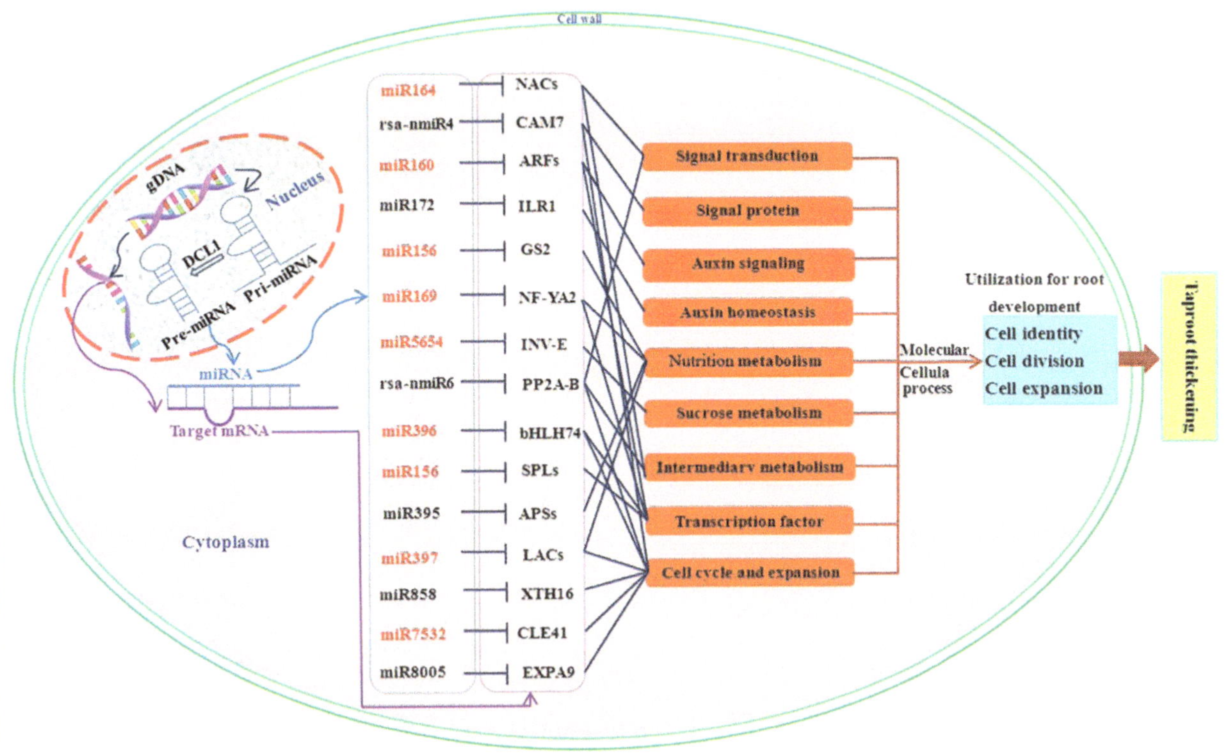

Figure 9 The hypothetical model of miRNAs-mediated regulatory network associated with radish taproot thickening. The red font represents the down-regulated miRNA in three libraries.

miR393 and miR399) were up-regulated, and two miRNA families (miR156 and miR169) were down-regulated [19]. As expected, some of these identified miRNAs also showed differential regulation during taproot thickening, such as miR156 and miR169 were down-regulated, suggesting these miRNAs may be involved in the regulatory networks during root development. Moreover, several miRNA families (miR167 and miR393) were reported to regulate root development in *Arabidopsis* [16,42], while in this study, they did not exhibit significant differences in expression during radish taproot thickening. Nevertheless, they also may be play crucial role during taproot thickening in radish, because these miRNAs expressed in all three libraries.

miRNA-mediated regulatory networks of taproot-thickening in radish

Although many functional studies have revealed that some miRNAs play crucial roles in plant root development [16,37,42], there is few studies on characterization of miRNAs and their target genes related to storage root formation to date. In this study, 191 targets for the differentially expressed miRNAs during radish taproot thickening were found to be involved in various biological functions including plant growth and development, metabolism, cell organization and biogenesis, signal sensing and transduction, and plant defense response. Of these predicted targets, many of them are transcription factors (*SPLs*, *NF-YA2*, *ILR1* and *bHLH74*) and regulate hormone accumulation (*ARFs*, *NACs*), and they were identified to be involved in plant root formation and development process [14,15,43-46]. For example, *NF-YA2* transcription factor was one of the targets of miR169, which acts in the control of primary root growth in *Arabidopsis* [15]. miR156 were predicted to target mRNA coding for SPL like family transcription factor. Previous studies have proven that miR156-mediated regulation of *SPLs* involves in plant development [44]. miR160 target gene encodes auxin response factors (*ARF16*, *ARF17*) which were found to affect primary and lateral root growth in *Arabidopsis*, rice and *Medicago* [14,16,20,47]. In addition, some targets may be involved in regulating substances and energy metabolism changes, and are widely considered to be important for root development in plants [48]. Sulfate adenylyltransferase and ATP sulfurylase 1 (*APSs*) genes targeted by miR395 were involved in sulfur assimilation and regulated root elongation by affecting root indole-3-acetic acid levels [49]. miR5654 targeting *INV* was thought to function in regulating sucrose metabolism and it has been proven that *INV* affects the root development [50]. Moreover, a number of target genes in this study seem to be involved in regulating cell cycle and cell expansion. For instance, xyloglucan endotransglucosylase/hydrolase protein

16 (*XTH16*, targeted by miR858) and expansin A9 (*EXPA9*, targeted by miR8005) were cell wall-related genes, which regulate the extension of cell wall during plant growth [51]. Laccases (*LACs*) targeted by miR397 were associated with thickening of the cell wall in secondary cell growth [52]. The protein CLAVATA3/ESR-related 41 gene (*CLE41*) was targeted by miR7532 and controls the rate and orientation of vascular cell division [53]. The protein phosphatase 2A regulatory B subunit gene (*PP2A B subunit*) targeted by rsa-nmiR6 could regulate cell grow in root of *Arabidopsis* [54]. Calmodulin 7 gene (*CAM7*), targeted by rsa-nmiR4, was found to be involved in the cell growth-promoting pathway, for calmodulin could bind to peptide phytosulfokine (PSK) receptor to cause cell growth [55]. Additionally, among the putative taproot thickening-related miRNAs, eight miRNAs (miR156, miR160, miR164, miR169, miR396, miR397, miR5654 and miR7532) were down-regulated, and five miRNAs (miR395, miR858, miR8005, rsa-nmiR4 and rsa-nmiR6) showed stage-specific pattern of expression during the taproot thickening (Figure 6 and Additional file 8). Therefore, it could be inferred that these miRNAs and their targets play important roles in regulating radish taproot thickening.

The thickening of taproot strongly influences the yield and quality in radish [1]. Thickening of underground storage root is a complex process with intricate pathways. Hormonal accumulation, transcription factor regulation, substances and energy metabolism changes, cell cycle and cell expansion, and others favor the thickening of taproot in radish [56-58]. In light of important functions of these differentially expressed miRNAs in regulating radish root development, a hypothetical model of miRNAs mediated regulation associated with taproot thickening in radish was put forward (Figure 9).

Conclusions

In summary, the small RNA population of radish taproot at different thickening stages were firstly identified using Solexa sequencing technology, a total of 175 known miRNAs and 107 novel miRNAs were found to be associated with radish taproot-thickening. Totally 98 differentially expressed miRNAs (85 known and 13 novel miRNAs) were identified and their 191 target genes were engaged in various biological functions, including plant growth and development, metabolism, cell organization and biogenesis, signal sensing and transduction, and plant defense response. Gene ontology categorization and enrichment analysis of the targets corresponding to the differentially expressed miRNAs revealed that a number of miRNA-targeted genes are required for radish taproot thickening. These findings provide significant insight into miRNA-mediated molecular regulatory mechanism underlying the taproot development and formation in radish.

Availability of supporting data

The RNA sequence dataset supporting the results of this article is available in the repository of NCBI Sequence Read Archive (SRA) with the GenBank accession No.: SRX707630.

Additional files

Additional file 1: RT-qPCR validated miRNAs primer sequences (A), and targets primer sequences (B).

Additional file 2: The distribution of sRNAs among different categories.

Additional file 3: Venn diagrams for analysis of Small RNAs. (A-F) Summary of common and specific unique (A, B and C) sRNAs and total (D, E and F) sRNAs between different libraries. (G) Known miRNAs among different libraries.

Additional file 4: Detailed information of the known miRNAs identified in stage1 library (A), stage2 library (B), stage3 library (C) and in all three libraries (D).

Additional file 5: Identification of novel miRNA candidates in three libraries. (A) Potential novel miRNA candidates, (B) novel miRNA candidates with complementary sequences.

Additional file 6: The secondary structures of novel *Raphanus sativus* miRNA precursors.

Additional file 7: Expression data of miRNAs differentially expressed during radish taproot thickening.

Additional file 8: The differentially expressed miRNAs in stage 1 to stage 2 and stage 2 to stage 3.

Additional file 9: The potential targets of differentially expressed miRNAs during radish taproot thickening.

Additional file 10: Significant GO terms for differentially expressed miRNA target genes.

Abbreviations

DAS: Day after sowing; RT-qPCR: Reverse transcription quantitative real-time PCR; TOC1: Timing of CAB Expression 1/two-component response regulator-like APRR1; TFIIIA: Ttranscription factor IIIA.

Competing interests

The authors declare that they have no competing interests.

Authors' contributions

YRG, LLW, GYQ and ZXW designed the experiments. YRG and ZW performed the radish cultivation and sample collection. YRG wrote the manuscript draft. LLW, ZXW, CL and XL edited and revised the manuscript. YRG, WY and WRH performed the experiments. All authors read and approved the final manuscript.

Acknowledgements

This work was in part supported by grants from the NSFC (31171956, 31372064, 30571193), Key Technology R & D Program of Jiangsu Province (BE2013429), JASTIF [CX (12)2006, (13)2007] and the PAPD.

Author details

[1]National Key Laboratory of Crop Genetics and Germplasm Enhancement; Engineering Research Center of Horticultural Crop Germplasm Enhancement and Utilization, Ministry of Education of P.R.China; College of Horticulture, Nanjing Agricultural University, Nanjing 210095, P.R. China. [2]School of Life Sciences, Huaibei Normal University, Huaibei, Anhui 235000, P.R. China. [3]Department of Plant Sciences, North Dakota State University, Fargo, ND 58108, USA.

References

1. Wang LZ, He QW. Chinese radish. Beijing: Scientific and Technical Documents Publishing House; 2005.
2. Pérez Gutiérrez RM, Perez RL. *Raphanus sativus* (Radish): their chemistry and biology. Sci World J. 2004;4:811–37.
3. Zaki HEM, Takahata Y, Yokoi S. Analysis of the morphological and anatomical characteristics of roots in three radish (*Raphanus sativus*) cultivars that differ in root shape. J Hortic Sci Biotech. 2012;87:172.
4. Ting FST, Wren MJ. Storage organ development in radish (*Raphanus sativus* L.). 2. Effects of growth promoters on cambial activity in cultured roots, decapitated seedlings and intact plants. Ann Bot. 1980;46:277–84.
5. Choi EY, Seo TC, Lee SG, Cho IH, Stangoulis J. Growth and physiological responses of Chinese cabbage and radish to long-term exposure to elevated carbon dioxide and temperature. Hortic Environ Biote. 2011;52:376–86.
6. Petricka JJ, Winter CM, Benfey PN. Control of *Arabidopsis* root development. Annu Rev Plant Biol. 2012;63:563.
7. Smith S, Smet ID. Root system architecture: insights from *Arabidopsis* and cereal crops. Phil Trans R Soc Lond B Biol Sci. 2012;367:1441–52.
8. Zhou XJ, Li Q, Chen X, Liu J, Zhang QH, Liu YJ, et al. The Arabidopsis *RETARDED ROOT GROWTH* gene encodes a mitochondria-localized protein that is required for cell division in the root meristem. Plant Physiol. 2011;157:1793–804.
9. Taramino G, Sauer M, Stauffer JL, Multani D, Niu X, Sakai H, et al. The maize (*Zea mays* L.) *RTCS* gene encodes a LOB domain protein that is a key regulator of embryonic seminal and post-embryonic shoot-borne root initiation. Plant J. 2007;50:649–59.
10. Zhi-Guo E, Ge L, Wang L. Molecular mechanism of adventitious root formation in rice. Plant Growth Regul. 2012;68:325–31.
11. Kitashiba H, Li F, Hirakawa H, Kawanabe T, Zou Z. Hasegawa Y, et al. Draft Sequences of the Radish (*Raphanus sativus* L.) Genome. DNA Res. 2014;5:481–90.
12. Wang Y, Pan Y, Liu Z, Zhu XW, Zhai LL, Xu L, et al. De novo transcriptome sequencing of radish (*Raphanus sativus* L.) and analysis of major genes involved in glucosinolate metabolism. BMC Genomics. 2013;14:836.
13. Cuperus JT, Fahlgren N, Carrington JC. Evolution and functional diversification of MIRNA genes. Plant Cell. 2011;23:431–42.
14. Guo HS, Xie Q, Fei JF, Chua NH. MicroRNA directs mRNA cleavage of the transcription factor NAC_1 to downregulate auxin signals for Arabidopsis lateral root development. Plant Cell. 2005;17:1376–86.
15. Sorin C, Declerck M, Christ A, Blein T, Ma L, Lelandais-Brière C, et al. A miR169 isoform regulates specific NF-YA targets and root architecture in Arabidopsis. New Phytol. 2014;202:1197–211.
16. Gutierrez L, Bussell JD, Păcurar DI, Schwambach J, Păcurar M, Bellini C. Phenotypic plasticity of adventitious rooting in *Arabidopsis* is controlled by complex regulation of AUXIN RESPONSE FACTOR transcripts and microRNA abundance. Plant Cell. 2009;21:3119–32.
17. Eyles RP, Williams PH, Ohms SJ, Weiller GF, Ogilvie HA, Djordjevic MA, et al. MicroRNA profiling of root tissues and root forming explant cultures in *Medicago truncatula*. Planta. 2013;238:91–105.
18. Kong X, Zhang M, Xu X, Li X, Li C, Ding Z. System analysis of microRNAs in the development and aluminium stress responses of the maize root system. Plant Biotechnol J. 2014;12:1108–21.
19. Liu P, Yan K, Lei YX, Xu R, Zhang YM, Yang GD, et al. Transcript profiling of microRNAs during the early development of the maize brace root via Solexa sequencing. Genomics. 2013;101:149–56.
20. Ma XX, Shao CG, Wang HZ, Jin YF, Meng YJ. Construction of small RNA-mediated gene regulatory networks in the roots of rice (*Oryza sativa*). BMC Genomics. 2013;14:510.
21. Lakhotia N, Joshi G, Bhardwaj AR, Katiyar-Agarwal S, Agarwal M, Jagannath A, et al. Identification and characterization of miRNAome in root, stem, leaf and tuber developmental stages of potato (*Solanum tuberosum* L.) by high-throughput sequencing. BMC Plant Biol. 2014;14:6.
22. Muvva C, Tewari L, Aruna K, Ranjit P. *In silico* identification of miRNAs and their targets from the expressed sequence tags of *Raphanus sativus*. Bioinformation. 2012;8:98–103.
23. Xu L, Wang Y, Xu YY, Wang LJ, Zhai LL, Zhu XW, et al. Identification and characterization of novel and conserved microRNAs in radish (*Raphanus sativus* L.) using high-throughput sequencing. Plant Sci. 2013;201:108–14.
24. Zhai LL, Xu L, Wang Y, Huang DQ, Yu RG, Limera C, et al. Genome-wide identification of embryogenesis-associated microRNAs in radish (*Raphanus sativus* L.) by high-throughput sequencing. Plant Mol Biol Rep. 2014;32:900–15.
25. Xu L, Wang Y, Zhai LL, Xu YY, Wang LJ, Zhu XW, et al. Genome-wide identification and characterization of cadmium-responsive microRNAs and their target genes in radish (*Raphanus sativus* L.) roots. J Exp Bot. 2013;64:4271–87.
26. Grabherr MG, Haas BJ, Yassour M, Levin JZ, Thompson DA, Amit I, et al. Full-length transcriptome assembly from RNA-Seq data without a reference genome. Nat Biotechnol. 2011;29:644–52.
27. Li R, Yu C, Li Y, Lam TW, Yiu SM, Kristiansen K, et al. SOAP2: an improved ultrafast tool for short read alignment. Bioinformatics. 2009;25:1966–7.
28. Zuker M. Mfold web server for nucleic acid folding and hybridization prediction. Nucleic Acids Res. 2003;31:3406–15.
29. Meyers BC, Axtell MJ, Bartel B, Bartel DP, Baulcombe D, Bowman JL, et al. Criteria for annotation of plant microRNAs. Plant Cell. 2008;20:3186–90.
30. Dai X, Zhao PX. psRNATarget: a plant small RNA target analysis server. Nucleic Acids Res. 2011;39:W155–9.
31. Xu YY, Zhu XW, Gong YQ, Xu L, Wang Y, Liu LW. Evaluation of reference genes for gene expression studies in radish (*Raphanus sativus* L.) using quantitative real-time PCR. Biochem Bioph Res Co. 2012;3:398–403.
32. Pantaleo V, Szittya G, Moxon S, Miozzi L, Moulton V, Dalmay T, et al. Identification of grapevine microRNAs and their targets using high-throughput sequencing and degradome analysis. Plant J. 2010;62:960–76.
33. Rajagopalan R, Vaucheret H, Trejo J, Bartel DP. A diverse and evolutionarily fluid set of microRNAs in *Arabidopsis thaliana*. Gene Dev. 2006;20:3407–25.
34. Wang ZJ, Huang JQ, Huang YJ, Li Z, Zheng BS. Discovery and profiling of novel and conserved microRNAs during flower development in *Carya cathayensis* via deep sequencing. Planta. 2012;236:613–21.
35. Xue LJ, Zhang JJ, Xue HW. Characterization and expression profiles of miRNAs in rice seeds. Nucleic Acids Res. 2009;37:916–30.
36. Li YF, Zheng Y, Jagadeeswaran G, Sunkar R. Characterization of small RNAs and their target genes in wheat seedlings using sequencing-based approaches. Plant Sci. 2013;203:17–24.
37. Khan GA, Declerck M, Sorin C, Hartmann C, Crespi M, Lelandais-Brière C. MicroRNAs as regulators of root development and architecture. Plant Mol Biol. 2011;77:47–58.
38. Sunkar R, Jagadeeswaran G. *In silico* identification of conserved microRNAs in large number of diverse plant species. BMC Plant Biol. 2008;8:37.
39. Meng FR, Liu H, Wang KT, Liu LL, Wang SH, Zhao YH, et al. Development-associated microRNAs in grains of wheat (*Triticum aestivum* L.). BMC Plant Biol. 2013;13:140.
40. Curaba J, Spriggs A, Taylor J, Li Z, Helliwell C. miRNA regulation in the early development of barley seed. BMC Plant Biol. 2012;12:120.
41. Kang MM, Zhao Q, Zhu DY, Yu JJ. Characterization of microRNAs expression during maize seed development. BMC Genomics. 2012;13:360.
42. Vidal EA, Araus V, Lu C, Parry G, Green PJ, Coruzzi GM, et al. Nitrate-responsive miR393/*AFB3* regulatory module controls root system architecture in *Arabidopsis thaliana*. Proc Natl Acad Sci. 2010;107:4477–82.
43. We S, Gruber MY, Yu B, Gao MJ, Khachatourians GG, Hegedus DD, et al. *Arabidopsis* mutant *sk156* reveals complex regulation of *SPL15* in a *miR156*-controlled gene network. BMC Plant Biol. 2012;12:169.
44. Wu G, Park MY, Conway SR, Wang J-W, Weigel D, Poethig RS. The sequential action of miR156 and miR172 regulates developmental timing in *Arabidopsis*. Cell. 2009;138:750–9.
45. Bao ML, Bian HW, Zha YL, Li FY, Sun YZ, Bai B, et al. miR396a-mediated basic helix-loop-helix transcription factor bHLH74 repression acts as a regulator for root growth in *Arabidopsis* seedlings. Plant Cell Physiol. 2014;7:1343–53.
46. Rampey RA, LeClere S, Kowalczyk M, Ljung K, Sandberg G, Bartel B. A family of auxin-conjugate hydrolases that contributes to free indole-3-acetic acid levels during Arabidopsis germination. Plant Physiol. 2004;135:978–88.
47. Bustos-Sanmamed P, Mao G, Deng Y, Elouet M, Khan GA, Bazin J, et al. Overexpression of miR160 affects root growth and nitrogen-fixing nodule number in *Medicago truncatula*. Funct Plant Biol. 2013;40:1208–20.
48. Wang ZR, Straub D, Yang HY, Kania A, Shen J, Ludewig U, et al. The regulatory network of cluster-root function and development in phosphate-deficient white lupin (*Lupinus albus*) identified by transcriptome sequencing. Physiol Plant. 2014;151:323–38.

49. Zhao Q, Wu Y, Gao L, Ma J, Li CY, Xiang CB. Sulfur nutrient availability regulates root elongation by affecting root IAA levels and the stem cell niche. J Integr Plant Biol. 2014;12:1151–63.
50. Yao Y, Geng MT, Wu XH, Liu J, Li RM, Hu XW, et al. Genome-wide identification, 3D modeling, expression and enzymatic activity analysis of cell wall invertase gene family from cassava (*Manihot esculenta Crantz*). Int J Mol Sci. 2014;15:7313–31.
51. Camacho-Cristóbal JJ, Herrera-Rodríguez MB, Beato VM, Rexach J, Navarro-Gochicoa MT, Maldonado JM, et al. The expression of several cell wall-related genes in Arabidopsis roots is down-regulated under boron deficiency. Environ Exp Bot. 2008;63:351–8.
52. Cai X, Davis EJ, Ballif J, Liang M, Bushman E, Haroldsen V, et al. Mutant identification and characterization of the laccase gene family in *Arabidopsis*. J Exp Bot. 2006;57:2563–9.
53. Etchells JP, Turner SR. The PXY-CLE41 receptor ligand pair defines a multifunctional pathway that controls the rate and orientation of vascular cell division. Development. 2010;137:767–74.
54. Tang W, Yuan M, Wang R, Yang Y, Wang C, Oses-Prieto JA, et al. PP2A activates brassinosteroid-responsive gene expression and plant growth by dephosphorylating BZR1. Nat Cell Biol. 2011;13:124–31.
55. Hartmann J, Fischer C, Dietrich P, Sauter M. Kinase activity and calmodulin binding are essential for growth signaling by the phytosulfokine receptor PSKR1. Plant J. 2014;78:192–202.
56. Jung JK, McCouch S. Getting to the roots of it: Genetic and hormonal control of root architecture. Front Plant Sci. 2013;4:186.
57. Montiel G, Gantet P, Jay-Allemand C, Breton C. Transcription factor networks. Pathways to the knowledge of root development. Plant Physiol. 2004;136:3478–85.
58. Rouhier H, Usuda H. Spatial and temporal distribution of sucrose synthase in the radish hypocotyl in relation to thickening growth. Plant Cell Physiol. 2001;42:583–93.

7

Transcriptome analysis of nitrogen-starvation-responsive genes in rice

Wenzhu Yang[1,2,3], Jinmi Yoon[1], Heebak Choi[1], Yunliu Fan[2,3], Rumei Chen[2,3*] and Gynheung An[1*]

Abstract

Background: Nitrogen (N), a critical macronutrient for plant growth and development, is a major limiting factor in most agricultural systems. Microarray analyses have been conducted to investigate genome-wide gene expression in response to changes in N concentrations. Although RNA-Seq analysis can provide a more precise determination of transcript levels, it has not previously been employed to investigate the expression of N-starvation-induced genes.

Results: We constructed cDNA libraries from leaf sheaths and roots of rice plants grown under N-deficient or -sufficient conditions for 12 h. Sequencing the libraries resulted in identification of 33,782 annotated genes. A comparison of abundances revealed 1,650 transcripts that were differentially expressed (fold-change ≥ 2) due to an N-deficiency. Among them, 1,158 were differentially expressed in the leaf sheaths (548 up-regulated and 610 down-regulated) and 492 in the roots (276 up, 216 down). Among the 36 deficiency-induced genes first identified via RNA-Seq analyses, 34 were subsequently confirmed by qRT-PCR. Our RNA-Seq data identified 8,509 multi-exonic genes with 19,628 alternative splicing events. However, we saw no significant difference in alternative splicing between N-sufficient and -deficient conditions. We found 2,986 novel transcripts, of which 192 were regulated under the N-deficiency.

Conclusion: We identified 1,650 genes that were differentially expressed after 12 h of N-starvation. Responses by those genes to a limited supply of N were confirmed by RT-PCR and GUS assays. Our results provide valuable information about N-starvation-responsive genes and will be useful when investigating the signal transduction pathway of N-utilization.

Keywords: N-starvation, *Oryza sativa*, Transcription factors, Transcriptome sequencing

Background

The macronutrient nitrogen (N) is an essential component of numerous important compounds, including amino acids, proteins, nucleic acids, chlorophyll, and some plant hormones. This element is a major limiting factor in most agricultural systems. Because the N-utilization efficiency strongly influences crop productivity, a vast amount of N fertilizers is applied to maximize yields. However, it is estimated that more than half of that N is lost from the plant–soil system, with unused N fertilizers severely polluting the environment [1]. Thus, N-uptake efficiency must be increased to improve productivity and reduce pollution.

During periods of N-starvation, various deficiency-responsive genes function to support plant survival by increasing the level of chlorophyll synthesis [2], altering root architecture [3], improving N-assimilation [4], enhancing lignin content [5], and changing the amounts of sugars and sugar phosphates [6]. Nitrate transporter genes (*NRTs*) are responsible for the high-affinity NO_3^- transport system and stimulate lateral root growth. *Arabidopsis NRT2.1* plays a major role in NO_3^- uptake and determines root architecture by controlling lateral root formation [7]. The ammonia transporter gene *AtAmt1.1*, which is highly expressed in the roots, also restructures this architecture under limited-N conditions [8]. The plant-specific Dof1 transcription factor (TF) from maize also functions to increase N-assimilation [9]. In Dof1-

* Correspondence: chenrumei@caas.cn; genean@khu.ac.kr
[2]Department of Crop Genomics and Genetic Improvement, Biotechnology Research Institute, Chinese Academy of Agricultural Sciences, Beijing 100081, China
[1]Department of Plant Molecular Systems Biotechnology and Crop Biotech Institute, Kyung Hee University, Yongin 446-701, Korea
Full list of author information is available at the end of the article

overexpressing *Arabidopsis* plants, genes are up-regulated under N-starvation to encode enzymes for carbon skeleton production [9]. Those transgenic plants also show markedly elevated amino acid contents, reduced levels of glucose, and improved growth during periods of N-deficient stress [9]. Overexpression of *glutaminesynthetase1* in tobacco and maize is associated with significant gains in plant heights, dry weights, and kernel numbers [10,11]. Overexpression of *NADH-glutamatesynthase* in rice and *alanine aminotransferase* in canola and rice also causes increases in grain weights [12] and biomass [13,14]. An early nodulin gene, *OsENOD93-1*, responds to both increases and reductions in N supplies. Furthermore, transgenic rice plants overexpressing *OsENOD93-1* have greater shoot dry biomass and seed yields [15].

Microarray analyses have been conducted to investigate genome-wide gene expression in response to changes in N conditions. Wang *et al.* [16] studied gene responses in *Arabidopsis* plants that were first grown for 10 d with ammonium as the sole N source, then treated with 250 mM nitrate for 20 min. That analysis identified 1,176 nitrate-responsive genes in the roots and 183 in the shoots. Peng *et al.* [17] monitored expression profiles from *Arabidopsis* plants grown under nitrate-limiting or -sufficient conditions. There, N-starvation altered transcript levels for 629 genes, with 340 being up-regulated and 289 down-regulated. Palenchar *et al.* [18] identified over 300 genes regulated by interactions between carbon and N signaling in *Arabidopsis*. Bi *et al.* [19] detected differential expression of genes under mild or severe chronic N stress. Plant responses were much more pronounced under severe conditions.

With 'Minghui 63' rice, Lian *et al.* [20] applied EST microarrays to examine expression profiles under low-N stress. In seedling roots, 473 responsive genes were identified, with 115 being up-regulated and 358 down-regulated. Beatty *et al.* [21] generated transgenic rice plants that overexpress alanine aminotransferase. Comparisons of transcriptomes between the transgenic plants and controls revealed that 0.11% and 0.07% of those genes were differentially regulated in the roots and shoots, respectively. Cai *et al.* [22] analyzed the dynamics of the rice transcriptome at 1 h, 24 h, and 7 d after N-starvation treatment. In all, 3,518 genes were identified, with most being transiently responsive to such stress.

Xu *et al.* [23] performed a genome-wide investigation to detect miRNAs that responded to either chronic or transient nitrate-limiting conditions in maize. They identified miRNAs showing overlapping or unique responses as well as those that were tissue-specific. Humbert *et al.* [24] reported that the concomitant presence of N and a water deficit affected expression much more than was anticipated in maize. This research group also revealed how the interaction between those two stresses shaped patterns of expression at different levels of water stress as well as during the recovery period. Finally, Brouillette and Donovan [25] identified five genes that had markedly different responses to nitrogen limitations in *Helianthus anomalus* when compared with *H. petiolaris* and *H. annuus*.

Although microarray analyses have been extensively used for the past few decades, RNA-Seq analysis can more precisely measure transcript levels and allow for the absolute quantification of gene expression. However, RNA-Seq has not previously been employed to investigate N-deficiency-induced genes. Here, we report transcriptome profiles for 1,650 N-starvation-responsive genes from rice for which expression was altered in the roots or shoots due to an N-limitation.

Results and discussion

RNA-Seq analysis of N-deficiency stress-responsive genes

Through microarray analyses, early-responsive genes have been detected in rice roots but not in leaves when sampled after 20 min, 1 h, and 2 h of N-starvation [20,22]. Cai *et al.* have monitored such genes after long-term (1- and 7-d) treatments with limited-N [22].

To identify additional responsive genes, we transferred rice seedlings at the six-leaf stage to an N-deficient hydroponic solution. Leaf sheaths and roots were harvested after 3 h, 6 h, 12 h, 1 d, and 2 d. We used two previously identified N-starvation-induced genes – *NRT2.3* and *AMT2.1* – to investigate induction kinetics (Figure 1). In both sheaths and roots, transcript levels were increased upon starvation, peaking at 12 h before declining to basal levels after 1 d. This trend was consistent with earlier reports [2,3,26]. Therefore, we selected the 12-h point for RNA-Seq analyses to distinguish between our results and those of studies that had investigated only very early- or late-responsive genes. Because expression of stress-responsive genes is mostly transient, we believed our data would be valuable for finding a new class of N-starvation-responsive genes. Leaf sheaths and roots were harvested from plants grown under deficient or sufficient conditions. RT-PCR analyses were used to determine the response of several N-metabolism genes, including *OsAMT1.1*, *OsAMT1.2*, *OsAMT2.1*, *OsAMT3.2*, *OsNAR2.2*, *OsNR*, *OsNRT2.2*, *OsNRT2.3*, *OsPEPC*, and *OsASN*. Significant changes in expression were revealed in the 12-h N-deficient samples (Figure 2).

We constructed eight cDNA libraries from two biological replicates of leaf sheaths and roots from plants grown under deficient or sufficient conditions. Sequencing those libraries resulted in the identification of 40,756,549 and 41,703,971 paired-end reads (202-

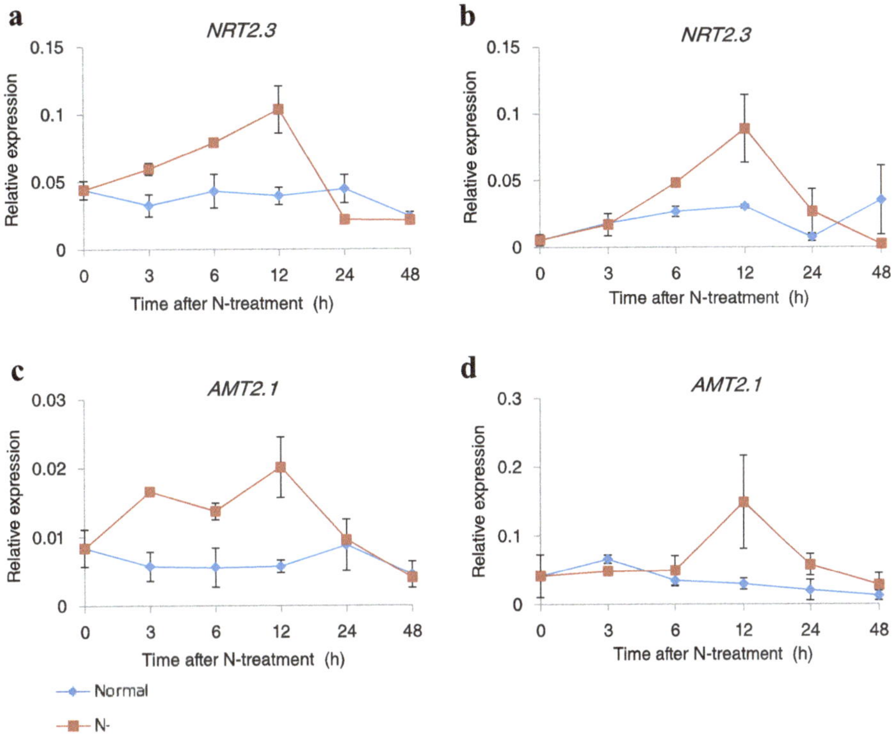

Figure 1 Induction kinetics of N-starvation-induced genes. Leaf sheaths **(a and c)** and roots **(b and d)** of rice seedlings at six-leaf stage were harvested at 3 h, 6 h, 12 h, 1 d, and 2 d after N-starvation and -sufficient treatments were applied. *NRT2.3* **(a and b)** and *AMT2.1* **(c and d)** were used to investigate induction kinetics.

nucleotide read length) from the sheaths and roots, respectively. The generated reads were then aligned to the rice genome (IRGSP/RAP build 5 data set) [27,28] by applying Bowtie [29] and TopHat2 programs [30]. In all, 86% of the reads from the sheaths and 69% from the roots were mapped to the reference genome, for which nearly 87% were correctly aligned and approximately 98% of them had unique locations in that genome (Table 1).

Transcript profiles of the RNA-Seq data were analyzed by calculating the reads per kilo base per million reads (RPKM). The sequenced RNA covered 33,782 annotated genes, accounting for 86.2% and 86.7% of those genes in the sheaths and roots, respectively. In addition, 2,986 novel transcripts were detected. Transcripts with low RPKM values were removed because they may not have been reliable due to low abundance or statistical faults. Among the 36,768 transcripts, 26,699 had RPKM ≥ 2. Of those, 22,992 were present in the leaf sheaths, 24,087 in the roots, and 18,319 in both. We identified 6,319 transcripts that were uniquely expressed in the sheaths (2,612) or roots (3,707).

Among the transcriptionally active transcripts, the top 500 most highly expressed were identified from the leaf sheath (Additional file 1) and roots (Additional file 2) under N-limited conditions. In both organ types, the most frequent transcripts functioned for protein synthesis, protein degradation, photosynthesis, stress responses, TFs, and DNA synthesis. Transcripts involved in lipid metabolism, transport, secondary metabolism, and amino acid metabolism were also common.

Differential expression of transcripts due to N-deficiency

Comparing transcript abundances revealed 1,650 transcripts that were differentially expressed (fold-change ≥ 2; $p \leq 0.05$) due to a deficient N supply (Additional files 3 and 4). Among them, 1,158 were differentially expressed in the leaf sheaths and 492 in the roots. Of those identified in the N-deficient sheaths, 548 transcripts were up-regulated and 610 transcripts were down-regulated. In the N-deficient roots, 276 transcripts were up-regulated and 216 were down-regulated. To gain insight into the effect of N status on transcript expression profiles, we illustrated expression patterns with a heat map obtained via hierarchical cluster analysis (Additional file 5). This clustering revealed the relatedness of the various transcripts.

Transcription factors are important for controlling the expression of other genes. Several TFs have been described in plants exposed to limited N. For example, an R2R3-type MYB TF, CmMYB1, is a central regulator of N-assimilation in *Cyanidioschyzon merolae* and

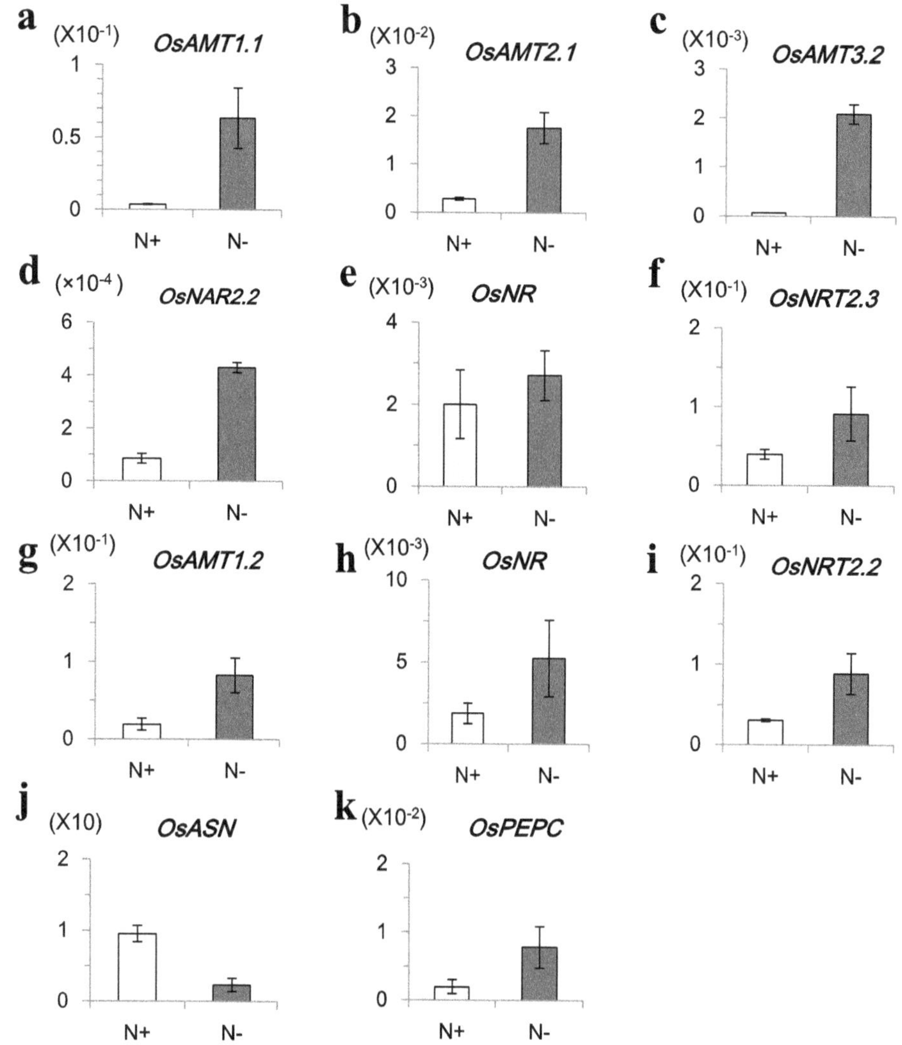

Figure 2 Analyses of N-metabolism genes by RT-PCR. (a-f) Transcript levels of *OsAMT1.1, OsAMT2.1, OsAMT3.2, OsNAR2.2, OsNR,* and *OsNRT2.3* were measured in leaf sheaths sampled from seedlings grown under N-sufficient (N+) or -deficient (N-) conditions. **(g-k)** Transcript levels of *OsAMT2.1, OsNR, OsNRT2.2, OsASN,* and *OsPEPC* were measured in roots sampled from seedlings grown under N+ or N- conditions. Levels were relative amounts against *OsUbi* expression.

enhances the expression of *CmNRT, CmNAR, CmNIR, CmAMT,* and *CmGS* under N-starvation [4]. A member of the *Arabidopsis* GATA TF gene family, At5g56860, is inducible by nitrate; loss-of-function mutants cause reduced chlorophyll levels and downregulation of the genes involved in carbon metabolism [2]. In *Arabidopsis*,

Table 1 Analysis of RNA-Seq data from rice seedlings

Category	Leaf sheaths	Roots
Total reads	40,756,549	41,703,971
Mapped reads[a]	34,863,681 (85.5%)	28,612,070 (68.6%)
Paired-end mapped reads[b]	29,808,452 (85.2%)	25,178,621 (88.1)
Uniquely mapped reads[c]	33,964,261 (97.4%)	28,057,411 (98.1%)

[a]Reads were aligned to the rice genome by Bowtie and TopHat2.
[b]Paired-end mapped reads.
[c]Reads were aligned to only one location in the genome.

ANR1 encodes a MADS box protein and is induced by nitrate. When expression of this gene is suppressed, lateral root proliferation is altered due to a reduction in sensitivity to NO_3^- [3]. Of the 1,650 transcripts that we found differentially expressed under an N-deficiency, 86 were identified as TFs, covering 28 families (Table 2; Additional file 6). This included one TF each from the GATA, Dof, and MADS families. The AP2/EREBP and WRKY TF families are the two largest families responsive to this deficiency. Here, six AP2/EREBP TF members were increased in the sheaths and seven in the roots under stress. Twelve WRKY members were induced in the sheaths versus none in the roots. It will be valuable in future investigations to determine whether these TFs also play a critical role in the N-starvation response and plant development.

Table 2 TFs differentially expressed in roots and leaf sheaths due to N-deficiency

TF Family	Leaf sheaths		Roots	
	Up	Down	Up	Down
ABI3/VP1		1		
AP2/EREBP	6	2	7	1
ARF		1	1	
AS2(LOB)			3	
bHLH	1	4		
bZIP	1			
C2H2	4	3	1	
C2C2-CO-like			1	
C2C2-GATA		1		
CPP		1		
DBB (Orphans)	1			
Dof		1		
E2F/DP		1		
FAR1		1		
GARP-G2-like		2	1	
GRAS		1		
GRF		2		
HB	2	1		
HSF	2	1		
LSD			1	
MADS	1			
MYB	3	1	3	
NAC	3		1	2
SBP		1		
TCP		1		
Trihelix			1	
WRKY	12			
ZF-HD	1			
Total	37	26	20	3

We classified the 1,650 differentially expressed genes into 54 functional groups by GO analysis (Figure 3). The dominant terms were 'cell part' (GO:0044464) in Cellular Component, 'binding' (GO:0005488) in Molecular Function, and 'cellular process' (GO:0009987) in Biological Process. In the third category, more than 30% of the genes for 'metabolic process' (GO:0008152), 'response to stimulus' (GO:0050896), and 'biological regulation' (GO:0065007) responded to N-starvation. 'Cellular process' accounted for 72.4% and 58.3% of the starvation-related genes in the leaf sheath and root, respectively. 'Metabolic process' genes made up 70.0% and 53% in the sheath and root, respectively; while those proportions were 46.7% (sheath) and 41.1% (root) for 'response to stimulus' and 45.7% (sheath) and 36.3% (root) for 'biological regulation'.

Confirmation by real-time PCR

Our RNA-Seq data appeared to be quite reliable for genes up-regulated by N-starvation, with 34 of the 36 deficiency-responsive genes first identified via RNA-Seq analyses subsequently being confirmed by qRT-PCR (Table 3). Only two could not be verified in that latter examination. By contrast, the identification of down-regulated genes by RNA-Seq was less reliable. Among 12 examined, eight were later confirmed through qRT-PCR (Table 4).

Validation by GUS assays

We used GUS assays of T-DNA gene trap lines to confirm the N-starvation-responsive TF genes. Those tagging lines were previously generated to have a translational fusion between the tagged gene and *GUS* [31]. Five *GUS*-positive lines displayed N-responsive GUS activity. Although this activity was weak when plants were grown in a standard N-sufficient medium, it was rapidly induced by N-starvation. Under low-N conditions, four lines (3A-60813, 3A-51694, 4A-02639, and 4A-01614) showed preferential GUS-staining in the sheaths (vascular bundles) while one (1B-11001) showed staining in the roots (vascular cylinder) (Figure 4, Table 5). In all five lines, GUS activity was higher for plants in the low-N medium than in the normal MS medium. The Os01g14440 and Os11g02480 genes encode a WRKY TF, Os12g07640 encodes a MYB TF, Os03g55220 encodes a bHelix-loop-helix TF, and Os02g43300 encodes the trihelix TF GTL1. Although the T-DNA vectors carry an intron with triple splicing donors/acceptors at the right border, only one pair of donors and acceptors is utilized that reduces the frequency of translational fusion between the tagged gene and *GUS* [32]. Nonetheless, the *GUS*-trapped TF lines are valuable for investigating their roles during N-starvation.

Analysis of alternative splicing

Alternative splicing (AS) is an important regulatory mechanism common in higher eukaryotes that results in a single gene coding for multiple proteins, thereby enhancing biological diversity [33]. Its products are efficiently identified using high-throughput sequencing techniques [34,35]. To investigate potential splicing junctions, we performed computational analyses that revealed 8,509 multi-exonic genes with 19,628 AS events (Figure 5). These events were categorized into six common types. 'Intron retention' was the dominant type (42.8%), which is consistent with previous observations from plants [36,37]. By contrast, 'exon skipping' is the

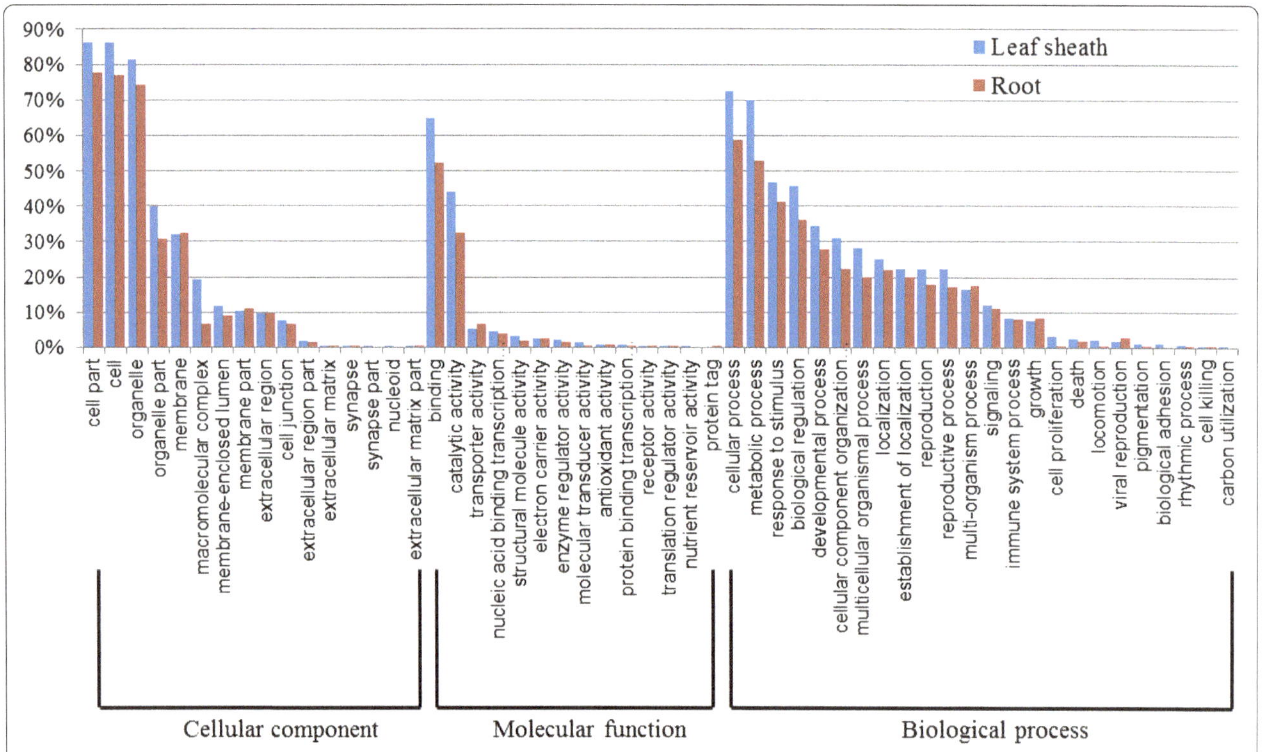

Figure 3 GO annotation clusters of differentially expressed genes. Gene Ontology functional enrichment analysis of differentially expressed genes in leaf sheaths and roots. Based on sequence homology, 1,650 genes were distributed among 3 main categories: Cellular Component (16 functional groups, dominated by 'cell part'), Molecular Function (14 groups, dominated by 'binding'), and Biological Process (24 groups, dominated by 'cellular process').

most prevalent mechanism in humans and yeast [38,39]. Here, 'alternative 3' site', 'exon skipping', and 'alternative first exon' accounted for 16.1%, 14.0%, and 13.2%, respectively, of all events. Frequencies were relatively low for 'alternative 5' site' (7.1%) and 'alternative last exon' (6.9%) (Figure 5a). These data were consistent with other recent reports for plants [36,37,40,41]. An example of the transcript isoforms is shown in Figure 5b.

Alternative splicing can occur because of environmental factors. For example, expression of *Wdreb2* is activated by cold, drought, salt, or exogenous ABA treatment; depending upon the source of the stress, three transcript forms may be produced [42]. However, we found no significant difference in AS between N-sufficient and -deficient conditions, which suggests that it is not involved in the low-N stress response.

Novel transcribed regions (NTRs) validated by RT-PCR

RNA-Seq technology has revealed novel transcripts that could not be identified previously [43]. Our RNA-Seq data contained 2,986 novel transcribed regions, of which 192 were regulated under the N-deficiency. To confirm their existence, we conducted semi-quantitative reverse transcription PCR with 13 NTRs (Figure 6). Among them, 10 (77%) were detected in the leaf sheath and 7 of those 10 (70%) showed expression patterns consistent with the sequencing data. The three exceptions, with inconsistent patterns, were NTR-1489, NTR-2195, and NTR-2240.

Conclusion

We performed deep transcriptomic investigations with rice plants and obtained detailed expression profiles for genes involved in responses to low-N stress. These data provide valuable information about the genes (1650 transcripts) induced by N-starvation, expecially the 86 TFs that are key regulators of growth and development. We then confirmed these RNA-Seq data by conducting qRT-PCR and GUS assays of T-DNA tagging lines. In all, 8,509 multi-exonic genes could be linked with 19,628 AS events. However, we found no significant difference in alternative splicing between N-deficient samples and controls. Our data will be useful for identifying N-deficiency-induced genes and investigating the signal transduction pathway of N-utilization.

Methods

Plant materials and growth conditions

Oryza sativa L. ssp. *japonica* cv. Dongjin rice was used in all experiments. Seeds were surface-sterilized and

Table 3 RNA-Seq results from leaf sheaths confirmed by real-time PCR

Locus ID	RNA-Seq log_2 (N-/N+)	RT-PCR log_2 (N-/N+)	Putative function
LOC_Os03g32230	4.35	2.25	Zinc finger protein
LOC_Os04g52090	2.32	1.56	Ethylene-responsive transcription factor 4
LOC_Os03g62200	2.26	2.66	Ammonium transporter protein
LOC_Os04g40410	2.25	4.89	High affinity nitrate transporter
LOC_Os02g08440	2.20	2.23	WRKY71
LOC_Os02g43790	1.89	1.27	Ethylene-responsive transcription factor 1A
LOC_Os03g55540	1.72	2.66	ZOS3-18 - C2H2 zinc finger protein
LOC_Os03g60560	1.67	1.08	Zinc finger protein ZAT12
LOC_Os03g60080	1.58	2.68	NAC domain-containing protein 67
LOC_Os01g50820	1.48	1.21	Transporter, major facilitator family
LOC_Os12g02440	1.47	1.89	WRKY transcription factor 46
LOC_Os12g07640	1.47	1.97	Myb-related protein
LOC_Os11g03540	1.46	1.63	Ethylene-responsive transcription factor
LOC_Os05g03900	1.43	1.85	WRKY109
LOC_Os06g41100	1.32	2.32	bZip, transcription factor
LOC_Os07g12340	1.32	1.38	NAC domain-containing protein 67
LOC_Os02g43330	1.30	1.43	Homeobox-associated leucine zipper
LOC_Os02g53130	1.26	2.12	Nitrate reductase
LOC_Os01g53220	1.26	−1.40	Heat stress transcription
LOC_Os02g43170	1.20	3.19	B-box zinc finger family protein
LOC_Os04g50770	1.20	2.21	MYB-related protein Zm1
LOC_Os04g45810	1.13	2.45	Homeobox-leucine zipper protein HOX22
LOC_Os02g49840	1.08	3.32	MADS-box transcription factor 57
LOC_Os01g54600	1.07	2.87	WRKY13
LOC_Os09g28354	1.07	1.07	Heat stress transcription factor B-1
LOC_Os04g43070	1.02	1.72	Ammonium transporter protein
LOC_Os03g42200	−2.00	−1.02	Dof zinc finger domain containing protein
LOC_Os03g08620	−1.60	−2.38	B3 DNA binding domain-containing protein
LOC_Os04g44440	−1.59	0.88	TCP family transcription factor
LOC_Os01g74540	−1.56	1.00	GATA zinc finger domain-containing protein
LOC_Os05g34110	−1.42	0.97	Homeodomain-related
LOC_Os10g42490	−1.26	−2.03	Homeobox and START domain-containing proteins
LOC_Os02g39140	−1.06	1.01	Helix-loop-helix DNA-binding domain containing protein
LOC_Os06g43220	−1.04	−2.12	AP2 domain-containing protein

Expression ratios for RNA-Seq data were calculated with the DESeq program. All ratios are presented as Log_2N-deficient/N-sufficient. Negative values indicate that expression was reduced under N-starvation.

germinated for two weeks in a Murashige and Skoog medium that lacked a nitrogen source. The seedlings were further grown in an N-sufficient nutrient solution at 28°C/25°C (day/night) under a 14-h photoperiod and 50 to 55% relative humidity. This hydroponic solution, refreshed every 3 d, contained 1.44 mM NH_4NO_3, 0.3 mM NaH_2PO_4, 0.5 mM K_2SO_4, 1.0 mM $CaCl_2$, 1.6 mM $MgSO_4$, 0.075 μM $(NH_4)_6Mo_7O_{24}$, 18.8 μM H_3BO_3, 9.5 μM $MnCl_2$, 0.16 μM $CuSO_4$, 0.15 μM $ZnSO_4$, 35.6 μM $FeCl_3$, and 74.4 μM citric acid (pH 5.0) [44]. At the six-leaf stage, the seedlings were divided into two groups: 1) N-starvation, with the amount of NH_4NO_3 in the solution reduced to 0.072 mM; and 2) N-sufficient, for which the nutrient solution contained the normal N concentration of 1.44 mM. At 12 h after the treatment began, the total roots and leaf sheaths were harvested from plants in both groups. Each biological replicate constituted a pool of three plants. Two of those replicates were subjected to RNA-sequencing.

Table 4 RNA-Seq results from roots confirmed by real-time PCR

Locus ID	RNA-Seq log_2(N-/N+)	RT-PCR log_2(N-/N+)	Putative function
LOC_Os02g02190	3.51	1.55	Transporter, major facilitator family
LOC_Os02g53130	2.57	2.97	Nitrate reductase
LOC_Os06g10780	2.01	3.76	Ethylene-responsive transcription factor ERF014
LOC_Os06g01480	1.66	1.91	NAC domain-containing protein 7
LOC_Os01g50720	1.41	2.23	MYB-related protein Hv33
LOC_Os02g57490	1.41	2.07	LOB domain-containing protein 16
LOC_Os02g41580	1.14	1.97	Phosphoenolpyruvate carboxylase kinase
LOC_Os10g07510	1.08	0.67	LOB domain-containing protein 18
LOC_Os03g32230	1.05	1.76	Zinc finger protein 1
LOC_Os02g40710	1.04	2.12	Ammonium transporter protein
LOC_Os01g64790	−4.97	−2.78	Ethylene-responsive transcription factor ERF110
LOC_Os03g18130	−1.73	−2.02	Asparagine synthetase
LOC_Os08g10080	−1.53	−1.01	NAC domain-containing protein 21/22
LOC_Os11g08210	−1.06	−1.84	NAC domain-containing protein 71

Expression ratios for RNA-Seq data were calculated with the DESeq program. All ratios are presented as Log_2N-deficient/N-sufficient. Negative values indicate that expression was reduced under N-starvation.

RNA extraction, preparation of cDNA library, and sequencing

Total RNA was prepared using RNAiso Reagent (Takara Bio Inc., Otsu, Japan). Quality was checked with the Agilent 2100 Bioanalyzer (Agilent Technologies, Santa Clara, CA, USA). Total RNA (30 μg) was used for synthesizing complementary DNA (cDNA). After the libraries were constructed, the cDNA was sequenced with the Illumina HiSeq™ 2000 according to the manufacturer's recommendations (http://www.illumina.com).

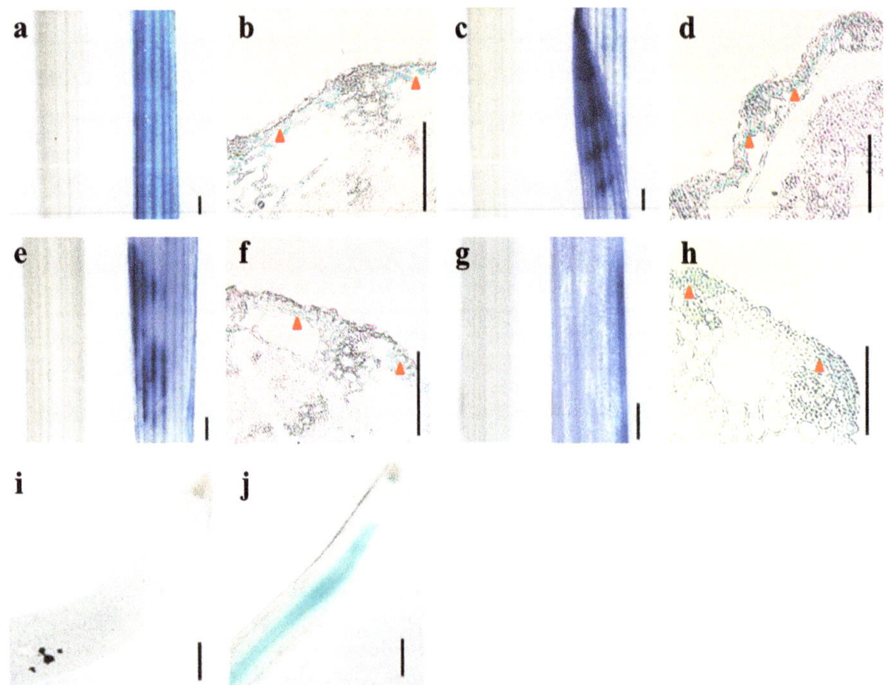

Figure 4 Patterns of *GUS* expression in 5 DAG seedlings. (A-H) Preferential expression of genes in leaf sheaths from Lines 3A-60813 **(a and b)**, 3A-51694 **(c and d)**, 4A-02639 **(e and f)**, and 4A-01614 **(g and h)**. **(a, c, e, and g)** Seedlings were grown under N-sufficient conditions (left) or N-deficient conditions (right). **(b, d, f, and h)** Cross section of leaf sheath under N-starvation. **(i and j)** Preferential expression in vascular cylinders of roots from Line 1B-11001. **(i)** N-sufficient conditions. **(j)** N-deficient conditions. Bar = 200 μm.

Table 5 Confirmation of RNA-Seq expression patterns by GUS assays

Line no.	Locus ID	Putative function	RNA-Seq log_2(N-/N+) (tissue)
3A-60813	01 g14440	WRKY1, expressed	1.14 (leaf sheath)
3A-51694	11 g02480	WRKY46, expressed	1.30 (leaf sheath)
4A-02639	12 g07640	MYB family transcription factor, putative, expressed	1.54 (leaf sheath)
4A-01614	03 g55220	bHelix-loop-helix transcription factor	1.38 (leaf sheath)
1B-11001	02 g43300	Trihelix transcription factor GTL1	1.13 (root)

Read alignment and assembly

RNA-Seq reads were aligned to the rice reference genomes by the TopHat2 program [30]. That program analyzes the RNA sequences to identify splice junctions between exons by using the ultra-high-throughput short-read aligner Bowtie [29]. Each read was mapped with Cufflinks, which assembled the alignments within the Sequence Alignment/Map file into transfrags [45]. The assembly files were then merged with reference transcriptome annotations into a unified annotation for further analysis [46].

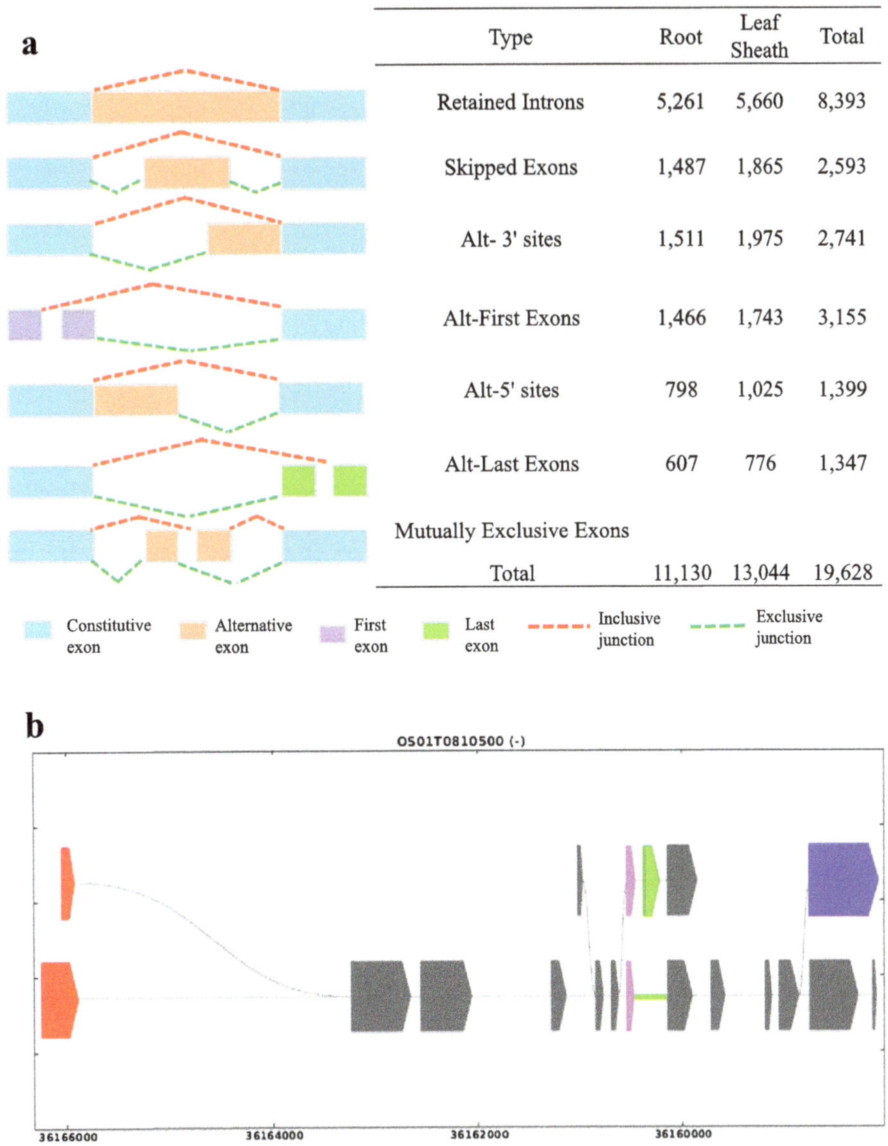

Figure 5 Analysis of alternative splicing. (a) Left, 6 types of AS events; right, total numbers of events, including annotated and newly identified splicings. **(b)** Example of NTR (Chromosome 1: 34,427,712-34,436,315). Four types of AS events are indicated. Blue frame, 'intron retention'; red, 'alternative first exon'; pink, '5′ site'; and green, 'skipped exon'.

NTR ID	RNA-seq $\log_2$(N-/N+)	N+ N-
NTR-2318	1.86	
NTR-1469	1.84	
NTR-1489	-1.41	
NTR-2139	2.98	
NTR-1840	2.4	
NTR-2195	-4.14	
NTR-1600	2.92	
NTR-1526	-2.17	
NTR-2385	1.26	
NTR-2240	1.16	
OsUbi	-0.15	

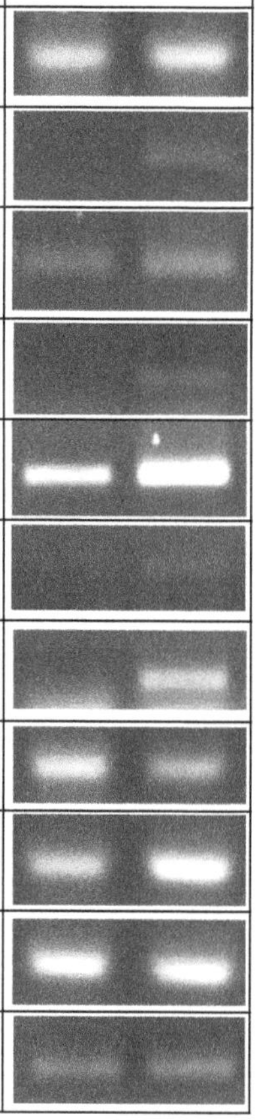

Figure 6 NTRs validated by RT-PCR. Semi-quantitative RT-PCR was performed to confirm existence of NTRs preliminarily identified from RNA-Seq analysis. Among 13 NTRs, 10 (77%) were detected in leaf sheath; 7 of those 10 (70%) showed expression patterns consistent with sequencing data.

Expression levels for each gene were calculated by quantifying the Illumina reads according to the RPKM method [47]. Replicates were examined independently for statistical analysis. Genes that were differentially expressed by at least two-fold were tested for False Discovery Rate correlations at p-values ≤ 0.05 [48]. We also selected any transcripts with RPKM ≥ 2 in at least one cDNA library. Heat maps illustrating patterns for differentially expressed genes were generated as described by Severin *et al.* [49].

Gene Ontology (GO) term analysis and discovery of alternatively spliced exons

Gene Ontology terms were examined by applying tools for GO enrichment (http://amigo.geneontology.org/cgi-bin/amigo/term_enrichment [50]) and Blast2GO [51], at p-values ≤ 0.05. Six basic modes of AS were identified by Cufflinks software, in which differentially spliced exons were detected by comparing pairs of gene models annotated to the same locus [46].

Identification of novel transcripts

Paired-end reads were mapped to the genome with a spliced-read mapper. Afterward, the reference annotations were used to generate faux-read alignments that covered the transcripts. Those alignments were used together with the spliced-read alignments to produce a reference genome-based assembly. Finally, this assembly was merged with the reference annotations and "noisy" read mappings were filtered, resulting in all reference annotation transcripts in the output as well as novel transcripts [52].

Real-time RT-PCR

Total RNA was isolated from seedling leaf sheaths and roots, using RNAiso Reagent. For first-strand cDNA synthesis, 1 μg of total RNA was reverse-transcribed in a total volume of 25 μL that contained 10 ng of oligo(dT) 12–18 primer, 2.5 mM dNTPs, and 200 units of AMV Reverse Transcriptase (Promega, Madison, WI, USA) in a reaction buffer. The samples were diluted 10 times prior to PCR. Gene-specific primers were designed using the Oligonucleotide Properties Calculator, or OligoCalc (http://basic.northwestern.edu/biotools/OligoCalc.html). Real-time PCR was performed with 3 μL of template cDNA, 1 μL of forward primer (5 pmol), 1 μL of reverse primer (5 pmol), and 5 μL of SYBR Green mix (Qiagen, Hilden, Germany). Conditions included 5 min of pre-denaturation at 95°C, then 45 cycles of 10 s at 95°C and 20 s at 60°C, followed by steps for dissociation curve generation (15 s at 95°C, 60 s at 60°C, and 15 s at 95°C). To examine the expression of novel transcripts, we performed semi-quantitative RT-PCR with *OsUbiquitin* as the internal reference to equalize the quantity of RNA. After 28 cycles of amplification, PCR products were resolved on a 2% agarose gel and stained with ethidium bromide. All primers are listed in Additional file 7.

GUS assays

Histochemical GUS-staining was performed according to the method of Jeon *et al.* [53]. Five-d-old seedlings were cut into approximately 1-cm pieces and submerged in a staining solution containing 0.5 M Na_2HPO_4 (pH 7.0), 0.5 M NaH_2PO_4 (pH 7.0), 0.1% TritonX-100, 0.5 M EDTA (pH 8.0), 1% DMSO, 0.1% X-gluc (5-bromo-4-chloro-3-indolyl-β-d-glucuronic acid/cyclohexylammonium salt), 1 mM $K_3[Fe(CN)_6]$, 1 mM $K_4[Fe(CN)_6]$, and 5% methanol. The samples were then incubated at 37°C for 12 h. Afterward, the staining solution was replaced with 70% (w/v) ethanol at 65°C to remove the chlorophyll.

Availability of supporting data

Illumina sequence data are available from NCBI under Short Read Archive accession SRP045923.

Additional files

Additional file 1: Table S1. The 500 most highly expressed transcripts in leaf sheaths under N-starvation.

Additional file 2: Table S2. The 500 most highly expressed transcripts in roots under N-starvation.

Additional file 3: Table S3. Leaf sheath transcripts differentially expressed between N-deficient and -sufficient conditions.

Additional file 4: Table S4. Root transcripts differentially expressed between N-deficient and -sufficient conditions.

Additional file 5: Figure S1. Hierarchical cluster analysis of differentially expressed transcripts due to N-deficiency. Heat map illustrates profiles of 1,650 transcripts differentially expressed due to N-starvation. Red, high expression; blue, low expression. Values on color scale (0 to 10) represent $\log_2$ (RPKM + 1) for each gene.

Additional file 6: Table S5. Transcription factors differentially expressed due to N-starvation.

Additional file 7: Table S6. Primer sequences for RT-PCR analyses.

Abbreviations

AS: Alternative splicing; cDNA: Complementary DNA; GO: Gene ontology; N: Nitrogen; NR: Nitrate reductase; NRT: Nitrate transporter; NTR: Novel transcribed region; PCR: Polymerase chain reaction; qRT-PCR: Quantitative reverse transcription polymerase chain reaction; RNA-Seq: RNA sequencing; RPKM: Reads per kilobase per million reads; TF: Transcription factor.

Competing interests

The authors declare that they have no competing interests.

Authors' contributions

Conceived and designed the experiments: YF, RC, GA. Performed the experiments: WY, JM, HC. Analyzed the data: WY, JM, HC. Contributed reagents/materials/analysis tools: WY, JM, HC, GA. Wrote the paper: WY, JM, GA. Revised the paper: WY, JM, GA. All authors read and approved the final manuscript.

Acknowledgements

We thank Ki-Hong Jung, Su-Zhen Li, Xiao-Jin Zhou, and Qiu-Xue Zhang for their valuable discussions. We also thank Kyungsook An for generating the transgenic lines and handling the seed stock, and Priscilla Licht for editing the English composition of the article. This work was supported in part by grants from the Next-Generation BioGreen 21 Program (No. PJ01108001); the Basic Research Promotion Fund, Republic of Korea (NRF-2007-0093862); and Kyung Hee University (20120227) to G. An.

Author details

[1]Department of Plant Molecular Systems Biotechnology and Crop Biotech Institute, Kyung Hee University, Yongin 446-701, Korea. [2]Department of Crop Genomics and Genetic Improvement, Biotechnology Research Institute, Chinese Academy of Agricultural Sciences, Beijing 100081, China. [3]National Key Facility for Crop Gene Resources and Genetic Improvement (NFCRI), Biotechnology Research Institute, Chinese Academy of Agricultural Sciences, Beijing 100081, China.

References

1. Socolow RH. Nitrogen management and the future of food: lessons from the management of energy and carbon. Proc Natl Acad Sci. 1999;96(11):6001–8.
2. Bi YM, Zhang Y, Signorelli T, Zhao R, Zhu T, Rothstein S. Genetic analysis of *Arabidopsis* GATA transcription factor gene family reveals a nitrate-inducible member important for chlorophyll synthesis and glucose sensitivity. Plant J. 2005;44(4):680–92.
3. Zhang H. An *Arabidopsis* MADS box gene that controls nutrient-induced changes in root architecture. Science. 1998;279(5349):407–9.
4. Imamura S, Kanesaki Y, Ohnuma M, Inouye T, Sekine Y, Fujiwara T, et al. R2R3-type MYB transcription factor, CmMYB1, is a central nitrogen assimilation regulator in *Cyanidioschyzon merolae*. Proc Natl Acad Sci. 2009;106(30):12548–53.
5. Peng M, Hudson D, Schofield A, Tsao R, Yang R, Gu H, et al. Adaptation of *Arabidopsis* to nitrogen limitation involves induction of anthocyanin synthesis which is controlled by the *NLA* gene. J Exp Bot. 2008;59(11):2933–44.
6. Watanabe CK, Hachiya T, Takahara K, Kawai-Yamada M, Uchimiya H, Uesono Y, et al. Effects of AOX1a deficiency on plant growth, gene expression of respiratory components and metabolic profile under low-nitrogen stress in *Arabidopsis thaliana*. Plant Cell Physiol. 2010;51(5):810–22.
7. Remans T, Nacry P, Pervent M, Girin T, Tillard P, Lepetit M, et al. A central role for the nitrate transporter NRT2.1 in the integrated morphological and physiological responses of the root system to nitrogen limitation in *Arabidopsis*. Plant Physiol. 2006;140(3):909–21.
8. Engineer CB, Kranz RG. Reciprocal leaf and root expression of AtAmt1.1 and root architectural changes in response to nitrogen starvation. Plant Physiol. 2007;143(1):236–50.
9. Yanagisawa S, Akiyama A, Kisaka H, Uchimiya H, Miwa T. Metabolic engineering with Dof1 transcription factor in plants: improved nitrogen assimilation and growth under low-nitrogen conditions. Proc Natl Acad Sci. 2004;101(20):7833–8.
10. Fuentes SI, Allen DJ, Ortiz-Lopez A, Hernandez G. Over-expression of cytosolic glutamine synthetase increases photosynthesis and growth at low nitrogen concentrations. J Exp Bot. 2001;52(358):1071–81.
11. Martin A, Lee J, Kichey T, Gerentes D, Zivy M, Tatout C, et al. Two cytosolic glutamine synthetase isoforms of maize are specifically involved in the control of grain production. Plant Cell. 2006;18(11):3252–74.
12. Yamaya T, Obara M, Nakajima H, Sasaki S, Hayakawa T, Sato T. Genetic manipulation and quantitative-trait loci mapping for nitrogen recycling in rice. J Exp Bot. 2002;53(370):917–25.
13. Good AG, Johnson SJ, De Pauw M, Carroll RT, Savidov N, Vidmar J, et al. Engineering nitrogen use efficiency with alanine aminotransferase. Can J Bot. 2007;85:252–62.
14. Shrawat AK, Carroll RT, DePauw M, Taylor GJ, Good AG. Genetic engineering of improved nitrogen use efficiency in rice by the tissue-specific expression of alanine aminotransferase. Plant Biotechnol J. 2008;6(7):722–32.
15. Bi YM, Kant S, Clarke J, Gidda S, Ming F, Xu J, et al. Increased nitrogen-use efficiency in transgenic rice plants over-expressing a nitrogen-responsive early nodulin gene identified from rice expression profiling. Plant Cell Environ. 2009;32(12):1749–60.
16. Wang R, Okamoto M, Xing X, Crawford NM. Microarray analysis of the nitrate response in *Arabidopsis* roots and shoots reveals over 1,000 rapidly responding genes and new linkages to glucose, trehalose-6-phosphate, iron, and sulfate metabolism. Plant Physiol. 2003;132(2):556–67.
17. Peng M, Bi YM, Zhu T, Rothstein SJ. Genome-wide analysis of *Arabidopsis* responsive transcriptome to nitrogen limitation and its regulation by the ubiquitin ligase gene *NLA*. Plant Mol Biol. 2007;65(6):775–97.
18. Palenchar PM, Kouranov A, Lejay LV, Coruzzi GM. Genome-wide patterns of carbon and nitrogen regulation of gene expression validate the combined carbon and nitrogen (CN)-signaling hypothesis in plants. Genome Biol. 2004;5(11):R91.
19. Bi Y-M, Wang R-L, Zhu T, Rothstein SJ. Global transcription profiling reveals differential responses to chronic nitrogen stress and putative nitrogen regulatory components in *Arabidopsis*. BMC Genom. 2007;8(1):281.
20. Lian X, Wang S, Zhang J, Feng Q, Zhang L, Fan D, et al. Expression profiles of 10,422 genes at early stage of low nitrogen stress in rice assayed using a cDNA microarray. Plant Mol Biol. 2006;60(5):617–31.
21. Beatty PH, Shrawat AK, Carroll RT, Zhu T, Good AG. Transcriptome analysis of nitrogen-efficient rice over-expressing alanine aminotransferase. Plant Biotechnol J. 2009;7(6):562–76.
22. Cai H, Lu Y, Xie W, Zhu T, Lian X. Transcriptome response to nitrogen starvation in rice. J Biosci. 2012;37(4):731–47.
23. Xu Z, Zhong S, Li X, Li W, Rothstein SJ, Zhang S, et al. Genome-wide identification of microRNAs in response to low nitrate availability in maize leaves and roots. PLoS One. 2011;6(11):e28009.

24. Humbert S, Subedi S, Cohn J, Zeng B, Bi YM, Chen X, et al. Genome-wide expression profiling of maize in response to individual and combined water and nitrogen stresses. BMC Genom. 2013;14:3.
25. Brouillette LC, Donovan LA. Nitrogen stress response of a hybrid species: a gene expression study. Ann Bot. 2011;107(1):101–8.
26. Suenaga A, Moriya K, Sonoda Y, Ikeda A, von Wirén N, Hayakawa T, et al. Constitutive expression of a novel-type ammonium transporter *OsAMT2* in rice plants. Plant Cell Physiol. 2003;44(2):206–11.
27. Sakai H, Lee SS, Tanaka T, Numa H, Kim J, Kawahara Y, et al. Rice Annotation Project Database (RAP-DB): an integrative and interactive database for rice genomics. Plant Cell Physiol. 2013;54(2):e6.
28. Kawahara Y, de la Bastide M, Hamilton JP, Kanamori H, McCombie WR, Ouyang S, et al. Improvement of the *Oryza sativa* Nipponbare reference genome using next generation sequence and optical map data. Rice. 2013;6(1):4.
29. Langmead B, Trapnell C, Pop M, Salzberg SL. Ultrafast and memory-efficient alignment of short DNA sequences to the human genome. Genome Biol. 2009;10(3):R25.
30. Kim D, Pertea G, Trapnell C, Pimentel H, Kelley R, Salzberg SL. TopHat2: accurate alignment of transcriptomes in the presence of insertions, deletions and gene fusions. Genome Biol. 2013;14(4):R36.
31. Jeong DH, An S, Kang HG, Moon S, Han JJ, Park S, et al. T-DNA insertional mutagenesis for activation tagging in rice. Plant Physiol. 2002;130(4):1636–44.
32. Kim SL, Choi M, Jung KH, An G. Analysis of the early-flowering mechanisms and generation of T-DNA tagging lines in Kitaake, a model rice cultivar. J Exp Bot. 2013;64(14):4169–82.
33. Black DL. Mechanisms of alternative pre-messenger RNA splicing. Annu Rev Biochem. 2003;72:291–336.
34. David CJ, Manley JL. The search for alternative splicing regulators: new approaches offer a path to a splicing code. Genes Dev. 2008;22(3):279–85.
35. Matlin AJ, Clark F, Smith CW. Understanding alternative sp licing: towards a cellular code. Nat Rev Mol Cell Biol. 2005;6(5):386–98.
36. Ner-Gaon H, Halachmi R, Savaldi-Goldstein S, Rubin E, Ophir R, Fluhr R. Intron retention is a major phenomenon in alternative splicing in *Arabidopsis*. Plant J. 2004;39(6):877–85.
37. Wang BB, Brendel V. Genomewide comparative analysis of alternative splicing in plants. Proc Natl Acad Sci. 2006;103(18):7175–80.
38. Sultan M, Schulz MH, Richard H, Magen A, Klingenhoff A, Scherf M, et al. A global view of gene activity and alternative splicing by deep sequencing of the human transcriptome. Science. 2008;321(5891):956–60.
39. Wang ET, Sandberg R, Luo S, Khrebtukova I, Zhang L, Mayr C, et al. Alternative isoform regulation in human tissue transcriptomes. Nature. 2008;456(7221):470–6.
40. Marquez Y, Brown JW, Simpson C, Barta A, Kalyna M. Transcriptome survey reveals increased complexity of the alternative splicing landscape in *Arabidopsis*. Genome Res. 2012;22(6):1184–95.
41. Zhang G, Guo G, Hu X, Zhang Y, Li Q, Li R, et al. Deep RNA sequencing at single base-pair resolution reveals high complexity of the rice transcriptome. Genome Res. 2010;20(5):646–54.
42. Egawa C, Kobayashi F, Ishibashi M, Nakamura T, Nakamura C, Takumi S. Differential regulation of transcript accumulation and alternative splicing of a *DREB2* homolog under abiotic stress conditions in common wheat. Genes Genet Syst. 2006;81(2):77–91.
43. Wang Z, Gerstein M, Snyder M. RNA-Seq: a revolutionary tool for transcriptomics. Nat Rev Genet. 2009;10(1):57–63.
44. Yoshida S, Forno DA, Cock JH, Gomez KA. Laboratory manual for physiological studies of rice. In: The International Rice Research Institute. 3rd ed. 1976. p. 61–6.
45. Trapnell C, Williams BA, Pertea G, Mortazavi A, Kwan G, van Baren MJ, et al. Transcript assembly and quantification by RNA-Seq reveals unannotated transcripts and isoform switching during cell differentiation. Nat Biotechnol. 2010;28(5):511–5.
46. Trapnell C, Roberts A, Goff L, Pertea G, Kim D, Kelley DR, et al. Differential gene and transcript expression analysis of RNA-seq experiments with TopHat and Cufflinks. Nat Protoc. 2012;7(3):562–78.
47. Mortazavi A, Williams BA, McCue K, Schaeffer L, Wold B. Mapping and quantifying mammalian transcriptomes by RNA-Seq. Nat Meth. 2008;5(7):621–8.
48. Anders S, Huber W. Differential expression of RNA-Seq data at the gene level-the DESeq package. Eur Mol Biol Lab. 2013.
49. Severin AJ, Woody JL, Bolon YT, Joseph B, Diers BW, Farmer AD, et al. RNA-Seq atlas of *Glycine max*: a guide to the soybean transcriptome. BMC Plant Biol. 2010;10:160.
50. Carbon S, Ireland A, Mungall CJ, Shu S, Marshall B, Lewis S, et al. AmiGO: online access to ontology and annotation data. Bioinformatics. 2009;25(2):288–9.
51. Conesa A, Gotz S. Blast2GO: a comprehensive suite for functional analysis in plant genomics. Intl J Plant Genom. 2008;2008:619832.
52. Roberts A, Pimentel H, Trapnell C, Pachter L. Identification of novel transcripts in annotated genomes using RNA-Seq. Bioinformatics. 2011;27(17):2325–9.
53. Jeon JS, Lee S, Jung KH, Jun SH, Jeong DH, Lee J, et al. T-DNA insertional mutagenesis for functional genomics in rice. Plant J. 2000;22(6):561–70.

8

Genome-wide SNP identification and QTL mapping for black rot resistance in cabbage

Jonghoon Lee[1], Nur Kholilatul Izzah[1,2], Murukarthick Jayakodi[1], Sampath Perumal[1], Ho Jun Joh[1], Hyeon Ju Lee[1], Sang-Choon Lee[1], Jee Young Park[1], Ki-Woung Yang[1,3], Il-Sup Nou[3], Joodeok Seo[4], Jaeheung Yoo[4], Youngdeok Suh[4], Kyounggu Ahn[4], Ji Hyun Lee[5], Gyung Ja Choi[5], Yeisoo Yu[6], Heebal Kim[7,8] and Tae-Jin Yang[1*]

Abstract

Background: Black rot is a destructive bacterial disease causing large yield and quality losses in *Brassica oleracea*. To detect quantitative trait loci (QTL) for black rot resistance, we performed whole-genome resequencing of two cabbage parental lines and genome-wide SNP identification using the recently published *B. oleracea* genome sequences as reference.

Results: Approximately 11.5 Gb of sequencing data was produced from each parental line. Reference genome-guided mapping and SNP calling revealed 674,521 SNPs between the two cabbage lines, with an average of one SNP per 662.5 bp. Among 167 dCAPS markers derived from candidate SNPs, 117 (70.1%) were validated as bona fide SNPs showing polymorphism between the parental lines. We then improved the resolution of a previous genetic map by adding 103 markers including 87 SNP-based dCAPS markers. The new map composed of 368 markers and covers 1467.3 cM with an average interval of 3.88 cM between adjacent markers. We evaluated black rot resistance in the mapping population in three independent inoculation tests using $F_{2:3}$ progenies and identified one major QTL and three minor QTLs.

Conclusion: We report successful utilization of whole-genome resequencing for large-scale SNP identification and development of molecular markers for genetic map construction. In addition, we identified novel QTLs for black rot resistance. The high-density genetic map will promote QTL analysis for other important agricultural traits and marker-assisted breeding of *B. oleracea*.

Keywords: Cabbage, Whole-genome resequencing, Genetic linkage map, Black rot, QTL

Background

Cabbage (*Brassica oleracea* L.) is one of the most important vegetable crops, and is consumed as a food worldwide due to its healthy compounds for humans. Besides its economic importance, cabbage is considered a valuable plant for the study of genome evolution because it contains a CC genome, which represents one of three basic diploid *Brassica* species in the U's triangle [1]. Recently, two draft genome sequences of *B. oleracea* were reported [2,3], and the availability of this reference genome enhances our understanding of the genome architecture of *B. oleracea* and the evolution of *Brassica* species, as well as facilitates identification of genes associated with important traits for breeding.

Black rot is one of the most devastating diseases to crucifers including *B. oleracea* and is caused by the vascular bacterium *Xanthomonas campestris* pv. *campestris* (Pammel) Dowson (*Xcc*). The disease infects the host plants through hydathodes, wounded tissue, insects and stomata [4,5]. The main disease symptoms are V-shaped chlorotic lesions at the margins of leaves, necrosis and darkening of leaf veins, which lead to serious production losses in vegetable crops [6]. Accordingly, development of cultivars resistant to black rot has been a priority for breeders.

Several methods have been attempted to control black rot disease, including crop diversification and rotation, production of disease-free seed, pre-treatment of seed with bactericide, elimination of potential pathogen sources

* Correspondence: tjyang@snu.ac.kr
[1]Department of Plant Science, Plant Genomics and Breeding Institute, and Research Institute of Agriculture and Life Sciences, College of Agriculture and Life Sciences, Seoul National University, Seoul 151-921, Republic of Korea
Full list of author information is available at the end of the article

such as infected crop debris and weeds, and planting of resistant cultivars [7]. Among these, utilization of resistant cultivars is one of the most effective and efficient ways to reduce disease incidence and crop loss. However, the development of commercially acceptable resistant varieties has proven to be extremely difficult due to the lack of studies on genetics and breeding for resistance in cabbage. Two major factors hinder black rot resistance breeding in *B. oleracea*: multigenic control of resistance and emergence of new races of the pathogen that overcome host resistance [8]. Nine races of *Xcc* have been identified [9], among which races 1 and 4 are the major pathogens causing black rot disease in *B. oleracea* crops [10]. Therefore, obtaining *B. oleracea* cultivars that have resistance to both races is considered a prerequisite to control black rot disease [11].

Molecular markers are highly useful for genomic analysis and allow exploration of heritable traits and the corresponding genomic variation [12]. DNA markers are now key components of crop improvement programs, and are applied to identify cultivars, analyze genetic diversity, construct linkage maps and identify quantitative trait loci (QTL) [13]. Advances in molecular markers have facilitated the identification of interesting traits via marker-assisted selection (MAS) in plant improvement. Marker-based approaches represent an effective and rapid strategy for identifying and transferring useful genes in breeding programs [14]. Furthermore, the identification of markers linked to QTL can allow analysis of the consistency of QTL effects across different environments and genetic backgrounds, and increase the frequency of favorable alleles during selection [15]. Several QTLs for black rot resistance in *B. oleracea* have been reported, including two on linkage groups 1 and 9, and two additional QTLs on linkage group 2 [15], as well as two other major QTLs on linkage groups 2 and 9, and two minor QTLs on linkage groups 3 and 7 [16]. Moreover, three QTLs analyzed using SNP markers in the F_2 mapping population derived from a cross between resistant cabbage and susceptible broccoli were found on linkage groups 2, 4 and 5, and exhibited significant effects in black rot resistance [4]. Recently, three further QTLs for black rot resistance were also detected in linkage groups 5, 8 and 9 [5]. In total, 14 QTLs with major and minor effects have been mapped on eight different *B. oleracea* chromosomes, suggesting that resistance to black rot disease is complex and quantitatively controlled by multiple genes in *B. oleracea*.

Successful QTL mapping requires a large number of genetic markers [17]. Markers based on simple sequence repeats (SSRs) and single nucleotide polymorphisms (SNPs) are commonly used due to their advantages over other types of genetic markers. SSR markers are highly reproducible, highly polymorphic, and amenable to automation. However, next-generation sequencing (NGS) technology makes SNP markers preferable to SSR markers [18]. SNPs have proved to be universal as well as the most abundant forms of genetic variation even among individuals of the same species [19]. Therefore, SNP markers exhibit higher polymorphism than SSR markers [20,21].

In this study, we have resequenced two parental cabbage lines up to 20× genome coverage and conducted a genome-wide survey for SNPs. We validated the SNPs and developed derived cleaved amplified polymorphic sequences (dCAPS) markers for resistance against black rot disease. The genome-wide catalog of SNPs, the high-density map derived from a mapping population generated from elite cabbage breeding lines with a narrow genetic background, and the QTLs reported herein all will be valuable for both breeding and genetic research in *B. oleracea*.

Results

Whole-genome resequencing of two cabbage parental lines and SNP detection

Whole genome sequencing data included about 114 million raw reads for C1184 and 113 million for C1234 (Table 1). The recently assembled *B. oleracea* genome sequence consists of 488.6 Mb, including 446.9 Mb in 9 pseudo-chromosomes and 41.2 Mb of unanchored scaffolds, and corresponding to almost 75% of the estimated genome size (648 Mb) [3]. Our new sequencing data represented approximately 18-fold genome coverage for both parental lines based on the estimated genome size. We mapped each set of paired reads onto the nine pseudo-chromosomes of reference genome sequence. In total, almost 94 million raw reads (82.1%) and 88 million (77.6%) from C1184 and C1234, respectively, were successfully aligned to the reference genome. The average mapping depth was 21.2- and 20-fold for C1184 and C1234, respectively.

The total number of SNPs relative to the reference sequence and average SNP densities were very similar in both parental lines. Approximately 1.20 and 1.24 million high-quality SNPs are identified in C1184 and C1234,

Table 1 Summary of whole-genome resequencing data for *B. oleracea* lines

	C1184	C1234
Raw reads	114,454,524	113,830,992
Raw bases	11,559,906,924	11,496,930,192
Coverage of *B.oleracea* genome	17.8 ×	17.7 ×
GC (%)	36.1	35.6
Mapped reads	93,956,750	88,382,752
Mapped percentage (%)	82.1	77.6
Mapped bases	9,489,631,750	8,926,657,952
Mapping depth (average)	21.2	20.0

respectively, by comparison to the reference genome. On average, a SNP was detected in each 372.8 bp in C1184, and each 360.0 bp in C1234. Chromosome C03 of both lines had the most SNPs, whereas the fewest SNPs were found on chromosome C06 of C1184 and chromosome C04 of C1234.

These SNPs were merged and used to detect SNPs between the two parental lines (Table 2). As a result, a total of 674,521 SNPs were found throughout nine chromosomes, with an average of 1 SNP per 662.5-bp interval. The highest density of SNPs was found on chromosome C03, with a SNP per 541 bp, while the lowest density was on chromosome C05, with one SNP per 818.9 bp. Analysis of the distribution of SNPs per 100 kb along the nine chromosomes revealed areas of high and low SNP density on each chromosome (Figure 1).

Development of dCAPS markers and construction of genetic map

We used the SNPs between C1184 and C1234 for development of dCAPS markers. Based on the physical positions of all markers used in a previous genetic map for *B. oleracea* [21], new dCAPS markers were designed for the regions of low marker density. Among 167 markers amplified, 117 (70.1%) were polymorphic between the two parental lines (Table 2 and Additional file 1: Table S1). Among the 117 polymorphic markers, 26 showed heterozygosity in one of parental lines (Table 2). We used 87 of these polymorphic dCAPS markers for genotyping of each individual in the F_2 population (Additional file 1: Table S1). Additionally, 16 other types of polymorphic markers including five EST-based dCAPS markers, five MIP markers, three IBP markers, two genomic SSR markers, and one INDEL marker were also genotyped with the same population. Among 103 newly analyzed markers, 25 markers showed a segregation pattern distorted from the 1:2:1 Mendelian ratio in the F_2 population, based on chi-square goodness of fit at the 0.05 probability level (Additional file 2: Table S2). There were six segregation distortion regions (SDRs) in the previous map [21], and all dCAPS markers designed from the SDRs of C01 and C05 showed the same distortion ratio.

The 103 novel polymorphic marker loci (Additional files 1 and 2: Tables S1 and S2) were added to the previous 265 markers [21] to develop a higher density genetic map. All 368 markers were placed on the map, and a linkage map was generated with nine linkage groups (LGs) in which each LG had more than 32 markers (Figure 2, Table 3). The improved *B. oleracea* genetic map spanned 1,467.3 cM, which is 135.4 cM more than the previous map, and the average distance between neighboring loci was reduced to 3.88 from 5.02 cM. Most of the new dCAPS markers were mapped to the originally estimated position of each chromosome sequence. The exceptions included BoRSdcaps1-35, which was designed on chromosome C01 but mapped to chromosome C02, and BoRSdcaps5-18, designed on chromosome C05 but mapped to chromosome C09.

Black rot resistance assays and QTL analysis

We performed three independent inoculation trials over three years. The final disease index for F_2 plants was determined by calculating the average value of the black rot disease indices for 10 ~ 15 $F_{2:3}$ progeny plants for each trial. Although all three inoculation tests were performed under the same conditions, the disease symptoms for each test were not consistent and tended to become more severe in later years (Additional file 3: Figure S1), possibly due to differences in plant growth or storage term for the F_3 seeds or to weather differences between years.

Table 2 Summary of SNPs detected from *B. oleracea* whole-genome resequencing data and development of dCAPS markers for validation

Ch.	Number of SNPs (average bp per SNP)						Validation		
	Ref vs. C1184		Ref vs. C1234		C1184 vs. C1234		Amplified/Designed	Polymorphic (h)[a]	%[b]
C01	122,191	(358.2)	114,778	(381.3)	66,197	(661.1)	31 / 35	20 (4)	64.5%
C02	149,730	(353.2)	161,246	(328.0)	74,741	(707.6)	14 / 17	10 (1)	71.4%
C03	196,150	(331.3)	205,306	(316.5)	120,115	(541.0)	13 / 22	11 (1)	84.6%
C04	136,815	(392.6)	132,144	(406.5)	86,999	(617.5)	14 / 20	8 (4)	57.1%
C05	130,557	(359.2)	132,887	(353.0)	57,417	(818.9)	15 / 18	14 (2)	93.3%
C06	87,712	(454.1)	102,422	(388.8)	63,017	(631.9)	28 / 34	16 (4)	57.1%
C07	119,275	(405.5)	128,978	(375.0)	69,905	(691.9)	8 / 15	5 (1)	62.5%
C08	108,586	(384.6)	113,956	(366.4)	68,361	(610.9)	21 / 26	18 (3)	85.7%
C09	147,866	(369.8)	149,581	(365.6)	67,768	(806.9)	23 / 35	15 (6)	65.2%
Total	1,198,882	(372.8)	1,241,298	(360.0)	674,521	(662.5)	167 / 222	117 (26)	70.1%

[a]h is the number of markers that showed heterozygous results.
[b]Percentage of the total amplified dCAPS markers that were polymorphic.

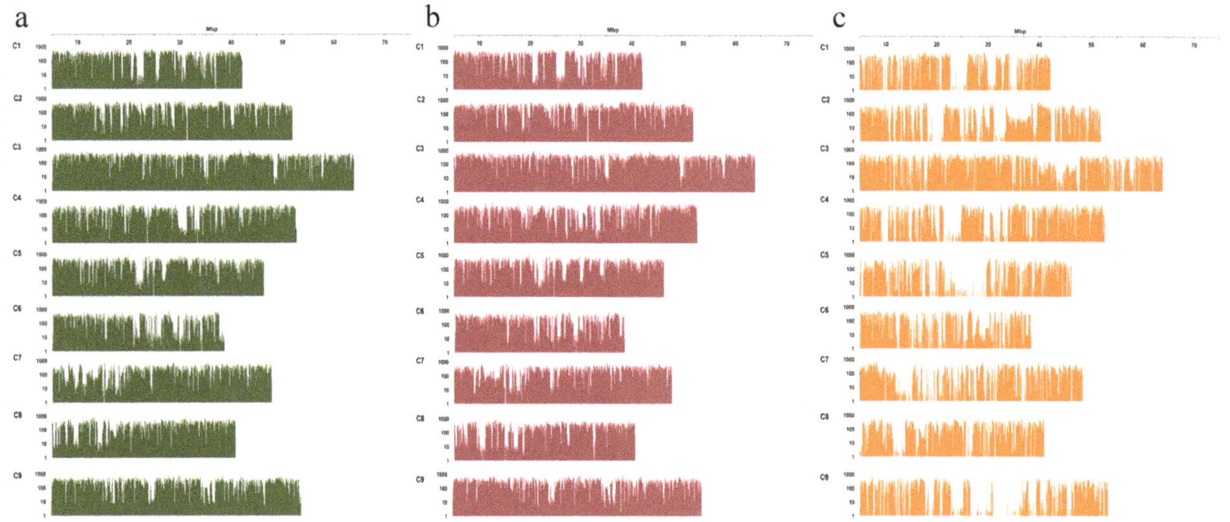

Figure 1 Distribution of SNPs in the pseudo-chromosomes of *B. oleracea*. SNPs within 100-kb intervals are shown for **(a)** Reference vs. C1184; **(b)** Reference vs. C1234; **(c)** C1184 vs. C1234.

QTL analyses were performed for each of three trials. We detected significant QTLs, based on higher LOD scores than the thresholds calculated in the permutation tests; LOD threshold values for the tests in 2012, 2013, and 2014 were 3.063, 2.912, and 2.906, respectively. In the first test performed in 2012, there were three significant QTL regions: BRQTL-C1_1 and BRQTL-C1_2 on chromosome C01, and BRQTL-C3 on chromosome C03 (Figure 2). Among these, BRQTL-C1_2 had the highest LOD score, additive effect, and variance explained (Table 4, Figure 2). The second test identified only a single QTL, which was included within BRQTL-C1_2 detected in the 2012 test, although this QTL had smallest LOD score among all QTLs identified in the three tests. The last test, carried out in 2014, identified BRQTL-C1_1 and 2 as well as a novel QTL in chromosome 6, BRQTL-C6. BRQTL-C1_2 in the 2014 test was identified as a smaller region than in 2012, but had the highest LOD score among all QTLs and accounted for 27.3% of the variation.

NBS-encoding genes in QTL regions

In most plants, disease resistance-related genes (R genes) encode proteins containing nucleotide binding sites (NBS) and a series of leucine-rich repeats (LRRs), termed NBS-LRR proteins. NBS-LRR proteins recognize and correspond to pathogen avirulence factors, and lead to defense responses and hypersensitive reactions [22]. Hence, we compared our genetic map to the pseudo-chromosome sequences [3] and searched for NBS-LRR genes within the QTL regions (Table 5). BRQTL-C1_1 was found between markers H073E22-3 and BoRSdcaps1-11, and BRQTL-C1_2 was between BoRSdcaps1-13 and BoEdcaps4 (Table 4). We identified eight NBS-LRR-encoding genes between H073E22-3 and BoEdcaps4 showing BRQTL-C1_1 and BRQTL-C1_2 QTLs. Seven NBS-LRR type R genes were detected within 1 Mb of the BoESSR291 marker, which is located near the BRQTL-C3 region. BRQTL-C6 contained five NBS-LRR type R genes.

We compared the sequences of these 21 candidate R genes against the *Brassica* Database (BRAD; http://brassicadb.org/) [23]. All 21 sequences showed similarity to disease resistance proteins, of which 19 and 11 sequences had syntenic genes in *B. rapa* and *A. thaliana*, respectively. According to the gene annotation, two candidate disease resistance genes (Bo1g094680 and Bo1g094710) in BRQTL-C1, and seven genes in BRQTL-C3 were found as gene clusters (Table 5).

Seven of nine NBS-LRR genes in BRQTL-C1 were syntenic with the R genes in the counterpart regions of chromosome A01 in *B. rapa* (Table 5). Orthologous genes of two NBS-LRR genes, Bo1g094680 and Bo1G094710, located within a 63-Kb portion of BRQTL-C1 appeared as tandem array at the counterpart syntenic region in *B. rapa* and *A. thaliana* (Figure 3a). All NBS-LRR genes in BRQTL-C6 also showed highly conserved syntenic relationships with counterpart regions in *B. rapa* and *A. thaliana*. However, a 72-Kb region near BRQTL-C3 contained a cluster of seven NBS-LRR genes, whereas the syntenic region in *B. rapa* contained a cluster of only three such genes, and the corresponding syntenic region in *A. thaliana* did not have any R genes (Figure 3b).

Discussion

Frequency and utility of SNPs revealed by whole-genome resequencing

An appropriate reference sequence allows whole-genome sequence data from individuals to be aligned, and thus

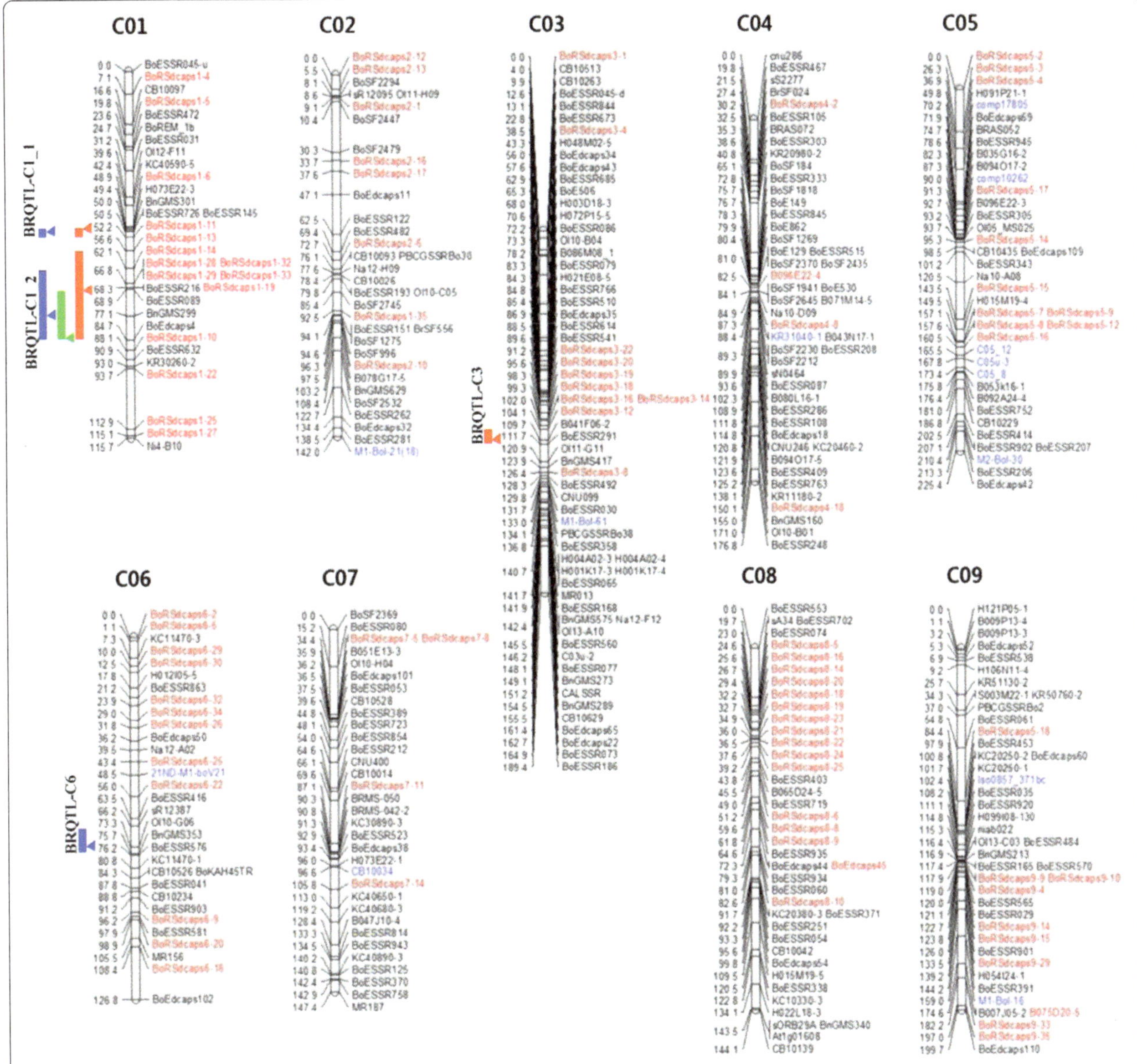

Figure 2 Genetic linkage map of cabbage constructed using 368 markers. Markers in red are newly developed dCAPS markers and markers in blue are EST-based dCAPs, MIP, IBP, genomic SSR, and INDEL markers. QTLs identified in inoculation tests in 2012, 2013, and 2014 are shown as red, green, and blue bars, respectively. The position of the peak LOD score in each QTL is indicated by an arrowhead.

Table 3 Distribution of molecular markers on the cabbage genetic map

Marker type		C01	C02	C03	C04	C05	C06	C07	C08	C09	Total
This study	dCAPS	15	8	10	3	11	12	4	15	9	87
	Other markers[a]	0	1	1	2	6	1	1	1	3	16
Previous study		18	25	52	43	23	19	29	25	31	265
Total		33	34	63	48	40	32	34	41	43	368
Length (cM)		115.7	142.0	189.4	176.8	225.4	126.8	147.4	144.1	199.7	1467.3
Average interval (cM)		3.51	4.18	3.01	3.68	5.64	3.96	4.34	3.51	4.64	3.88

[a]Including MITE insertion polymorphism, EST-based dCAPS, Intron based polymorphism, genomic SSR, and INDEL markers.

Table 4 QTLs identified for resistance to *Xcc* KACC 10377

Inoculation test	QTL name	Linkage group	Marker interval (cM)	Marker nearest to peak in LOD score	LOD[a]	Additive effect[b]	Variance explained (%)[c]
1st test (2012)	BRQTL-C1_1	C1	H073E22-3 - BoRSdcaps1-11 (2.8 cM)	BnGMS301	3.871	−0.714	17.8
	BRQTL-C1_2	C1	BoRSdcaps1-13 - BoEdcaps4 (28.1 cM)	BoESSR089	4.720	−0.697	21.2
	BRQTL-C3	C3	BoRSdcaps3-12 - BoESSR291 (7.6 cM)	B041F06-2	3.834	−0.661	17.6
2nd test (2013)	BRQTL-C1_2	C1	BoESSR089 - BoEdcaps4 (15.8 cM)	BoEdcaps4	3.051	−0.602	15.1
3rd test (2014)	BRQTL-C1_1	C1	H073E22-3 - BoRSdcaps1-11 (2.8 cM)	BoESSR726, BoESSR145	3.881	−0.912	19.8
	BRQTL-C1_2	C1	BoRSdcaps1-14 - BoEdcaps4 (22.6 cM)	BnGMS299	5.619	−0.987	27.3
	BRQTL-C6	C6	sR12387 - BnGMS353 (9.5 cM)	Ol10-G06	3.847	−0.868	19.6

Shown are position of the QTL on the map, LOD scores, additive and dominant effects, and percentage of variance explained.
[a]Peak LOD score of the QTL.
[b]Additive or dominant effect of C1234 allele.
[c]Percentage of variance explained at the peak of QTL.

resequencing can be used to detect genetic variation between different samples as high-confidence sequence differences [24]. Accordingly, we performed whole-genome resequencing for two cabbage lines to detect genome-wide SNPs for marker development to construct a high-density genetic map. There are two available draft genome sequences for *B. oleracea*; although the total assembled sequence of Liu et al. (539.9 Mb) [2] is larger than that of Parkin et al. (488.6 Mb) [3], the size of the nine pseudo-chromosomes of the latter (446.9 Mb) is larger than that of the former (388.8 Mb). Therefore, we chose the genome sequence of Parkin et al. [3] as a reference for this research because of its advantage for sequence-guided SNP marker development.

Approximately 80% of our newly generated PE sequence reads were successfully aligned to the reference genome. Almost 1.2 million SNPs were found in both lines compared to the reference, and 670,000 SNPs were found between C1184 and C1234. The number and density of SNPs between both parental lines were much lower than those detected by comparison with the reference. This could be related to the fact that the plant material used for the reference genome sequencing was kale-like *B. oleracea* [3], whereas the two parental lines used in this work were typical cabbages. In addition, we detected fewer SNPs between these lines than the 1.42 million SNPs (averaging one SNP in every 360 bp) previously reported between

Table 5 NBS-LRR-encoding genes in black rot resistance QTL regions identified for *B. oleracea* in this study, and syntenic orthologs in closely related species

QTL region in *B. oleracea*	Genes in *B. oleracea* (Parkin et al. 2014 [14])	Orthologs in *B. rapa*		Syntenic genes in *A. thaliana*
		Gene ID[b]	Position in *B. rapa*	
BRQTL-C1_1 C01: 14,884,502 - 16,579,946	Bo1g056920	Bra034079	A01: 25,091,903 - 25,095,843	
BRQTL-C1_2 C01: 18,227,386 – 37,119,290	Bo1g057060/070	Bra039560	A01: 11,678,267 - 11,687,802	AT4G14380
	Bo1g086130	Bra013691	A01: 7,172,559 - 7,175,366	AT4G23440
	Bo1g087610	Bra038144	Scaffold000140	AT1G57850
	Bo1g091560			
	Bo1g094680/710[a]	Bra031456/455[a]	A01: 17,128,737 – 17,140,522	AT1G61100/105[a]
	Bo1g103860			
BRQTL-C3 C03: 19,714,632 - 22,846,644	Bo3g060060/070/080/100/110/130/140[a]	Bra001160/161/162[a]	A03: 15,040,407 - 15,054,981	
BRQTL-C6	Bo6g025490	Bra004192	A07: 20,618,348 - 20,627,341	AT1G66840
C06: 7,423,787 - 10,466,894	Bo6g031330/350/360/380	Bra003997	A07: 19,462,054 - 19,467,133	AT1G69550

[a]Tandemly arrayed genes.
[b]Gene ID in BRAD database (http://brassicadb.org/).

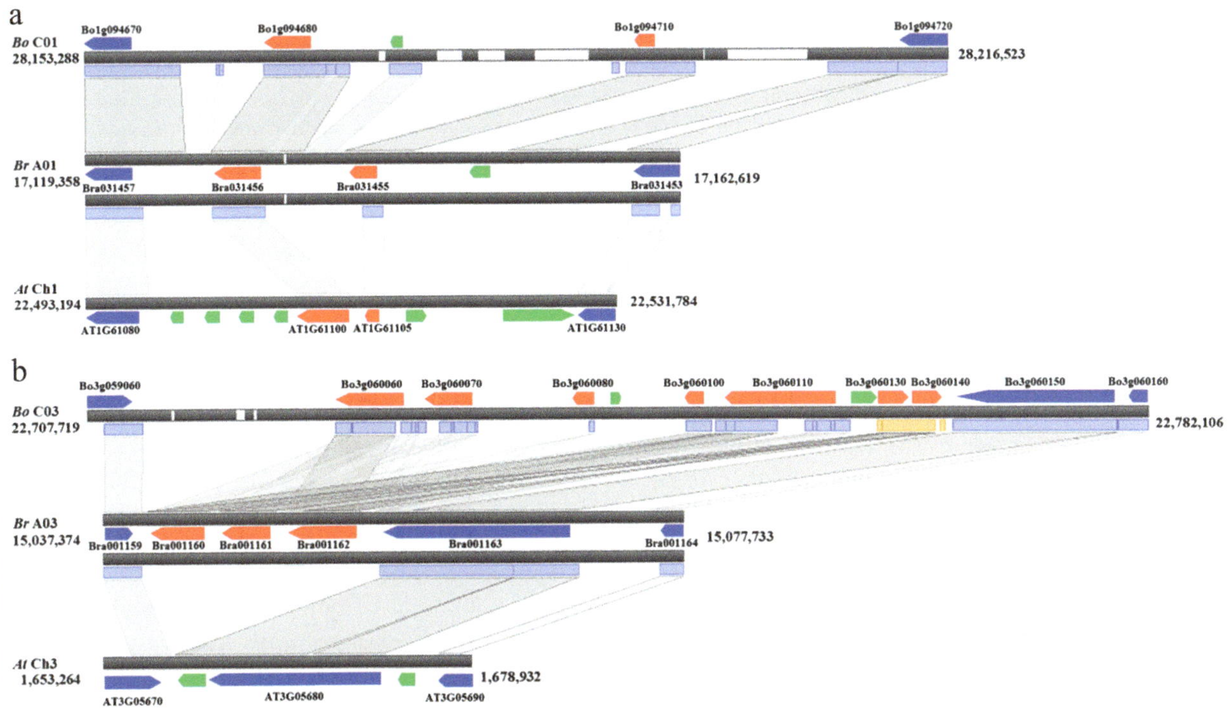

Figure 3 Syntenic relationships among crucifer species of QTL regions containing genes encoding NBS-LRR proteins. Black bars represent the chromosomal blocks and white regions are N-gaps. Red and blue indicate genes for which orthologous genes were found in relative species, with those encoding NBS-LRR genes in red and non-NBS-LRR genes in blue. Green denotes genes that were annotated only in one species; Bo-*Brassica oleracea*, Br-*Brassica rapa*, and At-*Arabidopsis thaliana*. **(a)** Syntenic regions that include disease resistance genes **(b)** Syntenic regions with different numbers of NBS-LRR-encoding genes clustered in *B. oleracea* and *B. rapa*.

two other cabbages [25]. Our plant materials have been used as elite breeding resources by a Korean company, and thus the genetic relationship between these two parental inbred lines is likely much closer than to the reference accession or the relationship between the two cabbages used by Liu et al. [25]; this close relationship likely underlies the relatively low number of SNPs we identified.

Among 167 dCAPS markers that successfully produced PCR products, 70.1% showed polymorphism. This rate was much higher than that of EST-derived dCAPS markers, in which the polymorphic rate was 58.44% when evaluated with low-sequencing depth 454 RNA-seq reads [21]. Paired-end read data generated by Illumina sequencing could allow more accurate alignment of raw reads compared to single reads from 454 RNA-seq, and thus the SNP calling process would also become more precise. The 29.9% of dCAPS markers showing no polymorphism might reflect false mapping of reads to paralogous regions, as there is high sequence similarity between the triplicated genomes and among recently duplicated chromosome segments [26-28]. This could also be the reason two dCAPS markers, BoRSdcaps1-35 and BoRSdcaps5-18, were mapped to unexpected chromosomes.

Collectively, our results demonstrate that whole-genome resequencing data generated by NGS techniques can be highly useful for large-scale discovery of SNPs and development of SNP-based molecular markers. Further study will enable high-throughput genotyping with SNPs detected here.

Improvement of the genetic map between cabbage breeding lines

By obtaining large numbers of reliable SNPs and utilizing them for development of DNA markers, we were able to improve the genetic map of cabbage. The genetic map now spans a total 1,467.3 cM after our addition of SNP markers developed for the relatively large gaps (greater than 20 cM) in the previous map [21]. Consequently, the 12 gaps in the previous genetic map are now reduced to 6 gaps and the average interval is smaller than before. The 368 markers used for the improved genetic map are promising for general cabbage breeding purposes because the map was built using a mapping population between two elite breeding lines with narrow genetic diversity. By contrast, most of previous genetic map was built using mapping populations derived between lines with wide genetic diversity for academic purposes, for example a cross between double-haploid (DH) lines derived from other subspecies [3,29] or DH lines selected based on simple morphological differences [20]. Therefore, the genetic map in this study will be helpful for molecular breeding associated not only with black rot resistance but also with many other important agricultural traits.

QTL mapping of black rot resistance

We identified four QTL regions that could contribute additively to resistance. The BRQTL-C1_2 QTL region was detected repeatedly in the three independent inoculation tests, had the highest LOD values and also accounted for the highest percentage of the variation in all tests. Accordingly, BRQTL-C1_2 is a strong candidate to be a major QTL for black rot resistance. BRQTL-C1_1, BRQTL-C3, and BRQTL-C6 seem to be minor QTLs, which could be influenced by plant conditions and environmental factors. Although two QTLs identified in chromosome C1 included SDRs, we retained all distorted markers for QTL analysis because distorted markers can also be helpful for QTL mapping when they are addressed properly [30].

The positions of our black rot resistance QTLs did not coincide with those of the 14 previously reported QTLs [4,5,15,16]. This lack of overlap is probably due to differences in disease resistance sources or inocula used. Some studies did not describe the races used [15,16], while some [4,5] used *Xcc* race 1. The exact race used in this study has not been classified yet. Integrated and standardized protocols for black rot disease races and testing would facilitate further research. However, even though the same *Xcc* race was used in our three inoculation tests, disease indices for same F_2 lineages were not consistent year to year and thus different QTLs were detected between tests. Resistance to *Xcc* has been reported to vary depending on accessions of *B. oleracea* and pathogen races [31,32]. Further, the resistance is likely also affected by complex polygenic control under different environmental conditions. Regardless of the race used (if the same as in previous studies or not), the different QTLs detected here should represent new regions.

Candidate genes for black rot resistance

The genomes of *B. oleracea*, *B. rapa*, and *A. thaliana* share a set of 24 conserved syntenic blocks, A to X, that can be identified among the ancestral karyotype [33]. The complete *B. oleracea* draft genome also demonstrates generally strong conservation with *B. rapa* in large segments at the pseudo-molecule level [2,3]. Comparative analysis revealed the presence of conserved R gene orthologs at the syntenic counterparts in *B. oleracea*, *B. rapa* and *A. thaliana*. In particular, the BRQTL-C1 region of C1 in *B. oleracea* showed large-scale conservation with A01 in *B. rapa*. Our analysis demonstrated that Bra038144, found in unanchored scaffold000140 of the *B. rapa* genome, is an ortholog of Bo1g087610 in *B. oleracea* (Table 5). Based on our finding that AT1G57850, the corresponding orthologous gene in *A. thaliana*, was also located in a syntenic region, the unanchored *B. rapa* scaffold000140 is likely derived from chromosome A01.

In plant genomes, hundreds of NBS-LRR genes are distributed as single genes or in tandem arrays as gene clusters, which arise from tandem gene duplications or homologous recombination and homogenization [34,35]. We detected 21 R genes in the four QTL regions, of which 9 were in gene clusters (Table 5). Most of the R genes showed conserved syntenic relationships in *Brassica* and *Arabidopsis* (Figure 3a). However, near BRQTL-C3 were NBS-LRR gene clusters that appear to be unique to the *Brassica* lineage (Figure 3b). This result implied that three-R-gene clusters arose by insertion in the *Brassica* lineage at *Br*A03 and subsequently amplified to a seven-R-gene cluster in *B. oleracea* over the 4.6 million years after divergence of the *Brassica* species [2].

Although genomes of *Brassica*-lineage species underwent whole-genome triplication events, the number of resistance genes was not proportionally increased in the *Brassica* genome [27,36]. Around 150 ~ 200 R genes were reported in the *A. thaliana* genome [2,35,37], and 206 [2] ~ 244 (http://brassicadb.org/) and 157 genes [2] were annotated as R genes in *B. rapa* and *B. oleracea* genomes, respectively. The 21 NBS-LRR genes found in the four QTL regions are proportionally higher density compared to other chromosomal regions, supporting the idea that some of these NBS-LRRs could be candidate to control black rot resistance in *B. oleracea*. Further analysis to reveal the function of these genes will be necessary for identification of the major resistance genes for *Xcc*.

Conclusion

We performed whole-genome resequencing of two cabbage inbred lines that are parental lines for black rot disease resistance and breeding lines with elite agricultural traits. Based on genome-wide SNP detection and validation with dCAPS markers, we report 670,000 SNPs with 70% accuracy between the parental lines. By combining SNP-based markers into the previous genetic map, we improved the genetic map and identified four QTL regions that contained 21 candidate R genes. We thus demonstrated that whole-genome resequencing can successfully be applied for detection of large-scale SNPs, development of molecular markers, and ultimately construction of a high-density genetic map for QTL analysis and marker-assisted breeding of *B. oleracea*.

Methods

Plant materials and whole-genome resequencing

Two cabbage (*Brassica oleracea L.* var. *capitata*) inbred lines, C1184 and C1234, were selected as parents to develop a mapping population. The two lines show different responses to black rot disease; C1184 is susceptible to *X. campestris* pv. *campestris* (*Xcc*), whereas C1234 is resistant. The mapping population consisted of 97 F_2 plants generated by crossing between C1184 and C1234, as described previously [21]. Furthermore, the 97 F_2 plants were vernalized and self-pollinated to produce seeds of F_3 progenies for inoculation tests. All plant materials examined in this study were obtained from Joeun Seeds Co. (Chungcheongbuk-Do, Korea).

Genomic DNAs were extracted from approximately 5 g samples of young leaves from the cabbage parental lines, following the modified cetyltrimethylammonium bromide (CTAB) protocol [38]. The quality and quantity of the DNA were examined using a NanoDrop ND-1000 (NanoDrop Technologies, Inc., USA). More than 5 μg extracted DNA was randomly sheared and quantified using DNA 1000 kit (Agilent Technologies, Inc., USA) according to the manufacturer's protocol. Sequencing with constructed shotgun libraries of C1184 and C1234 was performed by Illumina Hi-seq 2000. Fragmentation, library construction, and sequencing were carried out by the National Instrumentation Center for Environmental Management (NICEM; Seoul, Korea).

SNP discovery and dCAPS marker design

Overall process of SNP discovery was performed by following the framework described by DePristo et al. [39]. Briefly, Illumina paired reads from the parental lines were aligned to the reference sequence of *B. oleracea* [3] using Bowtie2 program [40]. Then, read grouping and removal of PCR duplicates were carried out using Picard (http://picard.sourceforge.net). Misalignments caused by INDELs were corrected by local re-alignment using Genome Analysis Toolkit (GATK) and candidate SNPs were called using Variant Caller, a utility in GATK [41]. To filter variants and avoid false positives, candidate SNPs exhibiting any of the following characteristics were removed: (1) mapping quality score lower than 4; (2) quality less than 30; (3) less than 10× or more than 45× mapping depth.

Initially, SNPs of C1184 and C1234 relative to the reference genome were called separately. All of the identified SNP positions from both parental lines were then merged and compared to each other, and promising SNPs for this research between C1184 and C1234 were identified. The selected SNPs were used to develop dCAPS markers using the dCAPS Finder 2.0 program (http://helix.wustl.edu/dcaps) for design of nearly-matched primers including SNP positions. After designing such mismatched primers for each SNP, the opposite primers were designed using the Primer3 program (http://primer3.wi.mit.edu/). All primers were synthesized by Macrogen (Seoul, Korea).

Molecular marker analysis

The newly developed dCAPS markers were validated by examining polymorphisms between the two parental lines C1184 and C1234. Additional expressed sequence tag (EST)-based dCAPS, intron-based polymorphic (IBP), genomic SSR, and INDEL markers that were not included in the previous genetic map [21] were also analyzed in this study. Furthermore, five polymorphic markers based on miniature inverted transposable element (MITE) insertion polymorphism (MIP) [42,43] were also used for genotyping the F_2 population.

PCR amplifications were performed in a total volume of 25 μL containing 20 ng genomic DNA template, 1 × PCR buffer, 20 pM each primer set, 0.2 mM each dNTP, 1 U Taq DNA polymerase (VIVAGEN, Korea). The amplicons of dCAPS markers were mixed with 3 U

Figure 4 Representative black rot disease symptoms on leaves of *B. oleracea* after spraying with *Xcc* suspension. Disease indices are: **(0)** less than 15%, **(1)** 15-30%, **(2)** 30-55%, **(3)** 55-75%, **(4)** more than 75% leaf area showing black rot symptoms.

appropriate restriction enzymes (New England Biolabs, USA), the corresponding 1 × buffer, and 1 × BSA when necessary, then incubated at 37°C for more than three hours. The digested fragments of dCAPS markers and amplicons of other markers stained by ethidium bromide were visualized on a UV trans-illuminator after electrophoresis using 9% non-denaturing polyacrylamide gels or 1% agarose gels depending on fragment size.

Inoculation test

Xanthomonas campestris pv. Campestris KACC 10366, obtained from the Korean Agricultural Culture Collection (KACC; Suwon, Korea) were used for the inoculation tests. Inoculum of the bacterium was scraped and cultured on tryptic soy agar (TSA) plates at 30°C for 48 h. Cultured bacteria were harvested using a spreader and diluted with distilled water to 0.125 OD at 600 nm to prepare bacterial suspension for inoculation.

Inoculation tests, carried out in 2012, 2013, and 2014 under the same conditions at the Korea Research Institute of Chemical Technology (Dae-jeon, Korea), were performed with 10 ~ 15 F_3 plants of each individual F_2 plants selected for genotyping analysis. The F_3 seeds were sown and grown on 5 × 8 plastic pots for 20 d in a greenhouse. Afterwards, 20-d-old plants, usually at a stage with two sufficiently developed true leaves, were inoculated by spraying bacterial suspension until adaxial and abaxial surfaces of leaves were sufficiently wet. Each plastic pot (40 plants) received 80 mL bacterial suspension, and the inoculated plants were moved into a dew chamber with the temperature set at 28°C. After 48 h incubation, all plants were transferred to a room maintained at 25°C and 80% humidity for further 7 d incubation with 12 h light/day, and disease symptoms on two inoculated leaves per each plant were surveyed. The severity of the black rot symptoms were recorded based on infected leaf area, with the following disease indices: (0) less than 15%, (1) 15-30%, (2) 30-55%, (3) 55-75%, (4) more than 75% leaf area showing black rot symptoms (Figure 4).

Map construction and QTL analysis

A total of 103 polymorphic markers were genotyped in the F_2 population, and the resulting scores were integrated into genotyping data used for a previous genetic map [21]. Linkage analysis and map construction were performed using JoinMap version 4.1 with the same parameters as in the previous study [21]. The Kosambi mapping function was used to convert recombination frequencies into genetic distances.

A disease index for each F_2 individual was calculated as the mean grade of 10 ~ 15 F_3 seedlings. QTLs for *Xcc* resistance were evaluated using composite interval mapping (CIM) analysis with QGene program. CIM was performed with LOD (logarithm of odds) threshold values that were estimated using 1,000 permutation tests at 5% significance with 0.5-cM scan intervals.

Additional files

Additional file 1: Table S1. Description of polymorphic markers between C1184 and C1234 used in this study.

Additional file 2: Table S2. Results of the chi-square goodness-of-fit tests for the observed segregation ratios with the genotyped markers among F_2 plants.

Additional file 3: Figure S1. Disease index distribution of F_2 population, evaluated by average scores from inoculated F_3 plants.

Abbreviations

QTL: Quantitative trait loci; SNP: Single nucleotide polymorphisms; Xcc: *Xanthomonas campestris* pv. *campestris*; SSR: Simple sequence repeat; NGS: Next-generation sequencing; NBS: Nucleotide binding sites; LRRs: Leucine-rich repeats; GATK: Genome analysis toolkit; EST: Expressed sequence tag; dCAPS: Derived cleaved amplified polymorphic sequences; IBP: Intron-based polymorphic; MITE: Miniature inverted transposable element; MIP: MITE insertion polymorphism; CIM: Composite interval mapping; MAS: Marker-assisted selection; CTAB: Cetyltrimethylammonium bromide.

Competing interests

The authors declare that they have no competing interests.

Authors' contributions

HJL, NKI, HJJ, and JL carried out the molecular experiments. MJ and SP carried out SNP discovery. JS, JY, YS, and KA developed the mapping population and maintained plant materials. JHL and GJC performed inoculation tests and investigated disease symptoms. K-WY, JYP, and S-CL collected plant materials, and provided technical assistance. JL interpreted the results, and wrote the manuscript. YY, HK, I-SN, and T-JY conceived of and managed the research. All authors critically read and approved the final version of the manuscript.

Acknowledgements

This research was supported by the Golden Seed Project (Center for Horticultural Seed Development, No. 213003-04-1-SB430), Ministry of Agriculture, Food and Rural Affairs (MAFRA), Ministry of Oceans and Fisheries (MOF), Rural Development Administration (RDA) and Korea Forest Service (KFS).

Author details

[1]Department of Plant Science, Plant Genomics and Breeding Institute, and Research Institute of Agriculture and Life Sciences, College of Agriculture and Life Sciences, Seoul National University, Seoul 151-921, Republic of Korea. [2]Indonesian Research Institute for Industrial and Beverage Crops (IRIIBC), Pakuwon, Sukabumi, Indonesia. [3]Department of Horticulture, Sunchon National University, Suncheon 540-950, Republic of Korea. [4]Joeun Seed, #174, Munbang-Ri, Cheonhan-Myun, 367-833 Goesan-Gu, Chungcheongbuk-Do, Korea. [5]Research Center for Biobased Chemistry, Korea Research Institute of Chemical Technology, Daejeon 305-600Yusong-Gu, Republic of Korea. [6]Arizona Genomics Institute, School of Plant Sciences, University of Arizona, Tucson, Arizona 85721, USA. [7]Department of Agricultural Biotechnology, Seoul National University, Seoul 151-921, Republic of Korea. [8]CHO & KIM genomics, Seoul National University Mt.4-2, Main Bldg. #514, SNU Research Park, NakSeoungDae, Seoul 151-919Gwanakgu, Republic of Korea.

References

1. UN. Genome analysis in *Brassica* with special reference to the experimental formation of *B. napus* and peculiar mode of fertilization. J Japanese Bot. 1935;7:389–452.

2. Liu S, Liu Y, Yang X, Tong C, Edwards D, Parkin IA, et al. The Brassica oleracea genome reveals the asymmetrical evolution of polyploid genomes. Nature Commun. 2014;5:3930.
3. Parkin IA, Koh C, Tang H, Robinson SJ, Kagale S, Clarke WE, et al. Transcriptome and methylome profiling reveals relics of genome dominance in the mesopolyploid *Brassica oleracea*. Genome Biol. 2014;15(6):R77.
4. Kifuji Y, Hanzawa H, Terasawa Y, Nishio T. QTL analysis of black rot resistance in cabbage using newly developed EST-SNP markers. Euphytica. 2013;190(2):289–95.
5. Tonu NN, Doullah MA-u, Shimizu M, Karim MM, Kawanabe T, Fujimoto R, et al. Comparison of positions of QTLs conferring resistance to xanthomonas campestris pv. campestris in Brassica oleracea. Am J Plant Sci. 2013;4:11.
6. Williams P. Black rot: a continuing threat to world crucifers. Plant Dis. 1980;64(8):736–42.
7. Jensen BD, Massomo SM, Swai IS, Hockenhull J, Andersen SB. Field evaluation for resistance to the black rot pathogen *Xanthomonas campestris* pv. *campestris* in cabbage (*Brassica oleracea*). Eur J Plant Pathol. 2005;113(3):297–308.
8. Hutton SF, Scott JW, Yang W, Sim S-C, Francis DM, Jones JB. Identification of QTL associated with resistance to bacterial spot race T4 in tomato. Theor Appl Genet. 2010;121(7):1275–87.
9. Fargier E, Manceau C. Pathogenicity assays restrict the species *Xanthomonas campestris* into three pathovars and reveal nine races within *X. campestris* pv. *campestris*. Plant Pathol. 2007;56(5):805–18.
10. Vicente J, Conway J, Roberts S, Taylor J. Identification and origin of Xanthomonas campestris pv. campestris races and related pathovars. Phytopathology. 2001;91(5):492–9.
11. Soengas P, Hand P, Vicente J, Pole J, Pink D. Identification of quantitative trait loci for resistance to Xanthomonas campestris pv. campestris in Brassica rapa. Theor Appl Genet. 2007;114(4):637–45.
12. Duran C, Appleby N, Edwards D, Batley J. Molecular genetic markers: discovery, applications, data storage and visualisation. Curr Bioinformatics. 2009;4(1):16–27.
13. Hendre PS, Aggarwal RK. DNA markers: development and application for genetic improvement of coffee. In: Varshney RK, Tuberosa R, editors. Genomics-assisted crop improvement. 2nd ed. Netherlands: Springer; 2007. p. 399–434.
14. Tanksley SD, McCouch SR. Seed banks and molecular maps: unlocking genetic potential from the wild. Science. 1997;277(5329):1063–6.
15. Camargo L, Williams P, Osborn T. Mapping of quantitative trait loci controlling resistance of Brassica oleracea to Xanthomonas campestris pv. campestris in the field and greenhouse. Phytopathology. 1995;85(10):1296–300.
16. Doullah M, Mohsin G, Ishikawa K, Hori H, Okazaki K. Construction of a Linkage Map and QTL analysis for Black Rot Resistance in *Brassica oleracea* L. Int J Nat Sci. 2011;1(1):1–6.
17. Yu Z, Guo X. Identification and mapping of disease-resistance QTLs in the eastern oyster, *Crassostrea virginica* Gmelin. Aquaculture. 2006;254(1):160–70.
18. Mammadov J, Aggarwal R, Buyyarapu R, Kumpatla S. SNP markers and their impact on plant breeding. Int J Plant Genomics. 2012;2012:11.
19. Rafalski JA. Novel genetic mapping tools in plants: SNPs and LD-based approaches. Plant Sci. 2002;162(3):329–33.
20. Wang W, Huang S, Liu Y, Fang Z, Yang L, Hua W, et al. Construction and analysis of a high-density genetic linkage map in cabbage (*Brassica oleracea* L. var. *capitata*). BMC Genomics. 2012;13(1):523.
21. Izzah NK, Lee J, Jayakodi M, Perumal S, Jin M, Park B-S, et al. Transcriptome sequencing of two parental lines of cabbage (*Brassica oleracea* L. var. *capitata* L.) and construction of an EST-based genetic map. BMC Genomics. 2014;15(1):149.
22. Hammond-Kosack KE, Jones JD. Plant disease resistance genes. Annu Rev Plant Biol. 1997;48(1):575–607.
23. Cheng F, Liu S, Wu J, Fang L, Sun S, Liu B, et al. BRAD, the genetics and genomics database for *Brassica* plants. BMC Plant Biol. 2011;11(1):136.
24. Bentley DR. Whole-genome re-sequencing. Curr Opin Genet Dev. 2006;16(6):545–52.
25. Liu J, Huang S, Sun M, Liu S, Liu Y, Wang W, et al. An improved allele-specific PCR primer design method for SNP marker analysis and its application. Plant Methods. 2012;8(1):34.
26. Lysak MA, Koch MA, Pecinka A, Schubert I. Chromosome triplication found across the tribe *Brassiceae*. Genome Res. 2005;15(4):516–25.
27. Yang T-J, Kim JS, Kwon S-J, Lim K-B, Choi B-S, Kim J-A, et al. Sequence-level analysis of the diploidization process in the triplicated flowering locus C region of *Brassica rapa*. Plant Cell Online. 2006;18(6):1339–47.
28. Wang X, Wang H, Wang J, Sun R, Wu J, Liu S, et al. The genome of the mesopolyploid crop species *Brassica rapa*. Nat Genet. 2011;43(10):1035–9.
29. Iniguez-Luy FL, Lukens L, Farnham MW, Amasino RM, Osborn TC. Development of public immortal mapping populations, molecular markers and linkage maps for rapid cycling *Brassica rapa* and *B. oleracea*. Theor Appl Genet. 2009;120(1):31–43.
30. Alheit KV, Reif JC, Maurer HP, Hahn V, Weissmann EA, Miedaner T, et al. Detection of segregation distortion loci in triticale (x *Triticosecale Wittmack*) based on a high-density DArT marker consensus genetic linkage map. BMC Genomics. 2011;12(1):380.
31. Taylor J, Conway J, Roberts S, Astley D, Vicente J. Sources and origin of resistance to Xanthomonas campestris pv. campestris in Brassica genomes. Phytopathology. 2002;92(1):105–11.
32. Vicente J, Taylor J, Sharpe A, Parkin I, Lydiate D, King G. Inheritance of race-specific resistance to Xanthomonas campestris pv. campestris in Brassica genomes. Phytopathology. 2002;92(10):1134–41.
33. Schranz ME, Lysak MA, Mitchell-Olds T. The ABC's of comparative genomics in the *Brassicaceae*: building blocks of crucifer genomes. Trends Plant Sci. 2006;11(11):535–42.
34. Holub EB. The arms race is ancient history in *Arabidopsis*, the wildflower. Nat Rev Genet. 2001;2(7):516–27.
35. Leister D. Tandem and segmental gene duplication and recombination in the evolution of plant disease resistance genes. Trends Genet. 2004;20(3):116–22.
36. Town CD, Cheung F, Maiti R, Crabtree J, Haas BJ, Wortman JR, et al. Comparative genomics of *Brassica oleracea* and *Arabidopsis thaliana* reveal gene loss, fragmentation, and dispersal after polyploidy. Plant Cell. 2006;18(6):1348–59.
37. Meyers BC, Kozik A, Griego A, Kuang H, Michelmore RW. Genome-wide analysis of NBS-LRR–encoding genes in *Arabidopsis*. Plant Cell. 2003;15(4):809–34.
38. Allen G, Flores-Vergara M, Krasynanski S, Kumar S, Thompson W. A modified protocol for rapid DNA isolation from plant tissues using cetyltrimethylammonium bromide. Nat Protoc. 2006;1(5):2320–5.
39. DePristo MA, Banks E, Poplin R, Garimella KV, Maguire JR, Hartl C, et al. A framework for variation discovery and genotyping using next-generation DNA sequencing data. Nat Genet. 2011;43(5):491–8.
40. Langmead B, Salzberg SL. Fast gapped-read alignment with Bowtie 2. Nat Methods. 2012;9(4):357–9.
41. Nekrutenko A, Taylor J. Next-generation sequencing data interpretation: enhancing reproducibility and accessibility. Nat Rev Genet. 2012;13(9):667–72.
42. Sampath P, Lee S-C, Lee J, Izzah NK, Choi B-S, Jin M, et al. Characterization of a new high copy stowaway family MITE, *BRAMI*-1 in *Brassica* genome. BMC Plant Biol. 2013;13(1):56.
43. Sampath P, Murukarthick J, Izzah NK, Lee J, Choi H-I, Shirasawa K, et al. Genome-wide comparative analysis of 20 miniature inverted-repeat transposable element families in *Brassica rapa* and *B. oleracea*. PLoS One. 2014;9(4):e94499.

Comparative genomics of grass EST libraries reveals previously uncharacterized splicing events in crop plants

Trees-Juen Chuang*, Min-Yu Yang, Chuang-Chieh Lin, Ping-Hung Hsieh and Li-Yuan Hung

Abstract

Background: Crop plants such as rice, maize and sorghum play economically-important roles as main sources of food, fuel, and animal feed. However, current genome annotations of crop plants still suffer false-positive predictions; a more comprehensive registry of alternative splicing (AS) events is also in demand. Comparative genomics of crop plants is largely unexplored.

Results: We performed a large-scale comparative analysis (ExonFinder) of the expressed sequence tag (EST) library from nine grass plants against three crop genomes (rice, maize, and sorghum) and identified 2,879 previously-unannotated exons (i.e., novel exons) in the three crops. We validated 81% of the tested exons by RT-PCR-sequencing, supporting the effectiveness of our *in silico* strategy. Evolutionary analysis reveals that the novel exons, comparing with their flanking annotated ones, are generally under weaker selection pressure at the protein level, but under stronger pressure at the RNA level, suggesting that most of the novel exons also represent novel alternatively spliced variants (ASVs). However, we also observed the consistency of evolutionary rates between certain novel exons and their flanking exons, which provided further evidence of their co-occurrence in the transcripts, suggesting that previously-annotated isoforms might be subject to erroneous predictions. Our validation showed that 54% of the tested genes expressed the newly-identified isoforms that contained the novel exons, rather than the previously-annotated isoforms that excluded them. The consistent results were steadily observed across cultivated (*Oryza sativa* and *O. glaberrima*) and wild (*O. rufipogon* and *O. nivara*) rice species, asserting the necessity of our curation of the crop genome annotations. Our comparative analyses also inferred the common ancestral transcriptome of grass plants and gain- and loss-of-ASV events.

Conclusions: We have reannotated the rice, maize, and sorghum genomes, and showed that evolutionary rates might serve as an indicator for determining whether the identified exons were alternatively spliced. This study not only presents an effective *in silico* strategy for the improvement of plant annotations, but also provides further insights into the role of AS events in the evolution and domestication of crop plants. ExonFinder and the novel exons/ASVs identified are publicly accessible at http://exonfinder.sourceforge.net/.

Keywords: Crop plants, Alternative splicing, Plant transcriptome evolution, Evolutionary rate, Comparative genomics, Bioinformatics

* Correspondence: trees@gate.sinica.edu.tw
Genomics Research Center, Academia Sinica, Taipei 11529, Taiwan

Background

Alternative splicing (AS) is a major post-transcriptional mechanism for producing multiple isoforms from the same precursor mRNA (pre-mRNA), thereby increasing the complexity of the transcriptome/proteome. AS is widespread in eukaryotes, and it has been suggested that over 95% of genes in human are alternatively spliced [1,2]. In contrast, 30% ~ 60% of genes in *Arabidopsis* or rice have been identified to undergo AS [3-11]. AS appears to be relatively less prevalent in plants than in mammals, but this may in part be due to limited detection of alternatively spliced variants (ASVs) in plants.

AS has been demonstrated to be involved in various biological functions [12-16] such as spatio-temporal regulation [17-20], disease resistance [21], and photosynthesis [22,23]. ASVs occur in both coding sequences (CDSs) and untranslated regions (UTRs). ASVs in CDSs can have influences on protein structure, subcellular localization, protein stability, post-translational modifications, enzymatic activity, and protein-protein interaction networks [24-26]. On the other hand, ASVs in 5′ UTRs (3′ UTRs) may include/exclude upstream open reading frames (premature termination codons), thereby altering translational stability/efficiency (nonsense-mediated decay pathway) [14,27]. Even so, a considerable number of ASVs are functionally irrelevant, or merely by-products during RNA splicing [28,29]. It remains challenging to determine whether an ASV is functionally important [30-33], not to mention that AS is less characterized in plants than in mammals, and that most plant ASVs have unknown functional consequences [10], but also that some of computationally-annotated genes/transcripts are subject to erroneous prediction. Although much effort to annotate plant transcripts produces several prominent databases [34-39], there still lacks an effective strategy to make use of public resources (e.g., EST traces) for better annotation of ASVs and accurate identification of novel isoforms in plant genomes.

In terms of molecular evolution, alternatively spliced exons and constitutively spliced exons are known to be under different evolutionary pressures. Previous studies reported that alternatively spliced exons tend to have higher nonsynonymous substitution rates (*dn*) and nonsynonymous-synonymous substitution rates (*dn/ds*) than constitutively spliced ones, indicating faster protein-level evolution in the former [40-47]. On the other hand, alternatively spliced exons were observed to have lower *ds* values than constitutively spliced ones due to the elevated synonymous rate in the latter [47]. This suggests that constitutively spliced exons are subject to weaker selection pressure than alternatively spliced ones at the RNA level. Therefore, the differences in evolutionary patterns may serve as an indicator to distinguish between these two types of exons.

In this study, we aimed to update the annotations of three crop plants, namely rice (*Oryza sativa*), maize (*Zea mays*), and sorghum (*Sorghum bicolor*). We designed a pipeline, ExonFinder, for the identification of novel exons/ASVs based on comparative genomics of the EST libraries of nine grass plants, including barley (*Hordeum vulgare*), maize, meadow ryegrass (*Festuca pratensis*), purple false brome (*Brachypodium distachyon*), rice, sorghum, sugarcane (*Saccharum officinarum*), switchgrass (*Panicum virgatum*), and wheat (*Triticum aestivum*). Such analysis resulted in the identification of a total of 2,963 ASV events (including cassette exons and retained introns) in rice, maize, and sorghum, with 2,879 novel exons that were cross-species conserved but not supported by prior Ensembl annotation or EST evidence from the same species. Evolutionary analysis reveals that though the novel exons are generally under more relaxed selection pressure than their flanking ones, some of them evolve at a similar evolutionary rate with their flanking exons. We reasoned that some of the previously-annotated isoforms that excluded the newly-identified exons may be subject to erroneous prediction. To test this possibility, we randomly selected rice exons of this kind, performed RT-PCR-sequencing, and found that over half (54%) of previously-annotated isoforms that excluded the novel exons were not detected in the same setting. The consistent results were observed in three rice cultivars (i.e., *O. sativa* L. ssp. *Indica* cv. 93-11, *O. sativa* L. ssp. *japonica* cv. Nipponbare, and *O. glaberrima*) and two wild rice species (i.e., *O. rufipogon* and *O. nivara*). Finally, we also discussed the functional potential of selected ASVs through the lens of evolution.

Results

Identification of novel exons in rice, maize, and sorghum

We introduced an *in silico* pipeline, ExonFinder, to identify previously unannotated exons/ASVs in target species (i.e., rice, maize, and sorghum) by comparative analysis of the EST library of non-target (designated as "subject") species against the genome of target species (Table 1 and Figure 1A). To achieve a better quality of cross-species alignment, we only considered grass plants in this study (Table 1). We supposed that the novel exons also represented novel AS events, since they were absent from known transcripts of the target species (Methods). ExonFinder identifies two types of novel exons: cassette exons and retained introns (Figure 1B). Authenticity and novelty of exons were considered through the following procedures. To eliminate false positives from accidental matches, we only considered EST matches that satisfied the following criteria: (1) a proper exon and its flanking exons must overlap with the same Ensembl-annotated transcript; (2) a proper cassette exon must be flanked by canonical splicing sites at its both ends; and (3)

Table 1 Summary of EST traces used in this study

Plant species	Abbreviation	UniGene version	Number of EST traces
Barley (*Hordeum vulgare*)	Hv	Build #59	517,604
Maize (*Zea mays*)	Zm	Build #81	1,705,606
Meadow ryegrass (*Festuca pratensis*)	Fp	Build #1	60,845
Purple false brome (*Brachypodium distachyon*)	Bd	Build #2	113,694
Rice (*Oryza sativa*)	Os	Build#86	1,202,546
Sorghum (*Sorghum bicolor*)	Sb	Build #30	199,401
Sugarcane (*Saccharum officinarum*)	Sof	Build #15	220,997
Switchgrass (*Panicum virgatum*)	Pv	Build #3	505,999
Wheat (*Triticum aestivum*)	Ta	Build #59	1,050,213

a proper exon that locates within CDS must not change the reading frame and must not result in any premature stop codon. Of note, Exonfinder also identifies novel cassette exons flanked by non-canonical splicing sites (Methods), although we only considered those flanked by canonical splicing sites for accuracy in the following analysis. To distinguish novel exons from currently-characterized exons, we removed the exons that were supported by Ensembl's annotation or EST traces from the target species (Methods). Of note, for each newly-identified transcript (or novel ASV), it must include at least one full-length novel exon and the flanking exons' segments of the novel exon(s) (Figure 1B). It is possible for a novel exon to be assigned to more than one novel ASV, in the case of uncertain boundaries of the flanking exons (Figure 1B). In addition, a novel ASV may also contain multiple novel exons (Case 2; Figure 1B). Consequently, we used ExonFinder to identify a total of 382 (381), 1,245 (1,150), and 1,336 (1,348) novel ASVs (novel exons) in rice, maize, and sorghum, respectively (Table 2 and Additional file 1).

Basic properties of the newly-identified exons/ASVs

As shown in Table 3, most of the identified exons/ASVs were supported by multiple EST traces, indicating these isoforms might not be rare. In addition, 14% ~ 30% of identified exons/ASVs were supported by EST traces from at least two non-target species, implying that they were widely expressed in grass plants. Since evolutionary conservation implies functional importance [33,48], these exons/ASVs may play an important role in grass plants, rather than random by-products during RNA splicing. Furthermore, the average length (~100 bp) of the novel cassette exons (Table 3) were considerably shorter than the average exon length (250 ~ 300 bp) of previously-annotated exons in rice, maize, and sorghum [3,26,49-51], reflecting a previous observation that conserved alternatively spliced exons tend to be shorter than non-conserved ones [48]. Next, we retrieved pure introns (i.e., constitutive introns; the Ensembl-annotated introns that do not contain any ExonFinder/Ensemble-identified alternatively spliced exons, and are flanked by two Ensemble-annotated constitutively spliced exons), and demonstrated that the average and median lengths of pure introns were significantly shorter than other known introns that contain the novel cassette exons (P value < 10^{-6} by the two-tailed t-test and Wilcoxon rank-sum test). This trends hold well across rice, maize, and sorghum, consistent with a previous observation that cassette exons tend to be flanked by longer introns than constitutively spliced exons [52].

We found that ExonFinder identified much more novel ASVs in maize and sorghum (both >1,000 ASVs) than in rice (382 ASVs). This was not unexpected, as the annotation of rice genome was more comprehensive than those of maize and sorghum. In addition, the number of exons identified by ExonFinder is related not only to the number of available EST traces but also to the level of divergence between the target and subject species. According to earlier phylogenetic analyses [53,54], the nine grass plants examined in this study can be classified into three groups: Ehrhartoideae (including rice), Pooideae (including purple false brome, meadow ryegrass, barley, and wheat), and Panicoideae (including switchgrass, maize, sorghum, and sugarcane), indicating a closer relationship between Ehrhartoideae and Pooideae (Figure 2A). In rice, the percentages of novel ASVs identified from non-rice grass plants were generally positively correlated with the quantities of Pooideae and Panicoideae EST traces, respectively (Figure 2B). However, the percentages of novel ASVs identified from Pooideae EST traces tended to be higher than those identified from Panicoideae EST traces. This tendency might reflect that the level of divergence between Ehrhartoideae (i.e., rice) and Pooideae is lower than that between Ehrhartoideae and Panicoideae (Figure 2A). For example, although the number of EST traces of maize (>1.7 million) is larger than that of wheat (~1 million), both data sets were used to identify similar percentages of novel exons in rice (Figure 2B). On the other hand, ExonFinder using Pooideae EST traces tended to identify fewer novel maize/sorghum exons (both of which belong to Panicoideae) than that using Panicoideae EST traces, even though EST traces from Pooideae (e.g., wheat) are about five times more than those from Panicoideae (e.g., sorghum in Figure 2C and sugarcane in Figure 2D). This indicates that ExonFinder is particularly powerful in the identification of novel exons/ASVs in poorly annotated species by using closely related species with abundant EST traces.

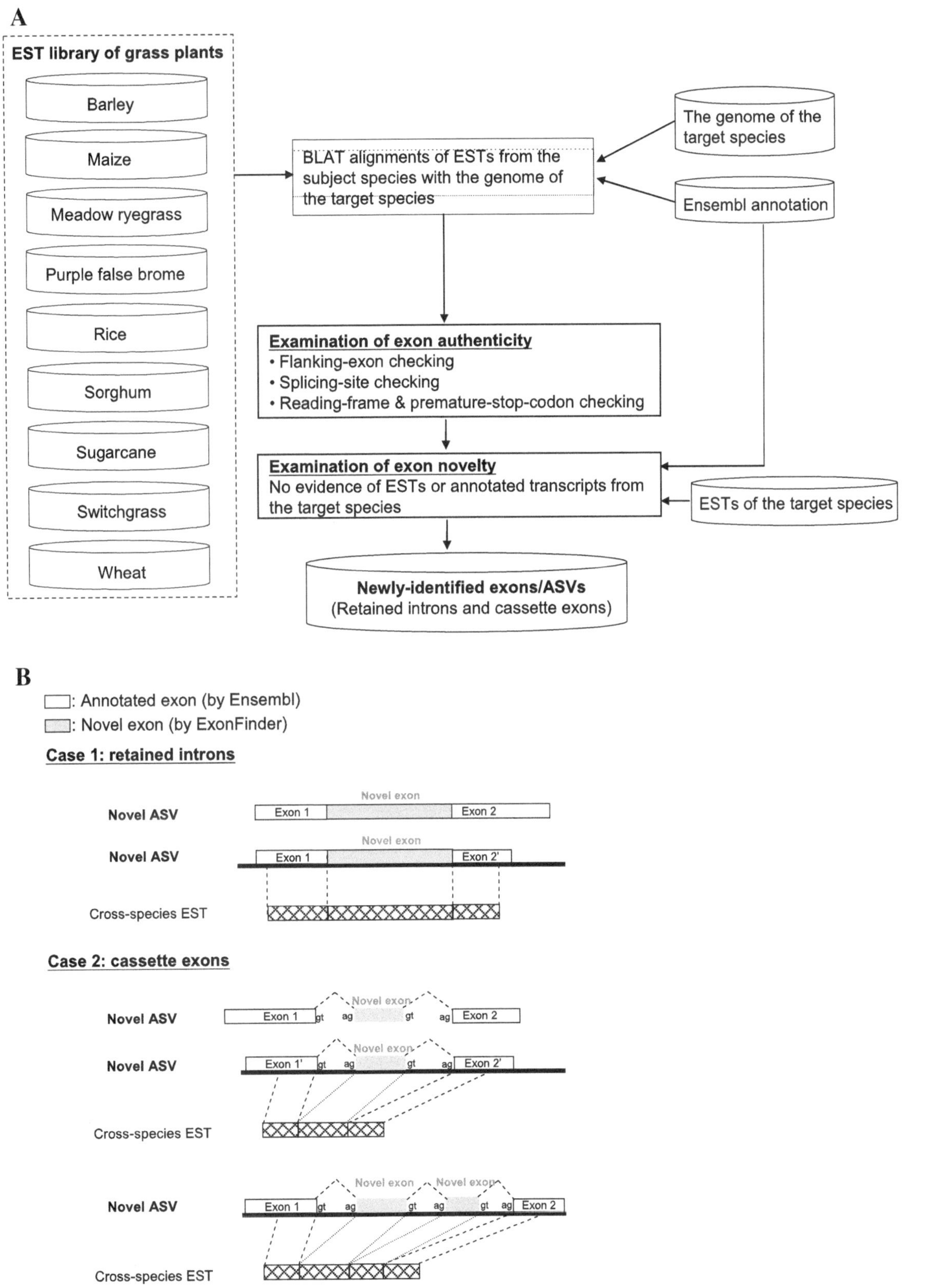

Figure 1 The ExonFinder process. (A) Flowchart of the identification of novel exons by ExonFinder. **(B)** Examples of newly-identified exons and ASVs, including retained introns (Case 1) and cassette exons (Case 2).

Table 2 Number of newly-identified exons/ASVs (including cassette exons and retained introns) in rice, maize, and sorghum

		Newly-identified exons (ASVs)		
Species	Genomic type	Cassette	Retained intron	Total
Rice	5′-UTR	36 (42)	8 (9)	44 (51)
	CDS	214 (216)	70 (70)	284 (286)
	3′-UTR	47 (38)	6 (7)	53 (45)
	Total	297 (296)	84 (86)	381 (382)
Maize	5′-UTR	272 (324)	95 (104)	367 (428)
	CDS	364 (367)	83 (85)	447 (452)
	3′-UTR	202 (209)	134 (156)	336 (365)
	Total	838 (900)	312 (345)	1,150 (1,245)
Sorghum	5′-UTR	242 (230)	53 (55)	295 (279)
	CDS	672 (684)	224 (227)	896 (913)
	3′-UTR	98 (78)	59 (62)	157 (140)
	Total	1,012 (992)	336 (344)	1,348 (1,336)

Newly-identified exons tend to have higher *dn* values and lower *ds* values than their flanking exons

To investigate the selection pressures imposed on the novel exons identified by comparative analysis of cross-species EST libraries, we calculated the evolutionary rates (*dn*, *ds*, and *dn/ds*) based on the alignments between the identified ASVs (including the novel exons and their flanking exons) in the target species and their corresponding EST sequences in the subject species (Methods). Since the novel exons are absent in the annotation (i.e., Ensembl annotation) of the target species, the inclusion level (the fraction of a gene's transcript isoforms that include a specific exon [55]) should be lower for the novel exons than for their corresponding flanking exons. Previous studies have demonstrated that alternatively spliced exons have higher *dn* and *dn/ds* values, but lower *ds* values, than constitutively spliced exons, and that the inclusion level of exons is negatively correlated with *dn* and *dn/ds* values, but positively correlated with *ds* values [44,47,56]. Therefore, we reasoned that the novel exons should exhibit higher *dn* and *dn/ds* values, but lower *ds* values, than their corresponding flanking exons. To test this hypothesis, we concatenated the flanking exons of each novel exon, and then calculated the evolutionary rates of the novel exon and its flanking exons, respectively (Methods). After that, we calculated the differences of *dn*, *ds*, and *dn/ds* values between each novel exon and its corresponding concatenated flanking exons. As expected, the differences in average evolutionary rates between novel exons and their flanking exons were higher than zero for *dn* and *dn/ds*, but lower than zero for *ds* (Figure 3A), indicating that the novel exons had higher *dn* and *dn/ds* values, but lower *ds* values, than their flanking exons. This result suggested that the novel exons were subjected to weaker selection pressure than their flanking exons at the protein level (*dn* and *dn/ds*), but the trend was reversed at the RNA level (*ds*), consistent with our hypothesis.

Interestingly, although the trend that the majority of novel exons (~80%) have higher *dn* values or lower *ds* values than their corresponding flanking exons was observed in rice, maize, and sorghum, only less than 50% of cases showed significant differences in *dn* or *ds* between these two types of exons (Methods) (Figure 3B). In other words, a considerable proportion of novel exons do not exhibit significant difference in evolutionary patterns as compared to their flanking exons. There are two possible scenarios for this consequence. First, the novel exon also represents a novel AS events. There may be some undetected transcript isoforms that include the novel exon, but exclude one or two of their flanking exons, resulting in the inclusion level of the novel exon being higher than or equal to those of its flanking exons. Second, the novel exon does not represent an AS event (in fact, it is a constitutively spliced exon), while the previously-annotated one that excludes the novel exon

Table 3 General properties of the newly-identified exons/ASVs

	Rice	Maize	Sorghum
Average number of supported EST traces (for cassette exons)	4.60	2.76	4.70
Average number of supported EST traces (for retained introns)	8.17	1.90	4.60
Percentage of the identified ASVs with EST evidence from ≥ 2 subject species	29.84	14.13	26.06
Average/median length of novel cassette exons (bp)	104.64/87	102.36/81	98.76/84
Average/median length of novel retained introns (bp)	105.40/81	108.72/92	112.19/93
Average/median length of the Ensembl-annotated introns that contain the novel cassette exons (bp)	1057.90/648	1665.04/556.5	932.09/547
Average/median length of pure introns (bp)	381.09/163	530.57/139	441.51/143
Significance test*	$P < 10^{-6}$; $P < 10^{-15}$	$P < 10^{-7}$; $P < 10^{-15}$	$P < 10^{-15}$; $P < 10^{-15}$

*Differences between the average/median lengths of previously-annotated introns that contain the newly-identified cassette events and those of pure introns were examined using the two-tailed *t*-test and Wilcoxon rank-sum test, respectively.

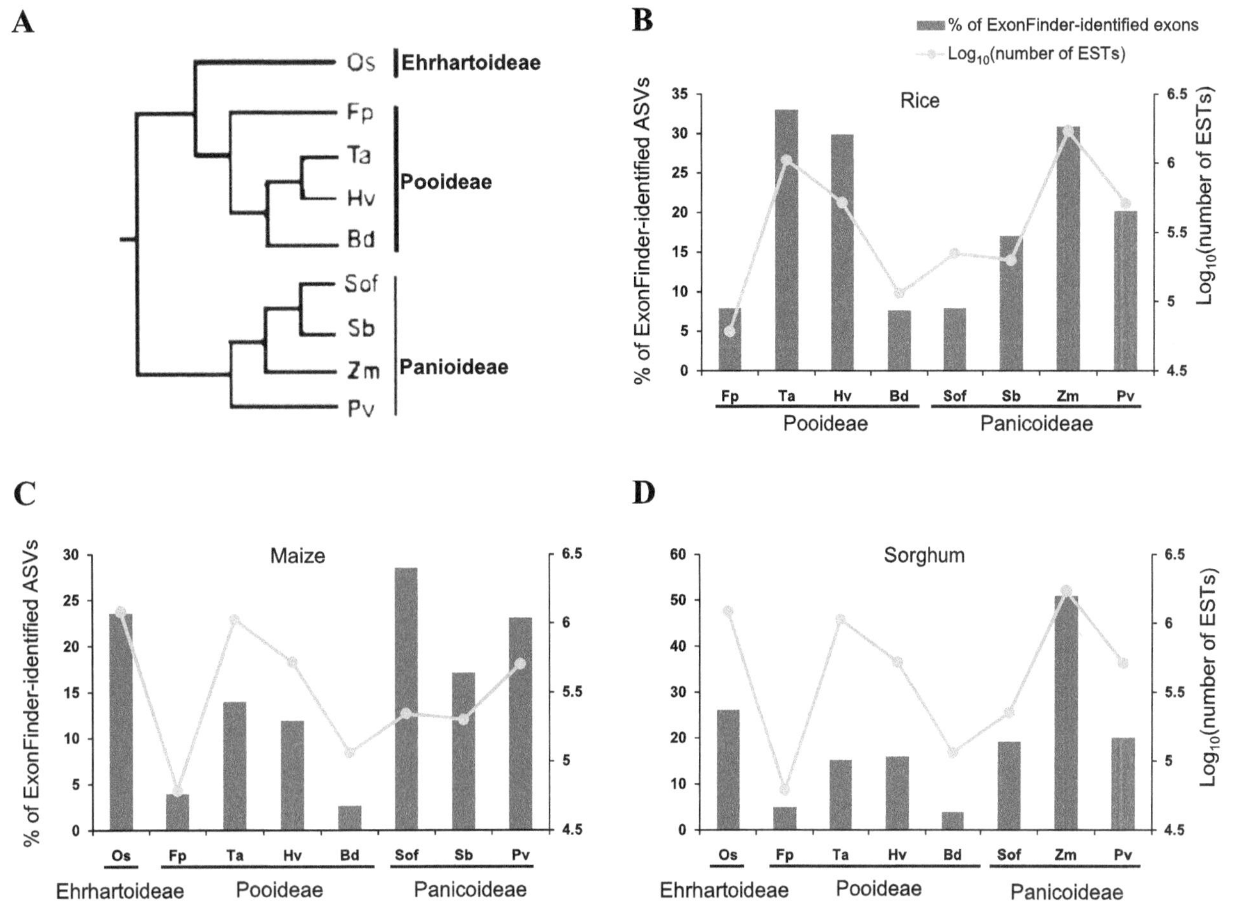

Figure 2 Comparative analysis of the AS events extracted from different subject species. (A) Phylogeny of the nine grass plants examined in this study [53,54]. These plants can be classified into three groups: Ehrhartoideae, Pooideae, and Panicoideae. **(B-D)** Comparison between the percentages of AS events identified from EST traces and the numbers of available EST traces of each subject species for Exonfinder identifications in three target species: rice **(B)**, maize **(C)**, and sorghum **(D)**. Os, rice; Fp, meadow ryegrass; Ta, wheat; Hv, Barley; Bd, purple false brome; Sof, sugarcane; Sb, sorghum; Zm, maize; Pv, switchgrass.

may be subject to erroneous prediction. Relatively, it is more important to examine these potentially erroneous predictions.

Certain previously-annotated isoforms remain non-evident by the existing transcript sequences

Taking rice as example, we then proceeded to confirm the authenticity of the newly-identified ASVs (i.e., the isoforms that include the novel exons and their flanking exons) and the previously-annotated ASVs (i.e., the isoforms that exclude the novel exons). Since the novel exons/ASVs identified here were based on the Ensembl annotation, we randomly selected 16 newly-identified ASVs and performed RT-PCR-sequencing experiments to examine their authenticity on a rice cultivar (i.e., *O. sativa* L. ssp. *japonica* cv. Nipponbare; Methods). The result showed that 13 of them (81%) were detected in *japonica* (Figure 4A and Additional file 2), supporting the effectiveness of ExonFinder. Intriguingly, while 13 novel AS isoforms were experimentally validated, more than half (54%; 7/13) of their previously-annotated isoforms were not detected (Figure 4A). We examined the alignments between rice EST traces (NCBI UniGene Database; Table 1) and the reference genome, and confirmed that no rice EST supported these previously-annotated isoforms. We further BLAST-aligned these previously-annotated transcript isoforms against the NCBI non-redundant database (Oct. 2014) and showed the absences of their homologous expressed sequences within other grass species. These results indicated that the previously-annotated isoforms were likely to be false positives. However, we cannot completely eliminate the possibility that these transcript isoforms are just absent in *japonica*, but are present in other cultivated or wild rice. To test this possibility, we attempted to detect these 13 newly-identified ASVs and their previously-annotated ASVs in other two cultivars (i.e., *O. sativa* L. ssp. *indica* cv. 93-11 and *O. glaberrima*) and two wild species (i.e., *O. rufipogon* and *O. nivara*)

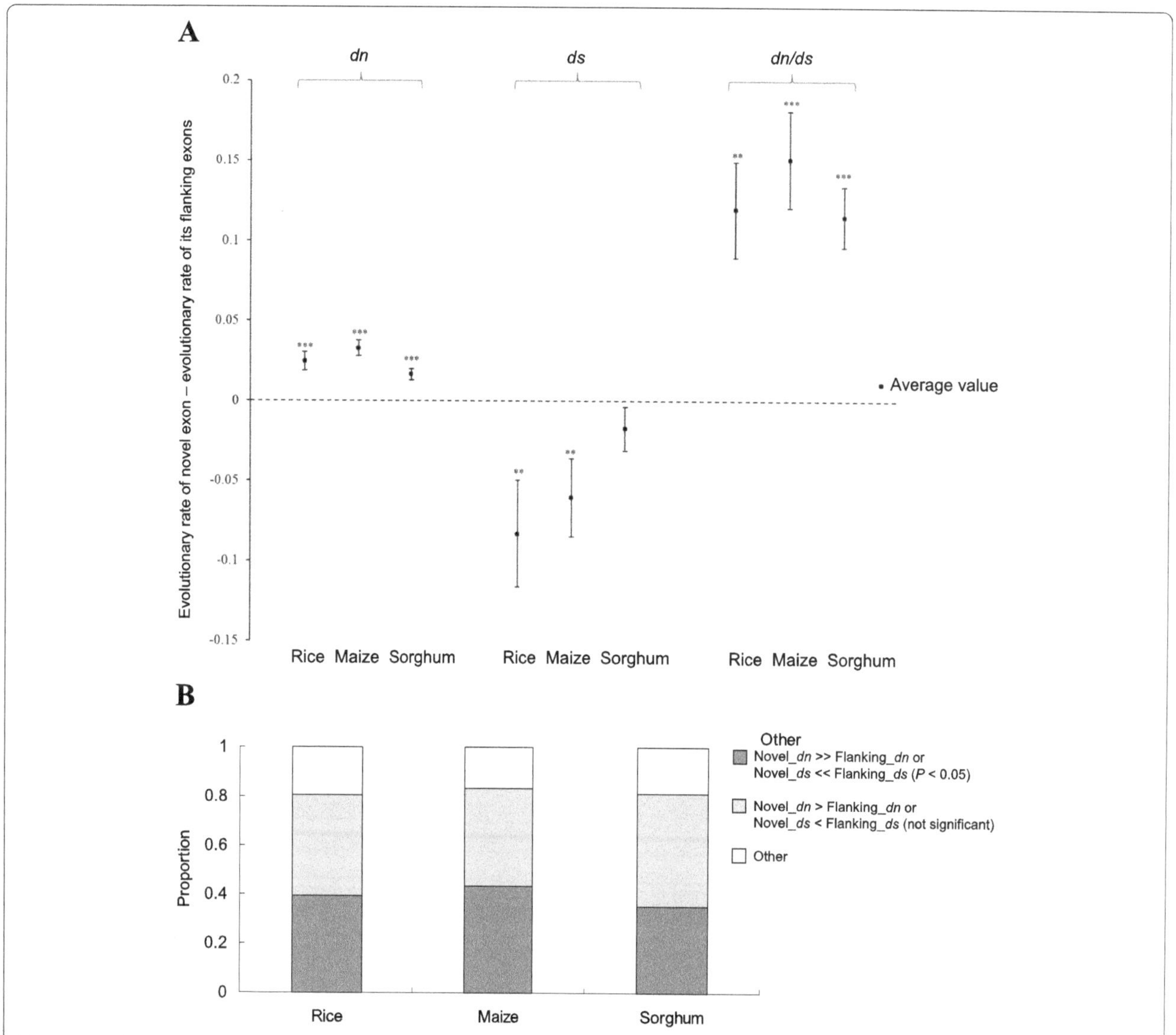

Figure 3 Evolutionary analysis of the newly-identified exons and their flanking exons. (A) Comparisons of evolutionary rates (*dn*, *ds*, and *dn/ds*) between the newly-identified exons and their flanking exons. Statistical significance was estimated by the paired two-tailed Wilcoxon signed rank-sum test. $^{**}P < 0.01$ and $^{***}P < 0.001$. Error bars represent the standard errors of the means. **(B)** Proportions of newly-identified ASVs with and without significant differences in evolutionary rates between the novel exons and their flanking exons ($P < 0.05$ by the two-tailed Fisher's exact test; Methods). Novel_*dn* and Novel_*ds* represent the *dn* and *ds* values of the novel exons; Flanking_*dn* and Flanking_*ds* represent the *dn* and *ds* values of their flanking exons, respectively.

(Methods). Our results revealed that the 13 novel isoforms were steadily detected in all of the rice species examined, but the previously-annotated isoforms that were not detected in *japonica* were also absent in other rice species examined (Figure 4B). These results support that certain previously-annotated ASVs may be subject to erroneous prediction. In fact, except for Os06g0472300, all the previously-annotated isoforms that were not detected in our experiments have not included in the mostly updated version of the Ensembl annotation (Release 23). Of note, the three newly-identified ASVs that could not be detected in *japonica* were also absent in the other rice species examined (Additional file 2). Although it is possible that these exons might be lost in rice and became pure introns during evolution, we observed that two of them (Os04g28460 and Os11g34120) had a *dn/ds* ratio significantly smaller than 1 (both P values < 0.05 by the Fisher's exact test). This indicates that these two newly-identified exons are subject to much stronger selective constrains on nonsynonymous changes than on synonymous ones [57-59], suggesting that they are more likely to be protein-coding exons.

Of the 13 experimentally-confirmed novel exons, 12 locate within CDS regions (Additional file 2). We observed

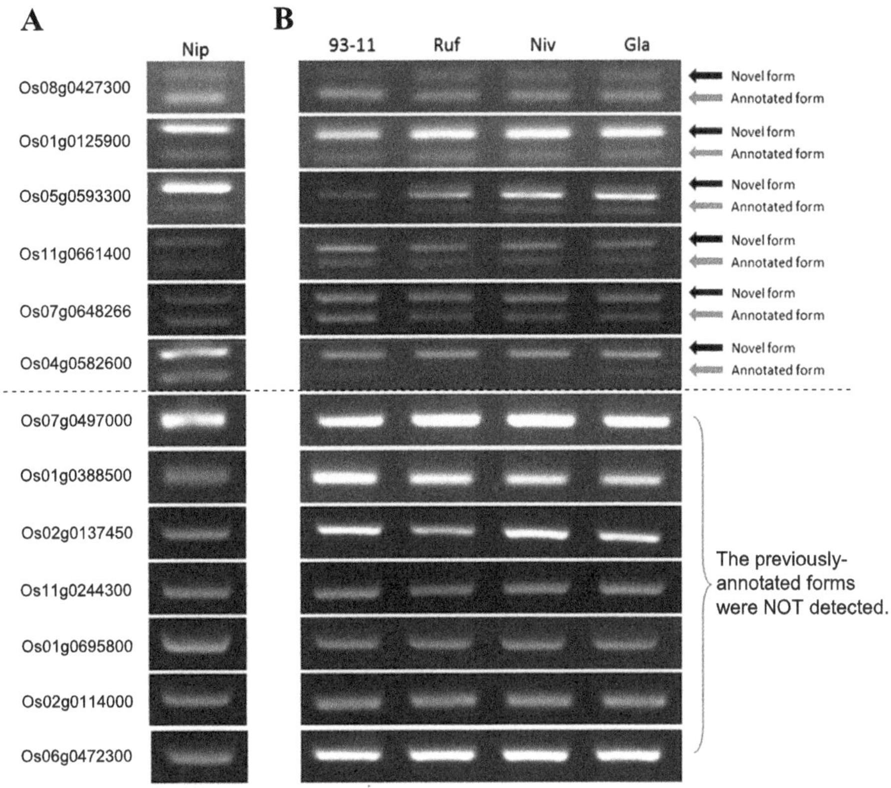

Figure 4 Experimental validations of the newly-identified exons/ASVs. Shown in the figure are. RT-PCR products of the newly-identified isoforms that include the novel exons and the previously-annotated isoforms that exclude the novel exons in **(A)** *O. sativa* L. ssp. *japonica* cv. Nipponbare (designated as "Nip") and **(B)** *O. sativa* L. ssp. *indica* cv. 93-11 (designated as "93-11"), *O. rufipogon* (designated as "Ruf"), *O. nivara* (designated as "Niv"), and *O. glaberrima* (designated as "Gla"). The black and gray arrows represent the newly-identified and previously-annotated isoforms, respectively.

that five exhibited significantly higher *dn* values or significantly lower *ds* values than their flanking exons, four of which were validated to be alternatively spliced (Figure 4 and Additional file 2). In contrast, the novel exons that exhibited neither higher *dn* values nor lower *ds* values than their flanking exons were not validated to be alternatively spliced (Additional file 2). This observation is consistent with the overall trend towards higher *dn* and lower *ds* values in alternatively spliced (or rarely utilized) exons as compared to constitutively spliced (or commonly utilized) exons, further suggesting that our evolutionary analysis is helpful for determining whether a newly-identified exon undergoes AS.

Implications of newly-identified ASVs for evolutionary studies

According to our experimental validation, there were six genes (i.e., Os08g0427300, Os01g0125900, Os05g0593300, Os11g0661400, Os07g0648266, and Os04g0582600) in which the previously-annotated isoforms that exclude the novel exons (designated as "ASV1") and newly-identified isoforms that include the novel exons (designated as "ASV2") were steadily detected in all rice species examined (Figure 4). Since both ASV1 and ASV2 were detected in Asian cultivated/wild rice and African cultivated rice, we hypothesized that both isoforms for each of the six genes might have been present in the common ancestral transcriptome of African and Asian rice species. Moreover, since the novel exons were derived from comparative analysis of non-rice EST traces, we speculated that ASV2 might also represent a common ancestral isoform of grass plants. As for ASV1, there are two possible scenarios. First, both ASV1 and ASV2 might be present in the common ancestral transcriptome of grass plants, inferring that the novel exons exhibited alternatively spliced exons (ASEs) in both rice and other grass plants (designated as "conserved ASEs") (Figure 5A). This implies that both AS isoforms are functionally important across grass plants. Second, ASV1 might represent a gain-of-ASV event that occurred after the divergence between rice and non-rice plants, inferring that the novel exons were constitutively spliced exons (CSEs) in the common ancestral transcriptome of grass plants (designated as "lineage-specific ASEs") (Figure 5B). This implies that ASV1 may play

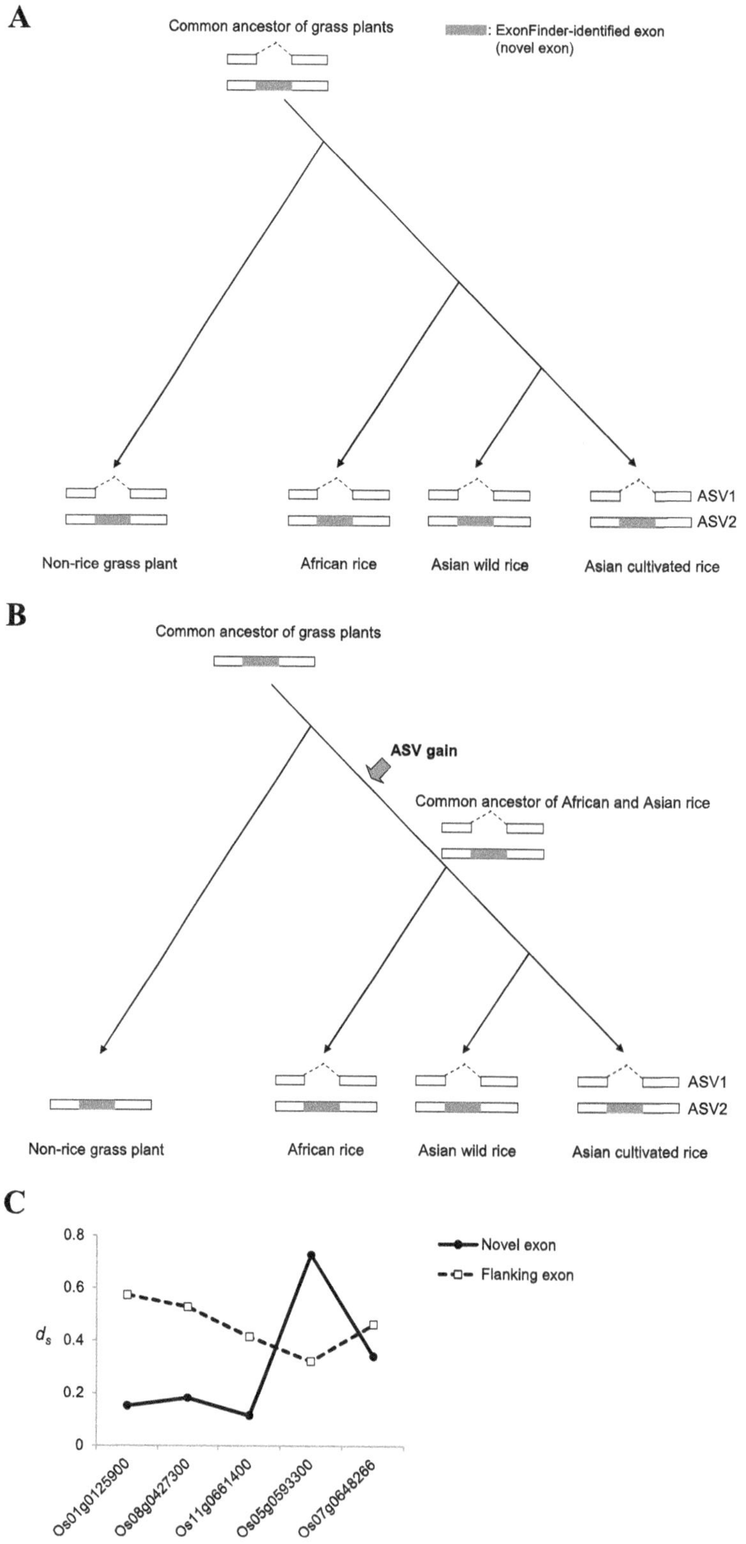

Figure 5 Possible evolutionary scenarios of the previously-annotated isoforms that exclude the novel exons (ASV1) and the newly-identified isoforms that include the novel exons (ASV2) during the evolution of rice transcriptome. (A) Both isoforms (ASV1 and ASV2) might have been present in the common ancestral transcriptome of grass plants. **(B)** A gain-of-ASV event might occur after the divergence of rice and non-rice plants. **(C)** Comparison of *ds* values of novel exons and their corresponding flanking exons.

a lineage-specific role in rice. Our previous study has showed that the *ds* values of conserved ASEs were markedly lower than those of both lineage-specific ASEs and CSEs [40], providing a possible way to examine whether the novel exons are conserved ASEs. To this end, on the basis of the rice-maize-sorghum orthologues (Additional file 3) and the phylogenetic context of these three species, we calculated the evolutionary rates of the rice transcript sequences and their orthologous sequences derived from the rice-maize-sorghum common ancestor using the CodeML program of PAML [60,61]. As shown in Figure 5C, the *ds* values of the novel exons were lower by three-fold or more compared with those of their flanking exons for Os08g0427300, Os01g0125900, and Os11g0661400, suggesting that the novel exons were subjected to be alternatively spliced in the rice-maize-sorghum common ancestral transcriptome. Meanwhile, for Os05g0593300 and Os07g0648266, the *ds* values of the novel exons were greater or insignificantly lower than those of their flanking exons (Figure 5C), inferring that the novel exons might be lineage- or rice-specific ASEs. Of note, Os04g0582600 was not considered due to the lack of the information of orthologues. We further aligned ASV1/ASV2 against currently-available non-rice transcripts and found that non-rice transcript evidence supported both ASV1 and ASV2 in Os08g0427300, Os01g0125900, and Os11g0661400, while non-rice evidence only supported ASV2 in Os05g0593300 and Os07g0648266 (Additional file 4). This result also supported the above speculation.

In summary, the above examples illustrate that the identified ASVs can serve a source for inferring the ancestral transcriptomes of rice and other grass plants. If the newly-identified ASVs (ASV2) were not considered in either of the above scenarios, one might speculate that ASV2 had been lost in rice, and the interpretation of transcriptome evolution could be incomplete or even misleading. The ASVs that were inferred from such a comparative analysis of cross-species EST library therefore provide new insights into evolutionary transcriptomic studies.

Implications of distinct ASVs for analysis of expression divergence

We then probed expression divergence of distinct ASVs (i.e., ASV1 and ASV2) among the five rice species examined. We analyzed the expression profiles of ASV1 and ASV2 for Os08g0427300, Os01g0125900, Os05g0593300, Os11g0661400, and Os07g0648266 by qRT-PCR (Figure 6). Of note, Os04g0582600 was not considered here because of difficulties in generating suitable primers for qRT-PCR. Two intriguing observations were made. First, ASV1 and ASV2 exhibited significantly different expression levels for all five genes in all rice species examined (all P values < 0.01 by the two-tailed t-test; Figure 6), suggesting that these two distinct AS isoforms might play different functional roles. Importantly, for Os05g0593300, Os11g0661400, and Os07g0648266, the expression levels of ASV2 were remarkably higher than those of ASV1 in all rice species examined, indicating that the newly-identified isoforms (i.e., ASV2) predominated over their previously-annotated counterparts (i.e., ASV1) for these genes. Second, the trend that ASV1 was more highly expressed than ASV2 for Os08g0427300 and Os01g0125900 but the reverse was true for Os05g0593300, Os11g0661400, and Os07g0648266 was observed in all five rice species examined (Figure 6). These results suggested that such ASV1 and ASV2 expression profiles for the five genes were present in the ancestral transcriptome before the domestication of Asian/African rice. Since *O. sativa* (such as *japonica* and *indica*; two Asian rice cultivars) and *O. glaberrima* (an African cultivated rice species) have independent histories of domestication [62,63], maintenance of such expression profiles may be of great importance during the domestication and evolution of rice transcriptome.

Discussion

In this study, we described an *in silico* pipeline ExonFinder to identify novel exons/ASVs based on comparative analysis of cross-species EST library. Using ExonFinder we identified 2,963 ASVs with 2,879 novel exons (including cassette exons and retained introns) that were previously unannotated in rice, maize, and sorghum. RT-PCR-sequencing confirmed the authenticity of 81% of the tested ASVs, supporting the effectiveness of the ExonFinder pipeline. Cross-species conservation of these exons/ASVs implies their biological importance and functional properties. In addition, a considerable proportion of newly-identified exons have no significant difference in evolutionary rates as compared to their flanking exons, suggesting that these novel exons and their flanking partners tend to co-occur in the transcripts (Figure 3B). While 13 novel ASVs were experimentally validated, 54% of their corresponding previously-annotated ASVs were not detected (Figure 4A and B). Such results were consistent across multiple rice species including cultivated and wild rice species (Figure 4A and B). This reveals that some of the previous annotations might be subject to erroneous prediction. These observations also indicate the capability and usefulness of ExonFinder for the curation and improvement of current plant genome annotations.

Regarding AS patterns, intron-retention events were observed to be the most prevalent AS event in plants such as rice and *Arabidopsis*, contributing to a higher proportion of all ASVs than cassette exons [10,14,26]. However, ExonFinder identified fewer retained introns than cassette exons (Table 2). There are several possibilities. First, the majority of retained introns are subject to nonsense-mediated mRNA decay [26,64], which tend to

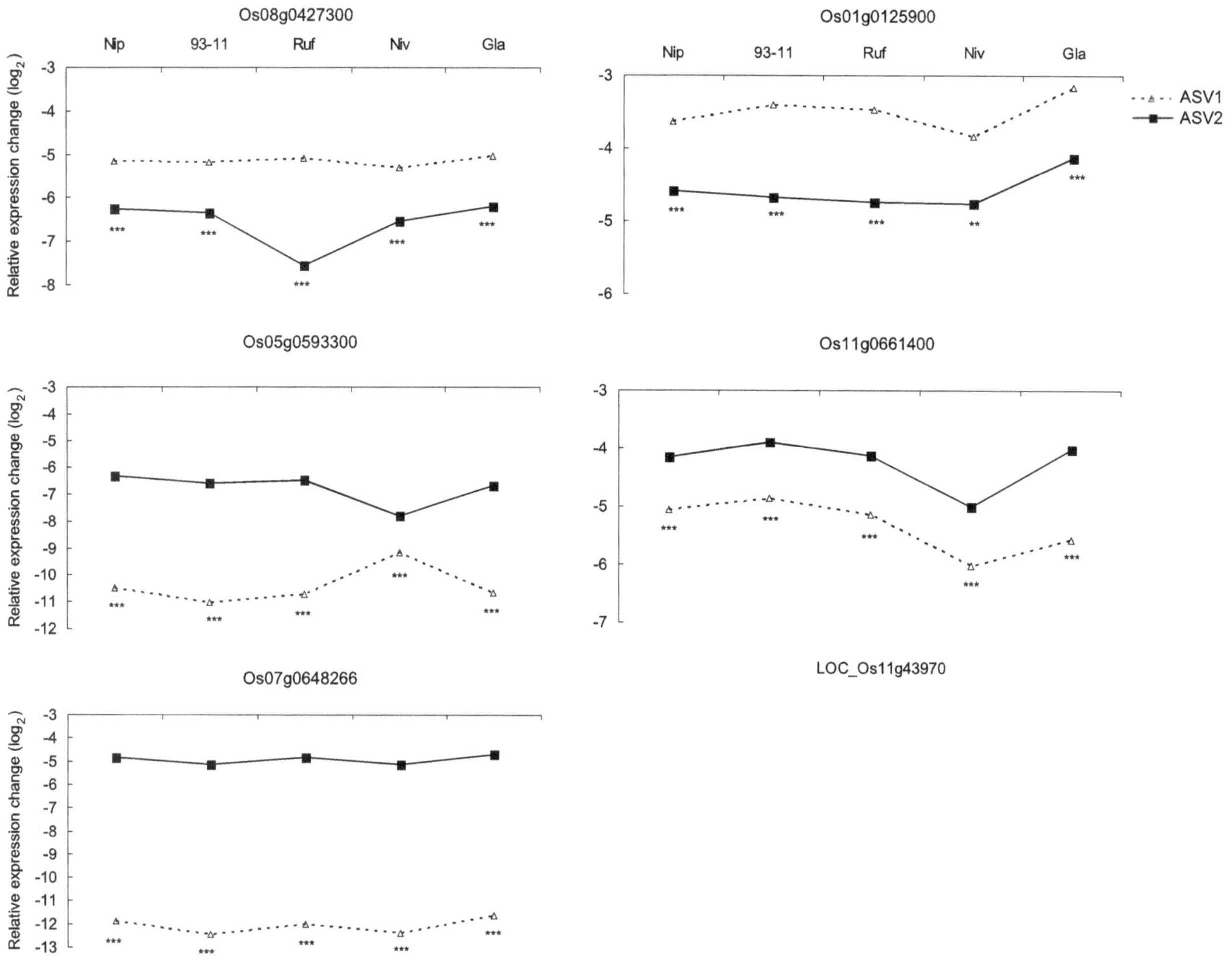

Figure 6 The qRT-PCR expression analysis of the previously-annotated isoforms that exclude the novel exons (ASV1) and the newly-identified isoforms that include the novel exons (ASV2) in the rice cultivars and wild species. Statistical significance was estimated by the two-sided *t*-test. $^{**}P < 0.01$ and $^{***}P < 0.001$.

locate at the 3′ terminal of transcripts. ExonFinder was designed for the accurate identification of internal exons, resulting in the limitation for the detection of exons close to the terminal of transcripts. This property also accounts for the observation that the number of UTR exons identified by ExonFinder is smaller than that of CDS exons (Table 2). Second, most of the retained introns in rice were supported by rice EST traces, and therefore were removed during the examination of exon novelty. In fact, before this examination, the number of identified retained introns (1,040) was much larger than that of identified cassette exons (533) (Additional file 5). Third, a substantial number of retained introns may be derived from spurious EST traces, such as genomic DNA contamination and incompletely-processed transcripts [26].

ExonFinder has the capability to identify "full-length" novel exons with cross-species EST evidence. For novel cassette exons, each identified ASV must have at least one EST from other species to support the corresponding exon-intron boundaries and two previously-annotated flanking exons (Figure 1B). For novel retained introns, ExonFinder requires EST evidence from other species to support the corresponding region along with the flanking exons (Figure 1B). We also demonstrated that ExonFinder is capable of inferring gain- and loss-of-ASV events, which may be valuable for evolutionary studies in the era of transcriptomics. Although the lack of an adequate quantity of EST traces in many plant species may hamper such a comparative transcriptome analysis, the rapid progress in high-throughput RNA sequencing (RNA-seq) technologies can provide alternative resources for this task. For example, transcript assemblers (e.g., Cufflinks [65] and Trinity [66]) can subsequently be used to generate preliminary multiple-exon transcript segments based on integration of long (e.g., PacBio-based) and short (e.g., Illumina-based) RNA-seq reads [67]. In this regard, ExonFinder can make a knowledge-based complement to these strategies, by revising the transcriptome annotation through the lens of cross-species comparative analysis.

For accuracy, ExonFinder does not consider the candidates that change the reading frame of the corresponding transcript or result in premature stop codons; however, some of these exons may be fragments of non-coding RNAs that are expressed in the target species. For example, we showed an exon candidate that resulted in a premature stop codon was experimentally-validated in multiple rice cultivars and wild species (Additional file 6), suggesting that this exon might be of biological significance in rice transcriptome. Thus, it is also possible to utilize ExonFinder to identify non-coding RNAs with cross-species conservations. Moreover, we showed that evolutionary rates might serve as an indicator for determining whether the identified exons were alternatively spliced. Integration of comparative analysis (ExonFinder) and evolutionary analysis (evolutionary rates) may enable the accurate identification of novel AS events by using publicly-available EST traces, without requiring costly experiments. In conclusion, this study not only presents an *in silico* pipeline for accurate identification of novel exons/ASVs, but also expands the discovery of AS events in the evolution of plant transcriptomes.

Conclusions

We have described a computational pipeline (ExonFinder) to identify previously-unannotated exons and ASVs using cross-species EST-to-genome comparisons. RT-PCR-sequencing confirmed 81% of the tested exons/ASVs, supporting the effectiveness of our *in silico* strategy. Exonfinder thus reannotated the rice, maize, and sorghum genomes, and identified many novel exons/ASVs that are cross-species conserved in grass plant ESTs. Evolutionary analysis further revealed the consistency of evolutionary rates between certain novel exons and their flanking exons, which provided evidence of their co-occurrence in the transcripts, suggesting that previously-annotated isoforms might be subject to erroneous predictions. This also indicated that evolutionary rates might serve as an indicator for determining whether the identified exons were alternatively spliced. Moreover, comparative analyses inferred the common ancestral transcriptome of grass plants and gain- and loss-of-ASV events, providing important targets for evolutionary and functional studies. Exonfinder can be applied to comparative analysis of other model organisms.

Methods

Data retrieval and availability

Genomic sequences and gene annotations of rice (*Oryza sativa*), maize (*Zea mays*), and sorghum (*Sorghum bicolor*) were downloaded from the EnsemblPlants genome browser at http://plants.ensembl.org (release 18). Of note, for rice, EnsemblPlants provided annotations from both the MSU Rice Genome Annotation Project and the Rice Annotation Project Database (RAP-DB)/International Rice Genome Sequencing Project. EST traces of nine grass species: rice, maize, sorghum, barley (*Hordeum vulgare*), meadow ryegrass (*Festuca pratensis*), purple false brome (*Brachypodium distachyon*), sugarcane (*Saccharum officinarum*), switchgrass (*Panicum virgatum*), and wheat (*Triticum aestivum*) were downloaded from the repository of NCBI UniGene Database at ftp://ftp.ncbi.nih.gov/repository/UniGene/ (Table 1).

The *in silico* pipeline of ExonFinder

ExonFinder was designed to identify novel exons in a target species on the basis of comparative analysis of the public EST library from one or more non-target species (subject species). As shown in Figure 1A, we first used BLAT [68] to align EST traces from subject species (see Table 1) against the genome of the target species. We considered the best hit in the case of multiple matches. We then extracted the genomic sequences of the target species that were previously annotated as introns and conserved in subject species ESTs, which represented the candidates of novel exons/ASVs. As shown in Figure 1B, two ASV types were identified: retained introns and cassette exons. Taking rice for example, we found approximately 2.5 million non-rice EST matches that located within introns and without overlapping with any annotated transcript (Figure 1). To guarantee the accuracy and novelty of the identified exons, we only considered the exons that satisfied the following four criteria: (1) the length of an exon candidate should be longer than 50 bp (in the case of cassette exons) and the lengths of both flanking EST segments of the exon candidate should overlap with the corresponding previously-annotated transcript by ≥ 50 bp; (2) a candidate of novel cassette exon must be flanked by canonical splicing sites (i.e., GT-AG, GC-AG, or AT-AC [69]) within a 10-bp window at boundaries of the candidate; non-canonical splicing sites (i.e., AT-AA, AT-AG, AT-AT, GT-AT, or GT-GG [69]) were considered, if the above-mentioned canonical splicing sites can not be detected within such a window; (3) an exon candidate that located within CDS must not alter the reading frame of the corresponding previously-annotated transcript and must not result in any premature stop codon; and (4) an exon candidate should not overlap with any EST trace of the target species or known transcripts from the Ensembl annotation. With these criteria, the number of candidates was largely reduced to several hundreds. All the programs of the ExonFinder pipeline were implemented in C language under Linux environment (i.e., Bio-Linux 6). Of note, ExonFinder also identifies novel exons/ASVs with EST evidence from the same species; users can choose whether or not to remain this information.

Comparison of *dn* and *ds* values between novel exons and their flanking exons

For each newly-identified ASV that located in CDS region and was supported by EST traces from more than one subject species, we selected the best hit of the matched EST traces. The two flanking CDS exons of the novel exon were concatenated. The reading frames of the novel exon and the concatenated sequence were determined according to the Ensembl annotation of the target species. Subsequently, the numbers of synonymous and non-synonymous sites, and the *dn*, *ds*, and *dn/ds* values of the novel exon and the concatenated sequence were respectively calculated using the YN00 program of the PAML package [60,61]. For detecting significant differences in *dn* values between the novel exon and the concatenated sequence (i.e., the flanking exons of the novel exon), we created a two-way contingency table, with rows comprised of the numbers of non-synonymous sites of the novel exon and the concatenated sequence, and columns comprised of the numbers of changed and unchanged sites; this table was used to test the independence of the number of changed nonsynonymous sites between the novel exon and the concatenated sequence using the two-tailed Fisher's exact test. The similar processes were applied to detecting significant differences in *ds* values between the novel exon and the concatenated sequence.

Rice material and growth conditions

Three rice cultivars (*O. sativa* L. ssp. *japonica* cv. Nipponbare, *O. sativa* L. ssp. *indica* cv. 93-11, and *O. glaberrima* [IRGC accession number 96717]) and two wild species (*O. rufipogon* [IRGC accession number 105491] and *O. nivara* [IRGC accession number 100897]) were used in this study. Seeds of each species were sterilized with 2.5% sodium hypochlorite for 20 minutes, and were cleaned with distilled water. Germination of seeds was triggered in petri dishes with wet papers at 37°C in the dark for two days, and then the germinated seeds were moved to culture in beakers (600 mL) containing half-strength Kimura B nutrient solution. Hydroponic cultivation was performed under natural sunlight at 30/25°C day/night and 70% relative humidity in the greenhouse of the Genomics Research Center, Academia Sinica, Taiwan. Nutrient solutions (pH 4.7-4.8) were replaced every three days. Once the third leaf was fully extended, the seedling was immediately collected for use in experiments.

RT-PCR screening

Total RNA was isolated from seedlings using TRIzol reagent (Invitrogen, CA, USA) according to the manufacturer's instructions. RNA was subsequently treated with Turbo DNase I (Ambion, TX, USA) to remove contamination of genomic DNA. For first strand cDNA synthesis, 200 ng total RNA was reverse transcribed with SuperScript II enzyme (Invitrogen, CA, USA). The primer pairs for novel exon screening were designed against the two flanking exons of each exon identified by ExonFinder (Additional file 7). The RT-PCR thermal cycle was as follows: denaturation at 95°C for 5 minutes; 40 cycles of 95°C for 30 seconds, 58°C for 30 seconds, and 72°C for 40 seconds; final extension at 72°C for 7 minutes; and final cooling to 16°C. The amplicons were separated by electrophoresis in a 2% agarose gel, and were visualized with ethidium bromide staining. Images of amplicons were captured using an E-box Vx2 system (Vilber Lourmat, Germany).

Quantitative RT-PCR (qRT-PCR) analysis

To quantify the expression of validated novel exons in rice species, primer pairs specifically targeting well-annotated and novel transcripts were designed (Additional file 8). Quantitative RT-PCR analysis was performed with a KAPA SYBR FAST qPCR Kit (KAPA Biosystems, USA) and a Roche LightCycler 480 Real-Time PCR System (Roche Diagnostics, USA). The amplification program consisted of the following steps: pre-incubation at 95°C for 3 minutes; 40 cycles of amplification (95°C for 10 seconds, 60°C for 20 seconds, and 72°C for 1 second); and melting curve analysis (95°C for 5 seconds, 65°C for 1 minute, and finally 5 - 10 continuous acquisitions to 97°C). The amplicons were further validated by electrophoresis and sequencing. Levels of each gene were normalized to the internal control OsUbiquitin (GENBANK/D12629); each result was shown as the mean of four independent experiments.

Availability of supporting data

The package of ExonFinder along with documentation, and the full list of novel exons/ASVs identified in this study are publicly available at http://exonfinder.sourceforge.net/ and Additional file 1, respectively. The multiple alignments of rice-maize-sorghum orthologous sequences of the ExonFinder-identified exons and their concatenated flanking exons for Os08g0427300, Os01g0125900, Os05g0593300, Os11g0661400, and Os07g0648266 are available in Additional file 3. The RT-PCR and Q-RT-PCR primer pairs used in this study are listed in Additional files 7 and 8, respectively.

Additional files

Additional file 1: The full list of novel rice, maize, and sorghum exons/ASVs identified in this study.

Additional file 2: Experimental validation and evolutionary examination of the previously-annotated isoforms that exclude the novel exons and the newly-identified isoforms that include the novel exons.

Additional file 3: The multiple alignments of rice-maize-sorghum orthologous sequences of novel exons and their concatenated flanking exons for Os08g0427300, Os01g0125900, Os05g0593300, Os11g0661400, and Os07g0648266.

Additional file 4: Non-rice EST evidence of the ASVs that include (novel ASVs) or exclude (annotated ASVs) the novel exons within the six genes: Os08g0427300, Os01g0125900, Os05g0593300, Os11g0661400, Os07g0648266, and Os04g0582600.

Additional file 5: Number of identified rice exons/ASVs without filtering out the exons/ASVs supported by rice EST traces.

Additional file 6: (A) Gene structure and (B) RT-PCR products of the ExonFinder-extracted exon containing a premature stop codon within Os01g0692300. Ruf, *O. rufipogon*; Niv, *O. nivara*; Nip, *O. sativa* L. ssp. *japonica* cv. Nipponbare; 93-11, *O. sativa* L. ssp. *indica* cv. 93-11; Gla, *O. glaberrima*.

Additional file 7: RT-PCR primer pairs used in this study.

Additional file 8: Q-RT-PCR primer pairs used in this study.

Abbreviations

AS: Alternative splicing; ASV: Alternatively spliced variant; EST: Expressed sequence tag; UTR: Untranslated region; CDS: Coding sequence; *dn*: non-synonymous substitution rate; *ds*: synonymous substitution rate; Nip: *Oryza sativa* L. ssp. *japonica* cv. Nipponbare; 93-11: *O. sativa* L. ssp. *indica* cv. 93-11; Gla: *O. glaberrima*; Ruf: *O. rufipogon*; Niv: *O. nivara*; Os: rice; Fp: Meadow ryegrass; Ta: wheat; Hv: Barley; Bd: Purple false brome; Sof: Sugarcane; Sb: Sorghum; Zm: Maize; Pv: switchgrass.

Competing interests

The authors declare that they have no competing interests.

Authors' contributions

TJC conceived/designed this study, interpreted the data, and wrote the manuscript. MYY and PHH collected the data, grew the rice species, and performed all experimental validations. CCL implemented the ExonFinder package. LYH helped to write the draft. All authors read and approve the final manuscript.

Acknowledgments

We thank Prof. Yue-ie Hsing, Dr. Yuan-Ching Tsai, Prof. Chwan-Yang Hong, and International Rice Germplasm Collection (IRGC) for providing seeds of three rice cultivars (i.e., *O. sativa* L. ssp. *Indica* cv. 93-11, *O. sativa* L. ssp. *japonica* cv. Nipponbare, and *O. glaberrima*) and two wild species (i.e., *O. rufipogon* and *O. nivara*). We also thank Prof. Wen-Hsiung Li and the Phytotron at the Genomics Research Center, Academia Sinica, for providing experimental facilities. This work was supported by the Genomics Research Center, Academia Sinica, Taiwan; and the Ministry of Science and Technology (MOST), Taiwan (under the contract MOST 103-2628-B-001-001-MY4, MOST 103-2911-I-001-507, and MOST 103-2911-I-001-502).

References

1. Pan Q, Shai O, Lee LJ, Frey BJ, Blencowe BJ. Deep surveying of alternative splicing complexity in the human transcriptome by high-throughput sequencing. Nat Genet. 2008;40(12):1413–5.
2. Wang ET, Sandberg R, Luo S, Khrebtukova I, Zhang L, Mayr C, et al. Alternative isoform regulation in human tissue transcriptomes. Nature. 2008;456(7221):470–6.
3. Chen FC, Wang SS, Chaw SM, Huang YT, Chuang TJ. Plant gene and alternatively spliced variant annotator. A plant genome annotation pipeline for rice gene and alternatively spliced variant identification with cross-species expressed sequence tag conservation from seven plant species. Plant Physiol. 2007;143(3):1086–95.
4. Zhang G, Guo G, Hu X, Zhang Y, Li Q, Li R, et al. Deep RNA sequencing at single base-pair resolution reveals high complexity of the rice transcriptome. Genome Res. 2010;20(5):646–54.
5. Liang C, Mao L, Ware D, Stein L. Evidence-based gene predictions in plant genomes. Genome Res. 2009;19(10):1912–23.
6. Lu T, Lu G, Fan D, Zhu C, Li W, Zhao Q, et al. Function annotation of the rice transcriptome at single-nucleotide resolution by RNA-seq. Genome Res. 2010;20(9):1238–49.
7. Filichkin SA, Priest HD, Givan SA, Shen R, Bryant DW, Fox SE, et al. Genome-wide mapping of alternative splicing in Arabidopsis thaliana. Genome Res. 2010;20(1):45–58.
8. Marquez Y, Brown JW, Simpson C, Barta A, Kalyna M. Transcriptome survey reveals increased complexity of the alternative splicing landscape in Arabidopsis. Genome Res. 2012;22(6):1184–95.
9. Z E, Wang L, Zhou J. Splicing and alternative splicing in rice and humans. BMB Rep. 2013;46(9):439–47.
10. Reddy AS, Marquez Y, Kalyna M, Barta A. Complexity of the alternative splicing landscape in plants. Plant Cell. 2013;25(10):3657–83.
11. Reddy AS, Rogers MF, Richardson DN, Hamilton M, Ben-Hur A. Deciphering the plant splicing code: experimental and computational approaches for predicting alternative splicing and splicing regulatory elements. Frontiers Plant Science. 2012;3:18.
12. Kelemen O, Convertini P, Zhang Z, Wen Y, Shen M, Falaleeva M, et al. Function of alternative splicing. Gene. 2013;514(1):1–30.
13. Singh RK, Cooper TA. Pre-mRNA splicing in disease and therapeutics. Trends Mol Med. 2012;18(8):472–82.
14. Carvalho RF, Feijao CV, Duque P. On the physiological significance of alternative splicing events in higher plants. Protoplasma. 2013;250(3):639–50.
15. Kalsotra A, Cooper TA. Functional consequences of developmentally regulated alternative splicing. Nat Rev Genet. 2011;12(10):715–29.
16. Keren H, Lev-Maor G, Ast G. Alternative splicing and evolution: diversification, exon definition and function. Nat Rev Genet. 2010;11(5):345–55.
17. Nilsen TW, Graveley BR. Expansion of the eukaryotic proteome by alternative splicing. Nature. 2010;463(7280):457–63.
18. McManus CJ, Graveley BR. RNA structure and the mechanisms of alternative splicing. Curr Opin Genet Dev. 2011;21(4):373–9.
19. Blencowe BJ. An exon-centric perspective. Biochem Cell Biol. 2012;90(5):603–12.
20. Vitulo N, Forcato C, Carpinelli EC, Telatin A, Campagna D, D'Angelo M, et al. A deep survey of alternative splicing in grape reveals changes in the splicing machinery related to tissue, stress condition and genotype. BMC Plant Biol. 2014;14(1):99.
21. Dinesh-Kumar SP, Baker BJ. Alternatively spliced N resistance gene transcripts: their possible role in tobacco mosaic virus resistance. Proc Natl Acad Sci U S A. 2000;97(4):1908–13.
22. Parry MA, Keys AJ, Madgwick PJ, Carmo-Silva AE, Andralojc PJ. Rubisco regulation: a role for inhibitors. J Exp Bot. 2008;59(7):1569–80.
23. Wu HP, Su YS, Chen HC, Chen YR, Wu CC, Lin WD, et al. Genome-wide analysis of light-regulated alternative splicing mediated by photoreceptors in Physcomitrella patens. Genome Biol. 2014;15(1):R10.
24. Stamm S, Ben-Ari S, Rafalska I, Tang Y, Zhang Z, Toiber D, et al. Function of alternative splicing. Gene. 2005;344:1–20.
25. Reddy AS. Alternative splicing of pre-messenger RNAs in plants in the genomic era. Annu Rev Plant Biol. 2007;58:267–94.
26. Barbazuk WB, Fu Y, McGinnis KM. Genome-wide analyses of alternative splicing in plants: opportunities and challenges. Genome Res. 2008;18(9):1381–92.
27. Hughes TA. Regulation of gene expression by alternative untranslated regions. TRENDS Genetics. 2006;22(3):119–22.
28. Sorek R, Shamir R, Ast G. How prevalent is functional alternative splicing in the human genome? TRENDS Genetics. 2004;20(2):68–71.
29. Lareau LF, Green RE, Bhatnagar RS, Brenner SE. The evolving roles of alternative splicing. Curr Opin Struct Biol. 2004;14(3):273–82.
30. McIntyre LM, Lopiano KK, Morse AM, Amin V, Oberg AL, Young LJ, et al. RNA-seq: technical variability and sampling. BMC Genomics. 2011;12:293.
31. Melamud E, Moult J. Stochastic noise in splicing machinery. Nucleic Acids Res. 2009;37(14):4873–86.
32. Tarazona S, Garcia-Alcalde F, Dopazo J, Ferrer A, Conesa A. Differential expression in RNA-seq: a matter of depth. Genome Res. 2011;21(12):2213–23.
33. Lu H, Lin L, Sato S, Xing Y, Lee CJ. Predicting functional alternative splicing by measuring RNA selection pressure from multigenome alignments. PLoS Comp Biol. 2009;5(12):e1000608.
34. Ouyang S, Zhu W, Hamilton J, Lin H, Campbell M, Childs K, et al. The TIGR Rice Genome Annotation Resource: improvements and new features. Nucleic Acids Res. 2007;35(Database issue):D883–7.

35. Yu J, Wong GK, Liu S, Wang J, Yang H. A comprehensive crop genome research project: the Superhybrid Rice Genome Project in China. Philos Trans R Soc Lond. 2007;362(1482):1023–34.
36. Sakai H, Lee SS, Tanaka T, Numa H, Kim J, Kawahara Y, et al. Rice Annotation Project Database (RAP-DB): an integrative and interactive database for rice genomics. Plant Cell Physiol. 2013;54(2):e6.
37. Kawahara Y, de la Bastide M, Hamilton JP, Kanamori H, McCombie WR, Ouyang S, et al. Improvement of the Oryza sativa Nipponbare reference genome using next generation sequence and optical map data. Rice (N Y). 2013;6(1):4.
38. Monaco MK, Stein J, Naithani S, Wei S, Dharmawardhana P, Kumari S, et al. Gramene 2013: comparative plant genomics resources. Nucleic Acids Res. 2014;42(Database issue):D1193–9.
39. Kersey PJ, Allen JE, Christensen M, Davis P, Falin LJ, Grabmueller C, et al. Ensembl Genomes 2013: scaling up access to genome-wide data. Nucleic Acids Res. 2014;42(Database issue):D546–52.
40. Chen FC, Chaw SM, Tzeng YH, Wang SS, Chuang TJ. Opposite evolutionary effects between different alternative splicing patterns. Mol Biol Evol. 2007;24(7):1443–6.
41. Chen FC, Chen CJ, Ho JY, Chuang TJ. Identification and evolutionary analysis of novel exons and alternative splicing events using cross-species EST-to-genome comparisons in human, mouse and rat. BMC Bioinformatics. 2006;7:136.
42. Chen FC, Chuang TJ. Different alternative splicing patterns are subject to opposite selection pressure for protein reading frame preservation. BMC Evol Biol. 2007;7(1):179.
43. Wu GC, Chen FC. Determinants of exon-level evolutionary rates in Arabidopsis species. Evol Bioinformatics Online. 2012;8:389–415.
44. Xing Y, Lee C. Evidence of functional selection pressure for alternative splicing events that accelerate evolution of protein subsequences. Proc Natl Acad Sci U S A. 2005;102(38):13526–31.
45. Gelly JC, Lin HY, de Brevern AG, Chuang TJ, Chen FC. Selective constraint on human pre-mRNA splicing by protein structural properties. Genome Biol Evol. 2012;4(9):966–75.
46. Chen FC, Liao BY, Pan CL, Lin HY, Chang AY. Assessing determinants of exonic evolutionary rates in mammals. Mol Biol Evol. 2012;29(10):3121–9.
47. Chen FC, Wang SS, Chen CJ, Li WH, Chuang TJ. Alternatively and constitutively spliced exons are subject to different evolutionary forces. Mol Biol Evol. 2006;23(3):675–82.
48. Darracq A, Adams KL. Features of evolutionarily conserved alternative splicing events between Brassica and Arabidopsis. New Phytol. 2013;199(1):252–63.
49. Yuan Q, Ouyang S, Wang A, Zhu W, Maiti R, Lin H, et al. The institute for genomic research osa1 rice genome annotation database. Plant Physiol. 2005;138(1):18–26.
50. Haberer G, Young S, Bharti AK, Gundlach H, Raymond C, Fuks G, et al. Structure and architecture of the maize genome. Plant Physiol. 2005;139(4):1612–24.
51. Paterson AH, Bowers JE, Bruggmann R, Dubchak I, Grimwood J, Gundlach H, et al. The Sorghum bicolor genome and the diversification of grasses. Nature. 2009;457(7229):551–6.
52. Yeo GW, Van Nostrand E, Holste D, Poggio T, Burge CB. Identification and analysis of alternative splicing events conserved in human and mouse. Proc Natl Acad Sci U S A. 2005;102(8):2850–5.
53. Kellogg EA. Evolutionary history of the grasses. Plant Physiol. 2001;125(3):1198–205.
54. Bouchenak-Khelladi Y, Salamin N, Savolainen V, Forest F, Bank M, Chase MW, et al. Large multi-gene phylogenetic trees of the grasses (Poaceae): progress towards complete tribal and generic level sampling. Mol Phylogen Evol. 2008;47(2):488–505.
55. Modrek B, Lee CJ. Alternative splicing in the human, mouse and rat genomes is associated with an increased frequency of exon creation and/or loss. Nat Genet. 2003;34(2):177–80.
56. Xing Y, Lee C. Alternative splicing and RNA selection pressure–evolutionary consequences for eukaryotic genomes. Nat Rev Genet. 2006;7(7):499–509.
57. Nekrutenko A, Chung WY, Li WH. An evolutionary approach reveals a high protein-coding capacity of the human genome. Trends Genet. 2003;19(6):306–10.
58. Nekrutenko A, Chung WY, Li WH. ETOPE: Evolutionary test of predicted exons. Nucleic Acids Res. 2003;31(13):3564–7.
59. Nekrutenko A, Makova KD, Li WH. The K(A)/K(S) ratio test for assessing the protein-coding potential of genomic regions: an empirical and simulation study. Genome Res. 2002;12(1):198–202.
60. Yang Z. PAML 4: phylogenetic analysis by maximum likelihood. Mol Biol Evol. 2007;24(8):1586–91.
61. Yang Z, Nielsen R. Estimating synonymous and nonsynonymous substitution rates under realistic evolutionary models. Mol Biol Evol. 2000;17(1):32–43.
62. Ma J, Bennetzen JL. Rapid recent growth and divergence of rice nuclear genomes. Proc Natl Acad Sci U S A. 2004;101(34):12404–10.
63. Sakai H, Ikawa H, Tanaka T, Numa H, Minami H, Fujisawa M, et al. Distinct evolutionary patterns of Oryza glaberrima deciphered by genome sequencing and comparative analysis. Plant J. 2011;66(5):796–805.
64. Maquat LE. Nonsense-mediated mRNA decay: splicing, translation and mRNP dynamics. Nat Rev Mol Cell Biol. 2004;5(2):89–99.
65. Trapnell C, Williams BA, Pertea G, Mortazavi A, Kwan G, van Baren MJ, et al. Transcript assembly and quantification by RNA-Seq reveals unannotated transcripts and isoform switching during cell differentiation. Nat Biotechnol. 2010;28(5):511–5.
66. Grabherr MG, Haas BJ, Yassour M, Levin JZ, Thompson DA, Amit I, et al. Full-length transcriptome assembly from RNA-Seq data without a reference genome. Nat Biotechnol. 2011;29(7):644–52.
67. Tilgner H, Grubert F, Sharon D, Snyder MP. Defining a personal, allele-specific, and single-molecule long-read transcriptome. Proc Natl Acad Sci U S A. 2014;111(27):9869–74.
68. Kent WJ. BLAT–the BLAST-like alignment tool. Genome Res. 2002;12(4):656–64.
69. Lewandowska D, Simpson CG, Clark GP, Jennings NS, Barciszewska-Pacak M, Lin CF, et al. Determinants of plant U12-dependent intron splicing efficiency. Plant Cell. 2004;16(5):1340–52.

10

Construction of a high-density genetic map and QTLs mapping for sugars and acids in grape berries

Jie Chen[1,2†], Nian Wang[3†], Lin-Chuan Fang[3], Zhen-Chang Liang[1], Shao-Hua Li[1*] and Ben-Hong Wu[1*]

Abstract

Background: QTLs controlling individual sugars and acids (fructose, glucose, malic acid and tartaric acid) in grape berries have not yet been identified. The present study aimed to construct a high-density, high-quality genetic map of a winemaking grape cross with a complex parentage (*V. vinifera* × *V. amurensis*) × ((*V. labrusca* × *V. riparia*) × *V. vinifera*), using next-generation restriction site-associated DNA sequencing, and then to identify loci related to phenotypic variability over three years.

Results: In total, 1 826 SNP-based markers were developed. Of these, 621 markers were assembled into 19 linkage groups (LGs) for the maternal map, 696 for the paternal map, and 1 254 for the integrated map. Markers showed good linear agreement on most chromosomes between our genetic maps and the previously published *V. vinifera* reference sequence. However marker order was different in some chromosome regions, indicating both conservation and variation within the genome. Despite the identification of a range of QTLs controlling the traits of interest, these QTLs explained a relatively small percentage of the observed phenotypic variance. Although they exhibited a large degree of instability from year to year, QTLs were identified for all traits but tartaric acid and titratable acidity in the three years of the study; however only the QTLs for malic acid and β ratio (tartaric acid-to-malic acid ratio) were stable in two years. QTLs related to sugars were located within ten LGs (01, 02, 03, 04, 07, 09, 11, 14, 17, 18), and those related to acids within three LGs (06, 13, 18). Overlapping QTLs in LG14 were observed for fructose, glucose and total sugar. Malic acid, total acid and β ratio each had several QTLs in LG18, and malic acid also had a QTL in LG06. A set of 10 genes underlying these QTLs may be involved in determining the malic acid content of berries.

Conclusion: The genetic map constructed in this study is potentially a high-density, high-quality map, which could be used for QTL detection, genome comparison, and sequence assembly. It may also serve to broaden our understanding of the grape genome.

Keywords: Berry quality, Genetic map, Next-generation sequencing (NGS), QTL analysis, Quantitative trait loci, Restriction-site associated DNA (RAD), *Vitis*

Background

The organoleptic quality of table grapes and the flavor and stability of wine depend strongly on the types of sugars and acids, as well as the total sugar and acid concentration, in the grapes. Generally, fructose and glucose are predominant in berries at maturity, and sucrose is present in smaller quantities [1-3]. They have different levels of sweetness: if sucrose is rated 1, then fructose is 1.75 and glucose 0.75 [4-6]. The main organic acids in grape berries are tartaric and malic acids, which typically account for 90% of total acids [7-9]. Malic acid is involved in many processes that are essential for the health and sustainability of the vine, and tartaric acid plays an important role in maintaining the chemical stability and the color of the wine. Tartaric acid has a stronger acidic flavor than malic (pKa: 3.04 vs. 3.40), and is also more sour [10].

* Correspondence: shhli@ibcas.ac.cn; bhwu@ibcas.ac.cn
†Equal contributors
[1]Beijing Key Laboratory of Grape Science and Enology, and CAS Key Laboratory of Plant Resources, Institute of Botany, Chinese Academy of Sciences, Beijing 100093, P. R. China
Full list of author information is available at the end of the article

Many studies have identified genomic loci that are linked to traits of interest in grapes. Modern strategies for the investigation of loci are based on the construction of genetic linkage maps, which was facilitated by the development of molecular markers. The first maps were constructed based mainly on RAPD [11] and AFLP [12] markers. Since then, a range of markers has been developed, and genetic maps of various grape cultivars and other *Vitis* species have been constructed [13-32]. One of these, a genetic map of a *V. vinifera* cross between Syrah and Pinot Noir, took into account most markers, including 483 SNP, 132 SSR and 379 AFLP markers [31]. Wang *et al.* [33] developed a genetic map with a total of 1 814 SNP markers. For a single SNP marker, the lowest integrity was ~85%. Of these 1 814 SNP markers, 1 545 were homozygous for one parent and heterozygous for the other (960 for lm×ll and 585 for nn×np), constituting 85.2% of all selected SNP markers. However, the other three types of markers that could be mapped on both female and male linkage maps amounted to 14.8% (ab×cd: 77, ef×eg: 171 and hk×hk: 21) [33]. Of these, 1 121 are on the female map, 759 are on the male map, and 1 646 are on the integrated map. This map was produced by combining next generation sequencing (NGS) and restriction-site associated DNA (RAD). Recently, Barba *et al.* [34] also used NGS to construct linkage maps for *V. rupestris* B38 and 'Chardonnay', with 1 146 and 1 215 SNPs each, covering 1 645 and 1 967 cM, respectively, and asserting that NGS was a powerful method for constructing a high-density, high-quality genetic map.

In grapes, quantitative trait loci (QTL) detection has mostly been used to investigate the genes related to resistance to diseases such as powdery and downy mildew and Pierce's disease [20,26,29,35-37], as well as pest resistance [19,20,38-41]. It has also been used to examine the genes related to a range of agronomic traits, e.g. berry size, seed number, mean and total seed fresh and dry weights, berry weight [14,17,20,27,39,42,43], inflorescence and flower morphology, number of inflorescences per shoot, flowering date [26], timing and duration of flowering and of veraison, veraison-ripening interval [14,44], architecture of the inflorescence [45], aroma profile [46], anthocyanin content [47], and number of clusters per vine [42]. In addition, the QTLs controlling sexual traits [26] and fertility [48] have been identified.

The genes controlling sugar and acid production in grapes are extremely complex, because of both the diverse chains of metabolic processes involved and the effect of environmental factors influencing these processes [49]. Viana *et al.* [50] have recently identified some QTLs involved in controlling soluble solid concentrations, pH, and titratable acidity in grape berries, but these explain a small amount of phenotypic variation in these traits. To our knowledge, no QTLs controlling the production of individual sugars and acids in grape berries have yet been identified. Some analyses of QTLs controlling soluble solid concentrations, titratable acidity, pH and the production of individual sugars and acids have, however, been conducted for other fruit tree species, such as peach [51,52], apple [53-56], sour cherry [57] and melon [58].

The aim of this work was to investigate the genetic determination of soluble solid concentrations, titratable acidity, and individual sugars and acids in grape berries. A high-density genetic map was constructed for the population, as described in Wang *et al.* [33]. The map was used in combination with phenotypic data to identify marker-linked loci, after which we identified loci related to phenotypic variability observed over three years. This population was derived from the interspecies cross of cultivars 'Beihong' (BH) and 'E.S.7-11-49' (ES).

Methods

Plant material

The population, which comprised 1 200 individuals, was obtained by crossing BH (*Vitis vinifera* 'Muscat Hamburg' × *V. amurensis*) with ES ((Minnesota 78 (*V. labrusca* 'Beta' × Witt) × *V. riparia*) × *V. vinifera* 'Chenin Blanc') in 2007. We randomly selected 249 individuals for our experiment, and used these to construct the genetic map. Due to plant mortality, poor fruit setting, and environmental factors (e.g. rainfall, hail storms), the number of individuals bearing fruits varied from year to year. Vines were planted in 2008, without replicate, in the vineyard at the Institute of Botany, Chinese Academy of Sciences, Beijing (39°90' N 116°30' E). They were trained to fan-shaped trellises and had single trunks, which facilitated protection during winter. The vines were spaced 1.0 m apart within the row and 2.5 m apart between rows, and rows were north–south oriented. They were maintained under routine cultivation conditions, including irrigation, fertilization, soil management, pruning and disease control.

A random set of fruiting genotypes and the two parents were used in each of the three years of the study (2011–2013). In total, 241 genotypes were used in 2011, 225 in 2012, and 197 in 2013 for phenotypic measurement. Of these, 187 were common to all three years. Three replicates of one or two berry clusters were harvested from each genotype and parent at maturity. Maturity date was estimated primarily by assessing the physical properties of the berries, the ease of removal of berries from pedicels (without berry tissue shriveling because of loss of water), and the change of seed color from bright green to tan-brown [59]. Date of maturity was also estimated partly based on previous records. In addition, by the same person was responsible for berry harvesting for the duration of the study, to ensure consistency in the estimation of date of maturity. Maturity date ranged from 15 August to 15 September in 2011, and from 20 August to 20 September in both 2012 and 2013, depending on the genotype. Harvested clusters were placed in plastic bags on ice and transported immediately to the laboratory, which took ~10 min. This mode of transportation did not result in significant change in tartaric acid concentration relative to normal transportation.

Measurement of sugars and acids

Each replicate was pressed using a hand juicer to extract berry juice. Soluble solids concentration (SSC, °Brix) of the juice was measured with a digital hand-held refractometer (Atago, Tokyo, Japan). A 2 mL sample of juice was diluted to 10 mL with deionized water, and titratable acidity was measured by titration up to pH 8.2 with 0.1 mol·L^{-1} NaOH, and expressed as g·L^{-1} of tartaric acid.

The remaining juice was centrifuged at 5 000 g for 15 min. The supernatants were decanted, passed through a SEP-C18 cartridge (Superclean ENVI C18 SPE), and filtered through a 0.22 μm Sep-Pak filter. The sugar and acid concentrations of the filtered supernatants were measured using a Dionex P680 HPLC system (Dionex Corporation, CA, USA).

Fructose and glucose concentrations were measured using a Shodex RI-101 refractive index detector with a Waters Sugar-Pak I column (300 mm × 6.5 mmI.D., 10 μm particle size) and a guard column cartridge (Sugar-Pak I Guard-Pak Insert, 10 μm particle size). The reference cell was maintained at 40°C. The column was maintained at 90°C using a Dionex TCC-100 thermostated column compartment. Degassed, distilled, deionized water at a flow rate of 0.6 mL·min^{-1} was used as the mobile phase. The injection volume was 10 μL.

Malic and tartaric acid concentrations were measured using a Dionex UltiMate3000 detector, with a Dikma PLATISIL ODS column (250 mm × 4.6 mmI.D., 5 μm particle size) and a guard column cartridge (DikmaSpursil C18 Guard Cartridge 3μm, 10 mm × 2.1 mm). The column was maintained at 40°C. Samples were eluted with 0.02 mol·L^{-1} KH_2PO_4 solution with pH 2.4, at a flow rate of 0.8 mL·min^{-1}. Eluted compounds were detected using UV absorbance at 210 nm.

The Chromeleon chromatography data system was used to integrate peak areas according to external standard solution calibrations [60] (reagents from Sigma Chemical Co. Castle Hill, NSW, Australia). Sugar and acid concentrations were expressed in mg·mL^{-1} juice.

DNA extraction

Young leaves (the second and third leaf from the apex) were harvested from each genotype and the two parents at the beginning of the vegetative period (late spring). The samples were immediately stored in liquid nitrogen and transferred to a freezer maintained at −80°C. Samples, weighing 0.5 g were ground in liquid nitrogen and genomic DNA was extracted using DNeasy plant mini prep kit (Qiagen). Briefly, 2 μg genomic DNA from each sample (249 F1 progeny and both parents) was treated with 20 units (U) MseI (New England Biolabs [NEB]) for 60 min at 37°C in a 50 μL reaction. A quick blunting kit (NEB) was used to convert 30 μL of the digested sample to 5'-phosphorylated, blunt-ended DNA in a 50 μL reaction mixture; the reaction was performed with 30 μL of digested sample, 5 μL 10× blunting buffer, 5 μL 1 mM dNTP mix, 2 μL blunting enzyme mix and 8 μL sterile dH_2O at room temperature for 30 min. A 3'-adenine overhang was added to the resulting samples in a 50 μL reaction with 32 μL blunt-ended DNA sample, 5 μL Klenow buffer (10×), 10 μL dATP (1 mM), 3 μL Klenow fragments (3'→5'exo-, 5 U·μL^{-1}) and sterile dH_2O to the final volume at 37°C for 1 h. Then 2 μL of 100 nM P1 and P2 adapter with a 3- to 5- bp plant-specific index (barcode) at the 5' end and a thymine overhang at the 3' end was added to each sample in a 50 μL reaction. A ligation reaction was carried out overnight at 16°C with T4 DNA ligase and 16 samples with different plant indices pooled into one. DNA fragments of 400–500 bp (including the ~120 bp adaptor) were separated on a 1.5% agarose gel and purified using a MiniElute gel extraction kit (Qiagen). Finally, all pooled samples were amplified with Phusion High-Fidelity PCR Master Mix (NEB) for 18 cycles in a 100 μL reaction including 20 μL Phusion master mix, 5 μL of 10 μM modified Solexa amplification primer mix (AP1 and AP1; 2006 Illumina, Inc., allright reserved) and sterile dH_2O to the final volume. The AP1 and AP2 primers contained Illumina paired end sequencing primer sites. DNA concentration was measured using a 2.0 fluorometer at BGI (Beijing Genomics Institute, China) [33].

High-throughput genotyping and map construction

High-density genetic maps for the two parents, BH and ES, were constructed using a slightly altered version of the method described by Wang *et al.* [33]. All experiments were performed at BGI. RAD-seq libraries for all 249 genotypes and the two parents were constructed according to Etter *et al.* (2011) [61], and sequenced using the Illumina HiSeq 2000 platform. The raw data produced were filtered to remove adaptors, indices and low-quality data (reads with > 15% of bases with quality score < 30). The cleaned data were analyzed using a standard RAD-seq analysis pipeline in the software package Stacks [62]. Genotypes for each plant in the population were assigned according to these results. Representative sequences for each SNP marker were obtained based on sequence clustering during the RAD-seq analysis pipeline. To manage the large quantity of data, a number of custom-programmed Perl scripts were also used to conduct the analysis.

To identify anchor markers for this study, we first identified a set of SNP markers, which we used to assign the 19 grapevine chromosomes to 19 linkage groups (LGs). This was done in two steps. Firstly, we marked the segregation patterns of all identified SNP markers as ab × cd, ef × eg, hk × hk, lm × ll, and nn × np. The first three of these pairs, which appeared in both parental linkage maps, were treated as candidate anchor markers. Secondly, because all alleles of each SNP marker had two nearly identical 100 bp sequences, the sequences from any allele could be taken as representative of the genotype of this SNP marker. These two representative sequences from the candidate anchor markers were aligned with the sequence of the 12× genomic assembly for *V. vinifera* PN40024, using local BLAST software with parameters set to −m 8 and −e 1E-5. The positions of each sequence for one SNP marker on the genome were identified based on their top hit. Three strict criteria were used to select anchor markers: 1) the marker had to show no significant segregation distortion among the 249 progeny genotypes in our population ($P < 0.001$); 2) both of the marker's end sequences had to align with the same chromosome position on the physical map for the reference PN40024 genome; and 3) the distance between the positions for the two end sequences on the reference genome had to fall between 200 and 500 bp (the expected size of the digested fragments was ~300–400 bp).

In constructing the map, the double pseudo-test cross strategy of Grattapaglia and Sederoff [63] was applied, using JoinMap4.0 (Kyazma). After data had been imported, a cross pollination (CP) model was used for data mining. The ratio of marker segregation was calculated using Chi-squared tests. Firstly, markers that showed significantly distorted segregation ($P < 0.001$) were excluded from further analyses; secondly, marker order on each linkage group was optimized by excluding markers with $\chi^2 > 3.0$. The genotypes of 1 826 SNP markers were analyzed for linkage and recombination, using the Kosambi function to estimate genetic map distances. Logarithm of odds (LOD) score thresholds ≥ 7 was used to group the markers. After the LGs had been computed, their number was assigned according to the anchor markers mapped on them.

QTL analysis

All trait data were Box-Cox transformed to unskew their distributions, and the normality of the distributions was tested using the Shapiro-Wilks test. The detection of QTLs using both the transformed and the original data yielded similar results in terms of number, location and contribution of QTLs, so the original data were henceforward used and reported.

QTLs for all traits in the population in the three separate years were analyzed for the parents only using the composite interval mapping (CIM) method in WinQTL Cartographer 2.5 [64,65]. CIM was used to scan the genetic map and estimate the likelihood of a QTL and its corresponding effect for every 1 cM. The forward regression algorithm was used to identify cofactors. A thousand permutations were performed using the CIM model within, and the thresholds for each environment were identified (almost all environments had thresholds at LOD ~3.0; $P \leq 0.05$). The 1-LOD confidence interval within the CIM model corresponded to the 95% confidence interval calculated by WinQTL Cartographer 2.5 for each QTL. The results showed that when LOD values were 3–3.2, the error rate was 5%. Threshold LOD value was therefore set to 3 for all traits. QTLs with peaks close to 5 cM were merged into one QTL, and each significant QTL was characterized by its maximum LOD score, the percentage of variation it explained and its confidence intervals in cM, corresponding to the maximum LOD score withinone unit's width either side of the LOD peak.

Search for candidate genes

For each QTL, the search for candidate genes was conducted in the genomic region corresponding to the confidence interval determined on the consensus map. The scrutinized sequence was limited by the most proximal SNP markers that were present in both the reference genome and the consensus map. The genes were selected based on the information available for the annotated reference genome (Genoscope 12×) of the quasi-homozygous line 40024 derived from Pinot noir (http://www.genoscope.cns.fr/externe/GenomeBrowser/Vitis/) [66]. They were classified according to their biological function as registered in the database. The genes catalogued as "unknown function" or equivalent were not considered in further analyses. In addition, a gene ontology (GO) enrichment analysis was performed, considering the genes identified in the physical genomic region that was associated with the confidence interval for each QTL. We also compared the frequency of each QTL vs. the complete reference genome, and searched for possible enrichment in gene functions. All enrichment analyses were done with the agriGO tool (http://bioinfo.cau.edu.cn/agriGO), using the options "singular enrichment analysis" and "complete GO". Significant GO terms ($P < 0.05$) were calculated using a hypergeometric distribution and the Yekutieli multi-test adjustment method [67].

Statistical analysis

Glucose-to-fructose ratio and β ratio (tartaric acid-to-malic acid ratio) were calculated, as these have been proposed as useful descriptors for evaluating the sugar and acid composition of grape berries [3,68]. For all further analyses, the means of the three replicates for each genotype and the parents were used.

All statistical analyses were performed using S-Plus (MathSoft Inc.). The frequency distribution of each trait was analyzed using the function 'hist', and the number of classes was determined using the Sturges method. Phenotypic correlations between traits within years and between years for each trait were calculated using the non-parametric Spearman correlation coefficient.

Results

Phenotypic characterization of parents and individuals

Averaged correlation coefficients between each pair of years were significant at $P < 0.001$ for almost all traits, ranging from 0.52 for the glucose-to-fructose ratio, to 0.74 for titratable acidity (Table 1).

Fructose, glucose, total sugar and SSC were positively correlated with each other. Fructose and glucose were strongly positively correlated, with a correlation coefficient of 0.93 ($P < 0.001$). The glucose-to-fructose ratio, however, was inconsistently correlated with fructose and glucose over the three years, and was not significantly correlated with total sugar, SSC or the acid-related traits. There were significant positive correlations between tartaric acid, malic acid, total acid and titratable acidity, from 0.36 between tartaric acid and malic acid to 0.88 between total acid and titratable acidity. The β ratio was significantly negatively correlated with malic acid and titratable acidity, but did not have consistent relationships with tartaric or total acid. The sugar-related and acid-related traits were, in general, negatively correlated, but the sugar-related traits were weakly positively correlated with the β ratio.

The traits examined showed approximately the same phenotypic data distributions for all three years (Figures 1 and Additional file 1: Figure S1). All traits exhibited continuous variation, which is typical of quantitatively inherited traits. Transgressive segregation was apparent in fructose, glucose, total sugar, SSC, glucose-to-fructose ratio and β ratio traits. For these traits, fewer than 12% of the genotypes had higher phenotypic values than the high-value parent (indeed only one genotype exceeded the parents' phenotypic value in 2011), and fewer than 29% of genotypes had lower phenotypic values than the low-value parent. Transgressive segregation was more apparent in the tartaric acid, malic acid, total acid and titratable acidity traits; for these traits, 37–88% of genotypes exceeded the high-value parent's phenotypic value, and 25–58% of genotypes were below the low-value parent.

Construction of genetic maps

A total of 1 826 SNP-based markers were used to construct the genetic maps. The lowest integrity for a single SNP marker was ~83.0%. Of the 1 826 SNP markers, 1 515 were homozygous for one parent and heterozygous for the other (803 for lm × ll and 712 for nn × np), constituting 83.0% of all selected SNP markers. The remaining 17.0% constituted the other three types of markers that could be mapped on both female and male linkage maps (ab × cd: 1, ef × eg: 109 and hk × hk: 201).

Table 1 Phenotypic correlation coefficients between the traits of grape berries produced by crossing 'Beihong' with 'E.S.7-11-49'

	Fructose	Glucose	Total sugar	SSC	G/F	Tartaric	Malic	Total acid	TA	β ratio
Fructose	**0.56*** **	0.93***	0.98***	0.86***	−0.27(ns)	−0.32(ns)	−0.52***	−0.49***	−0.57***	0.25(ns)
Glucose		**0.57*** **	0.98***	0.86***	ns(+)	−0.32 (ns)	−0.48***	−0.47***	−0.55***	0.20 (ns)
Total sugar			**0.56*** **	0.87***	ns	−0.32 (ns)	−0.52***	−0.49***	−0.57***	0.23 (ns)
SSC				**0.63*** **	ns	−0.32***	−0.54***	−0.54***	−0.56***	0.20 (ns)
G/F					**0.53*** **	ns	ns(+)	ns	ns	ns(−)
Tartaric						**0.63*** **	0.36***	0.76***	0.59***	0.39(ns)
Malic							**0.71*** **	0.88***	0.82***	−0.56***
Total acid								**0.68*** **	0.88***	−0.27(ns)
TA									**0.74*** **	−0.30***
β ratio										**0.64*** **

Correlation coefficients were averaged over three years, and over 241 genotypes in 2011, 225 in 2012, and 197 in 2013 (except for TA in 2013, for which there were 189 genotypes). The averages of the correlation coefficients between each two-year combination (2011 and 2012, 2011 and 2013, 2012 and 2013) for each trait are shown in the diagonal. SSC is the soluble solids content, G/F is the glucose-to-fructose ratio, TA is titratable acidity, and β ratio is the tartaric acid-to-malic acid ratio.

***Significant at $P<0.001$ in all three years.

ns: not significant and/or significant at $P<0.05$ in all three years.

ns (+/−): significant (+ = positive, − = negative) only in one year at $P< 0.001$ or $P< 0.01$.

ns (+) in diagonal: significant only between 2011 and 2012.

(ns): not significant only in one year; the correlation coefficients significant at $P< 0.001$ or $P< 0.01$ for the other two years were averaged.

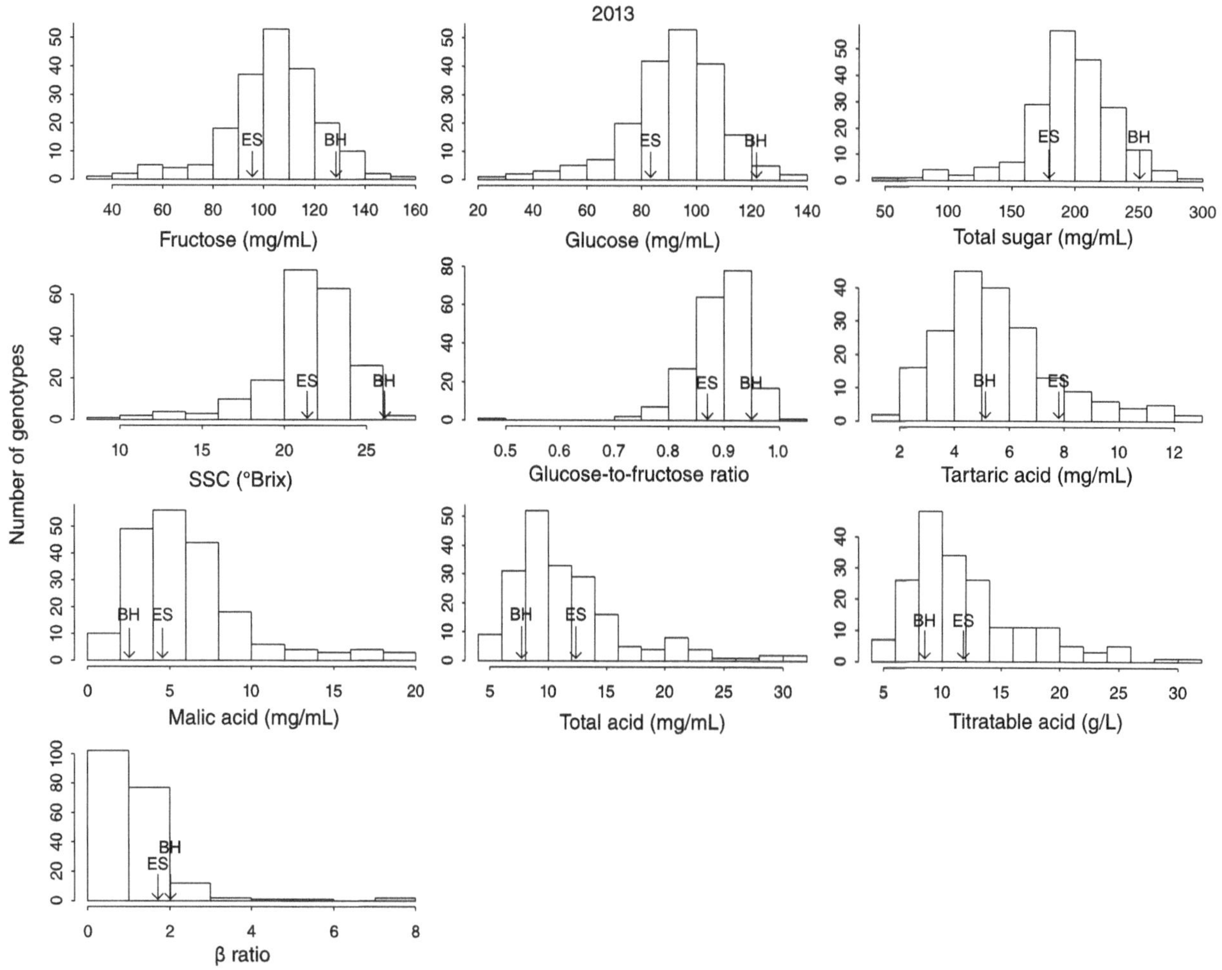

Figure 1 Distribution of traits of the F1 population derived from the cross 'Beihong' (BH) × 'E.S.7-11-49' (ES) in 2013. There were 197 genotypes in 2013 (189 for titratable acidity), using the averages of three replicates per genotype. The values for the maternal parent, BH, and the paternal parent, ES, are indicated by arrows. SSC and β ratio represent soluble solids content and the tartaric acid-to-malic acid ratio, respectively.

The minimum number of reads for an SNP marker to be accepted was five per allele; 181 distorted markers were removed. For the BH map, 621 markers were assembled into 19 LGs spanning 1 553.43 cM of map distance, with an average interval length of 2.50 cM. The ES map was based on 696 markers positioned in 19 LGs, and covered 1 381.02 cM, with an average interval length of 1.98 cM (Table 2, Additional file 2: Figure S2, Additional file 3: Table S1). The integrated map of maternal and paternal LGs included 1 254 markers, unevenly distributed between LGs. The total number of markers per LG ranged from 11 (LG16) to 66 (LG18) for the BH map, and from seven (LG05) to 63 (LG07) for the ES map. Each 1 000 kb of DNA sequence occupied an average of ~3.68 cM on the BH map and ~3.27 cM on the ES map. The average interval between two adjacent mapped markers was estimated at ~679 kb (2.50/3.68 × 1 000) for the BH map, and ~606 kb (1.98/3.27 × 1 000) for the ES map.

Comparison of genetic and reference sequences

Of the 1 254 markers used in the integrated genetic map, 1 055 were on the physical map for the reference PN40024 genome (Table 2), which suggests our genetic maps cover 84.1% of the reference genome. Of the 621 markers on the BH map, 480 (77.3%) were common to both the genetic and physical maps, and of the 696 markers on the ES map, 625 (89.8%) were shared (Table 2). The physical size of the corresponding chromosomes ranged from 16.5 Mb (LG17) to 30.1 Mb (LG14). In individual LGs, the number of markers common to both the genetic and physical maps ranged from eight (LG16) to 54 (LG18) for BH, and from seven (LG05) to 52 (LG07) for ES. The positions of the common markers on the genetic maps were compared with their physical positions on the reference genome (Additional file 4: Figure S3, Additional file 3: Table S1). Most of the markers showed good linear agreement

Table 2 Genetic map and number of common markers between genetic and physical maps for linkage groups

	Number of markers			Genetic size (cM)		Number of common markers		Chromosome size (Mb)
	BH	ES	Integrated	BH	ES	BH	ES	
LG01	25	43	66	81.688	78.187	23	41	22.8
LG02	49	30	75	106.868	67.662	37	25	18.6
LG03	16	30	45	96.402	69.187	15	27	19.2
LG04	46	41	83	106.634	104.325	40	39	23.7
LG05	39	7	46	79.219	22.831	13	7	24.8
LG06	30	47	75	79.630	71.633	27	45	21.3
LG07	13	63	73	25.035	98.122	9	52	20.9
LG08	53	41	88	94.558	73.718	41	41	22.3
LG09	24	33	54	86.867	71.318	23	32	23.0
LG10	45	20	60	81.571	79.890	23	10	17.7
LG11	24	40	60	79.699	74.507	23	40	19.3
LG12	34	53	75	66.618	52.297	29	47	22.3
LG13	33	54	81	79.181	82.256	30	41	24.4
LG14	28	38	65	83.639	90.139	26	36	30.1
LG15	29	22	49	87.052	58.515	14	21	20.1
LG16	11	19	27	50.544	56.048	8	18	21.7
LG17	27	34	60	63.685	58.969	26	34	16.5
LG18	66	52	117	92.401	101.306	54	44	29.3
LG19	29	29	54	112.139	70.110	19	25	23.8
Total	621	696	1254	1553.430	1381.020	480	625	421.8

The number of markers on the 19 linkage groups (LGs) of the 'Beihong' (BH) and 'E.S.7-11-49' (ES) genetic maps, their genetic sizes, and the number of markers common to both the genetic maps and the physical map for the reference PN40024 genome.

between the genetic and physical maps, with exceptions found on a few specific chromosomes (e.g. Chr05 and Chr16).

QTL identification

QTLs were analyzed separately on the parental maps for each of the three years (Table 3, Additional file 2: Figure S2). The CIM procedure detected 19 QTLs on the BH map, on LG02, LG03, LG06, LG09 and LG18, with 1, 1, 2, 1 and 14 QTLs, respectively. The average LOD value of the QTLs was 4.0, ranging from 3.0–8.1. On the ES map, 19 QTLs were detected on LG01, LG04, LG07, LG11, LG13, LG14, LG17 and LG18, with 1, 1, 1, 1, 1, 10, 3 and 1 QTLs, respectively. Here the average LOD value of the QTLs was 3.9, ranging from 3.0–6.1. The genomic threshold for both maps was ~3.0.

The number of QTLs identified for each trait varied between one and six, reflecting the quantitative nature of these traits, although no QTLs were detected for tartaric acid or titratable acidity. The QTLs that were identified were located within 13 of the 19 LGs. They each accounted for 5.28–17.31% of the total phenotypic variance in each trait.

Five QTLs for fructose were found in LG04, LG11, LG14, and LG17 of the ES map, each accounting for 5.58–9.71% of total variance. Three QTLs controlling glucose were found in LG14 of the ES map, contributing 6.04–8.16% of the variance. The QTLs for total sugar overlapped with those for fructose and/or glucose in LG14 of the ES map, individually contributing 5.85–8.37% of the variance. The QTL for SSC in LG14 of the ES map was the same as that for fructose, glucose and total sugar. A QTL for SSC was also identified in LG18 of the BH map, which explained 6.03% of the variance. A QTL for the glucose-to-fructose ratio and the α ratio, which was identified in LG03 and LG09 of the BH map, did not overlap with any of those for the individual sugars. Another QTL for the glucose-to-fructose ratio was found in LG02. QTLs for glucose-to-fructose ratio were also found in LG07 and LG17, on the ES map.

Malic acid, total acid and β ratio each had two to six QTLs in LG18 of the BH map, and there was another QTL for the β ratio in LG13 and LG18 of the ES map. There was also one QTL for both malic acid and total acid in LG06 of the BH map, which contributed 16.77–17.31% of the variance. However, no QTL could be identified for tartaric acid or titratable acidity.

Table 3 Summary of QTLs in F1 population derived from the cross 'Beihong' (BH) × 'E.S.7-11-49' (ES)

Trait	Year	LG	Parent	Marker name	Peak location	LOD score	R^2 (%)	95% confidence interval left	95% confidence interval right
Fructose	2011	14	ES	186084	17.51	3.52	8.24	16.90	20.90
	2011	14	ES	66968	23.11	4.20	7.63	22.40	24.50
	2011	17	ES	275393	30.71	3.47	6.27	29.90	32.50
	2012	4	ES	200017	85.31	3.11	5.58	81.90	88.90
	2013	11	ES	201825	72.71	4.21	9.71	72.20	73.90
Glucose	2011	14	ES	17727	16.51	3.00	6.18	16.10	18.00
	2011	14	ES	66968	23.11	4.44	8.16	22.40	24.30
	2011	14	ES	149275	46.21	3.31	6.04	42.90	49.30
Total sugar	2011	14	ES	17727	16.51	3.26	6.72	16.20	18.00
	2011	14	ES	66968	23.11	4.52	8.37	22.40	24.30
	2011	14	ES	149275	46.21	3.19	5.85	43.00	49.10
SSC	2011	14	ES	66968	23.11	6.13	11.42	22.80	25.70
	2011	14	ES	167058	29.81	3.58	7.17	28.10	31.90
	2012	18	BH	120619	47.41	3.08	6.03	45.40	47.80
	2013	1	ES	32402	61.01	3.58	6.96	54.60	65.00
Glucose-to-fructose ratio	2011	2	BH	125072	103.91	3.08	6.33	102.90	104.90
	2011	3	BH	182932	28.41	3.92	6.77	17.00	29.90
	2012	9	BH	117666	40.41	3.18	5.71	39.60	44.70
	2013	7	ES	296046	48.41	4.58	10.87	47.60	52.30
	2013	17	ES	250686	29.91	3.01	5.94	29.10	30.80
	2013	17	ES	76880	41.41	4.91	9.92	40.90	41.60
Malic acid	2011	6	BH	248487	74.11	3.68	17.31	72.00	76.90
	2011	18	BH	215808	24.31	5.45	9.49	23.90	26.40
	2011	18	BH	120619	47.41	5.07	9.63	45.50	47.60
	2012	18	BH	280521	23.21	3.68	7.13	22.80	26.80
	2012	18	BH	120619	47.41	6.09	11.83	45.60	47.60
	2012	18	BH	15694	53.51	4.78	8.49	53.20	55.40
Total acid	2011	6	BH	248487	74.11	3.20	16.77	72.10	75.30
	2012	13	ES	52766	54.81	3.31	6.25	51.20	54.90
	2012	18	BH	233088	38.21	3.00	16.05	37.10	39.40
	2012	18	BH	120619	46.91	3.75	7.90	44.90	47.40
	2012	18	ES	254573	34.91	3.29	6.04	32.70	37.60
β ratio	2011	18	BH	166745	23.61	8.10	16.06	23.20	24.40
	2011	18	BH	120619	47.41	4.20	7.91	44.90	47.60
	2011	18	BH	15694	53.51	3.03	5.29	53.20	55.80
	2012	18	BH	215808	24.31	3.10	5.69	23.90	27.10
	2012	18	BH	120619	47.41	4.22	8.34	45.10	47.60
	2012	18	BH	15694	53.51	3.23	6.09	53.20	55.40

Locations on linkage groups (LGs) of the BH and ES genetic maps, and contributions of the putative QTLs that control sugar- and acid-related traits, which were identified in at least two of three successive years (2011, 2012 and 2013). The locus is the marker showing the strongest association with the trait. The location of markers is given in cM, quoted from the top of each linkage group. R^2 represents the individual contribution of one QTL to the variation in a trait, and LOD is the logarithm of the odds ratio. SSC and β ratio represent soluble solids content and the tartaric acid-to-malic acid ratio, respectively. Traits in bold had QTLs detected in two years.

Candidate gene identification

In total, we identified 499 genes underlying the 19 QTLs of the BH map, and 724 genes underlying the 19 QTLs of the ES map. Of these, 835 (68.3%) were annotated and classified. However, only two QTLs (for malic acid, total acid and β ratio on LG18 of BH map) were stable across years (having been observed in two years). We therefore henceforward focused only on the candidate genes located within the confidence intervals of these two QTLs. For these two QTLs, 134 candidate genes were found. They were unevenly distributed, with 106 (22.8–26.8 cM) for one and 28 (45.5–47.6 cM) for the other QTL. Fifty of these genes were catalogued as having an "unknown protein function", and the others were classified into six major groups, namely cell, glycolysis, protein, RNA, TCA/org transformation, and transport. Of the 134 candidate genes, 10 that were probably related to TCA, acid metabolism or transport were listed, mainly (but not exclusively) based on their biological function as described in model plant species such as *Arabidopsis*, rice and poplar (Table 4).

Discussion

Phenotypic evaluation

The grape berries we analyzed displayed similar substantial variation in sugar and acid concentration across three successive years, which supports previous results showing that sugar and acid concentrations of grapes vary significantly by year [1]. For the study period, fructose, glucose, total sugar, tartaric acid, malic acid, and total acid concentration ranges were 7.5–136.7, 7.8–154.4, 15.3–291.9, 1.5–17.2, 0.8–21.3, and 4.9–34.7 mg·mL^{-1}, respectively. These ranges were greater than those found in other *Vitis* populations [49,69]. The reported ranges for fructose, glucose, and total sugar concentrations for these populations are 36.2–111.9, 38.5–104.4, and 78.9–216.3 mg·mL^{-1}, respectively, and those for tartaric acid, malic acid, and total acid concentrations are 1.1–6.0, 0.6–8.3 and 2.1–11.8 mg·mL^{-1}, respectively. Phenotypic correlations between sugar and acid concentrations were also relatively stable across the three years, although these may be affected by environmental factors.

Genetic map

Although genetic maps for grape cultivars have developed greatly in recent years, the number of markers in the LGs in existing maps is still generally less than 1 000, and some of the mapped markers have no sequence information. We recently identified 1 814 high-quality SNP markers for a population of 'Z180' (1 212 markers) × 'Beihong' (759 markers) [33]. In this study we used the same procedure to construct the genetic map, and the density of the resultant linkage map was similarly high. In total we identified 1 826 SNP markers, 621 of which were mapped on the female BH genetic map, and 696 on the male ES map. The difference between the number of markers we identified in this study and in the earlier one may be related to the different F1 population. On the BH map, the average size of LGs was 81.76 cM, ranging from 25.04 cM (LG07) to 112.14 cM (LG19). On the ES map, the average size was 72.69 cM, ranging from 22.83 cM (LG05) to 104.33 cM (LG04). There were 17 and 12 marker-free regions longer than 10 cM on the BH map (LG02, 03, 04, 05, 09, 11, 14, 16, 19) and the ES map (LG02, 04, 06, 09, 10, 11, 14, 15, 18), respectively.

The total physical size of the grape genome is ~470 Mb [66,70]. In most regions of the parental genetic and physical maps (for *V. vinifera*), the markers occurred in the same order, but not in all the chromosome regions. This indicates that on one hand the genome is well conserved among grape species, but that some changes in marker order have occurred during speciation. In this study, the maternal parent, BH, is bred from *V. vinifera* and *V. amurensis*, and the paternal parent, ES, is bred from *V. labrusca* × *V. riparia* and *V. vinifera*. Differences

Table 4 Genes in LG18 that may participate in acid regulation

Groups	Position	Gene ID	Gene symbol	Description	References
Cell	5462823-5465587	GSVIVT01009139001	CYCD4;1	CYCLIN D4;1	[87]
Glycolysis	5505643-5516683	GSVIVT01009147001	PGI,PGI1	phosphoglucose isomerase 1	[88]
protein	6442873-6452788	GSVIVT01009228001	ATCIPK8,CIPK8,PKS11,SnRK3.13	CBL-interacting protein kinase 8	[89]
protein	6652256-6660443	GSVIVT01009251001	ATDBR1,DBR1	debranching enzyme 1	[90]
RNA	6526891-6532321	GSVIVT01009238001	IAA9	indole-3-acetic acid inducible 9	[91,92]
TCA/org transformation	5663288-5667315	GSVIVT01009165001	ATBCA5,BCA5	beta carbonic anhydrase 5	[93]
transport	6771378-6774993	GSVIVT01009260001	AAP6	amino acid permease 6	[94]
transport	5605382-5606712	GSVIVT01009152001	ATPUMP5,DIC1,UCP5	uncoupling protein 5	[86]
transport	6688299-6690573	GSVIVT01009253001	ZF14	MATE efflux family protein	[80-85]
transport	10039446-10043834	GSVIVT01009629001		MATE efflux family protein	[80-85]

in the order of markers on some chromosomes have presumably resulted from different micro-structures on chromosomes in the various species. Alternatively, these differences might have arisen because of possible errors in mapping causing small inversions in marker order.

QTL detection

QTLs were analyzed separately for each of the traits on the parental maps for each of the three years, but were inconsistently detected. Some minor QTLs were detected only in a single year, such as a glucose QTL in LG14, for which R^2 = 6.04–6.18%, and a total sugar QTL in LG14, with R^2 = 5.85–6.72%. Other QTLs that contributed strongly to total variance were also detected in only one year, e.g. an SSC QTL in LG14 in 2011, for which R^2 = 11.42%, and a malic acid QTL in LG06 in 2011, with R^2 = 17.31%. In some cases, no QTLs were detected for a trait in a specific year, e.g. malic acid and total acid in 2013. Similar instability of QTLs across years has been widely reported for grapes [14,42,71], and also for other fruit tree species [52,57,72-74]. In contrast to crops such as maize, soya and rice, in which there may be many plants and biological replications of each genotype in one growth environment, there was only one vine per genotype in our trial. Phenotypic value assessment is potentially subject to bias which would increase the likelihood of error and affect the QTL analysis, with possible results including underestimated LOD values and overlap between QTLs across years. Furthermore, variation in climatic factors (such as rainfall and temperature) between years could bias assessment of fruit maturity, which would affect phenotypic evaluations. The observed low repeatability of QTL detection may thus have been exacerbated by the lack of replicate vines and the potential inconsistency in assessment of maturity. Addressing these problems in QTL studies on fruit species is difficult, however.

The percentage of variation explained by each QTL was small, and varied between 5.29% and 17.31%. This is consistent with the low R^2 values previously reported for some grape agronomic traits. Fanizza *et al.* [42] found that a QTL controlling berry weight had R^2 = 19%, but QTLs controlling the number of clusters per vine, cluster weight, number of berries per cluster, and berry weight had substantially lower R^2 values (1.2–10%). Similarly, Viana *et al.* [50] found that most QTLs accounted for less than 5.5% of the variance. QTLs with high R^2 values have generally been found to be related to properties such as veraison time/period, anthocyanin content (up to 48–62%) [47], and seed dry/fresh weight (up to 91.4%) [14,71]. It seems that agronomic traits, including sugar and acid concentration, are generally controlled by numerous QTLs, each with small effects. This might be due to the quantitative nature of these traits, as well as complicated metabolic pathways and regulatory networks.

Heritability for most traits is generally less than 50%, so the heritability associated with each QTL is a small fraction of this [75]. The more QTLs there are in the population, the smaller their individual contribution and the more difficult they are to detect [75]. As a result, precise map construction may be challenging, and maps may include some QTLs with very small R^2 values. Furthermore, the number of QTLs detected and the phenotypic variance they explain might be biased because of the limitations of the experiment itself, such as small sample size (as the effectiveness of marker loci increases with the number of individuals in a population) [76].

To our knowledge, no QTLs controlling the production of individual sugars and acids in grape berries have previously been identified. Viana *et al.* [50] reported one QTL in LG03 for SSC, one QTL in each of LG06, 13 and 19 for titratable acidity (% tartaric acid), and one QTL in each of LG01, 06, 11, 13 and 16 for pH, based on results for one year. We did not detect a QTL in LG03 for SSC in any of the three years of our study, or any QTLs for titratable acidity. This discrepancy between results may result from different genetic determinants of trait variation in the populations studied. Another cause might be differences in sampling strategies and the methods used for measuring traits. In this study, a QTL for malic acid in LG06 positioned at 74.11 cM explained a relatively large amount of variance (17.31%). Viana *et al.* [50] reported a QTL in LG06 positioned at 0.00 cM for pH, which explained 10.34% of variance. Although they were not the same QTL, these two regions might be worth exploring for genes controlling the quality of fruit acidity.

QTLs co-location

For breeding purposes, it is worth examining QTLs that are co-located. With respect to individual sugars, a QTL in LG14 affected both fructose and glucose, which explained the high correlation between them (r = 0.93). However, this QTL was detected only in 2011. In peaches, three QTLs, located in three different LGs are related to both glucose and fructose concentration [52]. The co-location of QTLs controlling fructose and glucose probably indicates a unique gene with a pleiotropic effect, or genes with close linkage, because glucose and fructose are absent from phloem sap and in grape berries are synthesized concurrently by sucrose hydrolysis [77]. The QTL in LG14 is potentially promising to work with to increase sugar concentrations, which may be beneficial for wine-making. Further study on candidate functional genes within the confidence intervals of this QTL may help to assess the mechanism for controlling hexose metabolism. The QTL for total sugar in LG14 overlapped with the QTL for fructose and glucose, which is probably related to the fact that these individual sugars contribute to total sugar. Similarly, a QTL for SSC in LG14 overlapped with one for fructose and glucose. However, the fact that the QTLs for the glucose-to-fructose ratio were not co-located with QTLs for either glucose or fructose is difficult to explain. With respect to the acid-related traits, QTLs for total acid and β ratio were co-located with those for malic acid, suggesting that malic acid contributes greatly to total acid and β ratio.

An important problem encountered when breeding for improved traits is negative correlations between favorable traits. For instance, as our results confirmed, malic acid and fructose are negatively correlated in 98 grape cultivars [1]. In some cases, this was caused by co-located QTLs with opposite horticultural effects. In tomatoes, fruit size and soluble sugar concentration are often negatively correlated, and the QTLs controlling them are located in the same LG [78,79]. In this study, although fructose, glucose, total sugar and SSC were negatively correlated with tartaric acid, malic acid, total acid and titratable acidity, in six LGs (01, 04, 11, 14, 17, 18) there were 15 QTLs related to sugars, and in three LGs (06, 13, 18) there were 11 QTLs related to acids. Viana *et al.* [50] also reported different LGs for sugar- and acid-related traits: one QTL in LG3 for SSC, and one QTL for pH and titratable acidity in each of LG01, 06, 11, 13, 16 and 19. The non-co-location of QTLs controlling sugar and acid concentration is favorable from the perspective of breeding. However, it is likely that some QTLs that have not yet been detected contribute to the remaining unexplained variance, and these should not be ignored.

Candidate genes for acid regulation

Of 134 genes located within the confidence intervals of the two QTLs controlling malic acid, total acid, and β ratio in LG18, 11 were probably involved in acid metabolism [80-94]. For example, beta carbonic anhydrase 5 is related to TCA/org transformation, and malic acid is involved in this process. Numerous studies reveal that the MATE family plays a key role in malate efflux from root apices in many plant species, including *Arabidopsis*, maize (*Zea mays*), wheat, rice (*Oryza sativa*), and rice bean (*Vignaumbellata*) [80-85]. Uncoupling protein 5 (DIC1) belongs to the mitochondrial carrier protein family. The *Arabidopsis* DIC protcins transport a wide range of dicarboxylic acids including malate, oxaloacetate and succinate [86], and these proteins might function as malate/oxaloacetate shuttles, which provide other cell components with reducing equivalents [86]. Expression levels of or changes in these candidate genes could be evaluated in further transcriptomic and gene-directed studies. The identification of the most relevant genes would help to reveal the molecular mechanisms operating in grape cultivars and could have a large impact on future breeding efforts.

Conclusions

The genetic map we have constructed for a winemaking grape cross is potentially a high-density, high-quality map, which could be used for QTL detection, genome comparison, and sequence assembly. In total, we mapped 1 254 markers with 60 bp sequences on the integrated map. These markers can be used as anchors to compare genetic and physical maps. This may facilitate the improved use of grape genomic resources. Moreover, this genetic map of a cross with a complex parentage, (*V. vinifera* × *V. amurensis*) × ((*V. labrusca* × *V. riparia*) × *V. vinifera*), will help to broaden our understanding of the grape genome. However, the stability and accuracy of QTLs are also affected by environmental factors. In our study, year and climatic conditions were the most important source of variability. Thus, controlling environmental factors can increase the likelihood that an observed phenotypic value is both accurate and repeatable.

Several QTLs controlling berry sugar and acid traits were detected in different LGs, suggesting that these traits are influenced by several genes that control different aspects of complex metabolic pathways. For example, we have identified a set of 10 candidate genes underlying the QTLs that are potentially related to malic acid. We anticipate that we will soon be able to narrow down these regions to the point where effects can be ascribed to specific genes. In addition, studies based on transcriptomics, proteomics, and metabolomics would help us to achieve a more accurate understanding of the molecular parameters involved in berry sugar and acid regulation. The long-term objective of this research is to provide information on the genetic basis of these traits, and to facilitate the selection of varieties to improve sugar and acid quality.

Additional files

Additional file 1: Figure S1. Distribution of traits of F1 population derived from the cross 'Beihong' (BH) × 'E.S.7-11-49' (ES) in 2011 and 2012.

Additional file 2: Figure S2. Genetic maps and QTL locations for 'Beihong' (BH, maternal parent) and 'E.S.7-11-49' (ES, paternal parent).

Additional file 3: Table S1. SNP-based markers and genetic positions on the 19 linkage groups (LGs).

Additional file 4: Figure S3. Positions of SNP-based markers on genetic maps, and their physical positions on the reference genome.

Competing interests
The authors declare that they have no competing interests.

Authors' contributions

JC, LCF and NW carried out the experiments. SHL, NW and ZCL conceived of the study, and participated in its design and coordination. BHW and JC performed the statistical analysis. BHW, JC, and NW drafted the manuscript. All authors read, revised and approved the final manuscript.

Acknowledgements

This work was funded by the National Natural Science Foundation of China (NSFC 31372030 and NSFC 31130047). The authors are grateful to Uni-edit for improving the English in this paper.

Author details

[1]Beijing Key Laboratory of Grape Science and Enology, and CAS Key Laboratory of Plant Resources, Institute of Botany, Chinese Academy of Sciences, Beijing 100093, P. R. China. [2]University of Chinese Academy of Sciences, Beijing 100049, P. R. China. [3]Key Laboratory of Plant Germplasm Enhancement and Speciality Agriculture, Wuhan Botanical Garden, Chinese Academy of Sciences, Wuhan 430074, China.

References

1. Liu HF, Wu BH, Fan PG, Li SH, Li LS. Sugar and acid concentrations in 98 grape cultivars analyzed by principal component analysis. J Sci Food Agric. 2006;86:1526–36.
2. Kliewer WM. Changes in concentration of glucose, fructose, and total soluble solids in flowers and berries of *Vitis vinifera*. Am J Enol Vitic. 1965;16:101–10.
3. Shiraishi M. Three descriptors for sugars to evaluate grape germplasm. Euphytica. 1993;71:99–106.
4. Doty TE. Fructose sweetness: a new dimension. Cereal Foods World. 1976;21:62–3.
5. Kulp K, Lorenz K, Stone M. Functionality of carbohydrates ingredients in bakery products. Food Technol. 1991;45:136–42.
6. Pangborn RM. Relative taste intensities of selected sugars and organic acids. J Food Sci. 1963;28:726–33.
7. Conde C, Silva P, Fontes N, Dias ACP, Tavares RM, Sousa MJ, et al. Biochemical changes throughout grape berry development and fruit and wine quality. Food. 2007;1:1–22.
8. Kliewer WM. Concentration of tartrates, malic acids, glucose and fructose in the fruits of genus *Vitis*. Am J Enol Vitic. 1967;18:87–96.
9. Kliewer WM, Howarth L, Omori M. Concentrations of tartaric acid and malic acids and their salts in *Vitis vinifera* grapes. Am J Enol Vitic. 1967;18:42–54.
10. Amerine MA, Roessler EB, Ough CS. Acids and the taste. I. The effect of pH and titratable acidity. Am J Enol Vitic. 1965;16:29–37.
11. Williams JGK, Kubelik AR, Livak KJ, Rafalski JA, Tingey SV. DNA polymorphisms amplified by arbitrary primers are useful as genetic markers. Nucl Acids Res. 1990;18:6531–5.
12. Vos P, Hogers R, Bleeker M, Reijans M, van de Lee T, Hornes M, et al. AFLP: a new technique for DNA fingerprinting. Nucl Acids Res. 1995;23:4407–14.
13. Adam-Blondon AF, Roux C, Claux D, Butterlin G, Merdinoglu D, This P. Mapping 245 SSR markers on the V vinifera genome: a tool for grape genetics. Theor Appl Genet. 2004;109:1017–27.
14. Costantini L, Battilana J, Lamaj F, Fanizza G, Grando MS. Berry and phenology-related traits in grapevine (Vitis vinifera L.): from quantitative trait loci to underlying genes. BMC Plant Biol. 2008;8:38.
15. Dalbó MA, Ye GN, Weeden NF, Steinkellner H, Sefc KM, Reisch BI. A gene controlling sex in grapevines placed on a molecular marker-based genetic map. Genome. 2000;43:333–40.
16. Di Gaspero G, Cipriani G, Adam-Blondon AF, Testolin R. Linkage maps of grapevine displaying the chromosomal locations of 420 microsatellite markers and 82 markers for *R*-gene candidates. Theor Appl Genet. 2007;114:1249–63.
17. Doligez A, Bouquet A, Danglot Y, Lahogue F, Riaz S, Meredith CP, et al. Genetic mapping of grapevine (*Vitis vinifera* L.) applied to the detection of QTLs for seedlessness and berry weight. Theor Appl Genet. 2002;105:780–95.
18. Doligez A, Adam-Blondon AF, Cipriani G, Di Gaspero G, Laucou V, Merdinoglu D, et al. An integrated SSR map of grapevine based on five mapping populations. Theor Appl Genet. 2006;113:369–82.
19. Doucleff M, Jin Y, Gao F, Riaz S, Krivanek AF, Walker MA. A genetic linkage map of grape, utilizing *Vitis rupestris* and *Vitis arizonica*. Theor Appl Genet. 2004;109:1178–87.
20. Fischer BM, Salakhutdinov I, Akkurt M, Eibach R, Edwards KJ, Töpfer R, et al. Quantitative trait locus analysis of fungal disease resistance factors on a molecular map of grapevine. Theor Appl Genet. 2004;108:501–15.
21. Grando MS, Bellin D, Edwards KJ, Pozzi C, Stefanini M, Velasco R. Molecular linkage maps of Vitis vinifera L and V riparia Mchx. Theor Appl Genet. 2003;106:1213–24.
22. Lamoureux D, Bernole A, Le Clainche I, Tual S, Thareau V, Paillard S, et al. Anchoring of a large set of markers onto a BAC library for the development of a draft physical map of the grapevine genome. Theor Appl Genet. 2006;113:344–56.
23. Lodhi MA, Daly MJ, Ye GN, Weeden NF, Reisch BI. A molecular marker based linkage map of *Vitis*. Genome. 1995;38:786–94.
24. Lowe KM, Walker MA. Genetic linkage map of the interspecific grape rootstock cross Ramsey (*Vitis champinii*) × Riparia Gloire (*Vitis riparia*). Theor Appl Genet. 2006;112:1582–92.
25. Mandl K, Santiago JL, Hack R, Fardossi A, Regner F. A genetic map of Welschriesling × Sirius for the identification of magnesium-deficiency by QTL analysis. Euphytica. 2006;149:133–44.
26. Marguerit E, Boury C, Manicki A, Donnart M, Butterlin G, Némorin A, et al. Genetic dissection of sex determinism, inflorescence morphology and downy mildew resistance in grapevine. Theor Appl Genet. 2009;118:1261–78.
27. Mejía N, Gebauer M, Muñoz L, Hewstone N, Muñoz C, Hinrichsen P. Identification of QTLs for seedlessness, berry size, and ripening date in a seedless × seedless table grape progeny. Amer J Enol Viticult. 2007;58:499–507.
28. Riaz S, Dangl GS, Edwards KJ, Meredith CP. A microsatellite marker based framework linkage map of *Vitis vinifera* L. Theor Appl Genet. 2004;108:864–72.
29. Riaz S, Krivanek AF, Xu K, Walker MA. Refined mapping of the Pierce's disease resistance locus, *PdR1*, and *Sex* on an extended genetic map of *Vitis rupestris* × *V. arizonica*. Theor Appl Genet. 2006;113:1317–29.
30. Salmaso M, Malacarne G, Troggio M, Faes G, Stefanini M, Grando MS, et al. A grapevine (*Vitis vinifera* L.) genetic map integrating the position of 139 expressed genes. Theor Appl Genet. 2008;116:1129–43.
31. Troggio M, Malacarne G, Coppola G, Segala C, Cartwright DA, Pindo M, et al. A dense single-nucleotide polymorphism-based genetic linkage map of grapevine (Vitis vinifera L.) anchoring Pinot Noir bacterial artificial chromosome contigs. Genetics. 2007;176:2637–50.
32. Zhang J, Hausmann L, Eibach R, Welter LJ, Töpfer R, Zyprian EM. A framework map from grapevine V3125 (*Vitis vinifera* 'Schiava grossa' × 'Riesling') × rootstock cultivar 'Börner' (*Vitis riparia* × *Vitis cinerea*) to localize genetic determinants of phylloxera root resistance. Theor Appl Genet. 2009;119:1039–51.
33. Wang N, Fang LC, Xin HP, Wang LJ, Li SH. Construction of a high-density genetic map for grape using next generation restriction-site associated DNA sequencing. BMC Plant Biol. 2012;12:148.
34. Barba P, Cadle-Davidson L, Harriman J, Glaubitz JC, Brooks S, Hyma K, et al. Grapevine powdery mildew resistance and susceptibility loci identified on a high-resolution SNP map. Theor Appl Genet. 2014;127:73–84.
35. Barker CL, Donald T, Pauquet J, Ratnaparkhe MB, Bouquet A, Adam-Blondon AF, et al. Genetic and physical mapping of the grapevine powdery mildew resistance gene, *Run1*, using a bacterial artificial chromosome library. Theor Appl Genet. 2005;111:370–7.
36. Coleman C, Copetti D, Capriani G, Hoffmann S, Kozma P, Kovács L, et al. The powdery mildew resistance gene *REN1* co-segregates with an NBS-LRR gene cluster in two Central Asian grapevines. BMC Genet. 2009;10:89.
37. Riaz S, Tenscher AC, Rubin J, Graziani R, Pao SS, Walker MA. Fine-scale genetic mapping of two Pierce's disease resistance loci and a major segregation distortion region on chromosome 14 of grape. Theor Appl Genet. 2008;117:671–81.
38. Marino R, Sevini F, Mandini A, Vecchione A, Pertot I, Serra AD, et al. QTL mapping for disease resistance and fruit quality in grape. Acta Hort. 2003;603:527–33.
39. Zyprian E, Eibach R, Töpfer R. Comparative molecular mapping in segregating populations of grapevine. Acta Hort. 2003;603:73–7.
40. Krivanek AF, Riaz S, Walker MA. Identification and molecular mapping of PdR1, a primary resistance gene to Pierce's disease in *Vitis*. Theor Appl Genet. 2006;112:1125–31.

41. Xu K, Riaz S, Roncoroni NC, Jin Y, Hu R, Zhou R, et al. Genetic and QTL analysis of resistance to *Xiphinema index* in a grapevine cross. Theor Appl Genet. 2008;116:305–11.
42. Fanizza G, Lamaj F, Costantini L, Chaabane R, Grando MS. QTL analysis for fruit yield components in table grapes (*Vitis vinifera*). Theor Appl Genet. 2005;111:658–64.
43. Cabezas JA, Cervera MT, Ruiz-Garcia L, Carreño J, Martínez-Zapater JM. A genetic analysis of seed and berry weight in grapevine. Genome. 2006;49:1572–85.
44. Duchêne E, Butterlin G, Dumas V, Merdinoglu D. Towards the adaptation of grapevine varieties to climate change: QTLs and candidate genes for developmental stages. Theor Appl Genet. 2012;124:623–35.
45. Correa J, Mamani M, Muñoz-Espinoza C, Laborie D, Muñoz C, Pinto M, et al. Heritability and identification of QTLs and underlying candidate genes associated with the architecture of the grapevine cluster (*Vitis vinifera* L.). Theor Appl Gene t. 2014;127:1143–62.
46. Eibach R, Hastrich H, Töpfer R. Inheritance of aroma compounds. Acta Hort. 2003;603:337–44.
47. Fournier-level A, Le Cunff L, Gomez C, Doligez A, Ageorges A, Roux C, et al. Quantitative genetic bases of anthocyanin variation in grape (Vitis vinifera L. ssp. sativa) berry: a quantitative trait locus to quantitative trait nucleotide integrated study. Genetics. 2009;183:1127–39.
48. Doligez A, Bertrand Y, Dias S, Grolier M, Ballester JF, Bouquet A, et al. QTLs for fertility in table grapes (*Vitis vinifira* L.). Tree Genet Genomes. 2010;6:413–22.
49. Liu HF, Wu BH, Fan PG, Xu HY, Li SH. Inheritance of sugars and acids in berries of grape (Vitis vinifera L.). Euphytica. 2007;153:99–107.
50. Viana AP, Riaz S, Walker MA. Genetic dissection of agronomic traits within a segregating population of breeding table grapes. Genet Mol Res. 2013;12:951–64.
51. Dirlewanger E, Moing A, Rothan C, Svanella L, Pronier V, Guye A, et al. Mapping QTLs controlling fruit quality in peach (*Prunus persica* (L.) Batsch). Theor Appl Genet. 1999;98:18–31.
52. Quilot B, Wu BH, Kervella J, Génard M, Foulongne M, Moreau K. QTL analysis of quality traits in an advanced backcross between *Prunus persica* cultivars and the wild relative species *P. davidiana*. Theor Appl Genet. 2004;109:884–97.
53. Kenis K, Keulemans J, Davey MW. Identification and stability of QTLs for fruit quality traits in apple. Theor Appl Genet. 2008;4:647–61.
54. King GJ, Maliepaard C, Lynn JR, Alston FH, Durel CE, Evans KM, et al. Quantitative genetic analysis and comparison of physical and sensory descriptors relating to fruit flesh firmness in apple (*Malus pumila* Mill.). Theor Appl Genet. 2000;100:1074–84.
55. Liebhard R, Kellerhals M, Pfammatter W, Jertmini M, Gessler C. Mapping quantitative physiological traits in apple (*Malus × domestica* Borkh.). Plant Mol Biol. 2003;52:511–26.
56. Maliepaard C, Alston FH, Van Arkel G, Brown LM, Chevreau E, Dunemann F, et al. Aligning male and female linkage maps of apple (*Malus pumila* Mill.) using multi-allelic markers. Theor Appl Genet. 1998;97:60–73.
57. Wang D, Karle R, Iezzoni AF. QTL analysis of flower and fruit traits in sour cherry. Theor Appl Genet. 2000;100:535–44.
58. Cohen S, Tzuri G, Harel-Beja R, Itkin M, Portnoy V, Sa'ar U, et al. Co-mapping studies of QTLs for fruit acidity and candidate genes of organic acid metabolism and proton transport in sweet melon (*Cucumis melo* L.). Theor Appl Genet. 2012;125:343–53.
59. Bisson L: In search of optimal grape maturity. Pract Winery Vineyard 2001, 23: 32, 34, 36, 38, 40, 42–43.
60. Kupiec T. Quality-Control Analytical Methods: High-Performance Liquid Chromatography. Int J Pharm Compound. 2004;8:223–7.
61. Etter PD, Bassham S, Hohenlohe PA, Johnson EA, Cresko WA. SNP Discovery and Genotyping for Evolutionary Genetics Using RAD Sequencing. Methods Mol Biol. 2011;772:157–78.
62. Catchen JM, Amores A, Hohenlohe P, Cresko W, Postlethwait JH. *Stacks*: building and genotyping Loci *de novo* from short-read sequences. G3. 2011;1:171–82.
63. Grattapaglia D, Sederoff R. Genetic linkage maps of Eucalyptus grandis and Eucalyptus urophylla using a pseudo-testcross: mapping strategy and RAPD markers. Genetics. 1994;137:1121–37.
64. Zeng ZB. Precision mapping of quantitative trait loci. Genetics. 1994;136:1457–68.
65. Wang SC, Basternand J, Zeng ZB (2006): Windows QTL Cartographer 2.5. North Carolina State University, Raleigh, NC [http://statgen.ncsu.edu/qtlcart/WQTLCart.htm].
66. Jaillon O, Aury JM, Noel B, Policriti A, Clepet C, Casagrande A, et al. The grapevine genome sequence suggests ancestral hexaploidization in major angiosperm phyla. Nature. 2007;449:463–7.
67. Du Z, Zhou X, Ling Y, Zhang Z, Su Z. agriGO: a GO analysis toolkit for the agricultural community. Nucleic Acids Res. 2010;38:W64–70.
68. Shiraishi M. Proposed descriptors for organic acids to evaluate grape germplasm. Euphytica. 1995;81:13–20.
69. Liang ZC, Sang M, Ma AH, Zhao SJ, Zhong GY, Li SH. Inheritance of sugar and acid contents in the ripe berries of a tetraploid diploid grape cross population. Euphytica. 2011;182:251–9.
70. Velasco R, Zharkikh A, Troggio M, Cartwright DA, Cestaro A, Pruss D, et al. A High Quality Draft Consensus Sequence of the Genome of a Heterozygous Grapevine Variety. PLoS One. 2007;12:e1326.
71. Grzeskowiak L, Costantini L, Lorenzi S, Stella Grando M. Candidate loci for phenology and fruitfulness contributing to the phenotypic variability observed in grapevine. Theor Appl Genet. 2013;126:2763–76.
72. Conner PJ, Brown SK, Weeden NF. Molecular-marker analysis of quantitative traits for growth and development in juvenile apple trees. Theor Appl Genet. 1998;96:1027–35.
73. Costa F, Peace CP, Stella S, Serra S, Musacchi S, Bazzani M, et al. QTL dynamics for fruit firmness and softening around an ethylene-dependent polygalacturonase gene in apple (*Malus × domestica* Borkh.). J Exp Bot. 2010;61:3029–39.
74. García MR, Asíns MJ, Carbonell EA. QTL analysis if yield and seed number in *Citrus*. Theor Appl Genet. 2000;101:487–93.
75. Kearsey MJ. The principles of QTL analysis (a minimal mathematics approach). J Exp Bot. 1998;49:1619–23.
76. Staub JE, Serquen FC, Gupta M. Genetic markers, map construction, and their application in plant breeding. HortSci. 1996;31:729–41.
77. Davies C, Robinson SP. Sugar accumulation in grape Berries (cloning of two putative vacuolar invertase cDNAs and their expression in grapevine tissues). Plant Physiol. 1996;111:275–83.
78. Causse M, Saliba-Colombani V, Lecomte L, Duffé P, Rousselle P, Buret M. QTL analysis of fruit quality in fresh market tomato: a few chromosome regions control the variation of sensory and instrumental traits. J Exp Bot. 2002;53:2089–98.
79. Fulton TM, Beck-Bunn T, Emmatty D, Eshed Y, Lopez J, Petiard V, et al. QTL analysis of an advanced backcross of *Lycopersicon peruvianum* to the cultivated tomato and comparisons with QTLs found in other wild species. Theor Appl Genet. 1997;95:881–94.
80. Liu JP, Magalhaes JV, Shaff J, Kochian LV. Aluminum-activated citrate and malate transporters from the MATE and ALMT families function independently to confer *Arabidopsis* aluminum tolerance. Plant J. 2009;57:389–99.
81. Ryan PR, Raman H, Gupta S, Horst WJ, Delhaize E. A second mechanism for aluminum resistance in wheat relies on the constitutive efflux of citrate from roots. Plant Physiol. 2009;149:340–51.
82. Yang XY, Yang JL, Zhou Y, Pineros MA, Kochian LV, Li GX, et al. A *de novo* synthesis citrate transporter, *Vigna umbellate* multidrug and toxic compound extrusion, implicates in Al-activated citrate efflux in rice bean (*Vigna umbellata*) root apex. Plant Cell Environ. 2011;34:2138–48.
83. Yokosho K, Yamaji N, Ma JF. An Al-inducible MATE gene is involved in external detoxification of Al in rice. Plant J. 2011;68:1061–9.
84. Maron LG, Guimarães CT, Kirst M, Albert PS, Birchler JA, Bradbury PJ, et al. Aluminum tolerance in maize is associated with higher *MATE1* gene copy number. Proc Natl Acad Sci U S A. 2013;110:5241–6.
85. Tovkach A, Ryan PR, Richardson AE, Lewis DC, Rathjen TM, Ramesh S, et al. Transposon-mediated alteration of *TaMATE1B* expression in wheat confers constitutive citrate efflux from root apices. Plant Physiol. 2013;161:880–92.
86. Palmieri L, Picault N, Arrigoni R, Besin E, Palmieri F, Hodges M. Molecular identification of three *Arabidopsis thaliana* mitochondrial dicarboxylate carrier isoforms: organ distribution, bacterial expression, reconstitution into liposomes and functional characterization. Biochem J. 2008;410:621–9.
87. Nieuwland J, Maughan S, Dewitte W, Scofield S, Sanz L, Murray James AH. The D-type cyclin CYCD4;1 modulates lateral root density in Arabidopsis by affecting the basal meristem region. PNAS. 2009;106:22528–33.
88. Sun JR, Wu Q, Zhou HX, Xu H, Cao Y, Xie L, et al. Correlation Between Polysaccharide Biosynthesis and Glycometabolism Related Enzymes of

Chlamydomonas sp. YB-204 in Different Conditions. J Pure Microbio. 2014;8:1923–8.
89. Hu HC, Wang YY, Tsay YF. AtCIPK8, a CBL-interacting protein kinase, regulates the low-affinity phase of the primary nitrate response. Plant J. 2009;57:264–78.
90. Khalid MF, Damha MJ, Shuman S, Schwer B. Structure–function analysis of yeast RNA debranching enzyme (Dbr1), a manganese-dependent phosphodiesterase. Nucleic Acids Res. 2005;33:6349–60.
91. Fujita K, Horiuchi H, Takato H, Kohno M, Suzuki S. Fujita. Auxin-responsive grape *Aux/IAA9* regulates transgenic Arabidopsis plant growth. Mol Biol Rep. 2012;39:7823–9.
92. Wang H, Jones B, Li ZG, Frasse P, Delalande C, Regad F, et al. The Tomato *Aux/IAA* Transcription Factor *IAA9* Is Involved in Fruit Development and Leaf Morphogenesis. Plant Cell. 2005;17:2676–92.
93. Wang M, Zhang Q, Liu FC, Xie WF, Wang GD, Wang J, et al. Family-wide expression characterization of Arabidopsis beta-carbonic anhydrase genes using qRT-PCR and Promoter::GUS fusions. Biochimie. 2014;97:219–27.
94. Okumoto S, Schmidt R, Tegeder M, Fischer WN, Rentsch D, Frommer WB, et al. High affinity amino acid transporters specifically expressed in xylem parenchyma and developing seeds of arabidopsis. J Biol Chem. 2002;277:5338–45346.

11

Linkage mapping of *Barley yellow dwarf virus* resistance in connected populations of maize

Frederike Horn[1], Antje Habekuß[2] and Benjamin Stich[1*]

Abstract

Background: With increasing winter temperatures, *Barley yellow dwarf virus* (BYDV) is expected to become an increasing problem in maize cultivation in Germany. Earlier studies revealed that BYDV has a negative impact on maize performance. Molecular markers would accelerate the development of BYDV resistant maize. Therefore, the objectives of this study were (i) the identification of quantitative trait loci (QTL) for BYDV resistance in five connected segregating maize populations in a field experiment and (ii) their comparison with the QTL detected under greenhouse conditions.

Results: In linkage analyses of the traits virus extinction, infection rate, and the symptom red edges, a highly associated major QTL was identified on chromosome 10. This QTL explained 45% of the phenotypic variance for the traits virus extinction and infection rate and 30% for the symptom red edges.

Conclusion: We could show that BYDV resistance traits are oligogenically inherited. The QTL on chromosome 10 could be observed in the connected linkage analyses and in the single population analyses. Furthermore, this QTL could also be confirmed in the greenhouse experiment. Our results let suggest that this QTL is involved in multiple virus resistance and the markers are promising for marker assisted selection.

Keywords: QTL analysis, BYDV, Connected populations, *Zea mays* L.

Background

Barley yellow dwarf (BYD) is one of the economically most important virus diseases in small grain cereals. *Zea mays* L. plays an important role as a summer host for the aphid-transmitted *Barley yellow dwarf virus* (BYDV) [1]. With increasing winter temperatures due to climate change, aphids overwinter anholocyclic. This leads to an earlier development of aphid populations with a higher number of aphids in spring. Consequently, the infestation pressure on maize by viruliferious aphids is increased [1]. Furthermore, aphids infect plants in early developmental stages [2], in which maize reacts with strong growth reduction on BYDV infection because of an incomplete organ development [1].

In earlier studies, a reduction of plant height, ear height, fresh yield, and grain yield as well as an earlier flowering was observed in BYDV inoculated maize compared to non-inoculated maize plants [3-5]. The control of the virus directly is not possible. The aphids could be controlled by application of an insecticide. However, in Germany the application of insecticides against aphids in maize is not allowed. Thus, genetic resistance is the only alternative. Furthermore, with resistant maize cultivars, the BYDV transmission cycle can be broken and with this, the situation in cereals could be improved as well.

Characteristic symptoms of BYDV infected maize are red bands at the edge of the leaves. These symptoms correlate positively with the virus extinction measured by double-antibody sandwich enzyme-linked immunosorbent assay (DAS-ELISA) and is a fair indicator for susceptibility [6]. Nevertheless, tolerant genotypes which do not show any symptoms but high virus extinction exist [6-8]. To identify resistant genotypes, it is therefore necessary to measure the virus extinction by DAS-ELISA. The evaluation of maize genotypes for their BYDV resistance by aphid inoculation and DAS-ELISA analyses, however, is very labor and cost intensive and therefore its integration in practical breeding programs is not reasonable. Therefore, the identification of genome regions which are linked to BYDV resistance and the application of this knowledge in marker assisted selection (MAS) programs of

*Correspondence: stich@mpipz.mpg.de
[1] Max Planck Institute for Plant Breeding Research, Carl-von-Linné Weg, 50829 Cologne, Germany
Full list of author information is available at the end of the article

maize would allow a faster progress in breeding of BYDV resistant maize.

In previous studies, resistance loci have been identified for various viruses in maize. On chromosome 10, three minor quantitative trait loci (QTL) were identified for *Sugarcane mosaic virus* (SCMV) resistances [9,10]. Zambrano et al., 2014 [11] identified a resistance locus on chromosome 1 for *Maize mosaic virus* (MMV) and two resistance loci on chromosome 10 for *Wheat strike mosaic virus* (WSMV) and *Maize dwarf mosaic virus* (MDMV). For WSMV, also McMullen and Simcox, 1995 [12] as well as Jones et al., 2011 [13] identified a resistance locus on chromosome 10. On chromosome 10, further resistance loci were found for *Maize chlorotic dwarf virus* (MCDV) [14]. But, to the best of our knowledge, no results from QTL mapping of BYDV resistance in maize are available. However, in a genome wide association study (GWAS) Horn et al., 2014 [15] identified single nucleotide polymorphisms (SNPs) on chromosome 4 and 10 in a diverse germplasm set, representing the world-wide maize diversity [16], which explained a high proportion of phenotypic variation for traits related to BYDV resistance.

The disadvantage of GWAS is that genes contributing to phenotypic variation which show a low allele frequency can remain undetected. Due to a balanced allele frequency in segregating populations, classical linkage mapping has the advantage of higher QTL detection power compared to GWAS [17]. In the GWAS of Horn et al., 2014 [15], population structure was taken into account to avoid the detection of spurious associations [16]. However, with this correction it is possible that QTL effects are absorbed in population effects and stay undetected in a GWAS [18]. Therefore, in this study we used linkage mapping to be able to detect additional QTL for BYDV resistance and furthermore, if possible, validate the genome regions identified previously by GWAS.

In this study, five connected segregating populations were used to map QTL for BYDV resistance. In connected QTL mapping populations, the probability to find alleles of interest is higher because more than two alleles can be considered in multiple genetic backgrounds in contrast to a single biparental population [19] and therefore this approach increases the probability that a QTL will be polymorphic in at least one population [20]. The objectives of this study were (i) the identification of QTL for BYDV resistance in five connected segregating maize populations in a field experiment and (ii) their comparison with the QTL detected under greenhouse conditions.

Material and methods

Phenotypic evaluation of five connected segregating populations

Five connected segregating maize populations with a total of 443 entries [A: n=85 ($F_{2:3}$), B: n=83 ($F_{3:4}$), C: n=77 ($F_{4:5}$), D: n=92 ($F_{4:5}$), E: n=106 ($F_{4:5}$)] derived from biparental crosses of five inbred lines were examined in our study (Table 1).

The field experiments with these populations were carried out at Borken and Wadersloh (both Germany) in 2011 and 2012. Each population was planted separately in a single trial and the experimental design of each trial was an α lattice design, where the five parental inbreds served as repeated checks.

Two weeks after sowing, the plants were inoculated with BYDV. For the transmission of the virus, aphids of the genus *Rhopalosiphum padi* were raised for three weeks at 20°C on plants of the *Triticum aestivum* cultivar "Tuareg" which were infected with the virus BYDV-PAV. For the

Table 1 Population means and adjusted means of the parental inbreds for the traits red edges (RE), virus extinction (EX), and infection rate (IR) from the field experiment, and for the traits EX and IR from the greenhouse experiment

	Field			Greenhouse	
Entry type	RE	EX	IR [%]	EX	IR [%]
Segregating populations					
Population A (P092 x FAP1360A)	1.42	0.66	54.57	0.68	54.04
Population B (P092 x Ky226)	1.50	0.49	41.21	0.87	80.46
Population C (Ky226 x FAP1360A)	1.17	0.21	13.57	0.50	37.70
Population D (D408 x W64A)	3.69	0.76	66.38	0.60	53.45
Population E (D408 x P092)	1.37	0.66	55.50	0.61	53.44
Parental inbreds					
Ky226 (resistant)	1.45	0.30	19.88	0.39	27.88
W64A (susceptible)	7.15	0.83	80.66	0.77	82.72
FAP1360A (resistant)	0.92	0.26	18.28	0.76	67.32
P092 (tolerant)	0.89	1.02	80.79	1.15	85.40
D408 (tolerant)	1.92	0.76	60.44	0.46	35.62

inoculation, a piece of wheat leaves with viruliferious aphids was placed in the leaf axil of each maize plant. Afterwards, the plants were covered with fleece *Climatex* (17g/qm) to prevent escape of the aphids. One week after inoculation, the fleece was removed and the insecticide "Biscaya" (Bayer, 300 ml in 200 to 400 l water/ha) was applied in the field.

Six weeks after inoculation, the BYDV symptom red edges (RE) was scored on a scale from 1 to 9 (1 = no symptoms, 9 = highest symptom expression) in the field experiments. We collected leaf material from the sixth leaf of each plant per row and measured virus extinction (EX) by DAS-ELISA as described by Horn et al., 2013 [6]. The infection rate (IR) was calculated as the percentage of plants of one plot with EX ≥ 0.5 [15].

In order to evaluate the five segregating populations under artificial conditions, controlling for light, temperature, humidity, soil nutrients, and water application, they were grown in a greenhouse experiment with two replicates, each with 10 plants per genotype. Each population was planted separately in a single trial and the experimental design of each trial was an α lattice design, where the five parental inbreds served as repeated checks. Experimental conditions (light, temperature, soil) were as described by Horn et al., 2014 [15] and in contrast to field experiment the insecticide "Lizetan Plus" (Bayer) was applied one week after inoculation. The evaluation of EX and IR in the greenhouse was assessed in the same way as described above for the field experiment.

Statistical analyses

Phenotypic evaluation of five connected segregating populations

We used the mixed model (1) to analyse the data collected for the checks in the field experiments. As no obvious year effect was observed, we considered each year-location combination as one environment:

$$Y_{ijk} = \mu + R_i + T_j + C_k + e_{ijk}, \tag{1}$$

where Y_{ijk} was the phenotypic observation for the kth check in the jth trial in the ith environment, and μ the general mean. In the analysis of the field experiment R_i was the effect of the ith environment, T_j the effect of the jth trial, C_k the effect of the kth check, and e_{ijk} the residual error. All effects, except e_{ijk} were regarded as fixed. The trial effect T_j was subtracted from the raw data of all entries of the corresponding trial to correct for the differences among the different trials. The adjusted data of all entries from the field experiment were then analyzed according to the following linear mixed model:

$$Y_{ijlmn} = \mu + R_i + D_l(TG)_{jm} + (RB)_{in} + e_{ijlmn}, \tag{2}$$

where Y_{ijlmn} was the phenotypic observation for the mth entry in the ith environment in the nth incomplete block of the jth trial corrected for T_j. G_m was the effect of the mth entry, B_n the effect of the nth block, and e_{ijlmn} the residual error. $(D_{1-5})_l$ was a indicator variable with $D_l =$ 04 for checks and $D_l = 1 - 5$ for the entry of the 1st - 5th trial which enabled the calculation of specific genotypic σ^2_{gj} and error σ^2_{ej} variances for the entries of the jth trial. R_i was regarded as fixed, whereas the $D_l(TG)_{jm}$ interactions and the $(RB)_{in}$ interaction were regarded as random.

Formula (1) and (2) were also used for the analysis of the greenhouse experiment where R_i was the effect of the ith replicate.

Broad-sense heritability H^2_j was calculated for each jth trial based on the formula

$$H^2_j = \frac{\sigma^2_{gj}}{\sigma^2_{gj} + \frac{\sigma^2_{ej}}{n}}, \tag{3}$$

where n was the number of environments.

Broad-sense heritability on a plot basis H^{*2}_j was calculated based on the formula

$$H^{*2}_j = \frac{\sigma^2_{gj}}{\sigma^2_{gj} + \sigma^2_{ej}}, \tag{4}$$

For each entry, an adjusted entry mean was calculated as:

$$M_m = \widehat{\mu} + \widehat{G_m}, \tag{5}$$

where $\widehat{\mu}$ was the estimate for the intercept and $\widehat{G_m}$ the estimate of the genetic effect of the mth entry calculated based on formula (2).

All mixed model analyses were performed with the software ASReml [21]. For each segregating population, the correlation coefficients among all pairs of traits were calculated. If not stated differently, all analyses were performed with the statistical software R [22].

Genotyping and consensus map construction

Plant material was collected from the leaves of each genotype and deoxyribonucleic acid (DNA) was isolated using *BioSprint 96*. The five parental inbreds were genotyped by the TraitGenetics GmbH (Gatersleben, Germany) with the MaizeSNP50 array [23]. Out of this set of 56.110 SNPs, we selected 163 SNPs which were equally distributed across the genome. Furthermore, we selected markers which were homozygote in the parental inbreds and were polymorphic in the highest number of segregating populations. These selected SNPs were genotyped for all five segregating populations with KASP marker technology by TraitGenetics GmbH.

A chi-square test was performed to test whether the SNPs deviate from the expected 1:1 ratio and SNPs which significantly ($\alpha = 0.001$) deviated from this ratio were excluded from further analyses [24]. For an improvement of the consensus map construction, further marker information of six connected segregating populations (Frey F, Stich B. Identfication of genome regions contributing to variation of heat tolerance in temperate maize by QTL mapping with multiple connected populations: Unpublished.), genotyped with the same marker set, were included. According to the position on the physical map, SNPs were assigned to their chromosomes and for each chromosome a consensus map was created using the software CarthaGène [25].

Linkage mapping

The linkage analyses with the software MCQTL [26] was based on the consensus map and the adjusted entry mean of each genotype. At first, we performed a single population analysis for each segregating population [19,20]. In the model we included additive and dominance effects because there are still heterozygous genotypes in our $F_{2:3}$, $F_{3:4}$, and $F_{4:5}$ populations. To consider relationships between the populations due to shared parental inbreds, we furthermore performed a connected analysis [19,20] including also additive and dominance effects. For QTL detection, 0.01 quantile F thresholds were determined for each trait by 1,000 permutations. F thresholds for the cofactor selection were fixed at 90% of the F threshold values for QTL detection as proposed by the MCQTL software. The QTL detection was performed using an iterative composite interval mapping approach (iQTLm) [27]. SNP markers associated with the respective trait were selected as cofactors by forward regression considering a minimal distance of 10 cM between two selected cofactors [19].

To test, if the dominance effects of the populations were significantly different from 0, we calculated significance ($\alpha = 0,05$) a posteriori from a normal distribution using a two-sided test (personal communication, B. Mangin, August, 2014). The difference of the additive effects among pairs of alleles was tested a posteriori using a multicomparison t-test (Tukey) with $\alpha = 0,05$.

All genes between the physical position of the flanking markers from significant ($\alpha = 0.01$) QTL were extracted from the filtered gene set of the maize genome sequence ZmB73_5b_FGS.

Using MCQTL, we performed tests among all pairs of marker loci to detect epistatic interactions with a model including additive, dominance, and epistatic effects and all identified QTL [28]. To make the allelic QTL (or cofactors) main effects estimable, we used the same constraints as in the QTL analysis for the single populations as well as for the connected analysis. For the detection of epistatic effects, 0.01 quantile F thresholds were determined by 1,000 permutation tests using the option "genowideforepistasy".

Results

In the field experiments, the parental inbred P092 showed no symptoms (RE) but the highest value for EX and IR (Table 1). The strongest symptoms, however, were observed in the parental inbred W64A, which showed the second highest EX values. The lowest values for EX were observed for the parental inbreds Ky226 and FAP1360A. Population D showed the highest adjusted means for the trait RE and also the highest EX and IR compared to the other segregating populations. Population C showed the lowest values for RE, EX, and IR.

The assessments made in the greenhouse and the field experiments correlated significantly ($\alpha = 0.01$) positive with $\rho = 0.43$ for EX and $\rho = 0.44$ for IR (Figure 1). For the single populations, the correlations varied for EX with ρ between 0.38 and 0.61 and for IR between 0.41 and 0.61. The lowest correlation was observed for population C for both traits.

Mean EX and IR values of the populations B and C were double as high in the greenhouse as in the field experiment (Table 1). The mean value of EX and IR were three times higher for the parental inbred FAP1360A in the greenhouse compared to the field experiment. On the other hand, the adjusted entry mean of both traits for the parental inbred D408 were double as high in the field experiment compared to the greenhouse experiment.

In the field experiment, we observed H_j^2 from 0.69 to 0.92 for the trait RE in the different segregating populations (Table 2). EX and IR showed high H_j^2 in all segregating populations (0.69-0.88). In the greenhouse experiment, the H_j^2 for EX (0.50-0.74) and IR (0.50-0.64) across the two replications was lower than the values observed in the field experiment for EX and IR.

In the connected linkage analysis of the traits assessed in the field experiment, we identified a QTL with significant additive effects on the bottom of chromosome 10 at the position of 46.90 cM for the traits EX, IR, and RE. This QTL with significant additive effects explains a high proportion of the phenotypic variance for the traits EX (45%), IR (46%), and RE (30%) (Table 3). In the single population analyses, the genome positions on chromosome 10, which showed the maximum LOD score, varied among the examined populations. However, the confidence intervals (CI) of the QTL with significant additive effects on chromosome 10 for EX and IR overlapped with the CI of the connected analysis, except in population C (Table 4 and Table 5). Furthermore, in the single population analysis a QTL with significant additive and dominance effects was identified on chromosome 3 for EX in population D.

A second QTL with significant additive effects for RE was identified on chromosome 2, explaining 11% of the phenotypic variance in the connected analysis. In the single population linkage analysis, this QTL CI with

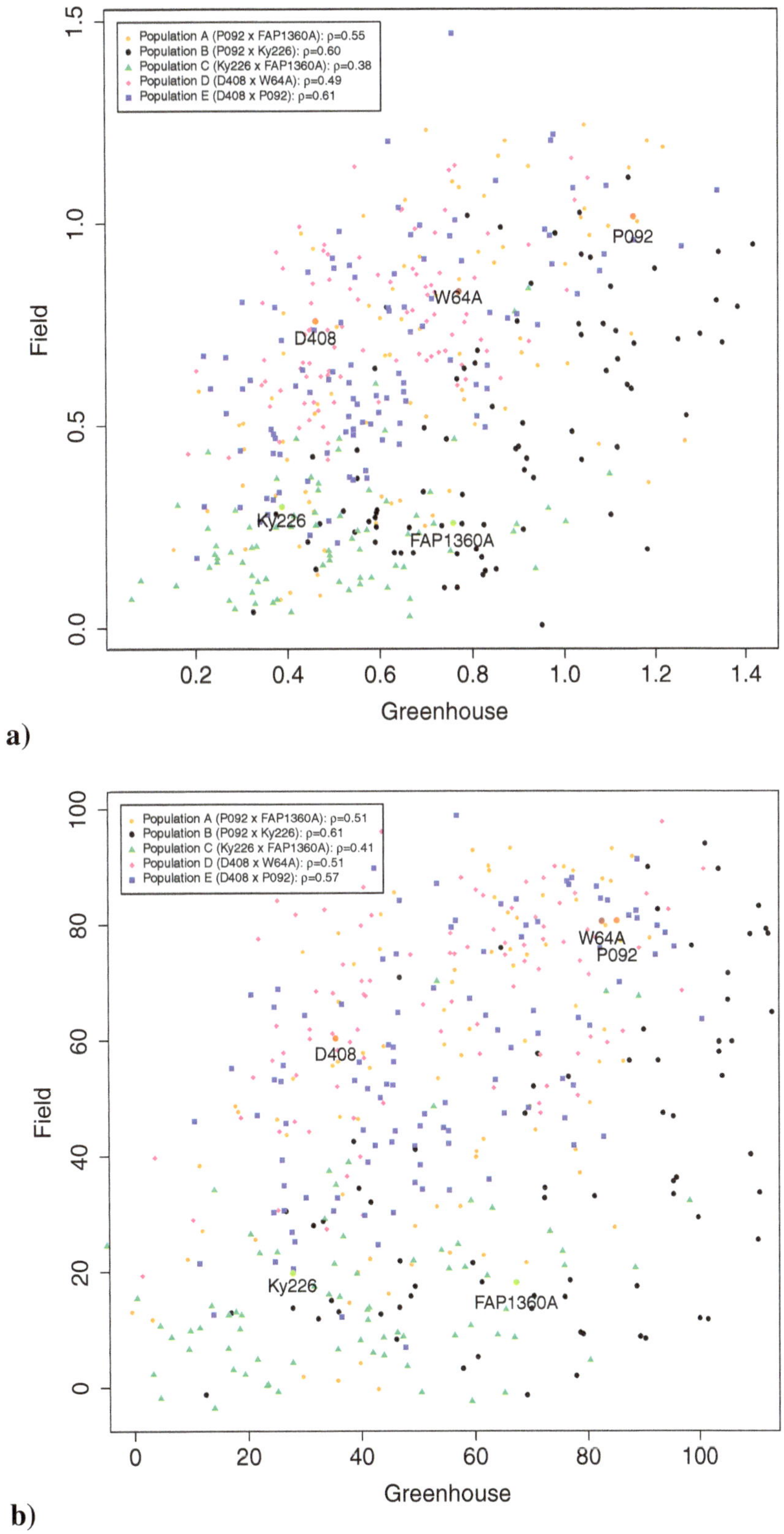

Figure 1 Correlation of (a) the traits virus extinction (EX) and (b) infection rate (IR) between greenhouse and field experiment. Each genotype is colored depending on the population it belongs to. Parental inbreds are lettered with their names and colored green (resistant), orange (tolerant) and red (susceptible). The legend shows the correlations of each population.

Table 2 Broad-sense heritabilities of the single populations for the traits traits red edges (RE), virus extinction (EX), and infection rate (IR) from the field experiment, and for the traits EX and IR from the greenhouse experiment

	Field			Greenhouse	
Entry type	RE	EX	IR	EX	IR
Segregating populations					
Population A (P092 x FAP1360A)	0.76	0.87	0.85	0.74	0.62
Population B (P092 x Ky226)	0.84	0.86	0.88	0.68	0.63
Population C (Ky226 x FAP1360A)	0.73	0.82	0.85	0.67	0.56
Population D (D408 x W64A)	0.92	0.69	0.59	0.50	0.50
Population E (D408 x P092)	0.69	0.83	0.75	0.68	0.64

significant additive effects was only significant in population D.

In the greenhouse experiment, we observed significant QTL with significant additive effects for EX and IR which colocalized with the QTL on chromosome 10 identified in the field experiment in the connected analysis. Furthermore, in the single population analyses the CI of the QTL with significant additive effects for EX and IR overlapped with the CI of the connected analysis, except in population C. A QTL with significant additive and dominance effects was identified on chromosome 9 for IR in the connected analysis and in the single population analysis for population B. In the single population analysis, we identified in population C a QTL with significant additive effects for IR at 176.10 cM and a QTL with significant additive effects for EX at 80.70 cM on chromosome 5. For the latter QTL on chromosome 5 at 80.7 cM, we observed a significant ($\alpha = 0.01$) epistatic interaction with a genome position on chromosome 6 at 60.8 cM for the trait EX, explaining 21% of the phenotypic variance (Table 6).

Discussion

Comparison of field and greenhouse experiments for the assessment of BYDV resistance

The high H_j^2 values observed for the field and greenhouse experiments indicated that the traits are under a strong genotypic control and, thus, an improvement of the genotypes regarding these traits is possible by breeding. The H_j^2 values observed for the traits EX and IR under greenhouse conditions were lower than these observed under field conditions. The reason could be that in the calculation of H_j^2 in the field experiments n=4 due to the four environments whereas in the greenhouse n=2 due to the two replicates. Therefore, H_j^2 on a plot basis $\left(H_j^{*2}\right)$ was calculated to be able to compare the H_j^{*2} values directly [29]. The H_j^{*2} was more similar between the field and greenhouse experiments but still slightly higher for EX in the field (0.36-0.62) compared to $\left(H_j^{*2}\right)$ for EX in the greenhouse (0.33-0.58). This finding can be explained by the higher number of plants per plot in the field (15 plants) compared to the greenhouse (10 plants). Furthermore, σ_{gj}^2

Table 3 Chromosomic locations and confidence intervals (CI) of QTL with their flanking markers and number of genes within the CI from the connected analysis for the traits red edges (RE), virus extinction (EX), infection rate (IR) from the adjusted means of the field experiment, and for the traits EX and IR from the adjusted means of the greenhouse experiment

Trait	Chr	Pos [cM]	LOD	R^2	CI [cM]	Flanking markers		Physical interval [bp]		Number of genes
Field										
EX	10	46.9	76.88	45	45-49	PUT-163a-60352819-2700	SYN15407	127005619	134855671	224
IR	10	46.9	80.26	46	45-51	PUT-163a-60352819-2700	SYN15407	127005619	134855671	224
RE	2	55.1	11.23	11	33-59	SYN29639	PZE-102117636	49294378	158105037	1173
	10	52.1	39.76	30	44-52	PUT-163a-60352819-2700	SYN15407	127005619	134855671	224
Simultaneous fit				36						
Greenhouse										
EX	10	46.9	37.15	29	44-52	PUT-163a-60352819-2700	SYN15407	127005619	134855671	224
IR	9	24.2	9.72	10	17-37	PZA03759.2	PZE-109068137	84335878	112017256	433
	10	46.9	38.51	30	44-52	PUT-163a-60352819-2700	SYN15407	127005619	134855671	224
Simultaneous fit				36						

Table 4 Chromosomic location and effects of QTL from the connected analysis for the BYDV resistance traits red edges (RE), extinction rate (EX), infection rate (IR) from the adjusted means of the field experiment, and for the traits EX and IR from the adjusted means of the greenhouse experiment

			Estimated additive allele effects					Estimated dominance allele effects				
Trait	**Chr**	**Pos [cM]**	**Ky226**	**W64A**	**FAP1360A**	**P092**	**D408**	**Pop A**	**Pop B**	**Pop C**	**Pop D**	**Pop E**
Field												
EX	10	46.9	-0.09^{B}	0.05^{A}	-0.09^{B}	0.18^{C}	-0.06^{B}	$-0.01^{n.s.}$	$-0.03^{n.s.}$	$0.07^{n.s.}$	$0.03^{n.s.}$	$-0.04^{n.s.}$
IR[%]	10	46.9	-9.27^{B}	7.19^{A}	-8.01^{B}	14.65^{A}	-4.57^{B}	$0.59^{n.s.}$	$-1.12^{n.s.}$	$5.44^{n.s.}$	$4.03^{n.s.}$	$-4.16^{n.s.}$
RE	2	55.1	0.04^{B}	-0.45^{A}	-0.01^{AB}	0.18^{B}	0.23^{B}	$0.31^{n.s.}$	$0.06^{n.s.}$	$-0.03^{n.s.}$	$-0.14^{n.s.}$	$0.05^{n.s.}$
	10	52.1	-0.20^{BC}	1.05^{A}	-0.42^{B}	-0.11^{C}	-0.33^{BC}	$0.12^{n.s.}$	$0.12^{n.s.}$	$-0.08^{n.s.}$	$0.10^{n.s.}$	$-0.09^{n.s.}$
Greenhouse												
EX	10	46.9	-0.07^{B}	0.11^{A}	-0.08^{B}	0.10^{A}	-0.06^{B}	$0.01^{n.s.}$	$-0.02^{n.s.}$	$0.03^{n.s.}$	$0.00^{n.s.}$	$0.00^{n.s.}$
IR [%]	9	24.2	-7.42^{A}	2.33^{ABC}	-3.74^{AB}	4.58^{C}	4.24^{BC}	-27.94*	15.58*	$7.74^{n.s.}$	$8.44^{n.s.}$	$-6.56^{n.s.}$
	10	46.9	-6.49^{B}	13.65^{A}	-8.08^{B}	8.75^{A}	-7.84^{B}	$4.04^{n.s.}$	$-3.02^{n.s.}$	$0.67^{n.s.}$	$-3.71^{n.s.}$	$0.13^{n.s.}$

*, **, ***, significant with significance level 0.05, 0.01 and 0.001, respectively.
^{ns}, not significant.
^{A}, ^{B}, ^{C}, populations with the same letter are not significantly ($\alpha = 0.05$) different from each other.

and σ^2_{ej} were calculated across four environments in the field compared to two replicates in the greenhouse. This could be the reason why σ^2_{ej} becomes smaller in the field leading to slightly higher H^{*2}_j in the field experiments.

We observed a significant ($\alpha = 0.01$) positive correlation between the BYDV resistance measured by EX and IR in the field and greenhouse experiments (Figure 1). This shows that our plant material reacts similarly to BYDV infection in the greenhouse and in the field. Nevertheless, the correlation was not tight with a correlation coefficient of 0.43 and 0.44. The difference between field and greenhouse results can be explained by the differences in growing conditions such as light, temperature, humidity, soil nutrients, water application and plant density. The different reaction of the genotypes to BYDV inoculation, resulting from different environmental conditions, are called genotype-environment interactions. Especially in population B and C the genotype-environment interaction led to higher EX and IR values in the greenhouse compared to the field experiments (Table 1 and Figure 1).

For EX and IR, the correlation between the field and greenhouse experiments was the lowest in population C compared to the other populations because the EX and IR values showed a much higher EX and IR values in the greenhouse experiment compared to the field experiments. Furthermore, we observed that one of the parental inbreds of population C, FAP1360A, which was resistant in the field experiment, showed high EX and IR values in the greenhouse experiment. In contrast, the parental inbred D408 showed lower EX and IR values in the greenhouse compared to the field experiment. In earlier studies, Grüntzig and Fuchs, 2000 [30] described the parental inbred D408 as resistant, which is in accordance to the greenhouse results. The reason for this finding could be that D408 reacts differently to BYDV infection, depending on the environmental conditions. However, this requires further research.

Consensus map

The advantage of a consensus map created with a higher number of populations compared to a linkage map created with a single biparental population is the availability of more genotypic information, which improves the construction of the genetic map. Furthermore, more alleles can be taken into account simultaneously and therefore, the probability that at least one population is polymorphic at a given marker locus is higher. In our study, the consensus map was constructed across all five segregating populations examined plus six additional segregating populations from a companion study of Frey and Stich, unpublished (Frey F, Stich B. Identfication of genome regions contributing to variation of heat tolerance in temperate maize by QTL mapping with multiple connected populations:), which were genotyped with the same marker set. With this we reached an even higher marker information and a better coverage of markers over the genome for the consensus map than only across five populations (Figure 2).

The plot of genetic positions vs. the physical positions of the markers showed a sigmoid curve (Figure 3). The same pattern was observed by Payseur and Nachman, 2000 [31] and occurs because the recombination rate is lower at centromeres. However, we observed that the genetic map calculation is reasonable because the order of markers on the genetic map is consistent with the order of the physical map positions.

Table 5 Chromosomic locations and effects of QTL from the single population analysis for the traits red edges (RE), virus extinction (EX), infection rate (IR) from the adjusted means of the field experiment, and for the traits EX and IR from the adjusted means of the greenhouse experiment

Trait	Chr	Pos	CI	R^2	Estimated additive allele effects parental inbred (1)	Estimated dominance allele effects
Population A (FAP1360A(1) x P092(2))						
Field						
EX	10	51.9	41-52	47	-0.12*	$0.02^{n.s.}$
IR [%]	10	51.9	41-52	46	-10.01*	$3.40^{n.s.}$
Greenhouse						
EX	10	51.9	39-52	31	-0.09*	$-0.06^{n.s.}$
IR [%]	10	52.1	39-52	31	-7.01*	$-3.78^{n.s.}$
Population B (Ky226(1) x P092(2))						
Field						
EX	10	51.9	50-52	68	-0.13*	$-0.01^{n.s.}$
IR [%]	10	52.1	50-52	71	-12.62*	$-1.18^{n.s.}$
Greenhouse						
EX	1	123.9	55-149	21	-0.05*	0.14*
	10	51.9	46-52	49	-0.10*	$-0.03^{n.s.}$
IR [%]	9	24.2	14-37	25	-6.99*	14.55*
	10	52.1	44-52	36	-8.53*	$-2.26^{n.s.}$
Population C (FAP1360A(1) x Ky226(2))						
Greenhouse						
EX	5	80.7	54-192	23	-0.18*	$-0.84^{n.s.}$
IR [%]	5	176.1	69-193	25	-11.03*	$11.09^{n.s.}$
Population D (W64A(1) x D408(2))						
Field						
EX	3	117.6	102-147	21	-0.05*	-0.19**
	10	44.8	33-52	30	0.06*	$0.04^{n.s.}$
IR [%]	10	44.8	34-52	30	6.06*	$4.19^{n.s.}$
RE	2	42.2	26-62	21	-0.37*	$-0.10^{n.s.}$
	10	52.1	39-52	47	0.69*	$0.12^{n.s.}$
Greenhouse						
EX	10	46.9	41-52	49	0.08*	$0.00^{n.s.}$
IR [%]	10	51.9	42-52	53	10.68*	$-2.91^{n.s.}$
Population E (P092(1) x D408(2))						
Field						
EX	10	39.8	37-47	48	0.13*	$-0.05^{n.s.}$
IR [%]	10	39.8	37-44	53	10.72*	$-5.22^{n.s.}$
Greenhouse						
EX	10	46.9	39-51	27	0.08*	$0.00^{n.s.}$
IR [%]	10	46.9	41-51	30	7.91*	$0.28^{n.s.}$

*, **, ***, significant with significance level 0.05, 0.01 and 0.001, respectively.

^{ns}, not significant.

Table 6 Virus extinction of genotypes from population C from the greenhouse experiment and (the epistatic additive-additive effects) at the allele combination at the epistatic interacting genome positions

	Genotype at marker within QTL on chromosome 5		
Allele (Effect)	**GG (-0.16)**	**GA (-1.15)**	**AA (0.16)**
Genotype at the epistatically interacting marker on chromosome 6			
TT (-0.02)	0.44 (0.11)	0.53	0.45 (-0.11)
TC (0.24)	0.56	0.53	0.68
CC (0.02)	0.38 (-0.11)	0.39	0.71 (0.11)

QTL for BYDV resistance

In the connected analysis of the traits EX and IR, we observed a huge peak at the end of chromosome 10 in the plot of the LOD scores (Figure 4, Additional file 1: Figure S1 and Figure S3) in both, the field and greenhouse experiments. This QTL with significant additive effects whose CI comprised the genome region between the markers *PUT-163a-60352819-2700* and *SYN15407*, explained in the connected analysis about 45% of the phenotypic variance. The identification of this single QTL with significant additive effects for EX and IR suggests that BYDV resistance is oligogenically inherited.

The QTL CI with significant additive effects on chromosome 10 overlapped with the CI of all individual populations, except population C and explained between 23% and 68% of the phenotypic variance (Table 5). This suggests that this QTL interval is involved in the inheritance of BYDV in all these populations and that this genome region could be validated in different genetic material. Population C was the only population in which the QTL on chromosome 10 could not be detected. This finding indicates that the parental inbred Ky226 and FAP1360A carry the identical allele at the highly linked marker *SYN4811* and therefore no QTL can be detected in this cross.

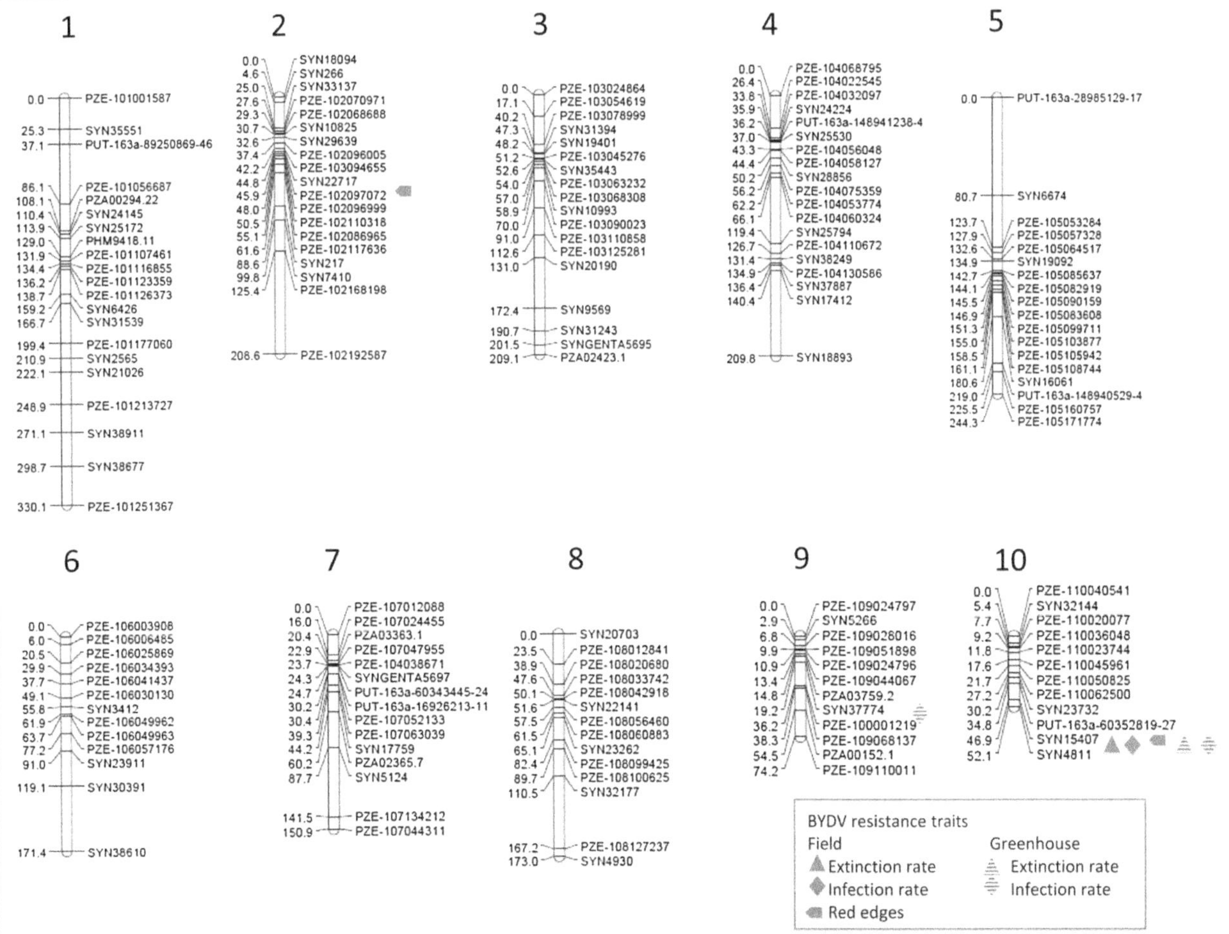

Figure 2 Projection of the QTL identified in this study on the genetic consensus map. QTL are illustrated by symbols for the traits red edges (RE), virus extinction (EX), and infection rate (IR) from the field experiment, and for the traits EX and IR, from the the greenhouse experiment.

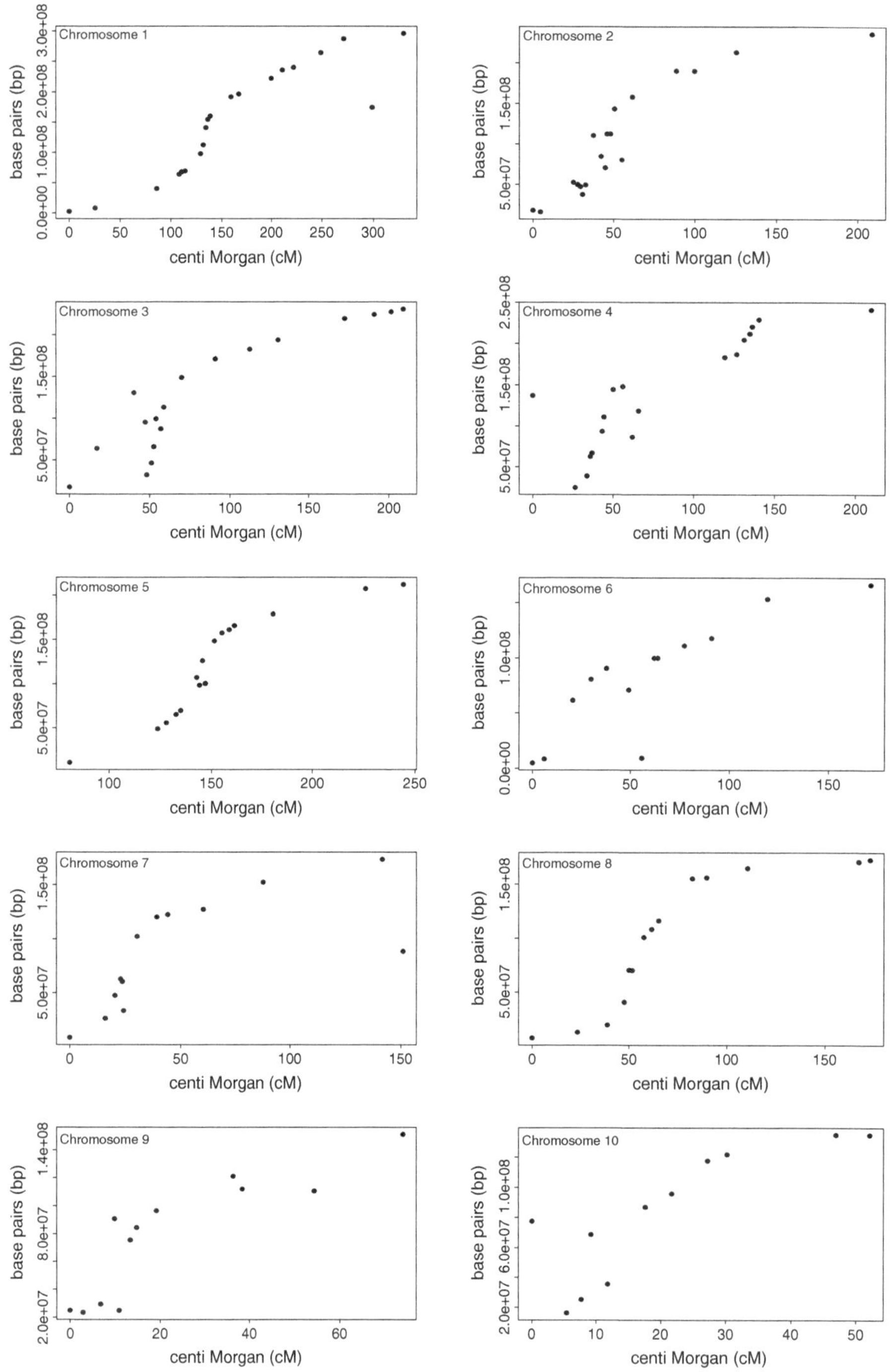

Figure 3 Plot of the physical map position [bp] of the markers versus their genetic position [cM] on the consensus map. The dots represent the markers used in our study on each of the 10 maize chromosomes.

These findings are in accordance with a previous GWAS by Horn et al., 2014 [15], where three SNPs in the same region on chromosome 10 explained 25% of the phenotypic variance for EX. This validation of the mentioned genome region on chromosome 10 in an independent group of genetic material leads to the assumption, that this major QTL for BYDV resistance in maize contributes to the resistance in various germplasm and might be broadly applicable in MAS projects of commercial breeding programs.

The QTL CI on chromosome 10 for EX and IR is located within the QTL CI for resistance to MDMV, identified

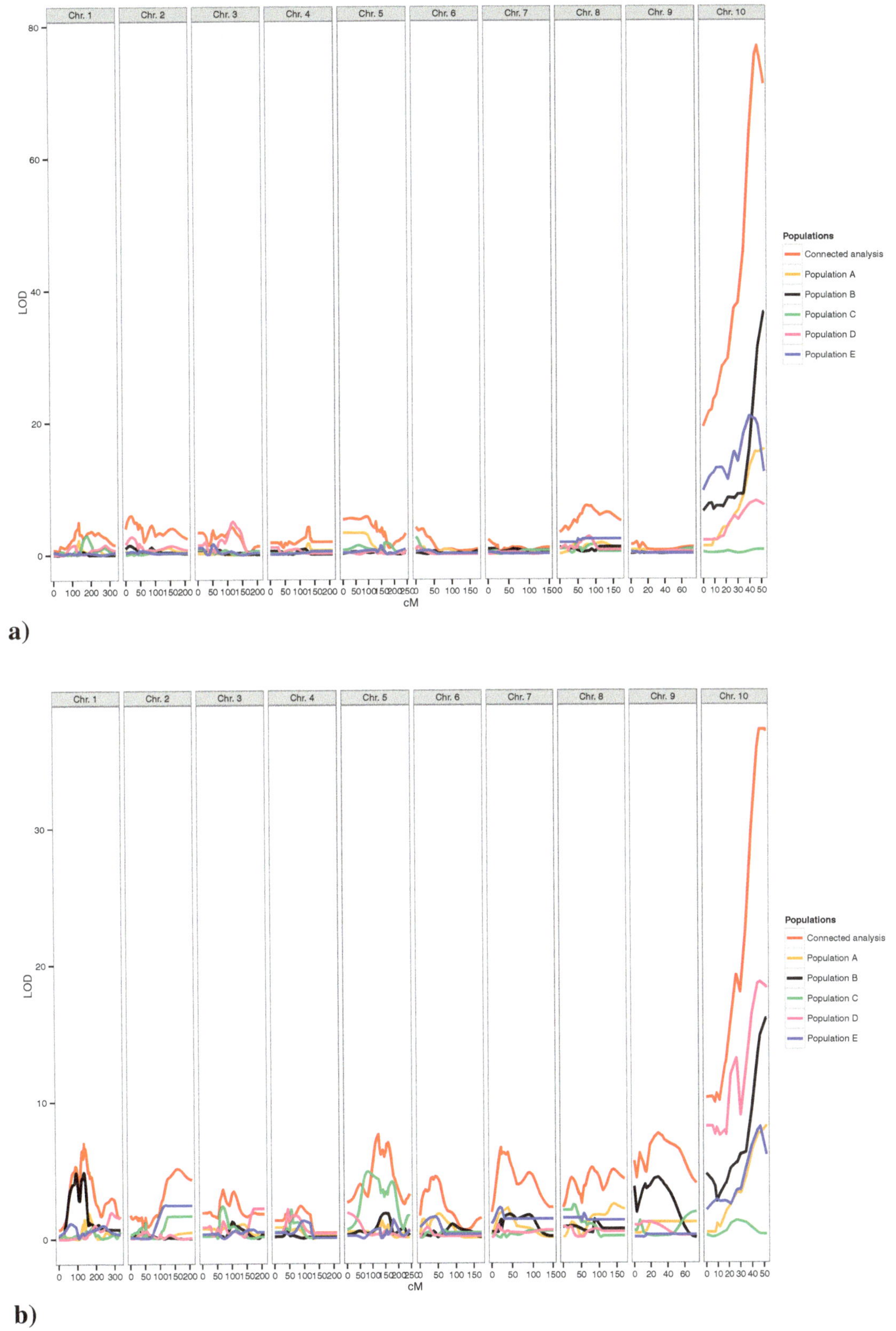

Figure 4 Plot of the LOD scores of the trait extinction rate (EX) across the 10 chromosomes in the (a) field experiment compared to the (b) greenhouse experiment. Each colored line represents a single population analysis, the red line represents the connected analysis.

by Zambrano et al., 2014 [11]. Furthermore, the locus *wsm3* for WSMV resistance [12,13] is located within the EX QTL on chromosome 10. The locus *mcd2* for MCDV resistance [14] and also three minor resistance QTL for SCMV [9,10] are reported to be associated with this genome region. This finding might suggest that the major QTL for BYDV on chromosome 10 possibly also leads to resistance of several other virus diseases. This, however, requires further research.

The QTL with significant additive effects for EX and IR on chromosome 10 colocalized with the QTL for the trait RE (Table 3, Additional file 1: Figure S2) which explained 30% of the phenotypic variance. Moreover, there was another QTL with significant additive effects on chromosome 2, explaining 11% of the phenotypic variance of RE. In the single population analysis, we identified these QTL with significant additive effects for RE on chromosome 2 and 10 only in population D. The reason could be that population D was the only population with the parental inbred W64A, which showed strong BYDV symptoms (RE).

In the greenhouse experiment, we identified a QTL with significant additive and dominance effects on chromosome 9 for IR which was not detected in the analyses of the data from the field experiments. Nevertheless, we could validate the QTL with significant additive effects from the field analysis on chromosome 10 for EX and IR in the greenhouse experiment in the connected analysis. And furthermore, in all single populations, except in population C, the QTL CI on chromosome 10 overlapped with the CI of the connected analysis. This confirms our findings from the field experiment under greenhouse conditions. In the single population analysis of the greenhouse experiment, a QTL with significant additive and dominance effects was identified on chromosome 1 in population B for EX. In the same region, Zambrano et al., 2014 [11] identified a QTL for MMV resistance. In this region a benzoxazionoid QTL in maize was identified [32] which causes aphid resistance and is associated with low levels of 2,4-dihydroxy-7-methoxy-1,4-benzoxazin-3-one glucoside methyltransferase leading to an increased aphid resistance by promoting callose deposition. This finding suggests that the QTL on chromosome 1 is possibly associated with aphid resistance but further research is needed to proof this hypothesis. The reason for the additional QTL in the greenhouse experiment could be the genotype-environment interaction which can lead to the detection of different QTL in the greenhouse compared to the field (Figure 4) [33].

A QTL with significant additive effects for EX was detected on chromosome 5 by the single population analysis of population C in which the major gene did not segregate. For this QTL on chromosome 5 at the position 60.8 cM (G/A), we detected a significant ($\alpha = 0.01$) epistatic interaction with the position at 80.7 cM on chromosome 6 (T/C) explaining 21% of the phenotypic variance (Table 6). We observed that all genotypes of population C which are homozygous (GG) at the QTL position on chromosome 5 and homozygous (TT) at the position on chromosome 6 showed a low (0.44) EX value. These genotypes have the same allele combination like the resistant parental inbred Ky226. In contrast, it can be observed that all genotypes which are homozygous at both positions AA/CC as the parental inbred FAP1360A showed high EX values.

Conclusions

A genome region on chromosome 10 was identified in a linkage mapping approach with five connected segregating populations explaining 45% of the phenotypic variance for BYDV resistance traits. This region could also be confirmed in single population analyses and under greenhouse conditions. This study shows that BYDV resistance is oligogenically inherited and influenced by one major QTL. Therefore, the BYDV resistance trait EX is the ideal candidate to apply for marker assisted selection as BYDV is phenotypically difficult to assess but genetically rather simply inherited.

Additional file

Additional file 1: Supplementary figures. Figure S1. QTL for the trait infection rate (IR) across the genome in the field experiment. **Figure S2.** QTL for the trait red edges (RE) across the genome in the field experiment. **Figure S3.** QTL for the trait infection rate (IR) across the genome in the greenhouse experiment.

Competing interests

The authors declare that they have no competing interests.

Authors' contributions

FH participated in the design of the experiment, carried out most of the experimental work, performed the statistical analyses, and drafted the manuscript. AH participated in the design of the experiment, experimental work, and drafted the manuscript. BS conceived the study, participated in its design and statistical analyses, and drafted the manuscript. All authors read and approved the final manuscript.

Acknowledgements

This work was supported by the Federal Ministry of Food and Agriculture (511-06.01-28-1-45.026-10) in the frame of the innovation program. We are grateful to "Syngenta Seeds GmbH" and "Monsanto Agrar Deutschland GmbH" for providing the experimental fields and carrying out the assessment of the plant performance traits. We would like to thank the "Gemeinschaft zur Förderung der privaten deutschen Pflanzenzüchtung e.V. (GFP)" for supporting the grant application and coordinating the project. For the technical assistance we thank Andrea Lossow, Dörte Grau, Gudrun Meißner, Isabell Scheibert, Jana Heinze, Katja Dlouhy, Nele Sylvester, and Nicole Kamphaus. For carrying out the genotyping of the five segregating populations we thank Jörg Plieske and Martin Ganal from TraitGenetics GmbH. We want to thank two anonymous reviewers for their valuable comments on an earlier version of this manuscript.

Author details

[1]Max Planck Institute for Plant Breeding Research, Carl-von-Linné Weg, 50829 Cologne, Germany. [2]Julius Kühn Institute, Erwin-Baur-Straße 27, 06484 Quedlinburg, Germany.

References

1. Huth W. Maisvirosen - Tendenz zunehmend. Pflanzenschutz-Praxis 1994;2:17–20.
2. Harrington R, Clark SJ, Welham SJ, Verrier PJ, Denholm CH, Hullé M, *et al.* Environmental change and the phenology of European aphids. Glob Change Biol 2007;13:1550–64.
3. Beuve M, Naibo B, Foulgocq L, Lapierre H. Irrigated hybrid maize crop yield losses due to barley yellow dwarf virus-PAV luteovirus. Crop Sci 1999;39:1830–4.
4. Loi N, Osler R, Lapierre H. Barley yellow dwarf associated to, BYDV-PAV virus In: Lapierre H, Signoret PA, editors. Virus and virus diseases of Poaceae (Gramineae). 1edition. Paris: INRA Editions; 2004. p. 618–620.
5. Panayotou PC. Effect of barley yellow dwarf on several varieties of maize. Plant Dis Rep 1977;61:815–9.
6. Horn F, Habekuß A, Stich B. Natural variation for BYDV resistance in maize. Maydica 2013;58:173–81.
7. Grüntzig M, Fuchs E, Werner M. Occurence and influence of barley yellow dwarf luteovirus (BYDV) on growth and yield of maize. Nachrichtenblatt des deutschen Pflanzenschutzdienstes 1997; 49:45–50.
8. Osler R, Loi N, Lorenzoni C, Snidaro M, Refatti E. Barley yellow dwarf virus infections in maize (Zea mays L.) inbreds and hybrids in northern Italy. Maydica 1985;30:285–99.
9. Xia X, Melchinger AE, Kuntze L, Lübberstedt T. Quantitative trait loci mapping of resistance to sugarcane mosaic virus in maize. Phytopathology 1999;89:660–67.
10. Zhang SH, Li XH, Wang ZH, George ML, Jeffers D, Wang FG, et al. QTL mapping for resistence to SCMV in chinese maize germplasm. Maydica 2003;48:307–12.
11. Zambrano JL, Jones MW, Brenner E, Francis DM, Tomas A, Redinbaugh MG. Genetic analysis of resistance to six virus diseases in a multiple virus-resistant maize inbred line. Theor Appl Genet 2014;127: 867–80.
12. McMullen MD, Simcox KD. Genomic organization of disease and insect resistance genes in maize. Mol Plant Microbe In 1995;8:811–5.
13. Jones MW, Boyd EC, Redinbaugh MG. Responses of maize (Zea mays L.) near isogenic lines carrying Wsm1, Wsm2, and Wsm3 to three viruses in the Potyviridae. Theor Appl Genet 2011;123:729–40.
14. Jones MW, Redinbaugh MG, Anderson RJ, Louie R. Identification of quantitative trait loci controlling resistance to maize chlorotic dwarf virus. Theor Appl Genet 2004;110:48–57.
15. Horn F, Habekuß A, Stich B. Genes involved in barley yellow dwarf virus resistance of maize. Theor Appl Genet 2014;127:2575–84.
16. Flint-Garcia SA, Thuillet AC, Yu J, Pressoir G, Romero SM, Mitchell SE, *et al.* Maize association population: a high-resolution platform for quantitative trait locus dissection. Plant J 2005;44:1054–64.
17. Stich B. Comparison of mating designs for establishing nested association mapping populations in maize and Arabidopsis thaliana. Genetics 2009;183(4):1525–34.
18. Stich B, Melchinger AE, Heckenberger M, Möhring J, Schechert A, Piepho HP. Association mapping in multiple segregating populations of sugar beet (Beta vulgaris L.) Theor Appl Genet 2008;117:1167–79.
19. Bardol N, Ventelon M, Mangin B, Jasson S, Loywick V, Couton F, et al. Combined linkage and linkage disequilibrium QTL mapping in multiple families of maize (Zea mays L.) line crosses highlights complementarities between models based on parental haplotype and single locus polymorphism. Theor Appl Genet 2013;126:2717–36.
20. Blanc G, Charcosset A, Mangin B, Gallais A, Moreau L. Connected populations for detecting quantitative trait loci and testing for epistasis: an application in maize. Theor Appl Genet 2006;113:206–24.
21. Gilmour AR, Gogel BJ, Cullis BR, Thompson R. ASReml User Guide Release 2.0. VSN International Ltd: Hemel Hempstead, HP1 1ES. UK; 2006.
22. R Core Development. Team R: a language and environment for statistical computing. Vienna: R Foundation for Statistical Computing; 2011.
23. Ganal MW, Durstewitz G, Polley A, Bérard A, Buckler ES, Charcosset A, *et al.* A large maize (Zea mays L.) SNP genotyping array: development and germplasm genotyping, and genetic mapping to compare with the B73 reference genome. PLoS One 2011;6:e28334.
24. Benke A, Urbany C, Marsian J, Shi R, Wirén NV, Stich B. The genetic basis of natural variation for iron homeostasis in the maize IBM population. BMC Plant Biol 2014;14:12.
25. Givry SD, Bouchez M, Chabrier P, Milan D, Schiex T. CarthaGene: multipopulation integrated genetic and radiation hybrid mapping. Bioinformatics 2005;21:1703–4.
26. Jourjon MF, Jasson S, Marcel J. MCQTL: multi-allelic QTL mapping in multi-cross design. Bioinformatics 2005;21:128–30.
27. Charcosset A, Mangin B, Moreau L, Combes L, Jourjon MF. Heterosis in maize investigated using connected RIL populations. In: Quantitative genetics and breeding methods: the way ahead. Paris, France: INRA; 2000. p. 89–98.
28. Mangin B, Cathelin R, Delannoy D, Escalière B, Lambert S, Marcel J, et al. MCQTL: A reference manual 2010. carlit.toulouse.inra.fr/MCQTL/.
29. Smalley M, Daub J, Hallauer A. Estimation of heritability in maize by parent-offspring regression. Maydica 2004;49:221–9.
30. Grüntzig M, Fuchs E. Occurence of luteoviruses of cereals in Zea mays L. J Plant Dis Protect 2000;107:523–38.
31. Payseur BA, Nachman MW. Microsatellite variation and recombination rate in the human genome. Genetics 2000;156:1285–98.
32. Meihls LN, Handrick V, Glauser G, Barbier H, Kaur H, Haribal MM, et al. Natural variation in maize aphid resistance is associated with 2,4-dihydroxy-7-methoxy-1,4-benzoxazin-3-one glucoside methyltransferase activity. Plant Cell 2013;25:2341–55.
33. Brachi B, Faure N, Horton M, Flahauw E, Vazquez A, Nordborg M, et al. Linkage and association mapping of Arabidopsis thaliana flowering time in nature. PLoS Genet 2010;6:e1000940.

Comparative transcriptome analysis of latex from rubber tree clone CATAS8-79 and PR107 reveals new cues for the regulation of latex regeneration and duration of latex flow

Jinquan Chao, Yueyi Chen, Shaohua Wu and Wei-Min Tian*

Abstract

Background: Rubber tree (*Hevea brasiliensis* Muell. Arg.) is the primarily commercial source of natural rubber in the world. Latex regeneration and duration of latex flow after tapping are the two factors that determine rubber yield of rubber tree, and exhibit a huge variation between rubber tree clones CATAS8-79 and PR107.

Results: To dissect the molecular mechanism for the regulation of latex regeneration and duration of latex flow, we sequenced and comparatively analyzed latex of rubber tree clone CATAS8-79 and PR107 at transriptome level. More than 26 million clean reads were generated in each pool and 51,829 all-unigenes were totally assembled. A total of 6,726 unigenes with differential expression patterns were detected between CATAS8-79 and PR107. Functional analysis showed that genes related to mass of categories were differentially enriched between the two clones. Expression pattern of genes which were involved in latex regeneration and duration of latex flow upon successive tapping was analyzed by quantitative PCR. Several genes related to rubber biosynthesis, cellulose and lignin biosynthesis and rubber particle aggregation were differentially expressed between CATAS8-79 and PR107.

Conclusions: This is the first report about probing latex regeneration and duration of latex flow by comparative transcriptome analysis. Among all the suggested factors, it is more important that the level of endogenous jasmonates, carbohydrate metabolism, hydroxymethylglutaryl-CoA reductase (HMGR) and *Hevea* rubber transferase (HRT) in mevalonate (MVA) parthway for latex regeneration while the level of endogenous ethylene (ETH), lignin content of laticifer cell wall, antioxidants and glucanases for the duration of latex flow. These data will provide new cues for understanding the molecular mechanism for the regulation of latex regeneration and duration of latex flow in rubber tree.

Keywords: *Hevea brasiliensis* Muell. Arg, RNA-Seq, Transcriptome, Latex regeneration, Duration of latex flow

Background

Rubber tree (*Hevea brasiliensis* Muell. Arg.) is the main source of natural rubber [1-4]. The natural rubber is synthesized and stored in laticifer, a specific tissue densely located in the secondary phloem of trunk [5]. By successive tapping, white or yellowish milky latex is expelled and collected. The latex is the cytoplasm of laticifer cells and used to refine natural rubber. It contains numerous rubber particles and lutoids as well as general eukaryotic organelles [6].

Latex regeneration and duration of latex flow after tapping are important factors that determine rubber yield of rubber tree. Sucrose, water and nitrogen sources supplying from the surrounding parenchyma cells act as raw materials for latex regeneration between two tappings [7,8]. Isopentenyl pyrophosphate (IPP) is the direct precursor for rubber biosynthesis and mainly derived from the MVA pathway although 2-C-methyl-D-erythritol 4-phosphate (MEP) pathway is suggested to be an alternative source [6]. Catalyzing by enzymes as prenyltransferase, the rubber

* Correspondence: wmtian@163.com
Ministry of Agriculture Key Laboratory of Biology and Genetic Resources of Rubber Tree/ State Key Laboratory Breeding Base of Cultivation and Physiology for Tropical Crops, Rubber Research Institute, Chinese Academy of Tropical Agricultural Sciences, Danzhou, Hainan 571737, PR China

transferase, IPP initiates the subsequent extensive prenyl chain elongation process for the formation of rubber macromolecules. It is well known that tapping can promote latex regeneration and there is obvious difference in the rubber content of latex among varieties upon ethrel stimulation [7]. The duration of latex flow is influenced by various factors, such as laticifer turgor, plug formation at the end of severed laticifer, and ethrel application. Plugging of severed laticifer end leads to the termination of latex flow from the wounded site of rubber trees and has been a key limiting factor for the yield of *Hevea* [9,10]. It is widely believed that the severed laticifers are plugged by rubber coagulum as a result of rubber particle aggregation (RPA) caused by the bursting of lutoids [11,12]. Inclusions and debris of lutoids from the burst lutoids are effective in rubber particle aggregation [13,14].

With the sequence technology development, digital gene expression tag profiling recently displays huge potential for exploring biological process [15-22]. By using next-generation massively parallel sequencing technologies, Triwitayakorn *et al.* sequenced 2,311,497 reads from rubber tree vegetative shoot apex transcriptome, generating 23 linkage groups covering 842.9 cM with a mean interval of 11.9 cM per linkage group [23]. Xia *et al. de novo* assembled 48,768 unigenes from transcriptome data of leave and latex of rubber tree in an effort to facilitate biological, biochemical and molecular researches on rubber biosynthesis [24]. Li *et al.* generated 22,756 unigenes from rubber tree bark transcritome and obtained 39,257 simple sequence repeats (SSRs) markers which may be benefit for marker-assisted selection in the cross breeding program of rubber tree [25]. Mantello *et al.* assembled 50,384 contigs with an average length of 400 bp from *H. brasiliensis* bark transcritome, and detected 17,927 SSRs and 404,114 single nucleotide polymorphisms (SNPs) [26]. In the present study, a comparative analysis of latex transcriptome between rubber tree clone PR107 and CATAS8-79 was performed to uncover the molecular mechanism for the regulation of latex regeneration and duration of latex flow.

Results

Difference in latex regeneration and duration of latex flow between rubber tree clone PR107 and CATAS8-79

Rubber tree clone CATAS8-79 and PR107 exhibited a huge difference in duration of latex flow (Figure 1A) and latex regeneration (Figure 1B) at each tapping. A total of about 260 ml of latex was gained from CATAS8-79 while only about 95 ml of latex from PR107 by four tappings (Figure 1B). The rubber content of latex, however, showed no difference between two clones at each tapping (Figure 1C). Consequently, rubber yield of CATAS8-79 was significantly higher than that of PR107 (Figure 1D), suggesting latex regeneration in CATAS8-79 was more effective than that in PR107 during the interval of successive tappings.

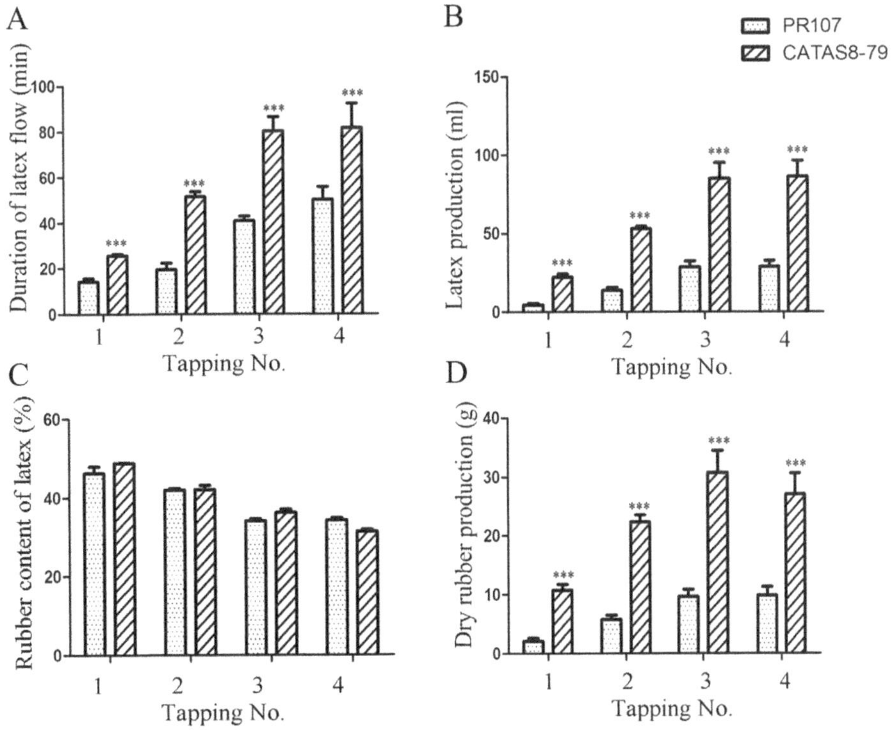

Figure 1 The difference in duration of latex flow **(A)**, latex regeneration **(B)**, rubber content of latex **(C)** and dry rubber production **(D)** between CATAS8-79 and PR107 upon successive tappings. Significant difference was indicated by the asterisks above the bars (***p < 0.01). Tapping No. represented Tapping Number.

Assembling, annotation of latex transcriptome

RNA were extracted from CATAS8-79 and PR107 at first tapping and sequenced with Illumina paired-end sequencing technology individually. After excluding low-quality reads such as empty adapters, 26 million clean reads were generated in each pool. Using SOAPdenove software, 296,736 and 308,262 contigs ranging from 100 bp to more than 3,000 bp were respectively assembled from CATAS8-79 and PR107 (NCBI accession numbers: GSE59981). By paired-end and gap-filling, contigs were further extended and finally assembled as a long sequence named "unigenes". In this way, 53,571 and 57,806 unigenes were generated from CATAS8-79 and PR107 individually. The unigenes were further integrated into 51,829 all-unigenes with an average length of 640 bp and a N50 of 526 bp by paired-end joining (Table 1, Additional file 1: Table S1). The analyses following were carried out with all-unigenes (reffered to unigenes).

All the integrated unigenes were used to match against both the NCBI Non-redundant (Nr) and Swissprot protein databases using BLASTx program with an E-value threshold of 1E-5. Of which, 40,373 (77.9%) and 23,387 (45.12%) unigenes were positively matched with Nr protein database and Swissprot protein database, respectively (Additional file 2: Figure S1). Clusters of Orthologous Group (COG) analysis showed that 16,242 unigenes could be divided into 25 categories (Figure 2A). Of which, "General function predicted only" represented the largest group (2,647), followed by "Transcription" (1,431), and "Posttranslational modification, protein turnover, chaperones" (1,427). Categories of "Extracellular structures" and "Nuclear structure" only included 4 and 8 unigenes whereas 926 unigenes were classified into the category of "Carbohydrate transport and metabolism".

By comparative analysis of transcriptome data of CATAS8-79 versus PR107 under condition of FDR ≤ 0.001 and |log2 Ratio| ≥ 1, there were 6,726 unigenes with differential expression. Of which, 3,018 were up-regulated while 3,708 were down-regulated. "Up-regulated" means the level of gene transcripts was higher in PR107 whereas "down-regulated" means the level of gene transcripts was higher in CATAS8-79. To validate the digital gene expression (DGE), 11 unigenes were selected to amplify by qRT-PCR in the latex samples from CATAS8-79 and PR107. These unigenes were different in abundance and expression pattern on the basis of DGE data. The expression pattern of most of them by qRT-PCR was similar to the corresponding DGE data (Figure 2B).

Table 1 *de novo* assembly of *H. brasiliensis* transcriptome

Sample	Number	N50 (bp)	Mean (bp)
CATAS8-79 unigene	53571	509	421
PR107 unigene	57806	427	375
All unigene	51829	640	526

Functional analysis of the enriched categories in CATAS8-79 and PR107

Kyoto Encyclopedia of Genes and Genomes (KEGG) analysis of all the differently expressed unigenes showed that mass of categories was differentially enriched between CATAS8-79 and PR107. Some of these differences may be related to the difference in latex regeneration and duration of latex flow between the two clones. The others may associate with the general difference in genetic background, considering that CATAS8-79 only contained one fourth descent of PR107.

Most of unigenes related to categories of stress-related proteins, cell wall biosynthesis and nitrogen requirement were significant up-regulated in PR107 (Figure 3). One unigene encoding for chitinase, eight unigenes for glucanase were detected in DGE data (Figure 4A), and unigene 49104 encoding beta-1,3-glucanase was analyzed by qRT-PCR (Figure 4B). Glutamate synthase (GS) plays a key role in integrating NH_4 into amino acid, which is crucial for nitrogen requirement [27]. Recently, there is evidence that GS can directly regulate lignin biosynthesis and deposition in rice seeding [28]. Nine unigenes encoding for enzymes involving in glutamine metabolism were up-regulated in PR107 (Figure 4A), and one of them, unigene19894, was analyzed by qRT-PCR (Figure 4B). Several unigenes involving cellulose and lignin metabolism were widely activated in rubber clone PR107. Lignin is a component of cell wall and synthesized through shikimate pathway [29]. Unigenes encoding 4-coumarate-coa ligase (unigene45298, 4CL), cinnamoyl-CoA reductase (unigene13817, CCR), the key enzymes in shikimate pathway were analyzed by qRT-PCR (Figure 4B).

By contrast, most of unigenes related to categories of carbohydrate metabolism, rubber biosynthesis, hormone, and antioxidation were up-regulated in CATAS8-79 (Figure 3). Carbohydrate metabolism could provide carbon skeleton for the formation of various organic compounds [30,31]. DGE analysis showed that plenty of unigenes associating with carbohydrate metabolism were up-regulated, and the expression pattern of several unigenes mediating sucrose transportation (unigene39426 encoding sucrose transporter, SUT), starch degradation (unigene21420 encoding beta-amylase, BAM) and glycolysis (Unigene8547 encoding pyruvate kinase, PK) were analyzed by qRT-PCR (Figure 5). Both MVA and MEP pathways provide IPP, a precursor for the formation of final rubber molecule [6]. In the present study, unigenes encoding 4-hydroxy-3-methylbut-2-enyl diphosphate reductase (HDR, unigene25794), 1-deoxy-D-xylulose 5-phosphate synthase (DXR, unigene3946) that is components of MEP pathway, and encoding hydroxymethylglutaryl-CoA reductase

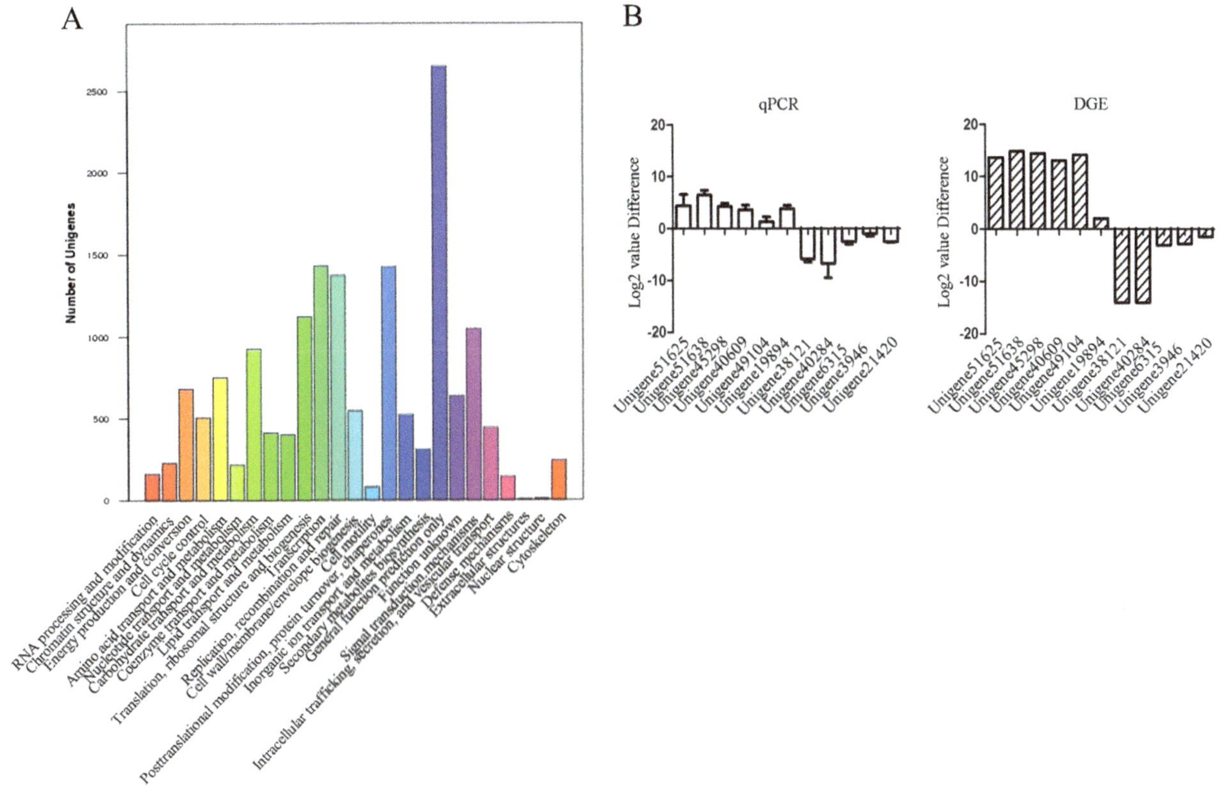

Figure 2 Classification and detection of differentially expressed unigenes. **(A)** Histogram of clusters of COG classifications of *H. brasiliensis* unigenes. Numbers on the Y-axis represent the uingenes numbers. **(B)** Validation of DGE by qRT-PCR. The left showed qRT-PCR and the right showed DGE data. Error bars for qRT-PCR show the standard deviation of three replicates. Unigene51625, BRASSINOSTEROID INSENSITIVE 1; Unigene51638, Cellulose synthase; Unigene45298, 4-coumarate-coa ligase; Unigene40609, aux/IAA protein; Unigene49104, beta-1,3-glucanase; Unigene19894, glutamine synthetase; Unigene38121, hydroxymethylglutaryl-CoA reductase; Unigene40284, Jasmonate O-methyltransferase; Unigene6315, 1-deoxy-D-xylulose 5-phosphate reductoisomerase; Unigene3946, 1-deoxy-D-xylulose 5-phosphate synthase; Unigene21420, Beta-amylase 2.

(HMGR, unigene38121) and acetyl coenzyme A acetyltransferase (AACT, unigene14484) that is the key enzymes of MVA pathway were notably up-regulated in CATAS8-79 (Figure 6). Of which, unigene38121 for HMGR, unigene3946 for DXR, unigene25794 for HDR were analyzed by qRT-PCR (Figure 5). Additionally, unigenes encoding farnesyl diphosphate synthase (FDPS, unigene33714), geranyl-diphosphate synthase (GPPS, unigene21852 and unigene4306), *Hevea* rubber transferase 2 (HRT2, Unigene37088) which directly participated in the formation of

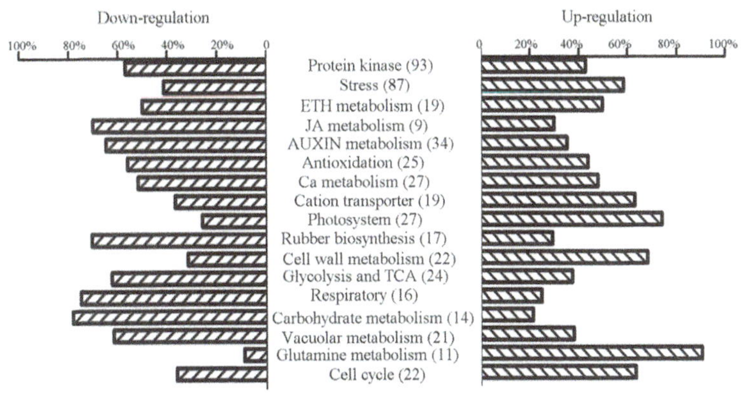

Figure 3 Histogram of metabolism categories enrichment in up-/down-regulation parts. Number in brackets showed corresponding unigenes detected in transcriptome.

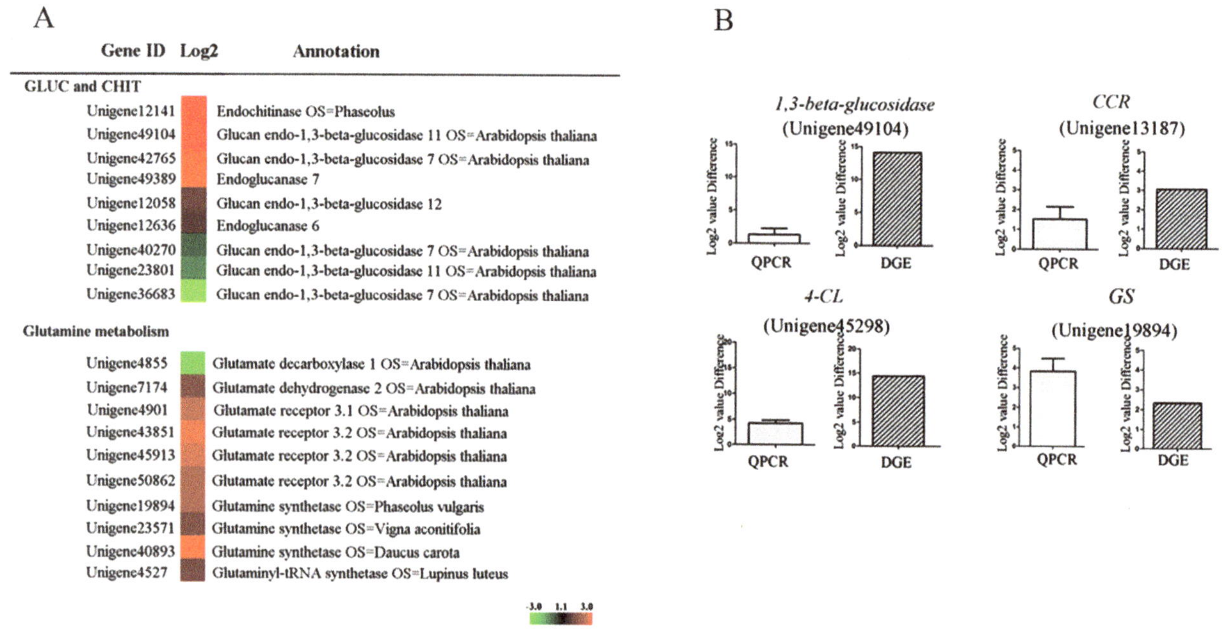

Figure 4 Differential expressed unigenes involving in cell wall biosynthesis, nitrogen metabolism and stress-related proteins. The left showed qRT-PCR **(A)** and the right showed heat map **(B)**. Error bars for qRT-PCR showed the standard deviation of three replicates. DGE values displayed as heat map. Colours bar represented expression levels of each gene which were either up-regulated (red) or down-regulated (blue).

the final high-molecular weight rubber molecule were also differentially expressed (Figure 6). FDPS and HRT2-related unigenes were up-regulated in CATAS8-79 whereas GPPS-related unigene was up-regulated in PR107, respectively (Figure 6). Jasmonates were pivotal to the secondary laticifer differentiation while ethylene was most effective in prolonging the duration of latex flow upon tapping in rubber tree [32,33]. In the present study, the up-regulated unigenes in JA signaling pathway in CATAS8-79 included unigene 34101 (*lipoxygenaseA, LOX*), 38638 (*lipoxygenaseA, LOX*), 24544 (*lipoxygenaseA, LOX*), 15932 (12-oxophytodienoate reductase3, *OPDR*), 40284 and 39621 (*jasmonate O-methyltransferase, JMT*), and 38882 (*Jasmonate ZIM-domain, JAZ8*). And the up-regulated unigenes in ethylene signaling pathway were unigene 23152 (*1-aminocyclopropane-1-carboxylate oxidase homolog 1, ACO*), 40400 (*1-aminocyclopropane-1-carboxylate oxidase, ACO*), 20082 and 17548 (*ethylene-overproduction protein 1*), 24000 and 20993 (*ethylene-insensitive protein 2*), 6206 and 17574 (*EIN3-binding F-box protein 1*) (Figure 6). Of which, unigene 38638 for LOX and unigene 23152 for ACO were analyzed by qRT-PCR (Figure 5). Whereas five of six unigenes (22954, 1657, 34623, 51430, 34785) encoding ethylene response factors were up-regulated in PR107. In addition, antioxidants antagonized reactive oxygen species -caused membrane lipid peroxidation and had a role in keeping the integrity of lutoid [16,17]. The unigene 18915 and 15723 encoding L-ascorbate peroxidase (APX) and superoxide dismutase (SOD) were up-regulated in CATAS8-79, respectively (Figure 5).

Expression pattern of unigenes related to latex regeneration and duration of latex flow upon successive tapping

Eleven unigenes with differential expression between the two clones have been demonstrated to participate in latex regeneration or duration of latex flow. For this purpose, their expression patterns upon successive tapping were analyzed by qRT-PCR (Figure 7). Six of the eleven unigenes exhibited a big difference in their expression patterns between CATAS8-79 and PR107. Of which, unigene 49104 (*beta-1,3-glucanase, HbGluc*), 51638 (*cellulose synthase, HbCS*) and 45298 (*4-coumarate-CoA ligase, Hb4CL*) were up-regulated and kept significant high level in PR107 while similar expression pattern of unigene 23152 (*HbACO*), 38121 (*HbHMGR1*) and 37088 (*HbHRT2*) occurred in CATAS8-79 (Figure 7). The expression pattern of the other five unigenes was similar between the two clones (Figure 7).

Discussion

The rubber tree clone PR107 is an original clone selected from Wickham germplasm in 1920s. The rubber tree clone CATAS8-79 was selected from hybrid offspring of CATAS88-13 and CATAS217. The CATAS88-13 is the product of RRIM600 and PilB84 cross while the CATAS217 is selected from the cross of RRIM513 and

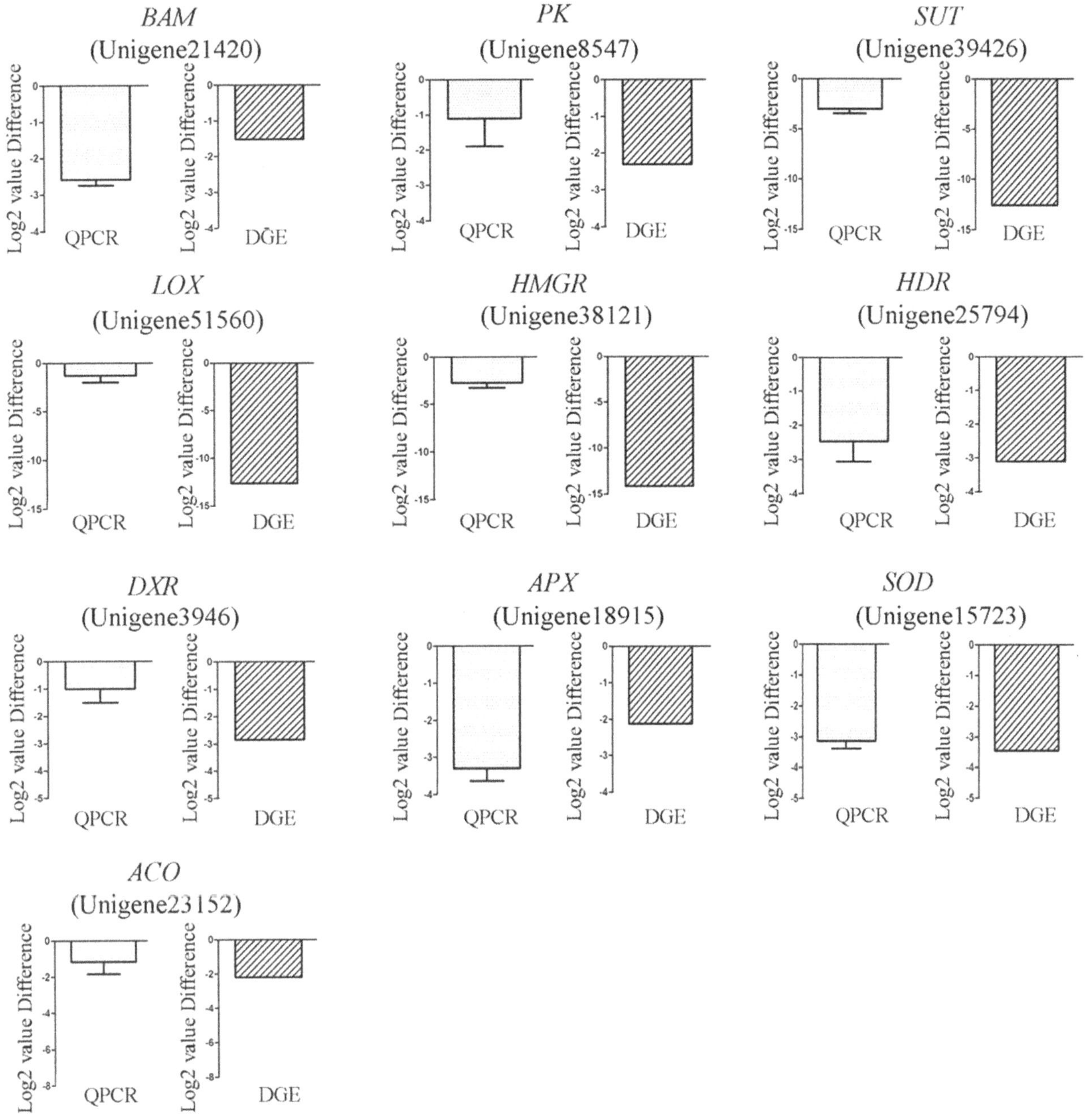

Figure 5 Differentially expressed unigenes involving hormone, carbohydrate and rubber biosynthesis processes. The left showed qRT-PCR and the right showed DGE data in each schematic. Error bars for qRT-PCR showed the standard deviation of three replicates.

PR107. The RRIM600 are selected from the cross of Tjir1 and PB86 while RRIM513 is the hybrid offspring of PilB16 and PilA44. As there is a significant difference in duration of latex flow and latex regeneration between CATAS8-79 and PR107, analysis of the transcriptome dataset of PR107 against that of CATAS8-79 will be prospect to dissect key genes mediating the process. Some of the differentially expressed unigenes should be related to the difference in latex regeneration and duration of latex flow although the others may associate with the general difference in genetic background, considering that CATAS8-79 only contained one fourth descent of PR107.

It has long been believed that the efficiency of sucrose transportation and metabolism and rubber biosynthesis closely associates with the ability of latex regeneration between interval of successive tappings [7]. Amylase, a member of glycosyl hydrolases, is activated at weakly alkaline pH [34,35]. One of the effects of ethrel is to enhance carbohydrate metabolism by alkalizing cytosol of laticifer cells [36], which may be ascribed to the activation of amylase. Besides, ethrel treatment can greatly up-regulate the expression of SUT genes in rubber tree [37]. In the present study, several unigenes encoding beta-amylases and SUTs are found to be expressed at higher level in

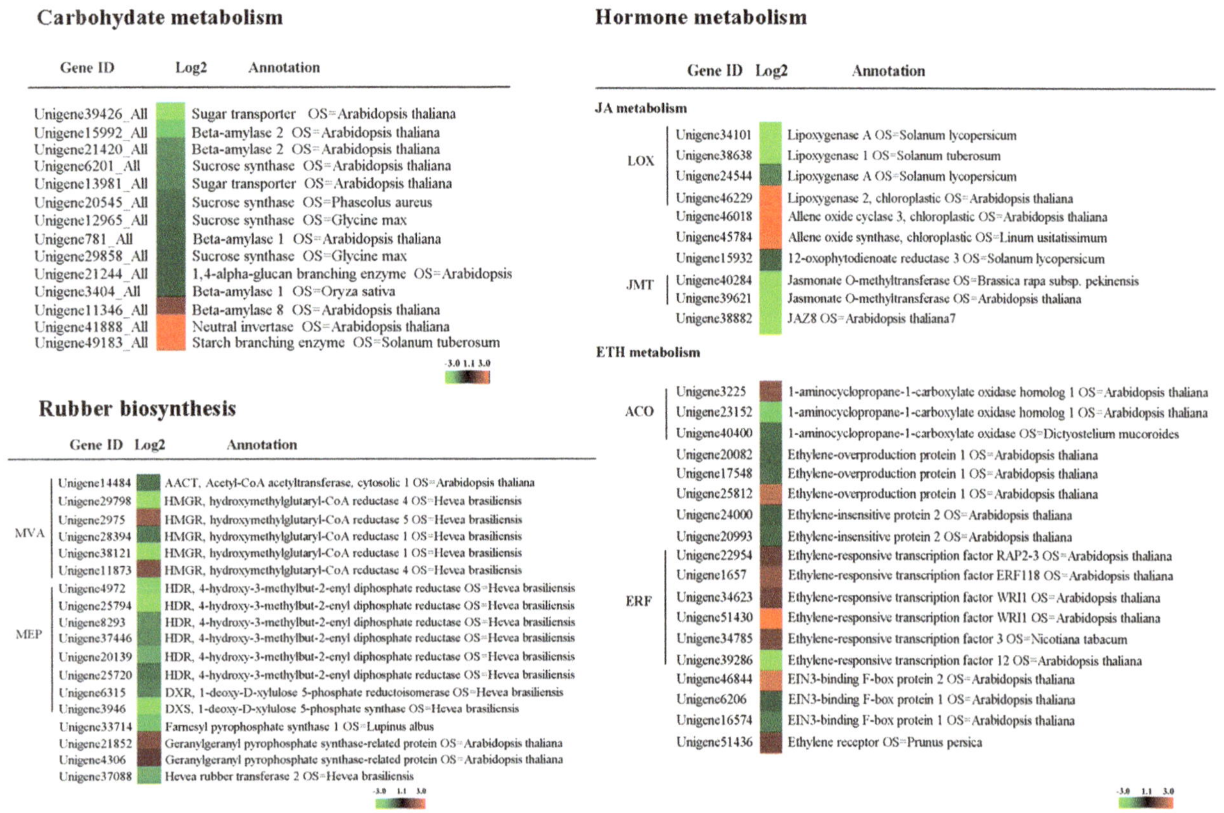

Figure 6 Heat map of genes involve in carbohydrate metabolism, rubber biosynthesis and hormone metabolism. DGE values displayed as heat map. Colours bar represented expression levels of each gene which were either up-regulated (red) or down-regulated (blue).

CATAS8-79 than that in PR107, suggesting a more efficiency of sucrose transportation and carbohydrate metabolism occurred in CATAS8-79 (Figure 6). IPP is biosynthesized though both MVA and MEP pathway and runs several enzymatic reactions to form the final high-molecular weight rubber molecule [7]. In present study, five unigenes encoding for HMGR and six unigenes encoding for HDR are detected with differently expression between two clones. Further sequence analysis shows four HDR-like members (unigene 4972, 25794, 8293, 37446) are perfectly blasted to one *HbHDR* with a Genbank number EU881977 whereas five HMGR-like unigenes are matched to three *HbHMGR* genes (unigene11873, unigene29798 for *HbHMGR4*, unigene28394, unigene38121 for *HbHMGR1*, unigene2975 for *HbHMGR5*) respectively. Previous work shows that the transcripts of *HbHMGR1* are most abundant in latex [38]. The present study shows that HMGR1-like unigenes are significantly up-regulated in CATAS8-79 while *HbHDR* displays a similar pattern between the two clones, suggesting HMGR1 in MVA pathway is critical for providing IPP.

Post-IPP processes include initiation and elongation of rubber macromolecules. FDPS family catalyzes the biosynthesis of farnesyl diphosphate (FDP). The FDP acts as the prime which is essential for initiating prenyl chain whereas HRT family is crucial for integrating IPP units into prenyl chain [39,40]. Two members of HRT family (HRT1 and HRT2) are reported in rubber tree [41]. *In vitro* analysis shows that only HRT2 has rubber transferase activity and may play a key role in extending prenyl chain [41]. In this study, unigenes related to HRT2 and FDPS are found to be expressed at higher level in CATAS8-79 than that in PR107, indicating that HRT2 and FDPS are crucial for enhancing rubber initiation and elongation. The higher expression level of HRT2 and FDPS like unigenes is in line with the higher rubber yield per tapping (Figure 1).

Jasmonate signaling plays a pivotal role in activating the secondary laticifer differentiation and activating the biosynthesis of secondary metabolites [42-45]. Although several members of COI1-JAZ-MCY module have been characterized in rubber tree and JA signaling is suggested to have an important role in regulating rubber biosynthesis in laticifer cells [46-48], the difference in the level of endogenous JAs in laticifer cells among *Hevea* germplasm remains largely unknown. The present study suggests that the level of endogenous JAs may be higher in the laticifer cells of CATAS 8-79 than that in the laticifer cells of PR107, considering that the expression of unigenes encoding enzymes such as LOX, OPDR, JMT is significantly up-regulated in CATAS8-79.

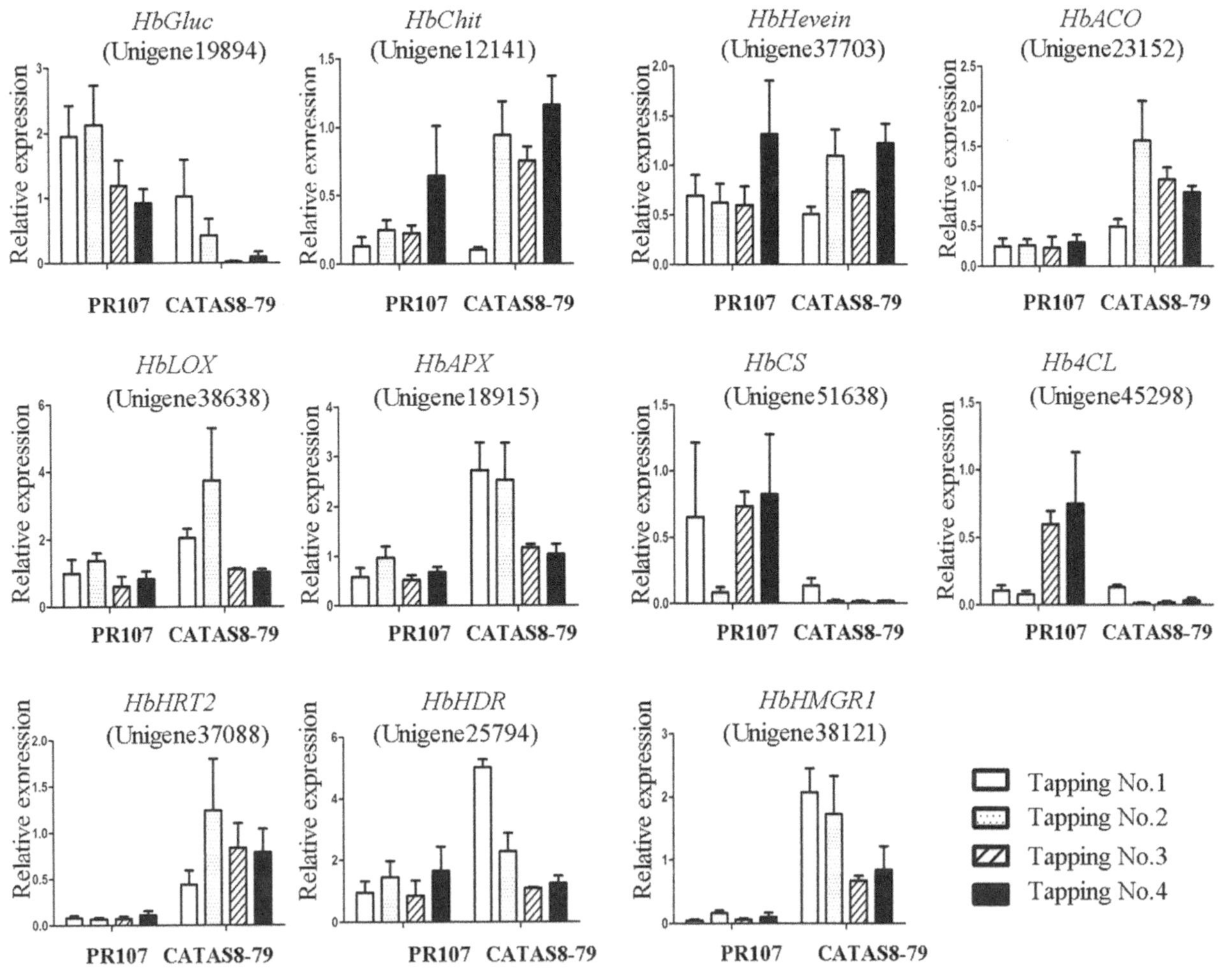

Figure 7 qRT-PCR analysis of the expression pattern of latex regeneration and expelling related genes upon successive tappings. Error bars for qRT-PCR showed the standard deviation of three replicates.

Ethrel, an ethylene releaser, is very effective in prolonging duration of latex flow [32]. One of the explanations for ethrel-induced prolongation of latex flow is its effect on maintaining the turgor pressure of laticifer and surrounding liber cells via its differential regulation on the aquaporins on the plasma membrane and tonoplast [49]. Considering that unigenes such as ACO, Ethylene overproduction protein and EIN3-binding F-box protein are up-regulated in CATAS8-79, the activity of ethylene biosynthesis and signal transduction in the laticifer cells of CATAS 8-79 may be higher than that of PR107. Besides, such factors as laticifer turgor, antioxidants, inclusions and debris of lutoid, and proteins in C-serum are suggested to influence the duration of latex flow [10-12]. After tapping, the bursting of lutoid particles leads to the release of hevein, chitinase, and glucanase. These protein inclusions are effective in rubber particle aggregation [5]. In the present study, it is the unigene encoding for glucanase other than the unigenes encoding for hevein and chitinase that is significantly up-regulated in PR107, suggesting that glucanase is more important in inducing rubber particle aggregation. The turgor pressure of laticifer in CATAS8-78 is significantly higher than that in PR107 [50]. In the present study, the unigenes enconding for enzymes (HbCS; Hb4CL) of cellulose and lignin biosynthesis are highly activated in RP107. This suggests that difference in turgor pressure may primarily associate with the difference in lignin content of laticifer cell wall between the two clones.

Conclusions

Taken together, comparative transcriptome analysis reveals new cues at molecular level for the difference in duration of latex flow and latex regeneration between rubber tree clone CATAS8-79 and PR107 (Figure 8). Up-regulated expression of unigenes *HbLOX* and *HbOPDR* in the pathway of JA biosynthesis, *HbSUT, HbBAM* and *HbPK* mediating carbohydrate metabolism, *HbHMGR1* in MVA parthway and *HbHR*T2 directly in rubber biosynthesis is important for enhanced latex regeneration. It may

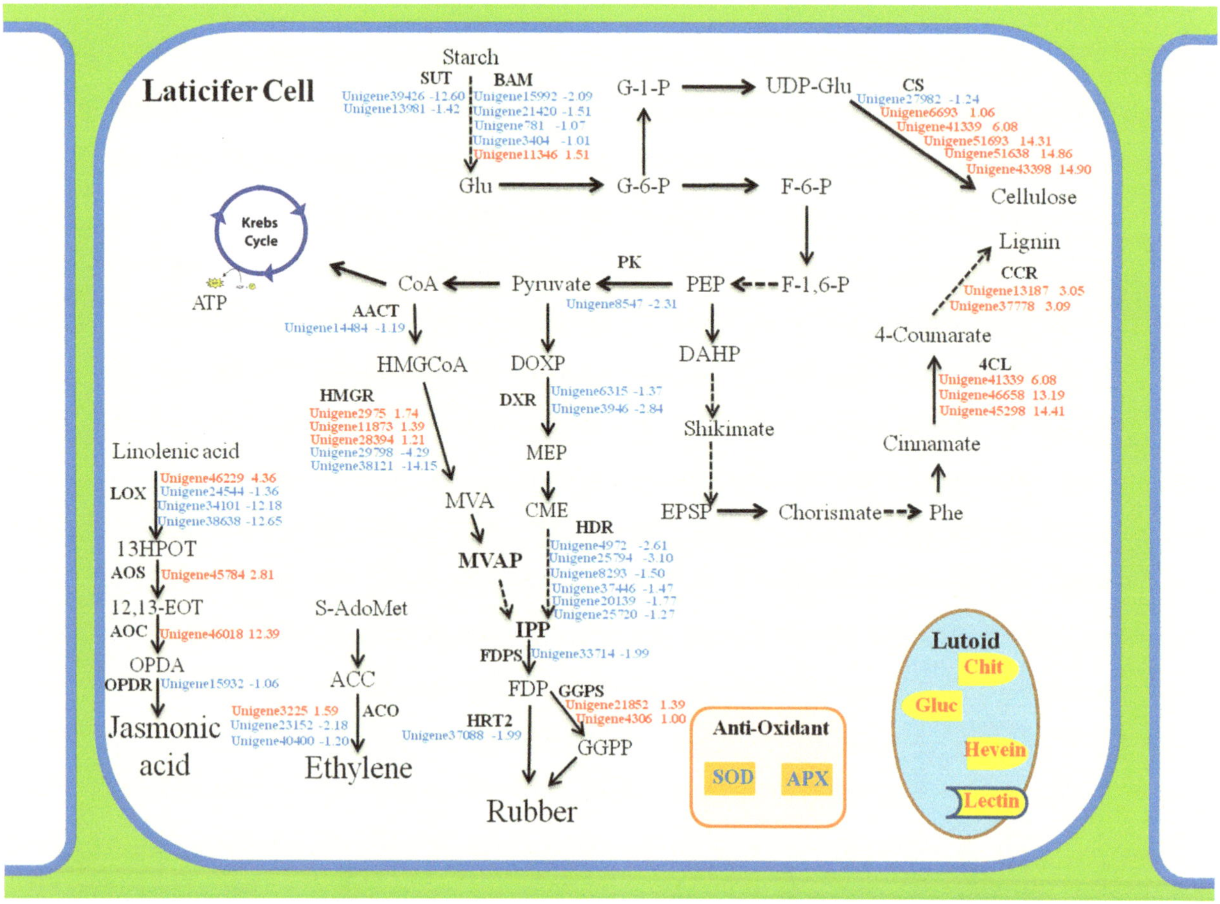

Figure 8 Schematic representation of differentially expressed unigenes related to latex regeneration and duration of latex flow in laticifer cell. Red colors showed unigenes up-regulated in PR107 while down-regulated in CATAS8-79. Blue colors showed unigenes up-regulated in CATAS8-79 while down-regulated in PR107. Numbers represented Log2 values in DGE data. The dashed arrows indicated multiple steps of enzymatic reactions.

be essential for prolonging duration of latex flow that up-regulated expression of unigenes *HbACO* mediating ethylene biosynthesis, *HbAPX* and *HbSOD* for antagonizing reactive oxygen species, but down-regulated expression of unigenes *Hb4CL* and *HbCCR* in the pathway of lignin biosynthesis and *HbGluc* for rubber particle aggregation.

Methods

Plant materials

Seven-year-old virgin trees of rubber tree clone CATAS8-79 and PR107 were grown at the Experimental Station of the Rubber Research Institute of the Chinese Academy of Tropical Agricultural Sciences in Danzhou city, Hainan province, P.R. China. The virgin trees with same circumference were seleced in this study. For RNA-Seq, latex from five individual trees by the first tapping was pooled for each clone. The samples were immediately stored at -80°C until RNA extraction. For real time-PCR and determination of physiological parameters, latex was individually collected from another batch of five trees for each clone upon the first, second, third and forth tapping, respectively. All the selected virgin trees were tapped with a tapping system of S/2, d/2 (a half spiral pattern, every two days) at 6:00 am in August, 2013.

Rubber content determination

To determine rubber content of latex, 100μl of acetic acid were dropped into 1g of fresh latex to obtain rubber coagula. The sample rubber coagula were washed in water for 2 h, then dried overnight at 55°C and weighted. The experiments were repeated three times.

RNA isolation and sequencing

Total latex RNA was extracted as described [7] and RNA integrity was evaluated by NanoDrop (Thermo Scientific Inc., USA). The double strand cDNA was synthesized using SuperScript® Double-Stranded cDNA Synthesis Kit (Invitrogen Inc., USA), and purified and added single nucleotide A (adenine) to the end with QiaQuick PCR extraction kit. Finally, sequencing adaptors were ligated to the cDNA fragments. The required fragments were purified by 2% agarose gel electrophoresis and enriched by PCR amplification. The library products were sequenced via Illumina HiSeq™ 2000 by Beijing Genomics Institute (Shenzhen, China). The original image datasets was transferred into sequence datasets by base calling. Clean reads were obtained by removing adaptor sequence, low quality sequences, empty tags, low complexity, and tags with only one copy.

Transcriptome de nove assembly, annotation and classification

Transcriptome *de novo* assembly was carried out using a *de* Bruijn graph and the SOAPdenovo as previously described [26]. Under a certain overlap length (k-mer = 29), SOAPdenovo combined overlapping reads into contigs. Adjacent contigs were constructed into scaffolds by read mate pairs. Within the scaffold, the connected contigs used 'N' to represent unknown sequences and insert size information. Finally, paired-end information was used to fill the gap of scaffolds to obtain the extended sequences with fewer Ns, which were defined as unigenes for further analysis.

All unigenes were used for BLAST searches (E-value < 1E-5) against databases as NCBI Nr (http://www.ncbi.nlm.nih.gov/), Swissprot (http://www.expasy.ch/sprot/), KEGG (http://www.genome.jp/kegg/) and COG (http://www.ncbi.nlm.nih.gov/cog/). The best aligning results were chosen for unigene annotation. The aligning results were selected with an order of Nr, Swiss-Prot, KEGG and COG.

To classify the unigenes, the Blast2GO program was used to get GO annotation based on molecular function, biological process and cellular component. All unigenes were also aligned to the COG database to predict possible functions and KEGG pathway database to perform pathway assignments.

Digital gene expression analysis

A rigorous algorithm was developed to identify differentially expressed genes between two different DGE libraries (CATAS8-79 versus PR107). Raw clean tags in each library were normalized to Tags Per Million (TPM) to obtain normalized gene expression level. Differential digital gene expression was deemed with FDR value ≤0.001 and |log2 Ratio| ≥ 1 in sequence counts across libraries. "Up-regulated" means the level of gene transcripts were higher in PR107 whereas "down-regulated" means the level of gene transcripts were higher in CATAS8-79.

Quantitative PCR analysis

Approximately 1 μg of RNA was used for reverse transcription based on the introduction of RevertAid™ First Strand cDNA Synthesis Kit (Thermo Scientific Inc., USA). qPCR was performed on the CFX96 System (Bio-Rad Laboratories Inc., USA) with SYBR PrimeScript RT-PCR Kit (TaKaRa Biotechnology, Japan). Nine housekeeping genes (*Hb18s, HbActin, HbELF1A, HbRH2B, HbRH8, HbYLS8, HbUBC2A, HbUBC2B, HbUBC4*) were selected to evaluate the stability by the software package NormFinder (version 0.953, http://www.mdl.dk/publicationsnormfinder.htm) [51]. *HbUBC2B* and *HbUBC4* were suitable for reference genes in the present study due to its stability in latex samples upon tapping (Additional file 3: Figure S2). Expression values were normalized for differences in cDNA input using parallel reactions employing primers designed against a reference gene *HbUBC2B* (HQ323247). All primer pairs used in this article were list in as an Additional file 4: Table S2.

Statistical analyses

The statistical analyses were ANOVAs carried out on raw data. A *t* test (Student's *t* test) was carried out for the significance analysis. *** indicates very significantly difference ($p < 0.01$).

Availability of supporting data

The authors confirm that all data underlying the findings are fully available without restriction. Data are available at the following URL: http://www.ncbi.nlm.nih.gov/geo/query/acc.cgi?acc=GSE59981.

Additional files

Additional file 1: Table S1. Statistics of DGE sequencing from CATAS8-79 and PR107 libraries.

Additional file 2: Figure S1. Species distribution of unigenes with matches in Nr (A) and Swissport (B) databases. The species distribution is shown as a total homologous sequences in the NCBI Nr (A) and Swissport (B) databases with an E-value < 10 – 5.

Additional file 3: Figure S2. Average expression stability values of the candidate reference genes evaluated by NormFinder. The least stable genes are on the left, and the two most stable genes are on the right.

Additional file 4: Table S2. Primers used in this paper.

Abbreviations

4CL: 4-coumarate-coa ligase; ACO: 1-aminocyclopropane-1-carboxylate oxidase; AOS: Allene oxide synthase; AOC: Allene oxide cycalse; BAM: Beta-amylase; CCR: Cinnamoyl-CoA reductase; Chit: Chitinase; COG: Clusters of Orthologous Group; CS: Cellulose synthase; DXR: 1-deoxy-D-xylulose 5-phosphate reductoisomerase; DGE: Digital gene expression; FDPS: Farnesyl diphosphate synthase; Gluc: Glucanase; GO: Gene Ontology; GPPS: Geranyl-diphosphate synthase; GS: Glutamate synthase; HDR: 4-hydroxy-3-methylbut-2-enyl diphosphate reductase; HMGR: Hydroxymethylglutaryl-CoA reductase; HRT: Hevea rubber transferase; IPP: Isopentenyl diphosphate; JA: Jasmonic acid; JMT: Jasmonate O-methyltransferase; KEGG: Kyoto Encyclopedia of Genes and Genomes; LOX: Lipoxygenase; MEP: 2-C-methyl-D-erythritol 4-phosphate; MVA: Mevalonate; OPDR: 12-oxophytodienoate reductase; PK: Pyruvate kinase; APX: L-ascorbate peroxidase; qRT-PCR: Quantitative reverse transcription-PCR; RPA: Rubber particle aggregation; SNP: Single nucleotide polymorphism; SOD: Superoxide dismutase; SUT: Sucrose transporter.

Competing interests

The authors declare that they have no competing interests.

Authors' contributions
Conceived and designed the experiments: WMT. Performed the experiments: JC, YC and SW. Analyzed the data: JC and SW. Wrote the paper: JC and WMT. All authors read and approved the final manuscript.

Acknowledgements
This work was supported by the National Natural Science Fundation of China (31170642) and the Special Program for Key Basic Research of the Ministry of Science and Technology, China (2012CB723005).

References

1. Mooibroek H, Cornish K. Alternative sources of natural rubber. Appl Microbiol Biotechnol. 2000;53:355–65.
2. Hagel JM, Yeung EC, Facchini PJ. Got milk? The secret life of laticifers. Trends Plant Sci. 2008;13:631–9.
3. Rahman AY, Usharraj AO, Misra BB, Thottathil GP, Jayasekaran K, Feng Y, et al. Draft genome sequence of the rubber tree *Hevea brasiliensis*. BMC Genomics. 2013;14:75.
4. van Beilen JB, Poirier Y. Establishment of new crops for the production of natural rubber. Trends Biotechnol. 2007;25:522–9.
5. Wang X, Shi M, Wang D, Chen Y, Cai F, Zhang S, et al. Comparative proteomics of primary and secondary lutoids reveals that chitinase and glucanase play a crucial combined role in rubber particle aggregation in *Hevea brasiliensis*. J Proteome Res. 2013;12:5146–59.
6. Chow KS, Mat-Isa MN, Bahari A, Ghazali AK, Alias H, Mohd-Zainuddin Z, et al. Metabolic routes affecting rubber biosynthesis in *Hevea brasiliensis* latex. J Exp Bot. 2012;63:1863–71.
7. Tang C, Huang D, Yang J, Liu S, Sakr S, Li H, et al. The sucrose transporter HbSUT3 plays an active role in sucrose loading to laticifer and rubber productivity in exploited trees of *Hevea brasiliensis* (para rubber tree). Plant Cell Environ. 2010;33:1708–20.
8. Tian WM, Zhang H, Yang SG, Shi MJ, Wang XC, Dai LJ, et al. Molecular and biochemical characterization of a cyanogenic β-glucosidase in the inner bark tissues of rubber tree (*Hevea brasiliensis* Muell. Arg.). J Plant Physiol. 2013;170:723–30.
9. d'Auzac J, Jacob JL, Chrestin H. Physiology of rubber tree latex. Florida: CRC Press; 1989. p. 314.
10. Kongsawadworakul P, Chrestin H. Laser diffraction: a new tool for identification and studies of physiological effectors involved in aggregation-coagulation of the rubber particles from *Hevea* latex. Plant Cell Physiol. 2003;44:707–17.
11. Wititsuwannakul R, Pasitkul P, Jewtragoon P, Wititsuwannakul D. *Hevea* latex lectin binding protein in C-serum as an anti-latex coagulating factor and its role in a proposed new model for latex coagulation. Phytochemistry. 2008;69:656–62.
12. Salomez M, Subileau M, Intapun J, Bonfils F, Sainte-Beuve J, Vaysse L, et al. Micro-organisms in latex and natural rubber coagula of *Hevea brasiliensis* and their impact on rubber composition, structure and properties. J Appl Microbiol. 2014;117:921–9.
13. Wahler D, Gronover CS, Richter C, Foucu F, Twyman RM, Moerschbacher BM, et al. Polyphenoloxidase silencing affects latex coagulation in *Taraxacum* species. Plant Physiol. 2009;151:334–46.
14. Leclercq J, Martin F, Sanier C, Clément-Vidal A, Fabre D, Oliver G, et al. Over-expression of a cytosolic isoform of the HbCuZnSOD gene in *Hevea brasiliensis* changes its response to a water deficit. Plant Mol Biol. 2012;80:255–72.
15. Wang Z, Fang B, Chen J, Zhang X, Luo Z, Huang L, et al. *De novo* assembly and characterization of root transcriptome using Illumina paired-end sequencing and development of cSSR markers in sweet potato (*Ipomoea batatas*). BMC Genomics. 2010;11:726.
16. Zenoni S, Ferrarini A, Giacomelli E, Xumerle L, Fasoli M, Malerba G, et al. Characterization of transcriptional complexity during berry development in *Vitis vinifera* using RNA-Seq. Plant Physiol. 2010;152:1787–95.
17. Zhang G, Guo G, Hu X, Zhang Y, Li Q, Li R, et al. Deep RNA sequencing at single base-pair resolution reveals high complexity of the rice transcriptome. Genome Res. 2010;20:646–54.
18. Garg R, Patel RK, Tyagi AK, Jain M. *De novo* assembly of chickpea transcriptome using short reads for gene discovery and marker identification. DNA Res. 2011;18:53–63.
19. Iorizzo M, Senalik DA, Grzebelus D, Bowman M, Cavagnaro PF, Matvienko M, et al. *De novo* assembly and characterization of the carrot transcriptome reveals novel genes, new markers, and genetic diversity. BMC Genomics. 2011;12:389.
20. Shi CY, Yang H, Wei CL, Yu O, Zhang ZZ, Jiang CJ, et al. Deep sequencing of the Camellia sinensis transcriptome revealed candidate genes for major metabolic pathways of tea-specific compounds. BMC Genomics. 2011;12:131.
21. Wong MM, Cannon CH, Wickneswari R. Identification of lignin genes and regulatory sequences involved in secondary cell wall formation in *Acacia auriculiformis* and *Acacia mangium* via de novo transcriptome sequencing. BMC Genomics. 2011;12:342.
22. Chow KS, Ghazali AK, Hoh CC, Mohd-Zainuddin Z. RNA sequencing read depth requirement for optimal transcriptome coverage in *Hevea brasiliensis*. BMC Res Notes. 2014;7:69.
23. Triwitayakorn K, Chatkulkawin P, Kanjanawattanawong S, Sraphet S, Yoocha T, Sangsrakru D, et al. Transcriptome sequencing of *Hevea brasiliensis* for development of microsatellite markers and construction of a genetic linkage map. DNA Res. 2011;18:471–82.
24. Xia ZH, Xu HM, Zhai JL, Li DJ, He CZ, Huang X. RNA-Seq analysis and *de novo* transcriptome assembly of *Hevea brasiliensis*. Plant Mol Biol. 2011;77:299–308.
25. Li D, Deng Z, Qin B, Liu X, Men Z. De novo assembly and characterization of bark transcriptome using Illumina sequencing and development of ESTSSR markers in rubber tree (Hevea brasiliensis Muell. Arg.). BMC Genomics. 2012;13:192.
26. Mantello CC, Cardoso-Silva CB, da Silva CC, de Souza LM, Scaloppi Junior EJ, de Souza GP, et al. *De novo* assembly and transcriptome analysis of the rubber tree (*Hevea brasiliensis*) and SNP markers development for rubber biosynthesis pathways. PLoS One. 2014;9:e102665.
27. Pujade-Renaud V, Clement A, Perrot-Rechenmann C, Prevot JC, Chrestin H, Jacob JL, et al. Ethylene-induced increase in glutamine synthetase activity and mRNA levels in *Hevea brasiliensis* latex cells. Plant Physiol. 1994;105:127–32.
28. Ohashi M, Ishiyama K, Kusano M, Fukushima A, Kojima S, Hanada A, et al. Lack of cytosolic glutamine synthetase1;2 in vascular tissues of axillary buds causes severe reduction in their outgrowth and disorder of metabolic balance in rice seedlings. Plant J. 2015;81:347–56.
29. Tzin V, Galili G. New insights into the shikimate and aromatic amino acids biosynthesis pathways in plants. Mol Plant. 2010;3:956–72.
30. Bolton MD. Primary metabolism and plant defense-fuel for the fire. Mol Plant-Microbe Interact. 2009;22:487–97.
31. Dieuaide-Noubhani M, Alonso AP, Rolin D, Eisenreich W, Raymond P. Metabolic flux analysis: recent advances in carbon metabolism in plants. Plant Syst Biol. 2007;97:213–43.
32. Zhu J, Zhang Z. Ethylene stimulation of latex production in *Hevea brasiliensis*. Plant Signal Behav. 2009;4:1072–4.
33. Hao BZ, Wu JL. Laticifer differentiation in *Hevea brasiliensis*: induction by exogenous jasmonic acid and linolenic acid. Ann Bot. 2000;85:37–43.
34. Monroe JD, Storm AR, Badley EM, Lehman MD, Platt SM, Saunders LK, et al. β-amylase1 and β-amylase3 are plastidic starch hydrolases in Arabidopsis that seem to be adapted for different thermal, pH, and stress conditions. Plant Physiol. 2014;166:1748–63.
35. Li J, Zhao H, Huang W. Mechanism of proanthocyanidins-induced alcoholic fermentation enhancement in *Saccharomyces cerevisiae*. J Ind Microbiol Biotechnol. 2014;41:1793–802.
36. Amalou Z, Bangratz J, Chrestin H. Ethrel (ethylene releaser)-induced increases in the adenylate pool and transtonoplast ΔpH within *Hevea* latex cells. Plant Physiol. 1992;98:1270–6.
37. Dusotoit-Coucaud A, Kongsawadworakul P, Maurousset L, Viboonjun U, Brunel N, Pujade-Renaud V, et al. Ethylene stimulation of latex yield depends on the expression of a sucrose transporter (HbSUT1B) in rubber tree (*Hevea brasiliensis*). Tree Physiol. 2010;30:1586–98.
38. Sando T, Takaoka C, Mukai Y, Yamashita A, Hattori M, Ogasawara N, et al. Cloning and characterization of mevalonate pathway genes in a natural rubber producing plant, *Hevea brasiliensis*. Biosci Biotechnol Biochem. 2008;72:2049–60.
39. Koyama T. Molecular analysis of prenyl chain elongating enzymes. Biosci Biotechnol Biochem. 1999;63:1671–6.
40. Tangpakdee J, Tanaka Y, Ogura K, Koyama T, Wititsuwannakul R, Wititsuwannakul D, et al. Isopentenyl diphosphate isomerase and prenyl

transferase activities in bottom fraction and C-serum from *Hevea brasiliensis*. Phytochemistry. 1997;45:261–7.

41. Ogura K, Koyama T. Enzymatic aspects of isoprenoid chain elongation. Chem Rev. 1998;98:1263–76.
42. Asawatreratanakul K, Zhang YW, Wititsuwannakul D, Wititsuwannakul R, Takahashi S, Rattanapittayaporn A, et al. Molecular cloning, expression and characterization of cDNA encoding cis-prenyltransferases from *Hevea brasiliensis*. A key factor participating in natural rubber biosynthesis. Eur J Biochem. 2003;270:4671–80.
43. Shitan N, Sugiyama A, Yazaki K. Functional analysis of jasmonic acid-responsive secondary metabolite transporters. Methods Mol Biol. 2013;1011:241–50.
44. Misra RC, Maiti P, Chanotiya CS, Shanker K, Ghosh S. Methyl jasmonate-elicited transcriptional responses and pentacyclic triterpene biosynthesis in sweet basil. Plant Physiol. 2014;164:1028–44.
45. Goldhaber-Pasillas GD, Mustafa NR, Verpoorte R. Jasmonic acid effect on the fatty acid and terpenoid indole alkaloid accumulation in cell suspension cultures of *Catharanthus roseus*. Molecules. 2014;19:10242–60.
46. Peng S-Q, Xu J, Li H-L, Tian W-M. Cloning and molecular characterization of *HbCOI1* from *Hevea brasiliensis*. Biosci Biotechnol Biochem. 2009;73:665–70.
47. Tian W-M, Huang W-F, Zhao Y. Cloning and characterization of HbJAZ1 from the laticifer cells in rubber tree (*Hevea brasiliensis* Muell. Arg.). Trees. 2010;24:771–9.
48. Pirrello J, Leclercq J, Dessailly F, Rio M, Piyatrakul P, Kuswanhadi K, et al. Transcriptional and post-transcriptional regulation of the jasmonate signalling pathway in response to abiotic and harvesting stress in *Hevea brasiliensis*. BMC Plant Biol. 2014;14:341.
49. Tungngoen K, Kongsawadworakul P, Viboonjun U, Katsuhara M, Brunel N, Sakr S, et al. Involvement of HbPIP2;1 and HbTIP1;1 aquaporins in ethylene stimulation of latex yield through regulation of water exchanges between inner liber and latex cells in *Hevea brasiliensis*. Plant Physiol. 2009;151:843–56.
50. An F, Cahill D, Rookes J, Lin WF, Kong LX. Real-time measurement of phloem turgor pressure in *Hevea brasiliensis* with a modified cell pressure probe. Bot Stud. 2014;55:19.
51. Li H, Qin Y, Xiao X, Tang C. Screening of valid reference genes for real-time RT-PCR data normalization in *Hevea brasiliensis* and expression validation of a sucrose transporter gene HbSUT3. Plant Sci. 2011;181:132–9.

13

VRN-1 gene- associated prerequisites of spring growth habit in wild tetraploid wheat *T. dicoccoides* and the diploid A genome species

Andrey B Shcherban*, Kseniya V Strygina and Elena A Salina

Abstract

Background: In order to clarify the origin of spring growth habit in modern domesticated wheat, allelic variability of the *VRN-1* gene was investigated in a wide set of accessions of the wild tetraploid species *Triticum dicoccoides* (BBAA), together with diploid species *T. monococcum*, *T. boeoticum* and *T. urartu*, presumable donors of the A genome to polyploid wheats.

Results: No significant variation was found at the *VRN-B1* locus of *T. dicoccoides*, whereas at *VRN-A1* a number of previously described alleles were found with small deletions in the promoter (*VRN-A1b*, *VRN-A1d*) or a large deletion in the first (1st) intron (*VRN-A1L*). The diploid A genome species were characterized by their own set of *VRN-1* alleles including previously described *VRN-A1f* and *VRN-A1h* alleles with deletions in the promoter region and the *VRN-A1ins* allele containing a 0.5 kb insertion in the 1st intron. Based on the CAPS screening data, alleles *VRN-A1f* and *VRN-A1ins* were species-specific for *T. monococcum*, while allele *VRN-A1h* was specific for *T. boeoticum*. Different indels were revealed in both the promoter and 1st intron of the recessive *VRN-A1u* allele providing specific identification of *T. urartu*, the proposed donor of the A genome to modern wheat. We found that alleles *VRN-A1b* and *VRN-A1h*, previously described as dominant, have either no or weak association with spring growth habit, while in some diploid accessions this habit was associated with the recessive *VRN-A1* allele.

Conclusions: Spring growth habit in diploid wheats was only partially associated with indels in regulatory regions of the *VRN-1* gene. An exception is *T. monococcum* where dominant mutations in both the promoter region and, especially, the 1st intron were selected during domestication resulting in a greater variety of spring forms. The wild tetraploid *T. dicoccoides* had a distinct set of *VRN-A1* alleles compared to the diploids in this study, indicating an independent origin of spring tetraploid forms that likely occurred after combining of diploid genomes. These alleles were subsequently inherited by cultivated polyploid (tetraploid and hexaploid) descendants.

Keywords: Allelic diversity, Vernalization, *VRN-1* gene, Promoter, First intron, Tetraploid, Diploid, *Triticum*

Background

Triticum dicoccoides or *T. turgidum* ssp. dicoccoides Thell is considered to be the wild form of all domesticated emmer wheats [1]. The species has two homologous sets of chromosomes (BBAA) thought to have arisen from spontaneous hybridization between two diploid donor species. The donor of the B genome is unknown but was likely a species closely related to modern *Aegilops speltoides* Tausch (2n = 14; SS) [2-4]. Potential sources of the A genome are represented by three closely related species: *T. urartu* Tum. ex Gandil., *T. boeoticum* Boiss. and *T. monococcum* L. These species are separated by crossing barriers [5], and differ in their plant morphology [6] and biochemical and molecular marker loci [3,7,8]. *T. boeoticum* is considered to be the progenitor of cultivated diploid einkorn wheat, *T. monococcum*. *T. urartu* is considered as the donor of A genome to the wild tetraploid species: *T. dicoccoides* (BBAA) and *T. araraticum* (GGAA) [9], which was subsequently inherited by other important polyploid wheat species including *T. durum* (BBAA), *T. timopheevii* (GGAA), and common wheat *T. aestivum* (BBAADD). Despite many studies of *Triticum*

* Correspondence: atos@bionet.nsc.ru
Institute of Cytology and Genetics, Lavrentiev ave. 10, Novosibirsk 630090, Russia

domestication, employing morphological, cytological [6,10] and molecular marker analysis techniques [7,11-14], many questions remain unanswered with respect to the phylogenetic relationships between diploid wheat species and between diploids and their polyploid descendants.

Adaptability of polyploid wheat to a wide range of environments has been at least partially facilitated by allelic diversity in *VRN* genes regulating growth habit and flowering time [15]. Wheats are categorized into two major forms: those requiring vernalization or exposure to cold to accelerate flowering (winter forms) and those which do not require vernalization (spring forms). Vernalization prevents floral initiation during winter, thereby protecting sensitive floral meristems from freezing temperatures. In recent years, researchers have significantly advanced our understanding of the molecular mechanisms of vernalization in wheat. The genomic locations and sequences of three vernalization genes (*VRN-1*, *VRN-2* and *VRN-3*) responsible for spring/winter growth habit have been determined [16-18].

The *VRN-1* gene is proposed as the main initiating factor of the regulatory cascade initiating flowering [16,19]. *VRN-1* encodes a MADS-box transcription factor which controls the transition of the vegetative shoot apical meristem to the reproductive phase [20,21]. In vernalization-requiring cereals, *VRN-1* expression is induced by vernalization, with the level of expression being dependent on the length of cold exposure [22]. *VRN-1* downregulates the floral repressor *VRN-2*, and allows long-day induction of the floral activator *VRN-3* to accelerate subsequent stages of floral development [17,19].

Changes in the growth habit of winter wheat to spring wheat are primarily due to dominant mutations in regulatory regions (promoter or intron 1) of *VRN-1* [15]. *VRN-A1* alleles of the hexaploid or common wheat *T. aestivum* L. (2n = 42, BBAADD) and the tetraploid *T. turgidum* L. (2n = 28, BBAA), have so far been identified containing nucleotide deletions as well as insertions of mobile elements in both regions [23,24]. By contrast, dominant mutations in *VRN-B1* and *VRN-D1* genes are predominantly caused by large deletions in the 1st intron [24]. Recently, a novel *VRN-B1* allele was identified as having a deletion coupled with sequence duplication within intron 1 [25-27].

Most of the wild *Triticeae* species have a winter growth habit, suggesting that the recessive *VRN-1* allele is the ancestral form. By contrast, there are many cultivated polyploid wheats with a spring growth habit and with at least one dominant *VRN-1* allele [28]. Reports demonstrate that different combinations of dominant and recessive *VRN-1* alleles (*VRN-1* genotype) significantly affect the time of flowering [29,30]. Spring growth habit determined by dominant *VRN-1* alleles could be either inherited from ancestral diploids or could result from selection of independent mutations appearing during the adaptation to different environments after domestication. To resolve this issue, a large scale screening of *VRN-1* polymorphisms among both diploid and polyploid wheat species is required. To date, a limited number of accessions of diploid and tetraploid progenitors of *T. aestivum* have been examined [16,23,24,31,32].

In the present study, molecular variability of *VRN-1* genes was analyzed based on a representative set of accessions of wild tetraploid wheat and diploid A genome species. The main objective was to identify and characterize different alleles of *VRN-1* genes and to analyze their distribution among accessions. We also examined the influence of different alleles and their combinations on vernalization requirement. This analysis sheds light on the evolution of *VRN-1* genes in diploids and during the first stages of wheat polyploidization. It contributes to our understanding of the phylogenetic relationships between diploid and polyploid wheat species and the evolutionary history of modern domesticated wheat.

Results

VRN-1 allelic variability in wild tetraploid wheat *T. dicoccoides*

Promoter region

Vernalization sensitivity in tetraploid wheat *T. dicoccoides* is controlled by alleles at the 2 homoeologous loci, *VRN-A1* and *VRN-B1*. The specific primers Vrn1AF and Int1R were used in order to identify variation in the promoter region of *VRN-A1* locus as described by Yan et al. [23] (Table 1). Almost all of the 80 studied accessions of *T. dicoccoides* yielded a PCR product of approximately 0.7 kb (Figure 1a). To further analyse these PCR products, we digested them with a frequently-cutting restriction endonuclease *Msp* I. The restriction patterns could be divided into two types (Figure 1b). The first type, characteristic of the vast majority of the studied accessions, contained two major fragments of ~140 and 200 bp long. The second type was found in only 3 of the 80 *T. dicoccoides* accessions and was characterized by a fragment of approximately 120 bp in place of the 140 bp band. To further analyze allelic variation at the *VRN-A1* promoter, we selected a set of accessions representing both restriction patterns and sequenced the PCR products obtained with VrnA1F/Int1R primers (Table 2).

The *VRN-A1* promoter sequences of accessions IG 113301, IG 139189, IG 46297, k-5198, belonging to the first restriction pattern type (Figure 1b) were completely identical to the known sequence of the recessive *VRN-A1* allele from *T. dicoccoides* (AY747598) and contained an *Msp I* restriction fragment of 138 bp. *VRN-A1* promoter sequences of all polyploid wheats both studied here and from databases contained the 8 bp insertion and an additional *Msp* I site characteristic of the respective

Table 1 PCR markers for determining the presence of different alleles of *VRN-A1, VRN-B1* in diploid and polyploid wheats

PCR marker	Name	Primer (5'→3')	Target allele(s)	Expected product size (bp)	Annealing temp. (°C)	Reference.
VRN-A1 marker*	Vrn1AF	GAAAGGAAAAATTCTGCTCG	*VRN-A1a*	876 and 965	55.0	[16,23,31]
	Int1R	GCAGGAAATCGAAATCGAAG	*VRN-A1b*	694		
			VRN-A1d	662		
			VRN-A1h	684		
			VRN-A1f	703		
			VRN-A1	704		
			VRN-A1u	705		
			VRN-A1u	713		
T. monococcum VRN-A1	Indel(-)F	CGCTCTTATATTTGTTTACCAGGG	*VRN-A1*	1025	50.0	–
	Indel(-)R	GGGTCAACTATTCTGTGGAG	*VRN-A1u, VRN-A1u'*	no product		
VRN-A1 deletion of 1.4 kb	Intr1/C/F	GCACTCCTAACCCACTAACC	*VRN-A1u, VRN-A1u'*	1068	56.0	[24]
	Intr1/AB/R	TCATCCATCATCAAGGCAAA				
VRN-A1 insertion of 0.5 kb	Intr 1	ATCATCTTCTCCACCAAGGG	*VRN-A1ins*	1980	50.0	–
	Intr1insR	AATGAACAGCACGGAAACAG	*VRN-A1*	1476		
VRN-A1 deletion of 7,2 kb	Ex1/C/F	GTTCTCCACCGAGTCATGGT	*VRN-A1L*	522	55.6	[24]
	Intr1/A/R3	AAGTAAGACAACACGAATGTGAGA				
VRN-B1 marker*	P2	TCATGCACGCACACACGGTA	*VRN-B1*	814	55.0	[25]
	P5	GGCCAACCCTACACCCCAAG				
VRN-B1 Non-deletion	Intr1/B/F	CAAGTGGAACGGTTAGGACA	*VRN-B1*	1149	56.4	[24]
	Intr1/B/R4	CAAATGAAAAGGAATGAGAGCA				

*These diagnostic markers detect allelic variation at the promoter regions. In other cases variation within intron 1 of corresponding genes is detected.

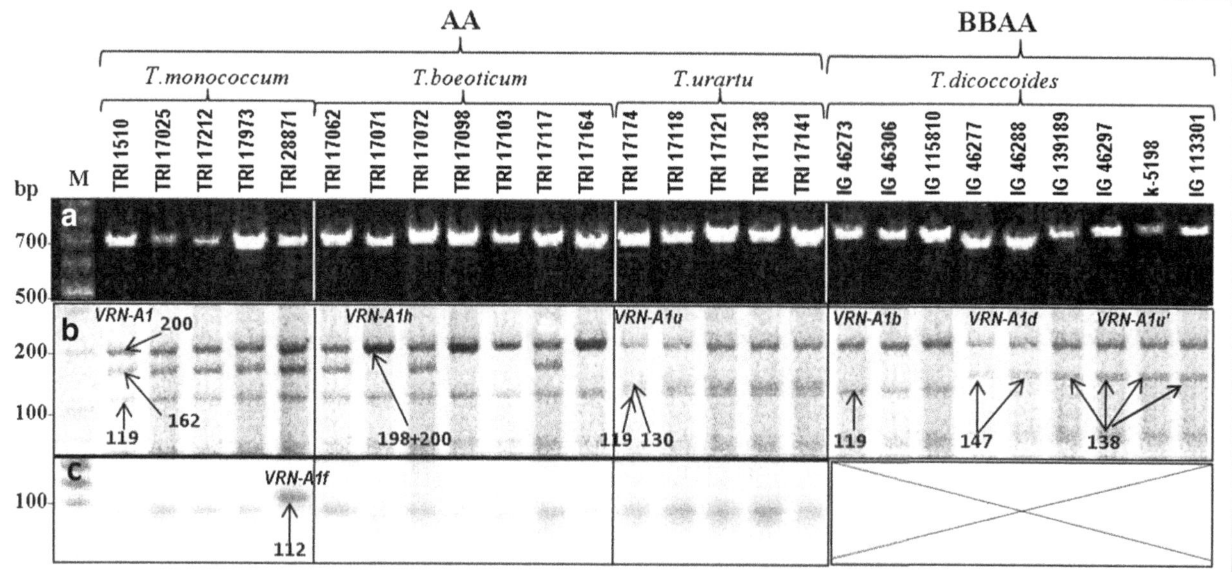

Figure 1 PCR amplification of *VRN-1* promoter region. **a**- PCR amplification with primers VrnA1F/Int1R. **b**- *Msp* I restriction digestion of corresponding PCR products. **c**- PCR amplification with primers VRN1F_F/VRN1F_R followed by *Taq* I restriction digestion. Accession numbers, species and genotypes are given at the top. The main restriction fragments more than 100 bp in length are indicated by arrows.

Table 2 Allelic variants of *VRN-A1* gene in a selected set of accessions of diploid *Triticum* species and tetraploid *T. dicoccoides*

Species (genome)	Accession number	Growth type (Winter/ Spring)	*VRN-A1* structural forms*			Alleles**	GenBank Ac.N.	Homologous sequences from GenBank
			Promoter	Intron 1				
				Insertion 0.5 kb	Deletions 1.4 kb/7.2 kb			
T. monococcum (AA)	TRI 1510	S	*VRN-A1*	+	–/–	*VRN-A1ins*	KM586653	AY188331
	TRI 17025	S	*VRN-A1*	-	–/–	*vrn-A1vrn2*	-//-	-//-
	TRI 17212	S	*VRN-A1*	+	–/–	*VRN-A1ins*	-//-	-//-
	TRI 17973	S	*VRN-A1*	-	–/–	*vrn-A1VRN-2*	-//-	-//-
	TRI 28871	S	*VRN-A1f*	+	–/–	*VRN-A1f VRN-A1ins*	KM586656	DQ146421
T. boeoticum (AA)	TRI 17062	W	*VRN-A1*	-	–/–	*vrn-A1*	KM586653	AY188331
	TRI 17071	S	*VRN-A1h*	-	–/–	*VRN-A1h*	KM586657	GQ451745
	TRI 17072	S	*VRN-A1*	-	–/–	*vrn-A1VRN-2*	KM586653	AY188331
	TRI 17098	S	*VRN-A1h*	-	–/–	*VRN-A1h*	KM586657	GQ451745
	TRI 17103	W	*VRN-A1h*	-	–/–	*VRN-A1h*	-//-	-//-
	TRI 17117	W	*VRN-A1*	-	–/–	*vrn-A1*	KM586653	AY188331
	TRI 17164	W	*VRN-A1h*	-	–/–	*VRN-A1h*	KM586657	GQ451745
T. urartu (AA)	TRI 17174	W	*VRN-A1u*	-	+/–	*vrn-A1u*	KM586659	GQ482970
	TRI 17118	W	*VRN-A1u*	-	+/–	*vrn-A1u*	-//-	-//-
	TRI 17121	W	*VRN-A1u*	-	+/–	*vrn-A1u*	-//-	-//-
	TRI 17138	W	*VRN-A1u*	-	+/–	*vrn-A1u*	-//-	-//-
	TRI 17141	W	*VRN-A1u*	-	+/–	*vrn-A1u*	-//-	-//-
T. dicoccoides (BBAA)	IG 113301	W	*VRN-A1u'*	-	+/–	*vrn-A1u´*	KM586660	AY747598
	IG 46273	W	*VRN-A1b*	-	+/–	*VRN-A1b*	KM586654	AY616461
	IG 46306	W	*VRN-A1b*	-	+/–	*VRN-A1b*	-//-	-//-
	IG 115810	W	*VRN-A1b*	-	+/–	*VRN-A1b*	-//-	-//-
	IG 46277	S	*VRN-A1d*	-	+/–	*VRN-A1d*	KM586655	AY616462
	IG 46288	S	*VRN-A1d*	-	+/–	*VRN-A1d*	-//-	-//-
	IG 139189	S	*VRN-A1u'*	-	?/+	*VRN-A1L*	KM586658	AY747598
	IG 46297	S	*VRN-A1u'*	-	?/+	*VRN-A1L*	-//-	-//-
	k-5198	S	*VRN-A1u'*	-	?/+	*VRN-A1L*	-//-	-//-

*In all cases the promoter region was sequenced; insertion and deletions in intron 1 were determined by PCR (see Table 1).
***vrn-A1, vrn-A1u, vrn-A1u´*- recessive alleles (initial structural forms); *VRN-A1*- structurally modified (mutated) forms.

VRN-A1u allele of *T. urartu* (see below). Taking into account these features, we designated the promoter sequence of the recessive *VRN-A1* allele of *T. dicoccoides* as *VRN-A1u´* (Figures 2 and 3; Table 2; Additional file 1).

Accessions IG 46273, IG 46306 and IG 115810 of the second type were identified to be a previously described *VRN-A1b* allele [23], with a 19 bp deletion in the promoter region as compared to the recessive allele (Figure 2). This deletion created the diagnostic *Msp I* restriction fragment of 119 bp (Figure 3).

Two out of 80 *T. dicoccoides* accessions (IG46277 and IG 46288) yielded a slightly smaller PCR product compared with the other accessions (Figure 1a). They were almost undistinguishable from the accessions of the first type by *Msp I* restriction (Figure 1b). Sequencing showed that these accessions contained a previously described *VRN-A1d* allele [23]. Compared with the *VRN-A1b* allele, this allele has an additional 32 bp deletion, generating a diagnostic 147 bp *Msp I* restriction fragment (Figures 2 and 3).

For analysis of the *VRN-B1* locus the primers P2/P5 were used to amplify an approximately 0.8 kb region of the promoter sequence (Table 1). No differences were revealed by PCR or subsequent *Msp* I digestion (see Additional file 2). Based on the known *VRN-B1* sequences of *T. aestivum* (AY747602-04, AY616453, AY616456), *Msp* I restriction fragments of 494, 104 and 128 bp were expected. We sequenced PCR products of 0.8 kb from 5 randomly selected accessions with diverse origins (see Additional file 1). All five sequences were highly similar to

```
                        TATA box                          CArG box
VRN-A1b   TCGGGGCCAGATCCCTTTAAAAACCCCTCCCCCCCTGCCGGAATCCTCGTTTTGGCCT
VRN-A1d   TCGGGGCCAGATCC--------------------------------TCGTTTTGGCCT
VRN-A1u'  TCGGGGCCAGATCCCTTTAAAAACCCCTCCCCCCCTGCCGGAATCCTCGTTTTGGCCT
VRN-A1u   TCGGGGCCAGATCCCTTTAAAAACCCCTCCCCCCCTGCCGGAACCCTCGTTTTGGCCT
VRN-A1h   TCGGGGCCAGATCCCTTTAAAAA--------------------CCCTCGTTTTGGCCT
VRN-A1f   TCGGGGCCAGATCCCTTTAAAAACCCCTCCCCCCCTGCCGGAACCCT-GTTTTGGCCT
VRN-A1    TCGGGGCCAGATCCCTTTAAAAACCCCTCCCCCCCTGCCGGAACCCTCGTTTTGGCCT
                                        AC
VRN-B1    TCGGGGCCAGATCCCTTTAAAAACCCCTCCCCCCTTGCCGGAAACCTCGTTTTGGCCT

           CAP signal                         8 bp ins.
VRN-A1b   GGCCATCCTC-------------------CACCTCACGTCCTCACCCAACCACCTGAT
VRN-A1d   GGCCATCCTC-------------------CACCTCACGTCCTCACCCAACCACCTGAT
VRN-A1u'  GGCCATCCTCCCTCTCCTCCCCTCTCTTCCACCTCACGTCCTCACCCAACCACCTGAT
VRN-A1u   GGCCATCCTCCCTCTCCTCCCCTCTCTTCCACCTCAC--------CCAACCACCTGAT
VRN-A1h   GGCCATCCTCCCTCTCCTCCCCTCTCTTCCACCTCAC--------CCAACCACCTGAT
VRN-A1f   GGCCATCCTCCCTCTCCTCCCCTCTCTTCCACCTCAC--------CCAACCACCTGAT
VRN-A1    GGCCATCCTCCCTCTCCTCCCCTCTCTTCCACCTCAC--------CCAACCACCTGAT

VRN-B1    GGCCATCCTCCCTCTCCTCCC-TCTCTTCCGCCTCAC--------CCAACCACCTGAC


VRN-A1b   AGCCATGGCTCCGCCGCCTCGCCTCCGCCTGCGCCAGTCGGAGTAGCCGTCGCGGTCT
VRN-A1d   AGCCATGGCTCCGCCGCCTCGCCTCCGCCTGCGCCAGTCGGAGTAGCCGTCGCGGTCT
VRN-A1u'  AGCCATGGCTCCGCCGCCTCGCCTCCGCCTGCGCCAGTCGGAGTAGCCGTCGCGGTCT
VRN-A1u   AGCCATGGCTCCGCCGCCTCGCCTCCGCCTGCGCCAGTCGGAGTAGCCGTCGCGGTCT
VRN-A1h   AGCCATGGCTCCGCCGCCTCGCCTCCGCCTGCGCCAGTCGGAGTAGCCGTCGCGGTCT
VRN-A1f   AGCCATGGCTCCGCCGCCTCGCCTCCGCCTGCGCCAGTCGGAGTAGCCGTCGCGGTCT
VRN-A1    AGCCATGGCTCCGCCGCCTCGCCTCCGCCTGCGCCAGTCGGAGTAGCCGTCGCGGTCT

VRN-B1    AGCCATGGCTCCGCCCCCCCGCCCCCGCCTGCGCCTGTCGGAGTAGCCGTCGCGGTCT

            ↓
VRN-A1b   GCCGGTGTTGGAGGGTAGGGGCGTAGGGTTGGCCCGGTTCTCGAGCGGAGATG
VRN-A1d   GCCGGTGTTGGAGGGTAGGGGCGTAGGGTTGGCCCGGTTCTCGAGCGGAGATG
VRN-A1u'  GCCGGTGTTGGAGGGTAGGGGCGTAGGGTTGGCCCGGTTCTCGAGCGGAGATG
VRN-A1u   GCCGGTGTTGGAGGGTAGGGGCGTAGGGTTGGCCCGGTTCTCGAGCGGAGATG
VRN-A1h   GCGGGTGTTGGAGGGTAGGGGCGTAGGGTTGGCCCGGTTCTCGAGCGGAGATG
VRN-A1f   GCGGGTGTTGGAGGGTAGGGGCGTAGGGTTGGCCCGGTTCTCGAGCGGAGATG
VRN-A1    GCGGGTGTTGGAGGGTAGGGGCGTAGGGTTGGCCCGGTTCTCGAGCGGAGATG
                           P5
VRN-B1    GCCGGTGTTGGAGGCTTGGGGTGTAGGGTTGGCC
```

Figure 2 Section (~0.2 kb) of a sequence alignment of the *VRN-1* promoter amplified with primers pairs VrnA1F/Int1R (*VRN-A1*; 0.7 kb product) and P2/P5 (*VRN-B1*; 0.8 kb product). The location of putative transcriptional signals (in bold) are given according to Yan et al. [23]. The ATG start codon is indicated in gray. The 8 bp insertion (8 bp ins.) is shown and nucleotide substitution leading to an additional *Msp* 1 restriction site (underlined) is indicated by an arrow. Polymorphic bases in *VRN-B1* as compared to *VRN-A1* are underlined. The *VRN-B1* sequence shown here is highly conserved among five diverse accessions of *T. dicoccoides*.

each other and to known *VRN-B1* sequences (98-100% homology). Minor variation between individual sequences was found including small deletions of up to 7 bp, however, almost all of this variation was upstream from the conserved region containing putative transcriptional signals (Figure 2). The *VRN-B1* sequences yielded in this study were deposited to Genbank under Ac.N. KM586661-65.

1st intron

The primer pair Intr1/C/F and Intr1/AB/R for the 1st intron of *VRN-A1* (Table 1; Figure 4) were previously used to confirm the presence of the recessive allele of *T. aestivum* [24]. This allele contains a 1.4 kb deletion compared with the respective alleles of *T. monococcum* and *T. boeoticum* (see below). Almost all the studied accessions of *T. dicoccoides* yielded a PCR product of approximately 1 kb, indicating the presence of the deletion, except for accessions k-5198, IG 139189 and IG 46297 which gave no products (Figure 5c). Previously, *VRN-A1* allele from the tetraploid variety 'Langdon' (AY747598) was determined as having a 7.2 kb deletion in intron 1. Using the primers pair Ex1/C/F and Intr1/A/R3 for detection of this allele (Table 1; Figure 4), the three *T. dicoccoides* accessions generated a PCR product of about 0.5 kb (Figure 5d) identical to the corresponding sequence from 'Langdon'. Here, we refer to this allele as *VRN-A1L* (Figure 4; Table 2).

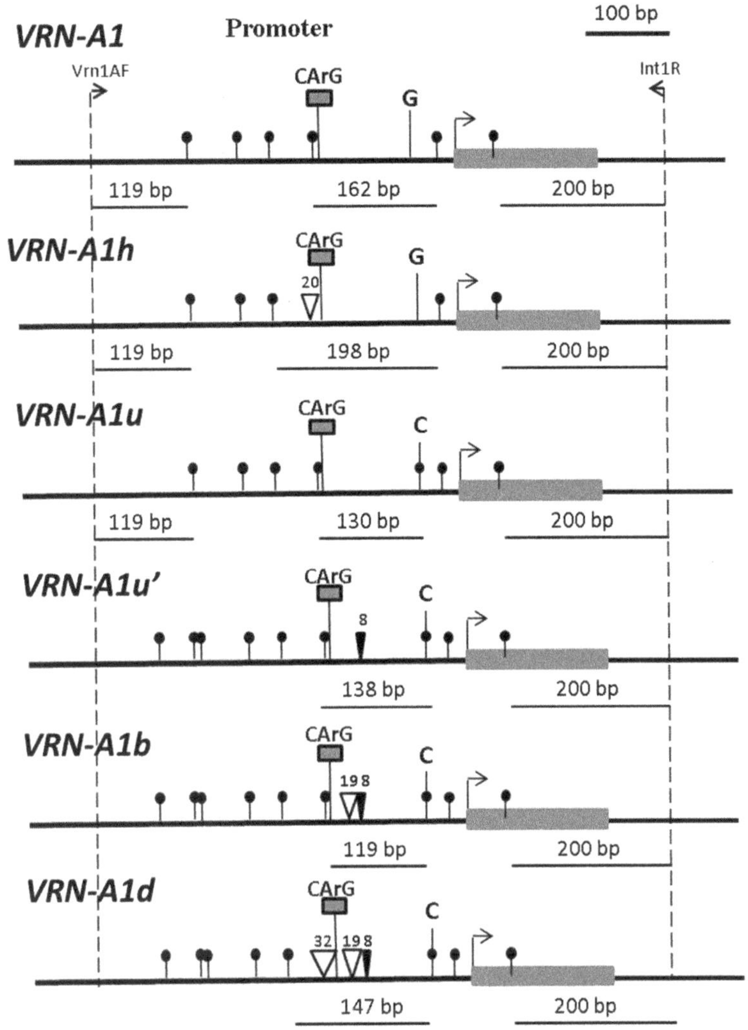

Figure 3 ***Msp*** **I restriction enzyme map of the promoter region of different *VRN-1* alleles.** Start codon and the 1st exon are indicated by the right-angled arrow flanking the grey rectangle. The positions of specific primers are shown at the top of the diagram. Dashes with black circles indicate *Msp* I restriction sites. Lengths of restriction fragments longer than 100 bp are indicated. Micro-deletions and insertions are shown by the empty and filled triangles, respectively, with sizes indicated. Positions of the CArG box and nucleotide G→C substitution are shown.

The pair of primers Intr1/B/F and Intr1/B/R4 was designed as a positive control for the absence of large mutations (deletions or insertions) in intron 1 of the *VRN-B1* locus [24] (Table 1). Using these primers, all accessions of *T. dicoccoides* yielded a 1149 bp PCR product indicating the presence of the wild recessive *VRN-B1* allele (see Additional file 2).

Thus, tetraploid *T. dicoccoides* displayed structural variation both at the promoter and 1st intron regions of *VRN-A1* locus, whereas at *VRN-B1* no significant differences were found in the studied accessions. To determine whether this variation originated at the tetraploid level or in diploid donors of the A genome, we conducted an analysis of diploid progenitor species.

VRN-1 allelic variability in diploid *Triticum* species

Promoter region

We analysed *VRN-1* promoter variability in 160 accessions of the diploid wheat species *T. monococcum*, *T. boeoticum* and *T. urartu* (Additional file 1). As for the tetraploid accessions, the diploids yielded PCR products of ~0.7 kb with primers Vrn1AF and Int1R (Figure 1a). *Msp* I- digestion yielded three patterns of restriction (Figure 1b). The first pattern contained bands of 119, 162 and 200 bp, consistent with the known recessive *VRN-A1* allele of *T. monococcum* (AY188331). This pattern was characteristic of all 61 accessions of *T. monococcum* and 39 accessions of *T. boeoticum* (Additional file 1). The remaining 20 accessions of *T. boeoticum* yielded a

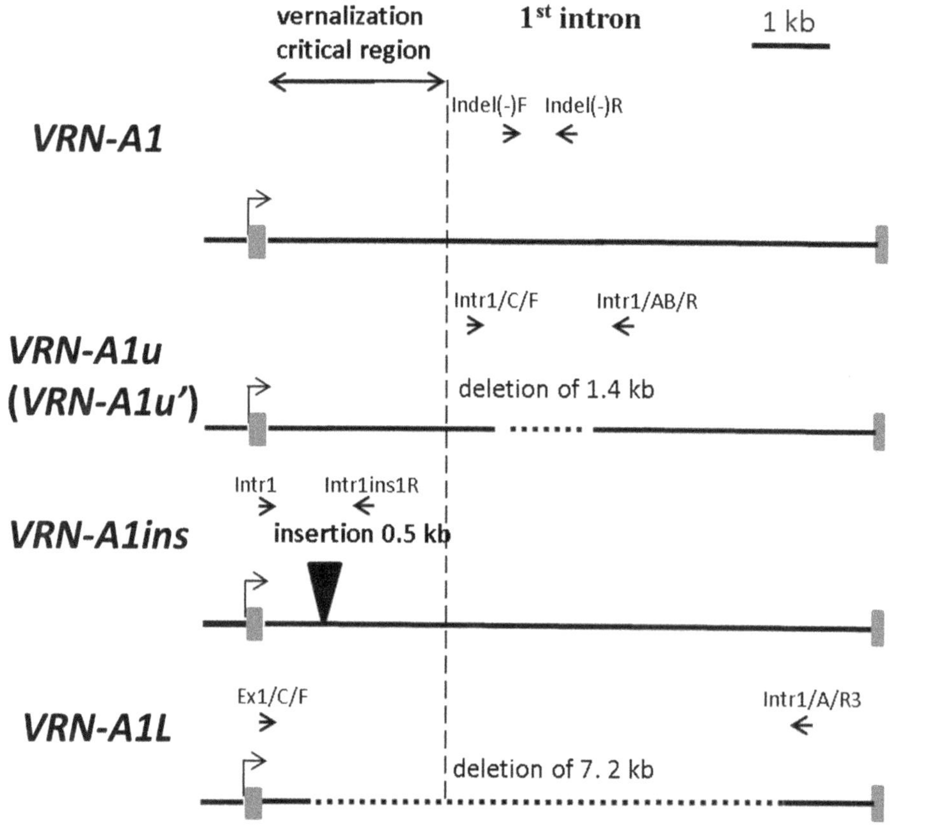

Figure 4 Schematic representation of the 1st intron region of different *VRN-1* alleles. The positions of specific primers are shown above each scheme. Large deletions and insertions are indicated.

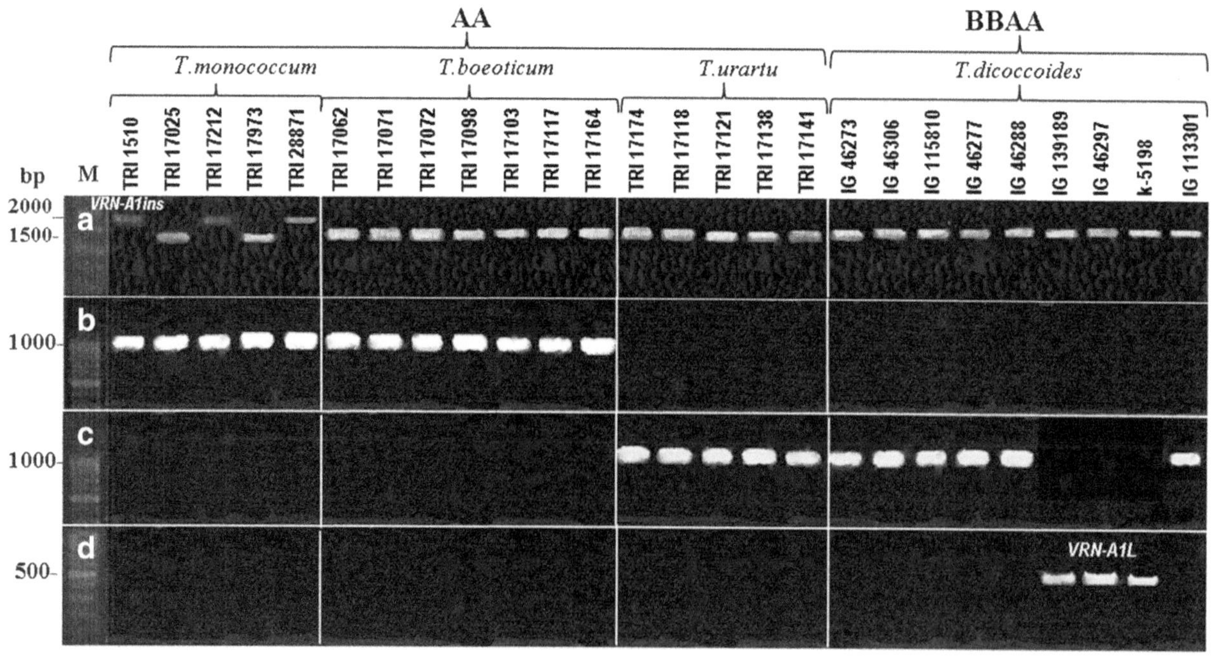

Figure 5 PCR amplification of *VRN-1* intron 1 region. PCR amplification with primers Intr1/Intr1insR **(a)**, Indel(–)F/Indel(–)R **(b)**, Intr1/C/F//Intr1/AB/R **(c)**, Ex1/C/F//Intr1/A/R3 **(d)**. Accession numbers, species and genotypes are given at the top.

pattern with no middle band at 162 bp, and intensive 200 bp band. All accessions of *T. urartu* were distinguished from the other two species by a double band of about 120–130 bp (Figure 1b). To further analyze allelic variation in the *VRN-A1* promoter, we selected a subset of accessions representing the three restriction patterns and sequenced the corresponding PCR products obtained with VrnA1F/Int1R primers (Table 2).

Among five selected accessions of *T. monococcum*, four had identical promoter *VRN-A1* sequences to the reported *VRN-A1* allele (AY188331) (Table 2). Accession TRI 28871 had a 1 bp deletion in a CArG-box motif in the promoter region (Figure 2). This allele has been previously identified in a spring accession of *T. monococcum* as *VRN-A1f* [16]. CAPS marker screening of all diploid accessions (see Methods) revealed a 112 bp band characteristic of *VRN-A1f* (Figure 1c) in three more accessions of *T. monococcum* (Additional file 1). Subsequent sequencing confirmed the presence of *VRN-A1f* in these accessions.

Of seven selected accessions of *T. boeoticum*, three had promoter sequences which were identical to the recessive *VRN-A1* allele of *T. monococcum* (Table 2). Sequences of the remaining four accessions were identical to the previously described *VRN-A1h* allele (GQ451745), which differs from *VRN-A1* by a 20 bp deletion near the CArG- box (Figure 2). This deletion resulted in a loss of one *Msp* I site (Figure 3), explaining the absence of the 160 bp band and increased staining intensity of the 200 bp band due to superposition of a 198 bp fragment (Figure 1b). We concluded that the 20 *T. boeoticum* accessions showing this restriction pattern carry *VRN-A1h*, whereas the remaining accessions are likely to have *VRN-A1* (Additional file 1).

The five selected accessions of *T. urartu* had an identical *VRN-A1* promoter sequence (Table 2), with 98% sequence identity to the respective sequence of *T. monococcum VRN-A1*. Of the few nucleotide substitutions, one (G → C, position −48 from the start codon) resulted in a new *Msp I* restriction site (Figure 2, 3). This explains the appearance of a 130 bp band coupled with the loss of the 160 bp band (Figure 1b). Based on our restriction analysis across the three diploid species, this substitution is specific for *T. urartu*, so we have designated the corresponding allele as *VRN-A1u* (Additional file 1).

1st intron

Alignment of the known *VRN-A1* sequences from databases at the 1st intron revealed two large indels (Figure 4). An insertion of 0.5 kb (*VRN-A1ins*) was found previously in two spring accessions of *T. monococcum* near the 5′ side of intron 1 [31]. A deletion of 1.4 kb located about 4 kb downstream from the start of intron 1 was found in all known *VRN-A1* sequences obtained from polyploid wheat species in comparison with *T. monococcum.*

We designed specific primers flanking both indels to study variation at the 1st intron of *VRN-A1* in three diploid species (Table 1; Figure 4). The primers Indel(−)F and Indel(−)R allow identification of the *VRN-A1* allele of *T. monococcum* without the deletion of 1.4 kb, with an expected PCR product of approximately 1 kb. All accessions of *T. monococcum* and *T. boeoticum* yielded this product, whereas no product was obtained with the *T. urartu* accessions (Figure 5b). The primers Intr1/C/F and Intr1/AB/R flanking the 1.4 kb deletion were used as a positive amplification control for the presence of this deletion in *T. urartu* (Table 1; Figure 4). As for *T. dicoccoides* (see above), all *T. urartu* accessions yielded the 1 kb product. This product was absent for all accessions of *T. monococcum* and *T. boeoticum*, confirming the previous result (Figure 5c). Hence, the 1.4 kb deletion in intron 1 is a specific marker for the *VRN-A1u* (*VRN-A1u′*) allele of *T. urartu* and polyploid species with an A-genome (Additional file 1).

Using primer pair Intr1/Intr1insR, 19 accessions of *T. monococcum* generated a 2 kb PCR product, demonstrating that these accessions carried *VRN-A1ins* (accessions TRI 1510, 17212 in Figure 5a; additional file 1). The remaining 42 *T. monococcum* accession yielded a fragment of about 1.5 kb indicating the absence of *VRN-A1ins.* Similarly, no *VRN-A1ins* was found in any of the *T. urartu* or *T. boeoticum* accessions, excepting a single *T. boeoticum* accession from Europe (which may have been mis-classified) (Additional file 1).

Evaluation of growth habit in diploid and tetraploid accessions

To assess the impact of *VRN-1* alleles on vernalization requirements in diploid and tetraploid wheats, we selected 26 accessions containing both initial (recessive), or structurally modified forms of *VRN-1* genes (Table 2). Noteworthy, in two accessions of *T. monococcum* and 1 accession of *T. boeoticum* the recessive form of *VRN-A1* (*VRN-A1* promoter, no insertion in intron 1) was associated with spring growth habit, implying an involvement of other genes in the determination of this habit. Four accessions of *T. monococcum* carrying a *VRN-A1f* allele was defined as a spring wheat. Out of four accessions of *T. boeoticum* carrying the *VRN-A1h* allele, two had a spring growth habit, while the other two were winter types. All five accessions of *T. urartu* with uniform *VRN-A1* sequences were winter types (Table 2).

With respect to the tetraploid *T. dicoccoides*, allele *VRN-A1b* was associated with winter growth habit in all three accessions containing this allele, whereas *VRN-A1d* and *VRN-A1L* were strongly associated with spring habit (Table 2). Accession IG 113301 with a recessive form of the *VRN-A1* gene (*VRN-A1u´* promoter, no changes in intron 1) was classified as a winter wheat.

Discussion

Previous studies of *VRN-1* allelic variability in wild *Triticum* species were restricted either by inclusion of a limited number of accessions for analysis [16,23,24,31], or by the use of a single regulatory region (*VRN-1* promoter) as a marker [32]. Here we have made an attempt to systematically study these species using a wide set of accessions representing wild tetraploid *T. dicoccoides* and three diploid A genome progenitor species (Additional file 1).

VRN-1 allelic diversity in diploid A genome species

A CAPS markers designed to the promoter *VRN-1* region allowed us to reveal the three most frequent allelic variants in diploids, while avoiding a large scale sequencing approach (Figure 1). The recessive *VRN-A1* allele was prevalent among the accessions of *T. monococcum* and *T. boeoticum* (Additional file 1; Table 2). Twenty accessions of *T. boeoticum* contained allele *VRN-A1h* which was absent from all accessions of *T. monococcum.* We found the previously described *VRN-A1f* allele (Figure 2) in four accessions of *T. monococcum* (Table 2; Additional file 1).

All accessions of *T. urartu* had a specific restriction pattern with the frequent base cutting enzyme, *Msp* I, allowing us to distinguish *T. urartu* from the other diploid species (Figure 1b). Sequencing determined that the corresponding *VRN-A1u* allele is highly homologous to the known *VRN-A1* alleles of polyploid wheat species. The majority of these alleles contain an 8 bp insertion which is absent in the respective sequences of diploids (Figures 2 and 3), except for a single previously studied sequence from *T. urartu* (GQ451737). We have classified the latter allele as *VRN-A1u´*, suggesting that it originated from the basic *VRN-A1u* allele of *T. urartu* and was subsequently inherited by polyploid descendants bearing an A-genome, including *T. dicoccoides* (Additional file 1).

Previously, molecular markers were designed for identification of large deletions in intron 1 affecting vernalization response [24]. Using these and our own markers (Table 1) we found two major alterations occurring at different evolutionary stages of diploid wheats (Figures 4 and 5). The insertion of 0.5 kb may have occurred relatively recently during the evolution of *T. monococcum,* since it was found only in some accessions of this species (Additional file 1). The deletion of 1.4 kb probably occurred in a progenitor of *T. urartu* after its divergence from *T. boeoticum.* The estimated divergence time between *T. boeoticum* and *T. urartu* and between *T. monococcum* and *T. boeoticum* is 570 000 and 290 000 YA, respectively [33]. Besides *T. dicoccoides,* the 1.4 kb deletion is characteristic of other polyploid species with an A genome.

Thus, the analysis of both the promoter and 1st intron regions of *VRN-1* gene in diploid wheat species allowed us to reveal species-specific alleles including *VRN-A1f* and *VRN-A1ins* for *T. monococcum, VRN-A1h* for *T. boeoticum* and *VRN-A1u* for *T. urartu.* The structural similarity of the latter allele with the respective alleles of wheat polyploids confirmed *T. urartu* as the donor of the A genome to wheat polyploids.

Tetraploid wheat genetic diversity at *VRN-1* loci; the beginning of the formation of a new allelic set

Wild emmer wheat *T. dicoccoides* as a species is about 360,000 years old, resulting from a spontaneous hybridization event which took place somewhere in the Fertile Crescent, a crescent-shaped area of fertile land in the Middle East [34]. *T. dicoccoides* belongs to the first cereals domesticated by humans and it is this domestication step which provided the key for subsequent bread wheat evolution [1].

Using the same approach as for the diploid wheats to study *T. dicoccoides,* we found two previously described alleles *VRN-A1b* and *VRN-A1d* with mutations in the promoter region (Figure 2 and 3; Table 2). These alleles occur in different polyploid wheats both tetra- and hexaploid but they have not been found in diploid wheat species [23,32]. We did not find the dominant *VRN-A1a* allele with insertion of a foldback element in the promoter region amongst the *T. dicoccoides* accessions (Table 1). This allele is abundant in hexaploid wheat *T. aestivum* [23,35,36].

PCR analysis of the 1st intron region revealed allele *VRN-A1L* in three accessions of *T. dicoccoides* (Figure 4 and 5). This allele is common among cultivated spring varieties of tetraploid *T. durum* [24,37]. *VRN-A1L* was not found in the diploid wheat species.

Unlike *VRN-A1,* locus *VRN-B1* displayed no significant variability within either the promoter or 1st intron regions (Figure 2, Additional file 2). Several large mutations, mainly deletions within the 1st intron of *VRN-B1* were previously found in spring varieties of *T. aestivum* [24,25,27,35]. But these mutations have not yet been detected in tetraploid wheat species, and are thus likely to have a later origin.

The analysis of *VRN-1* polymorphism in polyploid and diploid wheat accessions showed that the two groups differ with respect to the sets of *VRN-1* alleles found among accessions.

Effect of *VRN-1* gene structure on growth habit

Diploid wheat species

We determined the impact of *VRN-1* alleles on growth habit in a selected set of accessions representing three diploid species of the A genome (Table 2). Out of five tested accessions of *T. monococcum,* four were spring types despite the presence of an intact promoter region at *VRN-A1*. In two of them spring growth habit was associated with insertion in intron 1 (*VRN-A1ins*). Totally, we found the *VRN-A1ins* allele in 31% of *T. monococcum* accessions (Additional file 1).

Previous data support the importance of intron 1 in the vernalization response and heading time determination, and the proposed key regulatory region has been narrowed to a highly conserved 2.8 kb "vernalization critical region" located near the left side of intron 1 [24]. A number of mutations (deletions, insertions etc.) in this region have been associated with high levels of *VRN-1* transcripts and a spring growth habit in different crops [24,26,31,38,39]. It is suggested that these mutations may affect epigenetic chromatin states, resulting in a higher basal level of *VRN-1* expression [19,26,40,41].

Dubcovsky et al. suggested that mutation in CArG- box of the *VRN-A1f* allele of *T. monococcum* results in spring habit preventing interaction with unknown repressor dependent on photoperiod [31]. However, we can not discriminate the effect of this allele on vernalization response, because in our material it always occurs in combination with *VRN-A1ins* (Table 2; Additional file 1).

In two accessions of *T. monococcum* (TRI 17025, 17973) and one accession of *T. boeoticum* (TRI 17072) spring growth habit was associated with the wild recessive *VRN-A1* allele (*VRN-A1*; no insertion in intron 1; Table 2). Alignment of *VRN-1* sequence from accessions TRI 17072 and TRI 17062 of *T. boeoticum* (spring and winter types, respectively) showed that the first accession had a mutation that has not been described previously, the substitution T → C coupled with 1 bp deletion about 200 bp downstream from the start of intron 1. It remains unclear whether this minor alteration within "critical region" (see above) is the reason for spring growth habit.

Previous studies of vernalisation genes in wheat have identified loss-of function mutations at the *VRN-2* gene, the flowering repressor, which result in up-regulation of *VRN-1* regardless of its allelic status. [15,17,31]. We studied 3 above mentioned accessions using previously developed molecular markers for *VRN-2* gene [17] and found that accession TRI 17025 of *T. monococcum* contains a mutated recessive *VRN-2* allele (data not presented). We failed to define the reason for spring growth habit in accession TRI 17973 of *T. monococcum*. In previous study, 6% of accessions of this species were of spring type despite the presence of *vrn-1/VRN-2* genotype [17]. Based on these data, it was suggested that genes other than *VRN-1* and *VRN-2* influence vernalisation response in diploid wheat.

Out of four accessions of *T. boeoticum* containing *VRN-A1h* allele and studied for vernalization requirements, two were of spring type, while the other two were associated with the winter habit (Table 2).

One plausible hypothesis has been that the spring growth habit of *T. monococcum* originated in wild populations of *T. boeoticum* and was later selected by man through cultivation [32]. However, absence of the *VRN-1h* allele in *T. monococcum* coupled with the presence of specific alleles conferring spring growth habit (*VRN-A1f*, *VRN-A1ins*) in this diploid leads us to propose an alternative hypothesis, of selection of independent mutations that appeared in *T. monococcum* during domestication.

We found no spring forms in *T. urartu* (Table 2). None of the *T. urartu* accessions included in this study had changes in the promoter or 1st intron regions of *VRN-1* which might confer a spring growth habit (Additional file 1). According to Dorofeev et al. [6], all forms of *T. urartu* are winter forms. Goncharov [42] found only 2% of *T. urartu* accessions to be spring types. In another previous study, no differences in the promoter region of *VRN-1* relative to the recessive allele were found in four spring accessions of *T. urartu* [32].

Tetraploid wheat

In *T. dicoccoides* we found several *VRN-1* alleles which are widely distributed among cultivated wheat polyploids. Allele *VRN-A1b* is not always associated with spring growth habit in hexaploid wheat lines [43], in our study this allele was associated with winter habit in three accessions of *T. dicoccoides* (Table 2). Allele *VRN-A1d* influences growth habit more strongly, probably due to an additional deletion affecting the CArG box (Figures 2 and 3). It confers spring growth habit in all accessions of *T. dicoccoides* both studied here and from databases. Allele *VRN-A1L* previously found in many cultivated tetraploid forms of wheat [37] is also associated with dominant spring type (Table 2).

The question of appearance of spring growth habit in *T.diccocoides* remains open. This habit could be inherited from ancestral diploids, in particular, from ancestral form of *T.urartu*, precursor of A genome. However, as shown above, *T. urartu* is predominantly winter species. Kato et al. [44] studied geographical variation in the vernalisation requirements of *T. dicoccoides* and found that the distribution of spring forms was sporadic and restricted to warmer areas. The authors suggest that the spring type might have evolved from a previous winter prototype of *T. dicoccoides* as an adaptation to warmer conditions. Our data are consistent with this hypothesis implying that the *VRN-1* associated determinants of spring growth habit in polyploid wheat were formed after combining of diploid genomes.

Conclusions

In the present study we investigated variability in the promoter and 1st intron regions of the vernalization gene *VRN-1* in different accessions of wild tetraploid species

T. dicoccoides and its diploid A genome progenitors. Our results indicated that a number of *VRN-1* alleles characteristic of polyploid wheats (*VRN-A1b*, *VRN-A1d*, *VRN-A1L*) are found in *T. dicoccoides*, while others (*VRN-A1a*, *VRN-B1* alleles) probably arose during later stages of polyploid evolution and were selected during domestication. The occurrence of spring forms in the diploid A genome progenitor species is only partially attributed to *VRN-1* variation, since some of these forms contained an intact recessive *VRN-1* allele, while others displayed a winter habit despite the presence of a modified *VRN-1* allele (*VRN-A1h*). The most abundant allele conferring spring growth habit in diploids was *VRN-A1ins* of *T. monococcum*, containing a 0.5 kb insertion in the 1st intron. A higher variability of *VRN-1* loci in polyploid species in comparison with diploids may be explained by the hypothesis that stresses due to allopolyploidization may provoke genome instability [45]. Greater *VRN-1* loci variability gives polyploid wheat an advantage in conferring adaptation to a broader range of environments.

Methods

Plant material and DNA extraction

Plant material included 80 accessions of the wild tetraploid wheat species *T. dicoccoides* (BBAA) and the diploid A genome species *T. monococcum*, *T. boeoticum* and *T. urartu* (61, 59 and 40 accessions, respectively). These accessions were selected from collections in the Leibniz Institute of Plant Genetics and Crop Plant Research (IPK, Gatersleben, Germany), the ICARDA genbank (Syria) and the Vavilov All-Russian Institute of Plant Industry RAN (St Petersburg, Russia). Accession numbers, phenotypes and genotypes are given in Additional file 1.

Total genomic DNA was extracted from etiolated seedlings as described by Plaschke et al. [46] with modifications. Leaves (from 5 seeds per accession) were pooled, placed in racked collection tubes (24 per rack) and homogenised directly in extraction buffer using a FastPrep-24 (MP Biomedicals, USA).

PCR

PCR primers reported in Yan et al. [23], Fu et al. [24], Shcherban et al. [25] were used to detect the presence of dominant or recessive alleles of *VRN-A1* and *VRN-B1* loci in diploid and polyploid wheats (Table 1). To further discriminate different alleles in promoter regions of both loci we digested corresponding PCR products with restriction endonuclease *Msp* I followed by separation of DNA fragments in 2-3% high resolution agarose gel (HydraGene Co., China). The polymorphic 1-bp deletion (*VRN-A1f*) was detected as described by Dubcovsky et al. [31]. Using primers VRN1F_F (5′-ACAGCGGCTATGCTCCAGAC-3′) and VRN1F_R (5′-GGAGGATGGCCAGGCCAAATC-3′), the second nucleotide of the reverse primer (underlined T) was mutated to generate a *Taq* I restriction site for the *vrn-A1* allele that was absent in the *VRN-A1f* allele.

The sequences of the *T. monococcum VRN-A1* gene (AY188331, DQ146421, DQ146422, DQ146423) and respective sequences of *T.aestivum* (AY747600, AY747601) were used to design specific primers flanking an insertion of ~0.5 kb and deletion of 1.4 kb in the 1st intron of *VRN-A1* (Table 1; Figure 4).

PCR was performed using a DNA Thermal Cycler 480 (Perkin Elmer Cetus, USA). Reaction mixtures were in a volume of 20 μl containing 50–100 ng of genomic template DNA, 1 ng of each of primer, 0.25 mM of each dNTP, 1x reaction buffer (67 mM TrisHCl, pH 8.8; 2 mM $MgCl_2$; 18 mM $(NH_4)_2SO_4$; 0.01% Tween 20) and 1 unit *Taq* polymerase. After initial denaturation at 94°C for 2 min, 35 cycles were run at 94°C for 1 min, 55-61°C (depending on the primer pair used) for 1 min, and 72°C for 0.5-2 min, followed by a final extension at 72°C for 5 min. PCR products were separated on 1% agarose gel, stained with ethidium bromide and visualized under UV light.

Sequencing of PCR products

Amplified DNA fragments were excised following electrophoresis, purified using a QIAquick PCR purification kit (QIAGEN, Germany) and directly sequenced using an ABI PRISM Dye Terminator Cycle Sequencing ready reaction kit (Perkin Elmer Cetus, USA) and corresponding specific primers. Sequencing was performed in an ABI PRISM 310 Genetic Analyzer (Perkin Elmer Cetus). All obtained sequences were deposited to GenBank (Table 2).

Evaluation of growth habit

Growth habits of the diploid and tetraploid accessions containing different *VRN-A1* alleles were determined from two replicates in 2013–2014 (Table 2). Seeds were sown in the greenhouse (ICG, Novosibirsk) without vernalization under a long day photoperiod regime. Three months after sowing, when all spring standard varieties had headed, experimental materials were classified as either spring (ear emergence) or winter (no visible ear formation) types.

Additional files

Additional file 1: Polymorphisms at the promoter and 1st intron regions of *VRN-A1* gene in X accessions of diploid *Triticum* species and Y accessions of tetraploid *T. dicoccoides*.

Additional file 2: Variability at *VRN-B1* locus in *T. dicoccoides*. a- PCR amplification using specific primers P2/P5 to detect variation within the *VRN-B1* promoter region in different accessions of *T. dicoccoides*. b- *Msp* I restriction digestion of corresponding PCR products. c- PCR amplification with primers Intr1/B/F//Intr1/B/R4 to detect the absence of deletions in the 1st intron of *VRN-B1*. Asterisks mark the accessions for which the *VRN-B1* promoter region was sequenced (Genbank: KM586661-65).

Competing interests

The authors declare that they have no competing interests.

Authors' contributions

AS designed the study, carried out the molecular and bioinformatics experiments and greenhouse analyses, and prepared the manuscript. KS contributed to the molecular analysis. EAS helped with interpretation of the results and critically revised the manuscript. All authors read and approved the final manuscript.

Acknowledgements

We are grateful to Dr. A.Böerner and Dr. A.Graner (Leibniz Institute of Plant Genetics and Crop Plant Research, Gatersleben, Germany) for supplying seeds of wheat species. The study of diploid species was done in framework of the State Budget Programme (Project No VI.53.1.5.), the analysis of *T. dicoccoides* was supported by the Russian Scientific Foundation (Project No. 14-14-00161).

References

1. Ozkan H, Willcox G, Graner A, Salamini F, Kilian B. Geographic distribution and domestication of wild emmer wheat (*Triticum dicoccoides*). Genet Resour Crop Evol. 2011;58:11–53.
2. Dvorak J, Zhang HB. Variation in repeated nucleotide sequences sheds light on the phylogeny of the wheat B and G genomes. Proc Natl Acad Sci U S A. 1990;87:9640–4.
3. Kilian B, Ozkan H, Deusch O, Effgen S, Brandolini A, Kohl J, et al. Independent wheat B and G genome origins in outcrossing *Aegilops* progenitor haplotypes. Mol Biol Evol. 2007;24:217–27.
4. Salina EA, Lim YK, Badaeva ED, Shcherban AB, Adonina IG, Amosova AV, et al. Philogenetic reconstruction of *Aegilops* section Sitopsis and the evolution of tandem repeats in the diploids and derived wheat poliploids. Genome. 2006;49:1023–35.
5. Johnson BL, Dhaliwal HS. Reproductive isolation of *T. boeoticum* and *T. urartu* and the origin of the tetraploid wheats. Am J Bot. 1976;63:1088–96.
6. Dorofeev VF, Filatenko AA, Migushova EF, Udaczin RA, et al. Wheat. In: Flora of cultivated plants. Volume 1. Edited by Dorofeev VF, Korovina ON. Leningrad (St. Petersburg): Kolos (In Russian); 1979, 346 pp.
7. Konovalov FA, Goncharov NP, Goryunova S, Shaturova A, Proshlyakova T, Kudryavtsev A. Molecular markers based on LTR retrotransposons *BARE-1* and *Jeli* uncover different strata of evolutionary relationships in diploid wheats. Mol Genet Genomics. 2010;283:551–63.
8. Dvorak J, McGuire PE, Cassidy B. Apparent sources of the A genomes of wheats inferred from polymorphism in abundance and restriction fragment length of repeated nucleotide sequences. Genome. 1988;30:680–9.
9. Dvorak J, Di Terlizzi P, Zhang HB, Resta P. The evolution of polyploid wheats: identification of the A genome donor species. Genome. 1993;36:21–31.
10. Friebe B, Kim NS, Kuspira J, Gill BS. Genetic and intercytogenetic analyses of the A genome of *Triticum monococcum*. VI. Production and identification of primary trisomics using the C-banding technique. Genome. 1990;33:542–55.
11. Hammer K, Filatenko AA, Korzun V. Microsatellite markers – a new tool for distinguishing diploid wheat species. Genet Resour Crop Evol. 2000;47:497–505.
12. Baum BR, Bailey LG. The origin of the A genome donor of wheats (Triticum: Poaceae) – a perspective based on the sequence variation of the 5S DNA gene units. Genet Resour Crop Evol. 2004;51:183–96.
13. Heun M, Haldorsen S, Vollan K. Reassessing domestication events in the Near East: einkorn and *Triticum urartu*. Genome. 2008;51:444–51.
14. Kilian B, Ozkan H, Walther A, Kohl J, Dagan T, Salamini F, et al. Molecular diversity at 18 loci in 321 wild and 92 domesticate lines reveal no reduction of nucleotide diversity during *Triticum monococcum* (einkorn) domestication: implications for the origin of griculture. Mol Biol Evol. 2007;24:2657–68.
15. Distelfeld A, Li C, Dubcovsky J. Regulation of flowering in temperate cereals. Curr Opin Plant Biol. 2009;12:1–7.
16. Yan L, Loukoianov A, Tranquilli G, Helguera M, Fahima T, Dubcovsky J. Positional cloning of the wheat vernalization gene *VRN1*. Proc Natl Acad Sci U S A. 2003;100:6263–8.
17. Yan L, Loukoianov A, Blechl A, Tranquilli G, Ramakrishna W, SanMiguel P. The wheat *VRN2* gene is a flowering repressor down-regulated by vernalization. Science. 2004;303:1640–4.
18. Yan L, Fu D, Li C, Blechl A, Tranquilli G, Bonafede M, et al. The wheat and barley vernalization gene *VRN3* is an orthologue of FT. Proc Natl Acad Sci U S A. 2006;103:19581–6.
19. Trevaskis B, Hemming MN, Dennis ES, Peacock WJ. The molecular basis of vernalization induced flowering in cereals. Trends Plant Sci. 2007;12:352–7.
20. Murai K, Miyamae M, Kato H, Takumi S, Ogihara Y. *WAP1*, a wheat *APETALA1* homolog, plays a central role in the phase transition from vegetative to reproductive growth. Plant Cell Physiol. 2003;44:1255–65.
21. Shitsukawa N, Ikari C, Shimada S, Kitagawa S, Sakamoto K, Saito H, et al. The einkorn wheat (*Triticum monococcum*) mutant, maintained vegetative phase, is caused by a deletion in the *VRN-1* gene. Genes Genet Syst. 2007;82:167–70.
22. Trevaskis B, Hemming MN, Peacock WJ, Dennis ES. *HvVRN2* responds to day-length, whereas *HvVRN1* is regulated by vernalization and developmental status. Plant Physiol. 2006;140:1397–405.
23. Yan L, Helguera M, Kato K, Fukuyama S, Sherman J, Dubcovsky J. Allelic variation at the *VRN-1* promoter in polyploid wheat. Theor Appl Genet. 2004;109:1677–86.
24. Fu D, Szucs P, Yan L, Helguera M, Skinner JS, vonZitzewitz J, et al. Large deletions within the first intron in *VRN-1* are associated with spring growth habit in barley and wheat. Mol Gen Genomics. 2005;273:54–65.
25. Shcherban AB, Efremova TT, Salina EA. Identification of a new *Vrn-B1* allele using two near-isogenic wheat lines with difference in heading time. Mol Breeding. 2012;29(3):675–85.
26. Shcherban AB, Khlestkina EK, Efremova TT, Salina EA. The effect of two differentially expressed wheat *VRN-B1* alleles on the heading time is associated with structural variation in the first intron. Genetica. 2013;141:133–41.
27. Milec Z, Tomkova L, Sumikova T, Pankova K. A new multiplex PCR test for the determination of *Vrn-B1* alleles in bread wheat (*Triticum aestivum* L.). Mol Breeding. 2012;30(1):317–23.
28. Goncharov NP. Genetic resources of wheat related species:The *Vrn* genes controlling growth habit (spring vs. winter). Euphytica. 1998;100:371–6.
29. Stelmakh AF. Genetic effects of *Vrn* genes on heading date and agronomic traits in bread wheat. Euphytica. 1993;65:53–60.
30. Iwaki K, Haruna S, Niwa T, Kato K. Adaptation and ecological differentiation in wheat with special reference to geographical variation of growth habit and *Vrn* genotypes. Plant Breed. 2001;120:107–14.
31. Dubcovsky J, Loukoianov A, Fu D, Valarik M, Sanchez A, Yan L. Effect of photoperiod on the regulation of wheat vernalization genes *Vrn-1* and *Vrn-2*. Plant Mol Biol. 2006;60:469–80.
32. Golovnina KA, Kondratenko E, Blinov AG, Goncharov NP. Molecular characterization of vernalization loci *VRN1* in wild and cultivated wheats. BMC Plant Biol. 2010;10:168.
33. Middleton CP, Senerchia N, Stein N, Akhunov ED, Wicker T, Kilian B. Sequencing of chloroplast genomes from wheat, barley, rye and their relatives provides a detailed insight into the evolution of the Triticeae tribe. PLoS. 2014;9(3):e85761. doi:10.1371/journal.pone.0085761.
34. Dvorak J, Akhunov E. Tempos of gene locus deletions and duplications and their relationship to recombination rate during diploid and polyploid evolution in the *Aegilops–Triticum* alliance. Genetics. 2005;171:323–32.
35. Santra DK, Santra M, Allan RE, Campbell KG, Kidwell KK. Genetic and molecular characterization of vernalization genes *Vrn-A1*, *Vrn-B1* and *Vrn-D1* in spring wheat germplasm from the Pacific Northwest Region of the USA. Plant Breeding. 2009;128:576–84.
36. Shcherban AB, Emtseva MV, Efremova TT. Molecular genetical characterization of vernalization genes *Vrn-A1*, *Vrn-B1* and *Vrn-D1* in spring wheat germplasm from Russia and adjacent regions. Cereal Res Commun. 2012;40(3):425–35.
37. Oliveira HR, Campana MG, Jones H, Hunt HV, Leigh F, Redhouse DI, et al. Tetraploid Wheat Landraces in the Mediterranean Basin: Taxonomy, Evolution and Genetic Diversity. PLoS One. 2012;7(5):e37063.
38. von Zitzewitz J, Szucs P, Dubcovsky J, Yan L, Francia E, Pecchioni N, et al. Molecular and structural characterization of barley vernalization genes. Plant Mol Biol. 2005;59:449–67.
39. Hemming MN, Peacock WJ, Dennis ES, Trevaskis B. Low-temperature and day-length cues are integrated to regulate FLOWERING LOCUS T in barley. Plant Physiol. 2008;147:355–66.

40. Oliver SN, Finnegan EJ, Dennis ES, Peacock WJ, Trevaskis B. Vernalization-induced flowering in cereals is associated with changes in histone methylation at the VERNALIZATION1 gene. Proc Natl Acad Sci U S A. 2009;106:8386–91.
41. Oliver SN, Deng W, Casao MC, Trevaskis B. Low temperatures induce rapid changes in chromatin state and transcript levels of the cereal VERNALIZATION1 gene. J Exp Bot. 2013;64:2413–22.
42. Goncharov NP. Comparative genetics of wheats and their related species. Novosibirsk: Siberian University Press (In Russian); 2002. p. 251.
43. Pidal B, Yan L, Fu D, Zhang F, Tranquilli G, Dubcovsky J. The CArG-box located upstream from the transcription start of wheat vernalization gene *VRN-1* is not necessary for the vernalization response. J Heredity. 2009;100:355–64.
44. Kato K, Mori Y, Beiles A, Nevo E. Geographical variation in heading traits in wild emmer wheat, *Triticum dicoccoides*. I. Variation in vernalization response and ecological differentiation. Theor Appl Genet. 1997;95:546–52.
45. Levy AA, Feldman M. The impact of polyploidy on grass genome evolution. Plant Physiol. 2002;130:1587–93.
46. Plaschke J, Ganal MW, Roeder MS. Detection of genetic diversity in closely related bread wheat using microsatellite markers. Theor Appl Genet. 1995;91:1001–7.

14

Expression of catalase and retinoblastoma-related protein genes associates with cell death processes in Scots pine zygotic embryogenesis

Jaana Vuosku[1,3*], Suvi Sutela[1], Johanna Kestilä[1], Anne Jokela[1], Tytti Sarjala[2] and Hely Häggman[1]

Abstract

Background: The cell cycle and cellular oxidative stress responses are tightly controlled for proper growth and development of Scots pine (*Pinus sylvestris* L.) seed. Programmed cell death (PCD) is an integral part of the embryogenesis during which megagametophyte cells in the embryo surrounding region (ESR) and cells in the nucellar layers face death. In the present study, we show both the tissue and developmental stage specific expression of the genes encoding the autophagy related ATG5, catalase (CAT), and retinoblastoma related protein (RBR) as well as the connection between the gene expressions and cell death programs.

Results: We found strong *CAT* expression in the cells of the developing embryo throughout the embryogenesis as well as in the cells of the megagametophyte and the nucellar layers at the early embryogeny. The *CAT* expression was found to overlap with both the *ATG5* expression and hydrogen peroxide localization. At the late embryogeny, *CAT* expression diminished in the dying cells of the nucellar layers as well as in megagametophyte cells, showing the first signs of incipient cell death. Accumulation of starch and minor *RBR* expression were characteristic of megagametophyte cells in the ESR, whereas strong *RBR* expression was found in the cells of the nucellar layers at the late embryogeny.

Conclusions: Our results suggest that ATG5, CAT, and RBR are involved in the Scots pine embryogenesis and cell death processes. CAT seems to protect cells against hydrogen peroxide accumulation and oxidative stress related cell death especially during active metabolism. The opposite expression of *RBR* in the ESR and nucellar layers alongside morphological characteristics emphasizes the different type of the cell death processes in these tissues. Furthermore, the changes in *ATG5* and *RBR* expressions specifically in the megagametophyte cells dying by necrotic cell death suggest the genetic regulation of developmental necrosis in Scots pine embryogenesis.

Keywords: Autophagy, Catalase, Conifer, Developmental cell death, Embryogenesis, Megagametophyte, *Pinus*, Retinoblastoma-related protein, Scots pine, Seed development

Background

In multicellular organisms such as plants, the organized destruction of cells by programmed cell death (PCD) is essential for body plans and specific organ shapes as well as for removing damaged or infected cells [1-4]. In plants, different cell death types are often defined by morphological characteristics because precise molecular mechanisms behind the regulation and execution of cell death processes are still poorly known [5,6]. Two classes of cell death, vacuolar cell death and necrosis, have been suggested for plants but the terms are not unambiguous [6,7]. Examples of vacuolar cell death are found during embryo, organ and tissue morphogenesis and senescence, whereas necrosis is most usually induced by a variety of abiotic stresses and by successful recognition of a pathogen during the hypersensitive response [6,8]. However, necrosis is no longer considered to be an uncontrolled process, but necrotic cell death can also be a regulated event that contributes to the development and to the maintenance of tissue and organismal integrity [9-11]. In plants, the roles

* Correspondence: jaana.vuosku@luke.fi
[1]Genetics and Physiology Unit, University of Oulu, P.O. Box 3000, FI-90014 Oulu, Finland
[3]Current address: Natural Resources Institute Finland (Luke), Rovaniemi Unit, FI-96301 Rovaniemi, Finland
Full list of author information is available at the end of the article

of nucleases and proteases in PCD are evident [12,13] but the cell death processes also involve many other molecular mechanisms such as catalase [14] and the retinoblastoma related proteins [15] which are required to maintain cellular homeostasis and thus may be difficult to identify as actual cell death effectors.

The reactive oxygen species (ROS) contribute to the induction, signalling and execution of plant cell death [16-18]. Hydrogen peroxide (H_2O_2), the most stable ROS [19], has received particular attention as a signal molecule involved in the regulation of essential biological functions such as cell death, pathogen responses, and gene expression [16,20-23]. H_2O_2 is a toxic byproduct of cell metabolism and H_2O_2 generation from a variety of cellular processes increases in response to numerous developmental signals and biotic and abiotic stimulants [24]. While plants contain several types of H_2O_2-metabolizing proteins, catalases (CAT, hydrogen peroxide oxidoreductase, EC 1.11.1.6) are highly active peroxisomal enzymes which express high specificity towards H_2O_2 and protect cells from the toxic effects by the conversion of H_2O_2 into water and molecular oxygen [25]. It has been suggested that impact of H_2O_2 is strongly influenced by the extent of accumulation of H_2O_2 allowed by antioxidant enzymes such as CAT. This has been shown in *Chlamydomonas reinhardtii* where a reversible partial inactivation of CAT correlates with a transient increase in the level of H_2O_2. The concentration range seems to be necessary to activate H_2O_2-dependent signaling pathways stimulating the expression of H_2O_2 responsive genes [26]. Furthermore, the direct interaction of CAT with ROS could allow CAT to act as a molecular link between ROS and the promotion of autophagy-dependent cell death [14].

In plants, the retinoblastoma related proteins (RBR) regulate the progression of cell cycle and transcription via chromatin-modifiers and, furthermore, function in promotion of cell differentiation [27]. Recently, RBR1 was found to control also cell death in maize (*Zea mays* L.) endosperm cells [15]. The plant RBR proteins are homologs of retinoblastoma susceptibility gene products (pRB) firstly discovered in human cells [28]. The pRB proteins are activated by phosphorylation and their A/B pocket domain enables them to interact with more than 100 proteins [29]. The interaction of hypophosphorylated RBR with E2F and DP transcription factors prevents cell division by suppressing the transition from G1 to S phase in the cell cycle [27]. In plants, *RBR* genes are conserved and they have been found from green algae, bryophytes, lycophytes and from both monocot and dicot angiosperm species [27,30]. In our previous study, we showed that the RBR/E2F pathway operates also in conifers by presenting the expression of *RBR* and *E2F* genes in *in vitro* cultured Scots pine (*Pinus sylvestris* L.) embryogenic cells [31].

PCD is an integral part of the seed development in angiosperm and gymnosperm species (e.g. [32-35]). In gymnosperms, the Scots pine seed represents a well-documented embryogenesis with coordinated action of several distinct cell death programs. The fertilization of several eggs [36] and furthermore the cleavage of polyzygotic embryos [37] leads to the presence of several embryos in a young seed. However, only the dominant embryo survives, whereas the subordinate embryos die via PCD during the seed development [34]. Embryos grow and develop inside the corrosion cavity of the megagametophyte tissue which functions as a nutrient source of embryos i.e. being congruent with the endosperm of angiosperms [38]. Throughout the embryo development, the megagametophyte cells in the embryo surrounding region (ESR) are destroyed by morphologically necrotic cell death [35]. Additionally, the cells in the megagametophyte surrounding nucellar layers die during the seed development [35,39] providing nutrition for the surrounding tissues and later functioning as a barrier against water and fungi [40,41]. The importance of caspase-like VEIDase activity, type II metacaspases, and Tad-D nuclease have been perceived in cell death during Scots pine and Norway spruce (*Picea abies* (L.) Karst.) embryogenesis [35,42,43]. In conifers, the autophagy related (ATG) proteins are essential for vacuolar cell death and for instance, in Norway spruce, depletion of ATG5 and ATG6 causes aberrations in the development of suspensors [44].

Here, we study the progress of two distinct types of cell death programs and reveal the link between PCD and the expression of *CAT* and *RBR* genes which are typically considered as maintainers of cellular homeostasis. We focus on the cell death processes in the ESR and nucellar layers during the Scots pine seed development, and show that *ATG5* previously connected to the regulation of vacuolar cell death also expresses in the cells destroyed via necrotic-like cell death.

Methods

Immature and mature Scots pine seeds

One-year-old immature seed cones were collected during the growing period in July from an open-pollinated elite Scots pine (*Pinus sylvestris* L.) clone, K818, growing in the Scots pine clone collection in Punkaharju, Finland (61°48′ N; 29°17′ E). Immature Scots pine seeds were dissected from the developing cones. For the anatomical observations, for the detection of PCD and for the localization of proteins, starch, and mRNA transcripts of *ATG5, CAT, RBR* and β-glucosidase (*βG*, EC 3.2.1.21), the seed coats were removed and immature seeds were fixed immediately and embedded in paraffin as described below. For the gene expression studies, immature seeds were stored in liquid nitrogen until use.

Mature Scots pine cones were collected from the same grafts as the immature cones in late fall of the same growing season. Mature Scots pine seeds were surface sterilized with 3% Plant Preservative Mixture (Plant Cell Technology) overnight and placed thereafter on moist filter paper and let to imbibe at room temperature (RT) for two days. Thereafter, seeds were opened with a scalpel, seed coats were removed, and the embryos and megagametophytes were separated from each other for the gene expression analysis. For the localization of *ATG5* and *CAT* mRNA transcripts mature seeds were fixed as described below.

Anatomical and histochemical observations

Developing and imbibed mature Scots pine seeds with and without seed coats were dissected for the histochemical localization of H_2O_2 and peroxidase. The 3,5,3′5′-tetramethylbenzidine (TMB) is oxidised by peroxidases in a reaction where H_2O_2 acts as a hydrogen acceptor, and thus, TMB can be used to localize H_2O_2 and peroxidases [45]. For the H_2O_2 localization the seeds were incubated in 50 mM Tris-acetate-buffer (pH 5.0), which contained 0.1 mg/ml 3,5,3′,5′-TMB-HCl [45], at RT at least for 45 min. In addition, seeds were treated with 3% H_2O_2 in 1x phosphate-buffered saline (PBS) buffer (10 mM phosphate, 150 mM NaCl, pH 7.4) for 10 min prior to the histochemical staining for the peroxidase location. Negative controls were incubated in 50 mM Tris-acetate-buffer.

The DAB (3,3′-diaminobenzidine) Peroxidase Substrate kit (Vector Laboratories) was used in localizing peroxidase activity following manufacturer's instructions. The dissected seeds were incubated for two or three min at RT in reaction mixture. The DAB Peroxidase Substrate reaction mixture contains H_2O_2 which oxidizes DAB, and subsequently generates a dark brown reaction product when peroxidases are present. Negative controls were incubated in reaction mixture without DAB. The blocking of peroxidase activity was attempted by incubating seed material in 1% H_2O_2 at RT for 10 min. The sections were examined by a stereo microscope (Stemi DV4, Carl Zeiss) and imaged with a digital camera (Nikon Coolpix 950, Japan).

Developing and mature Scots pine seeds were fixed for studying the anatomical features of the embryos and for *in situ* mRNA hybridization analyses using the following protocol from fixation to coverslip mounting. Tissues were fixed in 4% (w/v) *p*-formaldehyde in 1x PBS. After dehydration with a graded series of ethanol, ethanol was replaced by tertiary butanol and then gradually by paraffin. Sections (5 and 7 μm) were cut from the embedded samples with a microtome, mounted on SuperFrost®Plus slides (Menzel-Glaser) and fixed by drying overnight at 40°C.

The paraffin sections were dewaxed in Histochoice (Sigma) and rehydrated through a graded series of ethanol. To study the developmental stage of the embryos, the preparates were stained with toluidine blue (0.05% toluidine blue in H_2O). According to Singh [46] Scots pine embryogenesis is divided into three different phases, proembyogeny, early embryogeny and late embryogeny. Here, we studied the early embryogeny which initiates with the elongation of the suspensor system and terminates with the appearance of the root meristem (Figure 1A) and the late embryogeny which contains the establishment of root and shoot meristems and the maturation of the leading embryo (Figure 1B). For the protein and starch visualization the sections were stained with 0.2% amido black [47] and with 0.5% potassium iodide-iodine (IKI) [48], respectively. All sections were studied with a light microscope (Nikon Eclipse E600) and photographed with a Qimaging Micropublisher 5.0 RTV digital camera. Adobe Photoshop CS was used to adjust contrast, brightness and colour uniformly to entire images.

TUNEL assay

Nuclear DNA fragmentation, i.e. DNA strand breaks leaving free 3′-OH termini, was shown by the TUNEL (terminal deoxyribonucleotidyl transferase (TdT)-mediated deoxyuridine triphosphate (dUTP) nick end labelling) assay. The dewaxed and rehydrated sections were digested with 10 μg/ml proteinase-K (Roche) for 30 min and two washings with PBS. The sections were labelled with TMR red (red fluorescence) using *in situ* cell death detection kit (Roche) according to the manufacturer's protocol. Prior to the labelling procedure, the positive control sections were incubated with DNase I recombinant (Roche) to induce DNA strand breaks. The label solution without terminal transferase, instead of the TUNEL reaction mixture, was used as a negative control. The labelled sections were examined with a confocal laser scanning microscope (LSM 5 Pascal, Carl Zeiss) with an HBO 100 mercury lamp and using the HeNe laser 543 nm line, dichroic beam splitter (HFT 488/543/633; Carl Zeiss) and LP 560 emission filter (Carl Zeiss).

RNA extraction, reverse transcription and cDNA cloning

Total RNA was extracted using the automated magnetic-based KingFisher™ mL method (Thermo Electron Corporation) with the MagExtractor® total RNA purification kit (Toyobo) according to the manufacturer's instructions. The RNA samples were treated with RNase-free DNase (Invitrogen) at RT for 15 min for the elimination of contaminating genomic DNA. Thereafter the RNA samples were purified with the NucleoSpin® RNA Clean-Up kit (Macherey-Nagel). The RNA yields were measured three times with OD_{260}

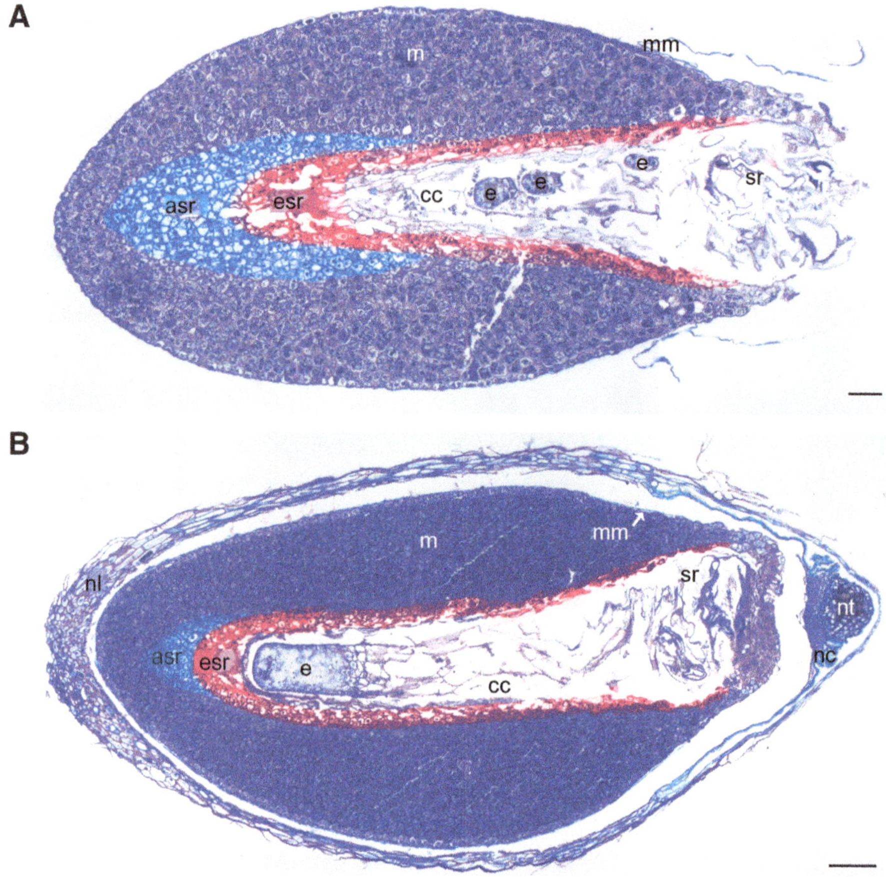

Figure 1 Early and late developmental stages of Scots pine embryogenesis. (A) The dominant embryo and subordinate embryos in the corrosion cavity at the early embryogeny surrounded by the embryo surrounding region (esr) of the megagametophyte highlighted with red color. The arrow-shaped region (asr) of the megagametophyte outside the esr highlighted with blue color. **(B)** The dominant embryo in the corrosion cavity at the late embryogeny. The esr of the megagametophyte highlighted with red color and asr highlighted with blue color. asr = arrow-shaped region, cc = corrosion cavity, e = embryo, esr = embryo surrounding region, m = megagametophyte, mm = megaspore membranes, nc = nucellar cap, nl = nucellar layers, nt = cellular nucellus, sr = suspensor remnants. Bars: 100 μm.

analysis using NanoDrop ND1000 spectrophotometer (NanoDrop Technologies), and 1 μg of each RNA sample was subsequently used for the cDNA synthesis. cDNA was reverse-transcribed from an anchored oligo-dT primer by SuperScript II reverse transcriptase (Invitrogen) using standard methods in a reaction volume of 20 μl. Fragments of the putative Scots pine *ATG5* and cell wall associated *βG* were amplified by standard PCR using cDNA from immature seeds as a template, gene specific primers (Additional file 1: Table S1) and DyNAzyme™EXT polymerase (Finnzymes). The fragments with appropriate length were gel-purified by Montage DNA Gel Extraction Kit (Millipore Corporation), cloned by TOPO TA Cloning Kit (Invitrogen) and sequenced by an Applied Biosystems 3730 DNA analyzer.

In situ mRNA hybridization analysis

The RNA probes for the *in situ* mRNA hybridization analyses of *ATG5*, *CAT*, *RBR* and *βG* transcripts in developing and mature seeds were prepared using a PCR-based technique [49]. The T7 RNA polymerase promoter sequence (TAATACGACTCACTATAGGG) was introduced at the 5′ends of the gene-specific primers (Additional file 1: Table S2). The downstream primers contained an artificially introduced T7 promoter at its 5′end, which enabled the synthesis of antisense transcripts. The upstream primers containing T7 promoters at the 5′ends was used for the synthesis of sense transcripts, i.e. as negative control.

The PCR fragment representing the coding region of either *ATG5*, *CAT*, *RBR* or *βG* was produced under standard PCR conditions by DyNAzyme™EXT polymerase (Finnzymes) using plasmid DNA containing the cDNA in question as a template. The PCR fragment was gel-purified (DNA Gel Extraction Kit, Millipore Corporation) and 250 ng was subsequently used as a template DNA for *in vitro* transcription by T7 RNA polymerase (Invitrogen), incorporating dig-UTP via DIG RNA labelling Mix (Roche

Molecular Biochemicals). Template DNA was digested with four units of RNase-free DNase (Invitrogen) in a reaction volume of 20 μl at 37°C for 10 min, the probe was precipitated and the yield of the DIG-labeled RNA probe was estimated by comparing the intensity of the sample to the defined controls made with DIG-labelled control RNA (Roche Molecular Biochemicals).

The sections were dewaxed, treated with Proteinase K (Finnzymes, 1 μg/ml in 100 mM Tris/HCl, 50 mM Na_2EDTA, pH 7.5) at 37°C for 30 min and dehydrated in a graded series of ethanol up to absolute. The slides were dried in a vacuum for 1 h. For the hybridization, 250 μl of hybridization mixture was added to the sections, mounted under coverslips and incubated in a water atmosphere at 55°C overnight. The hybridization mixture contained 0.8 μg/ml of DIG-labelled RNA antisense or sense probe, 50% (v/v) deionized formamide, 300 mM NaCl, 10 mM Tris/HCl (pH 7.0), 10 mM Na_3PO_4 (pH 7.0), 50 mM EDTA, 10% (w/v) dextran sulfate, 200 μg/ml tRNA, 1x Denhardt's solution and 10 units/ml RNase OutTM inhibitor (Invitrogen).

After the hybridization, the slides were washed in 2x NaCl/Cit (300 mM NaCl, 30 mM sodium citrate, pH 7.0) at RT for 10 min followed by 1x and 0.5x NaCl/Cit treatment by slow shaking at 37°C for 10 min. The washing was followed by treatment of the RNase A (Roche, 1 μg/ml in 10 mM Tris/HCl, 500 mM NaCl, 1 mM Na_2EDTA, pH 7.5) at 37°C for 60 min. Then the slides were washed four times in the same solution but without RNase A in a slow shaking at 37°C for 15 min and once in 2x NaCl/Cit at RT for 30 min.

For the detection of hybridized probe, the slides were washed in Tris/NaCl buffer (100 mM Tris/HCl, 150 mM NaCl, pH 7.5) for 5 min and blocked with 3% (w/v) blocking reagent (Dig Nucleic acid detection Kit, Roche) and 0.3% (v/v) Triton X-100 in Tris/NaCl buffer in a water atmosphere at RT for 30 min. 1 unit/ml of Anti-DIG-AP Fab fragments (Roche) in Tris/NaCl buffer was added to the sections, mounted under parafilm and incubated in a water atmosphere at RT for 2 h. The slides were washed at RT four times in Tris/NaCl buffer for 10 min and in AP buffer (100 mM Tris/HCl, 100 mM NaCl, 50 mM $MgCl_2$, pH 9.5) for 5 min. For colour development, 2% (v/v) NBT/BCIP substrate (Dig Nucleic acid detection Kit, Roche) and 1% (v/v) Triton X-100 in AP buffer was dispersed on the sections, mounted with coverslips and incubated in a water atmosphere at RT in the dark overnight. The slides were washed with water, dehydrated in a graded series of ethanol, air-dried and then mounted with immersion oil and covered with coverslips. The sections were examined with a light microscope (Nikon Optiphot 2, Japan) and imaged with an Infinity*1–3*C camera (Lumenera Corporatiom, Ottawa, Ontario, Canada), using the IMT iSolution Lite image-processing program (IMT i-Solution Inc., Vancouver, BC, Canada). The auto and manual tiling feature of image-processing program was used to combine separate images to one. Non-specific signal was observed in the remnants of the degenerated suspensors, ESR, and inner nucellar layers generated by fragmented nucleic acids as previously described in Vuosku *et al.* [50]. Adobe Photoshop CS5 was used to adjust contrast, brightness and colour uniformly to entire images.

Quantification of gene expression

The real-time reverse-transcription PCR analysis (Q-RT-PCR) was used for the quantification of the *CAT, RBR* and *βG* expression in immature seeds at the developmental stages of early and late embryogeny as well as in the embryos and megagametophytes of mature seeds. Being aware of the challenges in finding suitable endogenous reference genes for pine embryogenesis [51], both absolute and relative Q-RT-PCR analyses were used. For the determination of mRNA copy numbers in the absolute quantification, the standard curves were generated using serial 10-fold dilutions of synthesized RNA molecules to control variability during both RT and PCR steps of Q-RT-PCR runs. The DNA templates from which the RNA molecules could be transcribed were amplified by basic PCR procedure using gene-specific primers (Additional file 1: Table S3). The upstream primers contained T7 promoter sequence (TAATACGACTCACTA TAGGG) and the downstream primers contained poly(T) tail at their 5′end. The DNA molecules were subsequently used as templates for *in vitro* transcription by T7 RNA polymerase. The numbers of standard RNA molecules added to the reverse-transcription reactions were calculated using the molecular weights of the oligonucleotides and Avogadro's constant ($6.022 \cdot 10^{23}$ mol^{-1}). In the relative quantification, ubiquinone (*UBI*) and glyceraldehyde-3-phosphate dehydrogenase (*GAPDH*) were used as endogenous reference genes (see Additional file 2 for details).

In the absolute Q-RT-PCR analysis (see Additional file 2 for details for the relative quantification), the number of biological replicates was 5, 5, 6, and 4 for early stage immature seeds, late stage immature seeds, embryos, and megagametophytes, respectively. The PCR amplification conditions of the gene fragment were optimized for the LightCycler® 2.0 instrument (Roche Diagnostics), and the subsequent PCR runs showed a single PCR product during melting curve (The Tm Calling Analysis of Lightcycler® 480 Software release 1.5.0 SP3) and electrophoretic analysis. The real-time PCR amplifications were performed using the LightCycler® 480 SYBR Green I Master (Roche), 50 nM gene specific primers (Additional file 1: Table S4) and 2 μl cDNA in the reaction volume of 20 μl. The negative controls contained 2 μl of molecular grade water instead of cDNA template. The real-time PCR amplification was initiated by incubation at 95°C for 10 min followed by 45 cycles: 10 s at 95°C, 10 s at 58°C and 5 s at 72°C. Every PCR reaction was done as duplicate to control for the variability of PCR amplification and the coefficient variation, CV%, of technical replicates was < 2.5. The Abs Quant/2nd Derivative analysis of Lightcycler® 480 software release 1.5.0 SP3 was utilized to generate the crossing point and concentration values for each sample.

Statistical analysis

The mRNA copy numbers generated with the absolute Q-RT-PCR analysis were utilized in calculating the gene expression. The expression of studied genes was calculated by dividing the individual values with the mean number of transcripts in the early embryogeny, that is, value of one was given to the gene expression in the early embryogeny. The significance of difference between the early and late embryogeny and between the embryo and megagametophyte of mature seeds was examined using two sample *t* test with the graphical user interface R Commander [52] in R software package (2.11.0) [53]. Log 10 transformation was conducted to variables not normally distributed.

Results

Nuclear DNA fragmentation in ESR and nucellar layers

In our previous study, we showed that megagametophyte cells in the ESR are destroyed by morphologically necrotic cell death and also the cells in the megagametophyte surrounding nucellar layers die during the Scots pine seed development [35]. Likewise, TUNEL positive nuclei were detected in the ESR cells and in the nucellar layers in this study (Additional file 3: Figure S2).

Transient decrease of *CAT* expression in megagametophyte at late embryogeny

The Scots pine *CAT* (EU513163) was amplified using zygotic embryo cDNA as a template. Previously, three *CAT* genes have been found present in the angiosperm species tobacco (*Nicotiana tabacum* L.), Arabidopsis, maize, pumpkin (*Cucurbita* sp.) and rice (*Oryza sativa* L.) [54-58]. The angiosperm class I CATs are strongly expressed in photosynthetic tissues, whereas class II CATs are associated with vascular tissues. Class III CATs are expressed especially in seeds and reproductive tissues [25]. Scots pine *CAT* showed high sequence similarity (75%) with maize *CAT1* belonging to class III.

Based on the Q-RT-PCR analysis, the number of *CAT* mRNA transcripts decreased significantly in developing Scots pine seeds when the embryogenesis proceeded from the early to the late stage (P <0.001). In mature seeds, the number of *CAT* transcripts was significantly greater in megagametophytes than in embryos (P =0.02) (Figure 2A). At the early embryogeny, intense *CAT* expression was localized in the cells of both the dominant embryo and subordinate embryos (Figure 2B and D) as well as in the cells of the megagametophyte and the nucellar layers (Figure 2B and E). At the late embryogeny, the *CAT* expression was still strong in the cells of the dominant embryo (Figure 2C and G), whereas it was clearly diminished in the cells of the nucellar layers (Figure 2C and F) and in the megagametophyte (Figure 2C and G). In mature seeds the megagametophyte cells surrounding the corrosion cavity showed intense *CAT* expression whereas in the embryo, the expression was comparable to the late embryogeny (Figure 3). The specificity of the antisense *CAT* probe was confirmed by the absence of signals in the sections hybridized with the sense *CAT* probe (Additional file 4: Figure S3).

Peroxidase and H_2O_2 localization

At the early embryogeny, intense blue colour indicating H_2O_2 presence was observed in the seed coat, flight wing (Figure 4A), and cellular nucellus (Figure 4B). In addition, H_2O_2 was detected in the nucellar layers (Figure 4B), and megaspore membranes (Figure 4C). At the late embryogeny, H_2O_2 was still present in the seed coat and flight wing, but the extent was clearly diminished (Figure 4D). The seeds treated with Tris-acetate buffer were used as negative control (Figure 4E). Peroxidase activity was detected throughout the embryogenesis in the very same seed structures as H_2O_2 and, additionally, in the ESR cells and in the arrow-shaped region (ASR) of megagametophyte, suspensor cells, and subordinate embryos (Figure 5A and B, Additional file 5: Figure S4). For negative control DAB was excluded from the reaction buffer (Figure 5C). The peroxidase localization with DAB and TMB protocol supplemented with H_2O_2 also gave comparable results (Additional file 5: Figure S4).

Minute *RBR* expression in dying megagametophyte cells in ESR

The predicted Scots pine RBR protein showed 57% identity with Arabidopsis RBR 1 protein. In Arabidopsis, only one *RBR* gene has been found, and the presence of two distinct *RBR* genes have been suggested to a unique feature of grasses [59]. In developing Scots pine seeds, there was overall low variation in the number of *RBR* mRNA transcripts across the studied Scots pine seed material. The *RBR* transcript levels were comparable between early and late embryogenesis as well as between the embryos and megagametophytes of mature seeds when the expression was quantified with Q-RT-PCR (Figure 6A). The *RBR* expression was localized in both embryonic and megagametophyte cells throughout the embryogenesis excluding the dying megagametophyte cells in the ESR and the ASR cells (Figure 6B, C and D, Additional file 6: Figure S5). The *RBR* expression signal

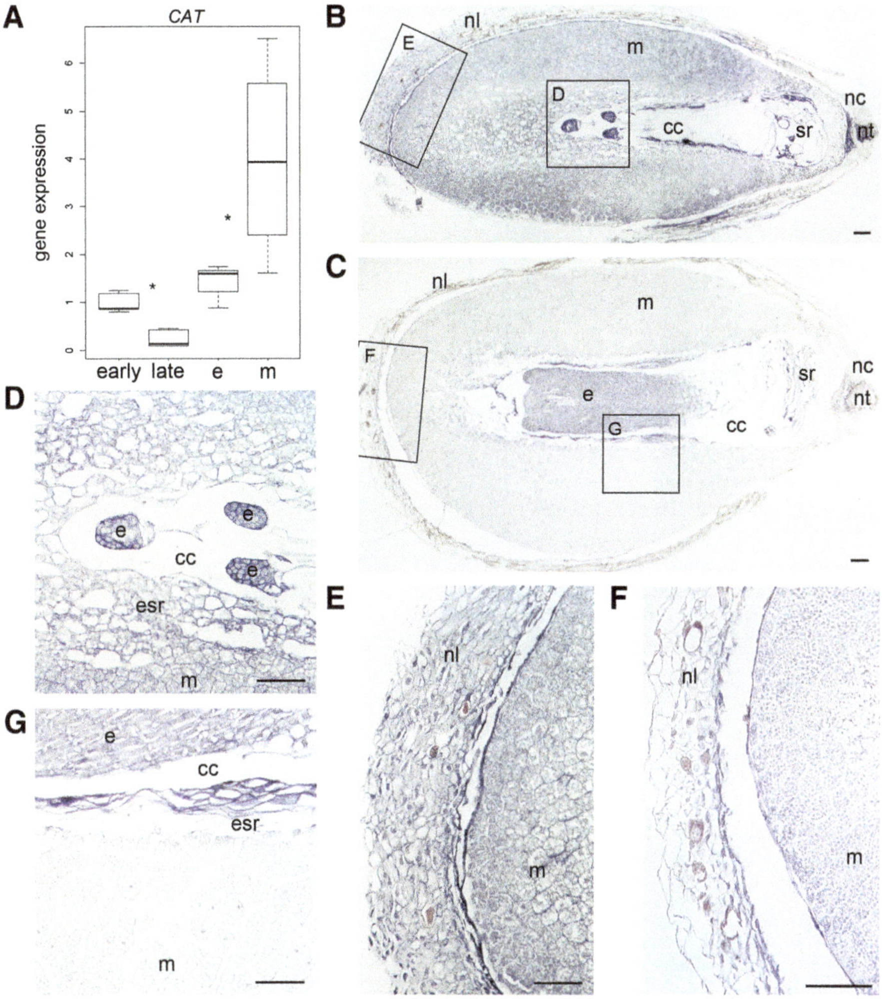

Figure 2 ***CAT*** **expression in developing and mature Scots pine seeds. (A)** The relative expression of *CAT* in developing seeds at the early and late embryogeny and in the embryos (e) and megagametophytes (m) of mature seeds. The expression was based on mRNA copy numbers generated with the absolute Q-RT-PCR analysis and values presented were normalized using the expression at the early embryogeny. A star denotes significant ($P < 0.05$) difference in the expression. **(B)** The localization of *CAT* mRNAs by *in situ* hybridization with DIG-labelled RNA-probes (blue signal) in a developing Scots pine seed at the early embryogeny. **(C)** The localization of *CAT* mRNAs at the late embryogeny. **(D)** Intense *CAT* expression in the developing embryos and in the megagametophyte at the early embryogeny. **(E)** Intense *CAT* expression in the nucellar layers at the early embryogeny. **(F)** Minor *CAT* expression in the nucellar layers at the late embryogeny. **(G)** Intense *CAT* expression in the cells of the leading embryo and minor *CAT* expression in the megagametophyte cells at the late embryogeny. cc = corrosion cavity, e = embryo, esr = embryo surrounding region, m = megagametophyte, nc = nucellar cap, nl = nucellar layers, nt = cellular nucellus, sr = suspensor remnants. Bars: 100 μm.

was found weak in nucellar cells at early embryogeny, but showed more intensity at later stages of embryogenesis (Figure 6E and F). The specificity of the antisense *RBR* probe was confirmed by the absence of signals in the sections hybridized with the sense *RBR* probe (Additional file 4: Figure S6).

ATG5 expression in dying cells of ESR and nucellar layers

In order to study the type of the cell death processes in the ESR of the megagametophyte and in the nucellar layers, we sequenced the complete coding sequence of the Scots pine *ATG5* gene (KM046993) and localized *ATG5* mRNA transcripts in developing Scots pine seeds at the early embryogeny and in mature seeds after two days of imbibition. The predicted Scots pine ATG5 protein showed 95% identity with the Norway spruce ATG5 protein which has previously shown to be essential for autophagy and vacuolar cell death [44]. At the early embryogeny, the *ATG5* expression was localized in both the ESR of the megagametophyte and the nucellar layers (Figure 7A). In a mature seed, *ATG5* expressed still in the ESR, but also in the maturating tracheids (Figure 7B). The specificity of the antisense *ATG5* probe was confirmed by the absence of signals in the sections hybridized with the sense *ATG5* probe (Additional file 4: Figure S7).

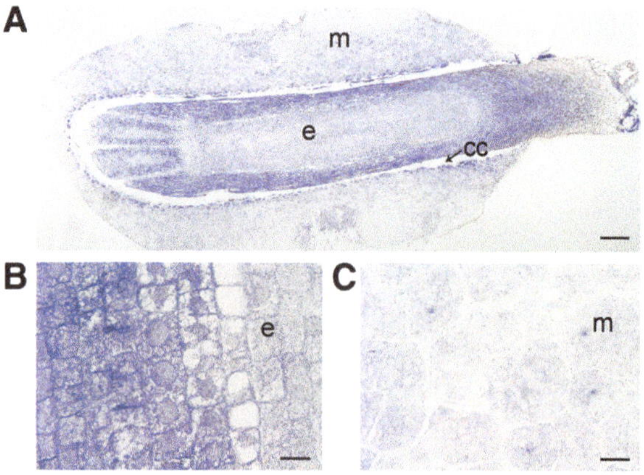

Figure 3 The localization of *CAT* mRNAs in a mature Scots pine seed. (A) Intense *CAT* expression (blue signal) in the embryo and in the megagametophyte cells surrounding the corrosion cavity. **(B)** Intense *CAT* expression in the embryonic cells. **(C)** Minor *CAT* expression in the cells in the inner part of the megagametophyte. cc = corrosion cavity, e = embryo, m = megagametophyte. Bars: **(B, C)** 20 μm and **(A)** 200 μm.

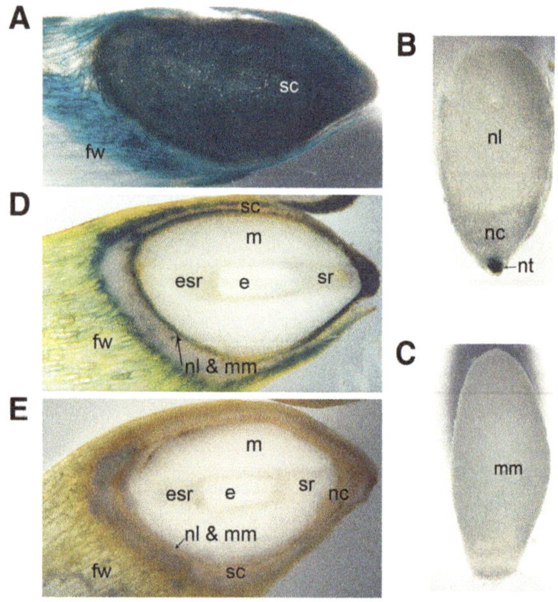

Figure 4 The localization of H_2O_2 in developing Scots pine seeds. (A) Intense blue colour from TMB indicates the presence of H_2O_2 in the seed coat and flight wing at the early embryogeny. **(B)** The localization of H_2O_2 in the nucellar cap, cellular nucellus, and in the nucellar layers at the early embryogeny. **(C)** The localization of H_2O_2 in the megaspore membranes at the early embryogeny. **(D)** The minor amount of H_2O_2 in the seed coat and flight wing and the large amount of H_2O_2 indicated by the dark blue colour in the nucellar layers at the late embryogeny. **(E)** The seed treated with Tris-acetate buffer without TMB. cc = corrosion cavity, e = embryo, esr = embryo surrounding region, fw = flight wing, m = megagametophyte, mm = megaspore membranes, nc = nucellar cap, nl = nucellar layers, nt = cellular nucellus, sc = seed coat, sr = suspensor remnants.

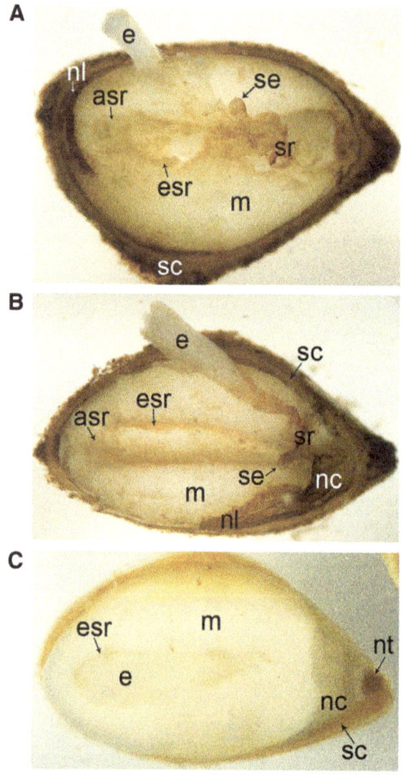

Figure 5 The localization of peroxidase activity in developing Scots pine seeds. (A, B) Oxidized DAB (brown colour) indicated peroxidase activity in the seed coat, nucellar layers, megagametophyte cells surrounding the corrosion cavity, suspensor cells and subordinate embryos at the early **(A)** and late **(B)** embryogeny. **(C)** The seed treated with the reaction buffer without DAB. asr = arrow-shaped region, e = embryo, esr = embryo surrounding region, m = megagametophyte, mm = megaspore membranes, nc = nucellar cap, nl = nucellar layers, nt = cellular nucellus, sc = seed coat, se = subordinate embryo, sr = suspensor remnants.

Accumulation and release of starch and storage proteins during Scots pine seed development

The Scots pine seed development comprises the accumulation of storage compounds such as starch and storage proteins. At the early embryogeny, the megagametophyte cells had only few storage proteins (Figure 8A) but during the late embryogenesis intensively stained protein bodies were found in the megagametophyte tissue (Figure 8B). However, the megagametophyte cells in ESR in addition to the cells in ASR did not pile up proteins during early embryogeny, but instead contained high quantity of starch (Figure 8C). As the embryo development proceeded the cell wall and plasma membrane of megagametophyte cells in ESR were destroyed with the release of starch grains into the corrosion cavity. In dying cells, the cell wall weakening and breakdown seemed to be connected with the increased intensity of the βG gene expression detected with *in situ* hybridization (Additional file 7: Figure S8, S9). In addition to the ESR and ASR, starch also

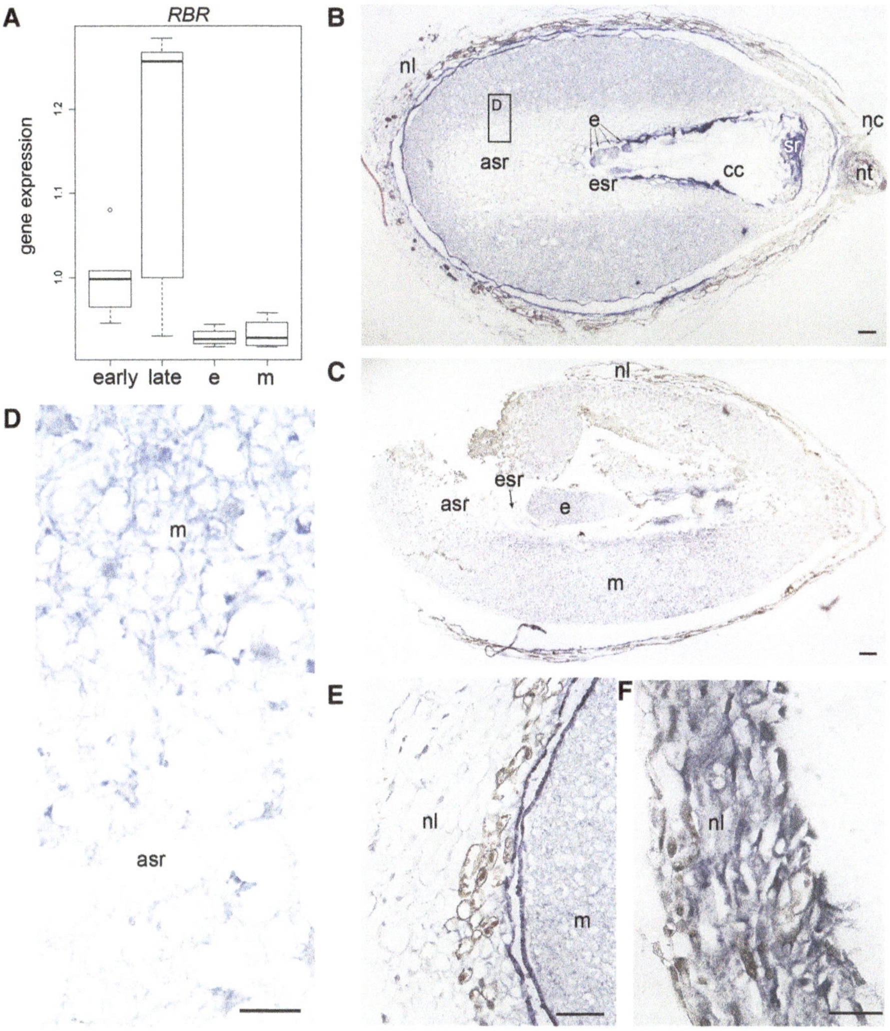

Figure 6 ***RBR*** **expression in developing Scots pine seeds.** **(A)** The relative expression of *RBR* in developing seeds at the early and late embryogeny and in the embryos (e) and megagametophytes (m) of mature seeds. The expression was based on mRNA copy numbers generated with the absolute Q-RT-PCR analysis and values presented were normalized using the expression at the early embryogeny. **(B)** The localization of *RBR* mRNAs (blue signal) in a developing Scots pine seed at the early embryogeny. **(C)** The localization of *RBR* mRNAs at the late embryogeny. **(D)** Minor *RBR* expression in the ESR of the megagametophyte at the early embryogeny. **(E)** Weak *RBR* expression in the cells of the nucellar layers at the early embryogeny. **(F)** Intense *RBR* expression in the cells of the nucellar layers at the late embryogeny. asr = arrow-shaped region, cc = corrosion cavity, e = embryo, esr = embryo surrounding region, m = megagametophyte, nc = nucellar cap, nl = nucellar layers, nt = cellular nucellus, sr = suspensor remnants. Bars: **(D)** 20 μm, **(F)** 50 μm, and **(B, C, E)** 100 μm.

accumulated into the cells of both the developing embryo and megagametophyte (Figure 8D).

Discussion

As an orthodox seed, a developing Scots pine seed experiences three characteristic features, embryo development, accumulation of reserve compounds and maturation/drying, which lead from a zygotic embryo to a mature, quiescent seed [60]. Both metabolic activity and moisture content fluctuate drastically during seed development [61-64]. Thus, the sources of ROS production, connected to basic cellular and specific seed developmental processes, in cells also vary [65]. During the seed development, the cell cycle is tightly controlled for proper growth and development [66] but also for cellular level oxidative stress responses [67]. Here, we show the tissue and developmental stage specific expression of the *ATG5*, *CAT*, and *RBR* genes as well as the connection between the gene expressions and cell death programs during the Scots pine seed development.

The development of a viable Scots pine seed includes the strictly coordinated action of several cell death programs (reviewed by Vuosku *et al.* [68]). The autophagic PCD of subordinate embryos as well as their suspensor

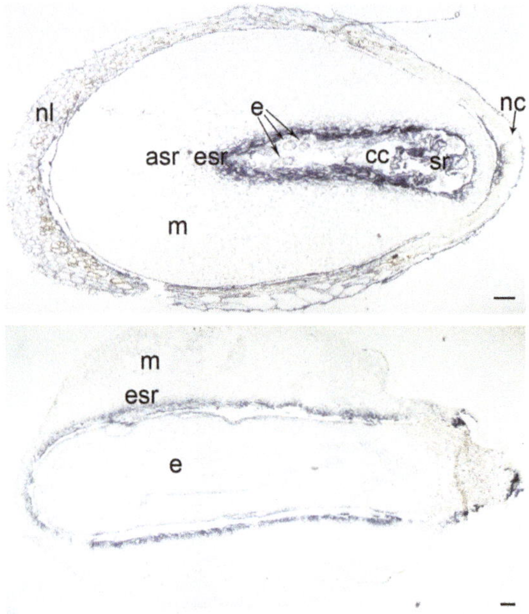

Figure 7 The localization of *ATG5* expression in developing and mature Scots pine seeds. (A) *ATG5* expression (blue signal) in the esr of the megagametophyte and in the nucellar layers at the early embryogeny. **(B)** *ATG5* expression in the esr of the megagametophyte and in the maturating tracheids of the embryo in a mature seed. asr = arrow-shaped region, cc = corrosion cavity, e = embryo, esr = embryo surrounding region, m = megagametophyte, nc = nucellar cap, nl = nucellar layers, sr = suspensor remnants. Bars: 100 μm.

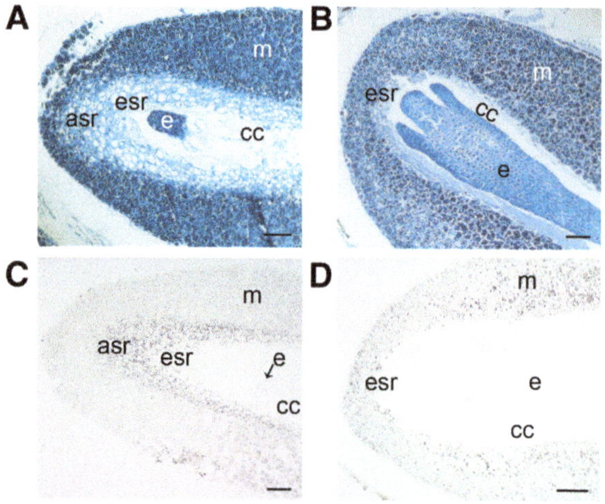

Figure 8 Histochemical localization of proteins by amido black and starch by potassium iodide-iodine in developing Scots pine seeds. (A) Megagametophyte cells with few protein storage vacuoles at the early embryogeny. **(B)** Megagametophyte cells stocked with protein storage vacuoles at the late embryogeny. **(C)** Megagametophyte cells with starch grains in the vicinity of the corrosion cavity at the early embryogeny. **(D)** Starch grains in the cotyledons and root meristem of a late embryo as well as in the megagametophyte. asr = arrow-shaped region, cc = corrosion cavity, e = embryo, esr = embryo surrounding region, m = megagametophyte. Bars: **(A, C)** 100 μm and **(B, D)** 200 μm.

tissues is initiated already at early embryogeny [34]. The nucellar layers face destruction at late embryogeny [35] to be used as a nutrition for the surrounding tissues of the seed and later on to form an efficient barrier to the passage of water and against fungal infections [40,41]. The necrotic-like death of the megagametophyte cells in the ESR continues throughout the embryogenesis. In the present study, the ESR cells differed already at the early embryogeny from the other megagametophyte cells in the expression of *ATG5, CAT, RBR* and *βG* genes as well as the storage compound distribution. The ESR cells had higher amount of starch and less storage proteins than the surrounding cells of the megagametophyte. The starch accumulation has been shown to precede PCD of the endosperm in several cereals [69-72] and in the perisperm of quinoa (*Chenopodium quinoa* Willd.) [73]. The role of starch accumulation may be more general in PCD: the endosperm cells of starch-deficient maize mutants contained more ethylene and faced accelerated PCD [69]. In cereals, the storage compounds are not evenly located in endosperm, but there is a gradient in the distribution of storage proteins and starch [69,74]. In a Scots pine seed, protein bodies and starch, however, filled all megagametophyte cells except the cells in the ESR at the late embryogeny.

Intense *βG* expression was found in the ESR of the megagametophyte and in the nucellar layers which suggests that during the cell death processes *βG* contributes to either weakening or breaking down of cell walls depending on the type and physiological function of the cell death. In Arabidopsis, cell wall associated *βG* breaks down polysaccharides to soluble sugars, and is induced by both starvation [75] and senescence [76]. In barley (*Hordeum vulgare* L.), hydrolysis of *β*-linked oligosaccharides results in cell wall degradation in endosperm during seed germination [77]. Hence, *βG* appears to contribute to both the degradation of oligosaccharides generated in cell wall turnover and release of monolignols from their glycosides for the stabilization of secondary cell wall by lignification [78].

Generally, CATs are considered as a sink for H_2O_2 [25,79]. In the present study, the tissue and developmental stage specific *CAT* expression suggested an important role for CAT in enzymatic ROS protection in Scots pine seeds. *CAT* showed intense expression in megagametophyte cells at the early embryogeny simultaneously with the accumulation of starch and storage proteins. Interestingly, the *CAT* expression signal faded out in the megagametophyte when the embryogenesis proceeded to the late stage but increased again during the imbibition phase of the seed germination. The transient decrease in the *CAT* expression co-occurred with slight DNA fragmentation in the nuclei of megagametophyte cells – that is to say, an incipient sign of cell death – detected in our previous study [35]. Unlike in the megagametophyte, *CAT* expression was intense in the cells of developing embryos throughout the embryogenesis and also in embryonic cells of mature seeds. Previously, CAT protein has been purified from the megagametophyte of loblolly pine (*Pinus taeda* L.) seeds [80], and the *CAT* gene has also been found to be active in *in vitro* cultured Scots pine embryogenic cells [31]. It has been suggested that in white spruce (*Picea glauca* (Moench.) Voss) the interplay of CAT and the ubiquitin mediated proteolytic system regulates ROS production and subsequent cell death in the megagametophyte during germination [81].

The H_2O_2 localization overlapped with the *ATG5* and *CAT* localizations at early embryogeny in the nucellar layers which undergo PCD later during Scots pine seed development [35,39]. Together the *CAT* expression and the presence of peroxidases in several seed structures indicate active H_2O_2 scavenging, and moreover, strict regulation of H_2O_2 levels during the embryogenesis. Possibly, the H_2O_2 detected in the nucellar layers exceeded the detoxifying capacity of ROS detoxifying enzymes, including CAT, thus triggering PCD, the mechanism also proposed by He and Kermode [81] for ROS mediated death of white spruce megagametophyte. Additionally, H_2O_2 can be involved in growth and developmental processes due to its role as an important signalling molecule [82]. Previously, high CAT activity has been detected in durum wheat (*Triticum durum* Desf.) kernels before the start of maturation drying [83], and CAT has also found to affect seed desiccation of sunflower (*Helianthus annuus* L.) by preventing dehydration-related oxidative damage [84]. Here, the reduced *CAT* expression in the megagametophyte cells at the late embryogeny, however, suggests that in Scots pine seeds the CAT protection against H_2O_2 damage was more connected to active metabolism than to dehydration-related oxidative stress during the seed desiccation.

We found that *RBR* expression remain quite stable throughout the Scots pine embryogenesis as well as in both the embryos and megagametophytes of mature seeds. In Arabidopsis, the loss of RBR function is gametophyte-lethal because the mature unfertilized mutant megagametophyte fails to arrest mitosis and undergoes excessive nuclear proliferation in the embryo sac [85]. In the maize endosperm, the down-regulation of *RBR1* by RNAi enhanced cell death and stimulated both the mitotic and endoreduplication cell cycles. Furthermore, DNA content was increased 43% by the down regulation of RBR1 during endosperm development, whereas storage protein content and kernel weight were essentially not affected [15]. In endoreduplication, a cell replicates chromosomal DNA without an intervening mitosis resulting in a higher ploidy level [86]. *Pinus* sp. exhibits the uniform haploid megagametophyte compared to the diploid DNA content of the corresponding embryo [87-89] indicating that no endoreduplication occurs in the megagametophyte cells. In the present study, the minor *RBR* expression found in the ESR suggests that *RBR* down regulation precedes cell death in Scots pine megagametophyte cells. The dying cells of the nucellar layers, however, showed intense *RBR* expression at the late embryogeny. The opposite *RBR* expression goes together with the different morphological features and different magnitudes of cell corpse processing in the ESR of the megagametophyte and in the nucellar layers emphasizing the specific developmental tasks of the cell death processes. Unlike in the ESR, cell death in the nucellar layers may be connected to oxidative stress leading to cell cycle arrest and preventing the replication of damaged DNA by intense *RBR* expression. Further protein level studies are, however, needed to confirm the possible role of RBR in the cell death processes during the Scots pine seed development. The stable *RBR* expressions detected in the present study is concordant with the previous findings revealing that in mammalian cells the pRB protein exhibits only slight variation at the gene expression level but significant differences in the protein phosphorylation status [90].

Conclusions

Starch accumulation precedes necrotic-like PCD in the ESR cells of the Scots pine megagametophyte, a phenomenon detected previously in the endosperm and perisperm of angiosperm seeds. Furthermore, the changes in the *RBR*, *ATG5* and *βG* expressions specifically in the cells doomed to death via morphologically necrotic cell death in the ESR and ASR of the megagametophyte suggest the genetic regulation of developmental necrosis. The results suggest that the function of CAT in the megagametophyte could be connected to the ongoing metabolism, whereas the continuous *CAT* expression in the embryo underlines the importance of CAT in the preventing of the transfer of H_2O_2 damaged components, especially DNA, to the next generation.

Additional files

Additional file 1: PCR primers for *in situ* mRNA hybridization assays and Q-RT-PCR.

Additional file 2: The determination of gene expression using relative quantification.

Additional file 3: Detection of nuclear DNA fragmentation by TUNEL assay.

Additional file 4 ***In situ*** **mRNA hybridizations with sense probes.**

Additional file 5: Verification of DAB visualized peroxidase location with H_2O_2 supplemented TMB.

Additional file 6: ***RBR*** **localization with** ***in situ*** **mRNA hybridization in the transition from early to late developmental stage of the Scots pine seed development.**

Additional file 7: ***βG*** **expression in developing Scots pine seeds.**

Abbreviations
ATG: Autophagy; βG: β-glucosidase; CAT: Catalase; ESR: Embryo surrounding region; PCD: Programmed cell death; RBR: Retinoblastoma related protein; ROS: Reactive oxygen species.

Competing interests
The authors declare that they have no competing interests.

Authors' contributions
JV conceived and designed the study. JV, AJ and SS carried out the histochemical stainings and JK and JV the *in situ* mRNA hybridizations. JV and SS conducted the Q-RT-PCR runs. JV and SS analysed the results and drafted the manuscript with TS and HH. All authors read and approved the final manuscript.

Acknowledgements
We thank our colleagues Dr. Anneli Kauppi from the University of Oulu and Dr. Eila Tillman-Sutela from Luke for their help and interest during the research. We are also grateful to Ms Eeva Pihlajaviita, Luke , Parkano Unit, for her skillful technical assistance, and to the personnel of Luke at the Punkaharju Unit for conducting the collections of the research material. The research was funded by the Academy of Finland (Project 121994 to TS), Thule Institute at University of Oulu (to HH) and by grants from the Taina Kuusi Foundation, the Finnish Cultural Foundation and the Niemi Foundation (to JV). The research was conducted as a part of the Bioeconomy Research Community of the University of Oulu.

Author details
[1]Genetics and Physiology Unit, University of Oulu, P.O. Box 3000, FI-90014 Oulu, Finland. [2]Natural Resources Institute Finland (Luke), Parkano Unit, Kaironiementie 15, FI-39700 Parkano, Finland. [3]Current address: Natural Resources Institute Finland (Luke), Rovaniemi Unit, FI-96301 Rovaniemi, Finland.

References

1. Greenberg JT. Programmed cell death: a way of life for plants. Proc Natl Acad Sci U S A. 1996;93(22):12094–7.
2. Jones AM. Programmed cell death in development and defense. Plant Physiol. 2001;125(1):94–7.
3. Lam E. Programmed cell death in plants: orchestrating an intrinsic suicide program within walls. Crit Rev Plant Sci. 2008;27(6):413–23.
4. Coll NS, Epple P, Dangl JL. Programmed cell death in the plant immune system. Cell Death Differ. 2011;18(8):1247–56.
5. Kroemer G, Galluzzi L, Vandenabeele P, Abrams J, Alnemri ES, Baehrecke EH, et al. Classification of cell death: recommendations of the nomenclature committee on cell death. Cell Death Differ. 2009;16(1):3–11.
6. van Doorn WG, Beers EP, Dangl JL, Franklin-Tong VE, Gallois P, Hara-Nishimura I, et al. Morphological classification of plant cell deaths. Cell Death Differ. 2011;18(8):1241–6.
7. van Doorn WG. Classes of programmed cell death in plants, compared to those in animals. J Exp Bot. 2011;62(14):4749–61.
8. Pennell RI, Lamb C. Programmed cell death in plants. Plant Cell. 1997;9(7):1157–68.
9. Zong WX, Thompson CB. Necrotic death as a cell fate. Genes Dev. 2006;20(1):1–15.
10. Berghe TV, Linkermann A, Jouan-Lanhouet S, Walczak H, Vandenabeele P. Regulated necrosis: the expanding network of non-apoptotic cell death pathways. Nat Rev Mol Cell Biol. 2014;15(2):135–47.
11. Golstein P, Kroemer G. Cell death by necrosis: towards a molecular definition. Trends Biochem Sci. 2007;32(1):37–43.
12. He X, Kermode AR. Proteases associated with programmed cell death of megagametophyte cells after germination of white spruce (*Picea glauca*) seeds. Plant Mol Biol. 2003;52(4):729–44.
13. Tsiatsiani L, Van Breusegem F, Gallois P, Zavialov A, Lam E, Bozhkov PV. Metacaspases. Cell Death Differ. 2011;18(8):1279–88.
14. Hackenberg T, Juul T, Auzina A, Gwizdz S, Malolepszy A, Van Der Kelen K, et al. Catalase and NO CATALASE ACTIVITY1 promote autophagy-dependent cell death in Arabidopsis. Plant Cell. 2013;25(11):4616–26.
15. Sabelli PA, Liu Y, Dante RA, Lizarraga LE, Nguyen HN, Brown SW, et al. Control of cell proliferation, endoreduplication, cell size, and cell death by the retinoblastoma-related pathway in maize endosperm. Proc Natl Acad Sci U S A. 2013;110(19):E1827–36.
16. Dat JF, Pellinen R, Beeckman T, Van De Cotte B, Langebartels C, Kangasjärvi J, et al. Changes in hydrogen peroxide homeostasis trigger an active cell death process in tobacco. Plant J. 2003;33(4):621–32.
17. Van Breusegem F, Dat JF. Reactive oxygen species in plant cell death. Plant Physiol. 2006;141(2):384–90.
18. de Pinto MC, Locato V, de Gara L. Redox regulation in plant programmed cell death. Plant Cell Environ. 2012;35(2):234–44.
19. Møller IM, Jensen PE, Hansson A. Oxidative modifications to cellular components in plants. Annu Rev Plant Biol. 2007;58:459–81.
20. Chamnongpol S, Willekens H, Moeder W, Langebartels C, Sandermann Jr H, Van Montagu M, et al. Defense activation and enhanced pathogen tolerance induced by H_2O_2 in transgenic tobacco. Proc Natl Acad Sci U S A. 1998;95(10):5818–23.
21. Rizhsky L, Hallak-Herr E, Van Breusegem F, Rachmilevitch S, Barr JE, Rodermel S, et al. Double antisense plants lacking ascorbate peroxidase and catalase are less sensitive to oxidative stress than single antisense plants lacking ascorbate peroxidase or catalase. Plant J. 2002;32(3):329–42.
22. Vandenabeele S, Van Der Kelen K, Dat J, Gadjev I, Boonefaes T, Morsa S, et al. A comprehensive analysis of hydrogen peroxide-induced gene expression in tobacco. Proc Natl Acad Sci U S A. 2003;100(26):16113–8.
23. Vanderauwera S, Zimmermann P, Rombauts S, Vandenabeele S, Langebartels C, Gruissem W, et al. Genome-wide analysis of hydrogen peroxide-regulated gene expression in Arabidopsis reveals a high light-induced transcriptional cluster involved in anthocyanin biosynthesis. Plant Physiol. 2005;139(2):806–21.
24. Apel K, Hirt H. Reactive oxygen species: metabolism, oxidative stress, and signal transduction. Annu Rev Plant Biol. 2004;55:373–99.
25. Mhamdi A, Queval G, Chaouch S, Vanderauwera S, Van Breusegem F, Noctor G. Catalase function in plants: a focus on Arabidopsis mutants as stress-mimic models. J Exp Bot. 2010;61(15):4197–220.
26. Michelet L, Roach T, Fischer BB, Bedhomme M, Lemaire SD, Krieger-Liszkay A. Down-regulation of catalase activity allows transient accumulation of a hydrogen peroxide signal in *Chlamydomonas reinhardtii*. Plant Cell Environ. 2013;36(6):1204–13.
27. Gutzat R, Borghi L, Gruissem W. Emerging roles of RETINOBLASTOMA-RELATED proteins in evolution and plant development. Trends Plant Sci. 2012;17(3):139–48.
28. Friend SH, Bernards R, Rogelj S, Weinberg RA, Rapaport JM, Albert DM, et al. A human DNA segment with properties of the gene that predisposes to retinoblastoma and osteosarcoma. Nature. 1986;323(6089):643–6.
29. Morris EJ, Dyson NJ. Retinoblastoma protein partners. In Advances in Cancer Research. Edited by Anonymous 2001:1–54. [Anonymous, vol 82].
30. Sabelli PA, Larkins BA. Regulation and function of retinoblastoma-related plant genes. Plant Sci. 2009;177(6):540–8.
31. Vuosku J, Suorsa M, Ruottinen M, Sutela S, Muilu-Mäkelä R, Julkunen-Tiitto R, et al. Polyamine metabolism during exponential growth transition in Scots pine embryogenic cell culture. Tree Physiol. 2012;32:1274–87.
32. Fath A, Bethke P, Lonsdale J, Meza-Romero R, Jones R. Programmed cell death in cereal aleurone. Plant Mol Biol. 2000;44(3):255–66.
33. Young TE, Gallie DR. Programmed cell death during endosperm development. Plant Mol Biol. 2000;44(3):283–301.
34. Filonova LH, von Arnold S, Daniel G, Bozhkov PV. Programmed cell death eliminates all but one embryo in a polyembryonic plant seed. Cell Death Differ. 2002;9(10):1057–62.
35. Vuosku J, Sarjala T, Jokela A, Sutela S, Sääskilahti M, Suorsa M, et al. One tissue, two fates: different roles of megagametophyte cells during Scots pine embryogenesis. J Exp Bot. 2009;60(4):1375–86.

36. Buchholz JT. Origin of cleavage polyembryony in conifers. Bot Gaz. 1926;81:55–71.
37. Sarvas R. Investigations on the flowering and seed crop of *Pinus silvestris*. Comm Inst Forest Fenn. 1962;53:1–198.
38. Costa LM, Gutierrez-Marcos JF, Dickinson HG. More than a yolk: the short life and complex times of the plant endosperm. Trends Plant Sci. 2004;9(10):507–14.
39. Hiratsuka R, Yamada Y, Terasaka O. Programmed cell death of *Pinus* nucellus in response to pollen tube penetration. J Plant Res. 2002;115(1118):141–8.
40. Tillman-Sutela E, Kauppi A. The significance of structure for imbibition in seeds of the Norway spruce, *Picea abies* (L.) Karst. Trees-Struct Funct. 1995;9(5):269–78.
41. Tillman-Sutela E, Kauppi A. The morphological background to imbibition in seeds of *Pinus sylvestris* L. of different provenances. Trees-Struct Funct. 1995;9(3):123–33.
42. Bozhkov PV, Filonova LH, Suarez MF, Helmersson A, Smertenko AP, Zhivotovsky B, et al. VEIDase is a principal caspase-like activity involved in plant programmed cell death and essential for embryonic pattern formation. Cell Death Differ. 2004;11(2):175–82.
43. Suarez MF, Filonova LH, Smertenko A, Savenkov EI, Clapham DH, Von Arnold S, et al. Metacaspase-dependent programmed cell death is essential for plant embryogenesis. Curr Biol. 2004;14(9):R339–40.
44. Minina EA, Filonova LH, Fukada K, Savenkov EI, Gogvadze V, Clapham D, et al. Autophagy and metacaspase determine the mode of cell death in plants. J Cell Biol. 2013;203:917–27.
45. Barceló AR. Hydrogen peroxide production is a general property of the lignifying xylem from vascular plants. Ann Botany. 1998;82(1):97–103.
46. Singh H. Embryology of gymnosperms. Berlin: Borntrager; 1978.
47. Fisher DB. Protein staining of ribboned epon sections for light microscopy. Histochemie. 1968;16:92–6.
48. Jensen WA. Botanical histochemistry. San Francisco: W.H. Freeman; 1962.
49. Young ID, Ailles L, Deugau K, Kisilevsky R. Transcription of cRNA for *in situ* hybridization from polymerase chain reaction-amplified DNA. Lab Invest. 1991;64(5):709–12.
50. Vuosku J, Sutela S, Sääskilahti M, Kestilä J, Jokela A, Sarjala T, et al. Dealing with the problem of non-specific *in situ* mRNA hybridization signals associated with plant tissues undergoing programmed cell death. BMC Plant Methods. 2010;6:7.
51. Gonçalves S, Cairney J, Maroco J, Oliveira MM, Miguel C. Evaluation of control transcripts in real-time RT-PCR expression analysis during maritime pine embryogenesis. Planta. 2005;222(3):556–63.
52. Fox J. The R commander: a basic-statistics graphical user interface to R. J Stat Softw. 2005;14(9).
53. Ihaka R, Gentleman R. R: a language for data analysis and graphics. J Comput Graph Stat. 1996;5(3):299–314.
54. Willekens H, Inzé D, Vanmontagu M, van Camp W. Catalases in plants. Mol Breed. 1995;1(3):207–28.
55. Frugoli JA, Zhong HH, Nuccio ML, McCourt P, McPeek MA, Thomas TL, et al. Catalase is encoded by a multigene family in *Arabidopsis thaliana* (L) Heynh. Plant Physiol. 1996;112(1):327–36.
56. Guan LQ, Scandalios JG. Molecular evolution of maize catalases and their relationship to other eukaryotic and prokaryotic catalases. J Mol Evol. 1996;42(5):570–9.
57. Esaka M, Yamada N, Kitabayashi M, Setoguchi Y, Tsugeki R, Kondo M, et al. cDNA cloning and differential gene expression of three catalases in pumpkin. Plant Mol Biol. 1997;33(1):141–55.
58. Iwamoto M, Higo H, Higo K. Differential diurnal expression of rice catalase genes: the 5'-flanking region of *CatA* is not sufficient for circadian control. Plant Sci. 2000;151(1):39–46.
59. Sabelli PA, Larkins BA. Grasses like mammals? Redundancy and compensatory regulation within the retinoblastoma protein family. Cell Cycle. 2006;5(4):352–5.
60. Berjak P, Pammenter NM. Orthodox and recalcitrant seeds. In Vozzo JA, editor. Tropical tree seed manual. Washington DC, USA: USDA Forest Service; 2003:137–47.
61. Kapik RH, Dinus RJ, Dean JFD. Abscisic acid and zygotic embryogenesis in *Pinus taeda*. Tree Physiol. 1995;15(7–8):485–90.
62. Carrier DJ, Kendall EJ, Bock CA, Cunningham JE, Dunstan DI. Water content, lipid deposition, and (+)-abscisic acid content in developing white spruce seeds. J Exp Bot. 1999;50(337):1359–64.
63. Silveira V, Balbuena TS, Santa-Catarina C, Floh EIS, Guerra MP, Handro W. Biochemical changes during seed development in *Pinus taeda* L. Plant Growth Regul. 2004;44:147–56.
64. Vuosku J, Jokela A, Läärä E, Sääskilahti M, Muilu R, Sutela S, et al. Consistency of polyamine profiles and expression of arginine decarboxylase in mitosis during zygotic embryogenesis of Scots pine. Plant Physiol. 2006;142(3):1027–38.
65. Bailly C, El-Maarouf-Bouteau H, Corbineau F. From intracellular signaling networks to cell death: the dual role of reactive oxygen species in seed physiology. Comptes Rendus - Biol. 2008;331(10):806–14.
66. Inzé D, De Veylder L. Cell cycle regulation in plant development. Annu Rev Genet. 2006;40:77–105.
67. Reichheld J, Vernoux T, Lardon F, Van Montagu M, Inzé D. Specific checkpoints regulate plant cell cycle progression in response to oxidative stress. Plant J. 1999;17(6):647–56.
68. Vuosku J, Sutela S, Tillman-Sutela E, Kauppi A, Jokela A, Sarjala T, et al. Pine embryogenesis - many licences to kill for a new life. Plant Sign Behav. 2009;4(10):928–32.
69. Young TE, Juvik JA, DeMason DA. Changes in carbohydrate composition and a-amylase expression during germination and seedling growth of starch-deficient endosperm mutants of maize. Plant Sci. 1997;129(2):175–89.
70. Young TE, Gallie DR. Analysis of programmed cell death in wheat endosperm reveals differences in endosperm development between cereals. Plant Mol Biol. 1999;39(5):915–26.
71. Li CY, Li WH, Li C, Gaudet DA, Laroche A, Cao LP, et al. Starch synthesis and programmed cell death during endosperm development in triticale (×*Triticosecale* Wittmack). J Integr Plant Biol. 2010;52(7):602–15.
72. Xu SB, Yu HT, Yan LF, Wang T. Integrated proteomic and cytological study of rice endosperms at the storage phase. J Proteome Res. 2010;9(10):4906–18.
73. López-Fernández MP, Maldonado S. Ricinosomes provide an early indicator of suspensor and endosperm cells destined to die during late seed development in quinoa (*Chenopodium quinoa*). Ann Bot. 2013;112(7):1253–62.
74. Tosi P, Gritsch CS, He J, Shewry PR. Distribution of gluten proteins in bread wheat (*Triticum aestivum*) grain. Ann Bot. 2011;108(1):23–35.
75. Lee EJ, Matsumura Y, Soga K, Hoson T, Koizumi N. Glycosyl hydrolases of cell wall are induced by sugar starvation in Arabidopsis. Plant Cell Physiol. 2007;48(3):405–13.
76. Mohapatra PK, Patro L, Raval MK, Ramaswamy NK, Biswal UC, Biswal B. Senescence-induced loss in photosynthesis enhances cell wall ß-glucosidase activity. Physiol Plantarum. 2010;138(3):346–55.
77. Leah R, Kigel J, Svendsen I, Mundy J. Biochemical and molecular characterization of a barley seed ß- glucosidase. J Biol Chem. 1995;270(26):15789–97.
78. Cairns JRK, Esen A. ß-Glucosidases. Cell Mol Life Sci. 2010;67(20):3389–405.
79. Mhamdi A, Noctor G, Baker A. Plant catalases: peroxisomal redox guardians. Arch Biochem Biophys. 2012;525(2):181–94.
80. Mullen RT, Gifford DJ. Purification and characterization of catalase from loblolly pine (*Pinus taeda* L.) megagametophytes. Plant Physiol. 1993;103(2):477–83.
81. He X, Kermode AR. Programmed cell death of the megagametophyte during post-germinative growth of white spruce (*Picea glauca*) seeds is regulated by reactive oxygen species and the ubiquitin-mediated proteolytic system. Plant Cell Physiol. 2010;51(10):1707–20.
82. Petrov VD, Van Breusegem F. Hydrogen peroxide-a central hub for information flow in plant cells. AoB Plants. 2012;pls014.
83. De Gara L, De Pinto MC, Moliterni VMC, D'Egidio MG. Redox regulation and storage processes during maturation in kernels of *Triticum durum*. J Exp Bot. 2003;54(381):249–58.
84. Bailly C, Leymarie J, Lehner A, Rousseau S, Côme D, Corbineau F. Catalase activity and expression in developing sunflower seeds as related to drying. J Exp Bot. 2004;55(396):475–83.
85. Ebol C, Mariconti L, Gruissem W. Plant retinoblastoma homologues control nuclear proliferation in the female gametophyte. Nature. 2004;429(6993):776–80.
86. Lee HO, Davidson JM, Duronio RJ. Endoreplication: polyploidy with purpose. Genes Dev. 2009;23(21):2461–77.
87. Bino RJ, Lanteri S, Verhoeven HA, Kraak HL. Flow cytometric determination of nuclear replication stage in seed tissues. Ann Bot. 1993;72(2):181–7.
88. Wakamiya I, Newton RJ, Johnston JS, Price HJ. Genome size and environmental factors in the genus *Pinus*. Am J Bot. 1993;80(11):1235–41.
89. Pichot C, El Maataoui M. Flow cytometric evidence for multiple ploidy levels in the endosperm of some gymnosperm species. Theor Appl Genet. 1997;94(6–7):865–70.
90. Miskolczi P, Lendvai Á, Horváth GV, Pettkó-Szandtner A, Dudits D. Conserved functions of retinoblastoma proteins: from purple retina to green plant cells. Plant Sci. 2007;172(4):671–83.

New insights into the genetic networks affecting seed fatty acid concentrations in *Brassica napus*

Xiaodong Wang[1,4†], Yan Long[2,3†], Yongtai Yin[1†], Chunyu Zhang[2], Lu Gan[1], Liezhao Liu[5], Longjiang Yu[1], Jinling Meng[2] and Maoteng Li[1*]

Abstract

Background: Rapeseed (*B. napus*, AACC, 2n = 38) is one of the most important oil seed crops in the world, it is also one of the most common oil for production of biodiesel. Its oil is a mixture of various fatty acids and dissection of the genetic network for fatty acids biosynthesis is of great importance for improving seed quality.

Results: The genetic basis of fatty acid biosynthesis in *B. napus* was investigated *via* quantitative trail locus (QTL) analysis using a doubled haploid (DH) population with 202 lines. A total of 72 individual QTLs and a large number pairs of epistatic interactions associated with the content of 10 different fatty acids were detected. A total of 234 homologous genes of *Arabidopsis thaliana* that are involved in fatty acid metabolism were found within the confidence intervals (CIs) of 47 QTLs. Among them, 47 and 15 genes homologous to those of *B. rapa* and *B. oleracea* were detected, respectively. After the QTL mapping, the epistatic and the candidate gene interaction analysis, a potential regulatory pathway controlling fatty acid biosynthesis in *B. napus* was constructed, including 50 enzymes encoded genes and five regulatory factors (*LEC1*, *LEC2*, *FUS3*, *WRI1* and *ABI3*). Subsequently, the interaction between these five regulatory factors and the genes involved in fatty acid metabolism were analyzed.

Conclusions: In this study, a potential regulatory pathway controlling the fatty acid was constructed by QTL analysis and *in silico* mapping analysis. These results enriched our knowledge of QTLs for fatty acids metabolism and provided a new clue for genetic engineering fatty acids composition in *B. napus*.

Keywords: *Brassica napus*, Fatty acid composition, QTL, Epistatic interaction, Regulatory pathway

Background

Oilseed rape (*Brassica napus* L., AACC, 2n = 38) is one of the most important oil crops producing multi-purpose oil for food and biofuel in many parts of the world. In 2007, biodiesel production accounted for 7% of the global vegetable oil supplies, in which 68% were used for biofuels in the EU [1]. As the global requirements for rapeseed oil are growing rapidly, increasing the oil content and improving the oil composition are important ways to meet the demands of agricultural feed stocks.

The fatty acid composition of rapeseed oil is considered to be genetically more variable than any other major vegetable oils [2]. Rapeseed oil is a mixture of seven main fatty acids [3]. Fatty acid biosynthetic pathways are generally controlled by multiple genes and considered as quantitative traits regulated by QTLs. So far, a number of QTLs controlling oil composition were identified in *B. napus*. Ecke *et al.* identified two QTLs for erucic acid distributed on chromosomes A6 and C2 [4]. Four QTLs for erucic acid distributed across chromosomes A1, A2, A8 and C3 were reported and three of these coincided with QTLs for the accumulation of oil content [5]. Burns *et al.* observed 13 QTLs affecting composition of 10 fatty acids, and seven also affected oil content [6]. Hu *et al.* identified two QTLs for oleic (on A1 and A5) and linolenic acids (on A4 and C4), respectively [7]. One to eight QTLs were detected for seven individual fatty acids by Zhao *et al.*, and eight of these also affected oil content [8]. Recently, Smooker *et al.* identified 34 QTLs for five major fatty acids [9], and Yan *et al.* detected a total of 40 QTLs for six fatty acids, which were most clustered on chromosomes A8, A9 and C3 [10].

* Correspondence: limaoteng426@mail.hust.edu.cn
†Equal contributors
[1]College of Life Science and Technology, Huazhong University of Science and Technology, Wuhan 430074, China
Full list of author information is available at the end of the article

The allotetraploid *B. napus* has two progenitor species, *B. rapa* and *B. oleracea*, which shared their last common ancestor with *A. thaliana* about 20 million years ago [11,12]. Both a high degree of sequence similarities and chromosomal colinearities between *Brassica* species and *Arabidopsis* were reported [13-15]. Parkin *et al.* reported 21 conserved blocks within the *Arabidopsis* genome shared with *B. napus* [16], and Schranz *et al.* proposed a set of 24 conserved chromosomal blocks in *B. napus* [17]. Furthermore, all the genome sequence of *B. rapa*, *B. oleracea* and *B. napus* have been released [18-20]. It is feasible to predict the *Arabidopsis* orthologous genes for specific agronomic traits within the *Brassica* genome. For example, a number of candidate genes were mapped to CIs of QTLs for flowering time by *in silico* mapping [21], and a total of 14 lipid-related candidate gene loci were located in the CIs of six QTLs for seed oil content [22]. In fact, many important genes involved in fatty acid metabolism were identified in *Arabidopsis*, such as *FAB2, FAD2, FAD3* and *FAE1* [23-26], and the orthologs of these genes in *B. napus* were also reported and mapped. *BnaFAD2* was mapped on A1, A5, C1 and C5 chromosomes [27-29], and one major QTL *BnaA.FAD2.a* located on A5 was responsible for high C18:1 [27]. *BnaFAD3* was mapped on A3, A4, A5, C3 and C4 [9,27], and two major QTLs *BnaA.FAD3.b* and *BnaC.FAD3.b* were both responsible for low C18:3 [27]. *BnaFAE1* was mapped on both A8 and C3 [9], and two *FAE1* homologous genes on A8 and C3 linkage groups were also found by Qiu *et al.* [5] and Fourmann *et al.* [30]. Collectively, although genes or QTLs for fatty acid biosynthesis have been identified, the genetic network for all these metabolic pathways in *B. napus* needs to be elucidated.

In *Arabidopsis*, more than 120 enzymatic reactions and at least 600 genes are involved in acyl-lipid metabolism [31]. Li-Beisson *et al.* gave metabolic pathways associated with the biosynthesis and degradation of acyl-lipids in *Arabidopsis* [31]. The genome of polyploid *B. napus* may typically contain six distinct alleles for each gene present in *Arabidopsis* [32], the fatty acid biosynthesis and the gene regulation in *B. napus* might have a more complex pathway than that in *Arabidopsis*. Though much attention was given to genes and regulatory factors involved in acyl-lipid metabolism in *Arabidopsis* [31,33-37], similar questions concerning the genetic basis of fatty acid biosynthesis in *B. napus* remain open, mainly due to the lack of integrative studies at a population scale. Moreover, the interaction of genes involved in acyl-lipid metabolism has not yet been studied based on co-location of mapped candidate genes with QTLs in *B. napus*. To determine these key steps in relevant complex metabolic pathways of acyl-lipids in *B. napus*, it is first necessary to identify QTLs or genes associated with fatty acids composition.

In this paper, we describe the genetic bases of seed fatty acid composition through QTL mapping in *B. napus*. The aims of this study were as follows: (1) to add knowledge concerning QTL mapping of the fatty acid composition in *B. napus*; (2) to predict candidate genes of major QTLs for different fatty acids' biosynthesis by comparative genome analysis; and (3) to construct a regulatory pathway for fatty acids metabolism in *B. napus*.

Results

Variation and single QTL analysis of fatty acid composition in the 'Tapidor' × 'Ningyou7' cross (TN) DH population

Means of all traits measured from the TN DH population over six environments were close to the mid-parent values (Table 1). There was a wide range of variations and transgressive segregations for the concentration of each fatty acid (Figure 1). The population appeared to have a normal or near-normal distribution for C16:0, C18:0, C18:2, C18:3, C20:0, C22:0 and FAS (Saturated Fatty Acid), suggesting complexity of their genetic networks. However, C18:1, C20:1 and C22:1 showed bi-modal distributions, indicating that they might be controlled by few major genes with a relatively large effect. The distribution patterns of 10 fatty acids' compositions showed that they were genetically stable but also affected by environment. The correlation between different fatty acid compositions showed great differences (Table 2). Erucic acid (C22:1) content was highly and positively correlated with the level of C20:0, C20:1 and C22:0 (Coefficients 0.30–0.63), but was negatively correlated with other fatty acids (−0.92 to −0.03), especially C18:1 (−0.92) (Table 2). C18:1 showed a high positive correlation with C16:0 and C18:0 (0.70–0.75) and moderate positive correlation with C18:2 and C18:3 (Table 2), but showed a high negative correlation with other fatty acids (−0.92 to −0.33).

For QTL mapping analysis, Wincart_2.5 detected a total of 139 QTLs distributed across 15 chromosomes (except for C1, C2, C4 and C7) and individual QTL for any given trait explained 1.27–47.56% of phenotypic variance (PV) (Additional file 1). QTLNetwork_2.0 detected a total of 44 QTLs across 15 chromosomes (Additional file 2). After combining QTLs for different traits clustered in the same regions indicated by the same close-linked molecular markers, a total of 72 QTLs controlling fatty acid composition were identified (Figure 2, Table 3), and 10, 44 and 18 QTLs were detected by using QTLNetwork_2.0 only, Wincart_2.5 only and both QTLNetwork_2.0 and Wincart_2.5, respectively (Table 3).

Individual saturated fatty acids were only analyzed in three environments (07D, 07 W1 and 08 W2). For C16:0, 18 QTLs were detected across seven chromosomes (Table 3). Among these QTLs, seven (39%) were detected by two types of software. The additive effect ranged from −0.29 to 1.72, and explained 2.74–37.99%

Table 1 Means and ranges for seed fatty acids of TN DH population evaluated in six environments

Traits		16:0	C18:0	C18:1	C18:2	C18:3	C20:0	C20:1	C22:0	C22:1
Tapidor	Mean[a]	4.82 ± 0.03	1.89 ± 0.1	57.43 ± 2.46	17.5 ± 0.84	7.76 ± 0.52	0.67 ± 0.01	2.41 ± 1.24	0.35 ± 0.01	2.83 ± 0.91
Ningyou7	Mean	3.15 ± 0.01	1.11 ± 0.04	17.45 ± 2.85	12.92 ± 1.45	8.54 ± 0.48	0.85 ± 0.04	8.82 ± 0.08	0.73 ± 0.01	45.55 ± 0.27
DH	Mean	3.94 ± 0.53	1.63 ± 0.41	34.12 ± 15.63	15.44 ± 3.3	8.22 ± 0.97	0.77 ± 0.3	10.88 ± 5.62	0.4 ± 0.16	24.3 ± 15.56
	Max	5.87	4.92	73.54	27.24	12.02	6.88	19.66	1.86	52.22
	Min	2.84	0.89	9.4	8.61	1.31	0.14	1.08	0	0

[a]Mean value ± SE.

of PV (Additional files 1 and 2). For C18:0, 12 QTLs distributed across seven chromosomes had −0.24 to 0.10 additive effect and explained 3.05–37.73% of PV (Table 3, Additional files 1 and 2). Three of them (on A1, A4 and C3) were environment-specific QTLs detected only in one environment. A total of 13 QTLs controlling C20:0 explained 4.02–47.55% of PV (Table 3, Additional files 1 and 2). Nine QTLs on five chromosomes for C22:0 were detected, and three were environment-specific, their additive effect ranged from −0.06 to 0.11, and explained 4.02–47.56% of PV (Table 3, Additional files 1 and 2). In 09D, 09 W2 and 09 W3, all saturated fatty acid compositions were considered as one trait named FAS, and a total of 10 QTLs were detected with additive effect ranged from −0.28 to 0.24 and explaining 3.96–31.33% of PV (Table 3, Additional files 1 and 2).

For the other four unsaturated fatty acid components, phenotypic data were obtained from six different environments, except for C20:1 from three environments (07D, 07 W1 and 08 W2). Twelve QTLs for C18:1 were distributed across seven chromosomes and their additive effect ranged from −10.96 to 3.18 and explained 1.88–42.43% of PV (Table 3, Additional files 1 and 2). Nine QTLs on six chromosomes were associated with C18:2, with the additive effect ranging from −1.69 to 0.56 and explaining 3.10–36.26% of PV (Table 3, Additional files 1 and 2). Twenty-one QTLs on 10 chromosomes were significantly associated with C18:3, which had −1.64 to 0.42 additive effect and explained 5.70–22.29% of PV (Table 3, Additional files 1 and 2). For C20:1, 10 QTLs were detected across A1, A8, A9, C1, C3 and C8 in three environments (Table 3). Six of them were found in two or more environments, and the remaining four were only in one environment. Nine QTLs were observed for C22:1 distributed across A4, A7, A8, C3, C6 and C9, and singly explained 1.27–45.79% of PV (Table 3, Additional files 1 and 2).

Co-localization of mapped candidate genes of *B. rapa* and *B. oleracea* with single-locus QTL

A total of 932 molecular markers were mapped to the new version of the TN map. This map covered a total length of 2116.73 cM with an average marker interval of 2.27 cM. The length of the 19 linkage groups varied from 65.83 (C1) to 154.53 (C3) cM (Additional file 3). Thirty-four synteny blocks (28 for A genome, 6 for C genome) and 149 insertion fragment islands (95 for A genome, 54 for C genome) were identified between *Arabidopsis* pseudochromosomes and TN DH genetic linkage groups by the *in silico* mapping approach (Additional file 3). More than twenty-nine thousand homologous genes were found to underline the CIs of 61 QTLs associated with the concentrations of 10 different fatty acids, by comparative analysis between the linkage map of TN DH and the genome of *Arabidopsis* (Additional file 4). Among them, a total of 111 key genes involving fatty acid metabolism were used as candidate genes in the present study (Additional file 4). A number of important genes were found, such as *FAB2, FAD2, FAD3* and *FAE1*, but also five regulatory factors, *LEC1, LEC2, FUS3, WRI1* and *ABI3*. As the polyploid *B. napus* genome may typically contain six loci for each gene present in *Arabidopsis*, these 111 genes were found to have 824 homologous genes mapped on the TN DH linkage map in total, including 97 key genes of 234 homologs underlying the CIs of 47 QTLs (Additional file 4). All of the QTLs with CIs containing homologous genes were separately compared to the physical genomic regions of *B. rapa* (A genome) and *B. oleracea* (C genome). To compare with the *B. rapa* genome, 32 QTLs containing candidate gene(s) distributed on the A genome of the TN DH linkage map were used for analysis, accounting for 65.3% of the total QTLs on the A genome. In total, 47 genes in *B. rapa* matched those in *Arabidopsis* underlying 24 QTL CIs (Additional file 4). For the C genome of the TN DH linkage map, 23 QTLs were detected and 15 (65.2%) of them contained candidate gene(s) in *Arabidopsis*. Comparison of the candidate genes showed 15 genes in *B. oleracea* matched those in *Arabidopsis* underlying 7 QTL CIs (Additional file 4). For example, on the C3 linkage group, 33 candidate genes (40 homologous genes) of *Arabidopsis* were located in the CIs of eight QTLs, and 10 candidate genes of *B. oleracea* were found based on the candidate genes of *Arabidopsis*, including the regulatory factor *ABI3* of previous studies (Figure 3).

Epistatic QTLs and interaction analysis of candidate genes for fatty acid compositions

QTLNetwork_2.0 and Genotype Matrix Mapping ver2.1 (GMM) software were used to identify the epistatic

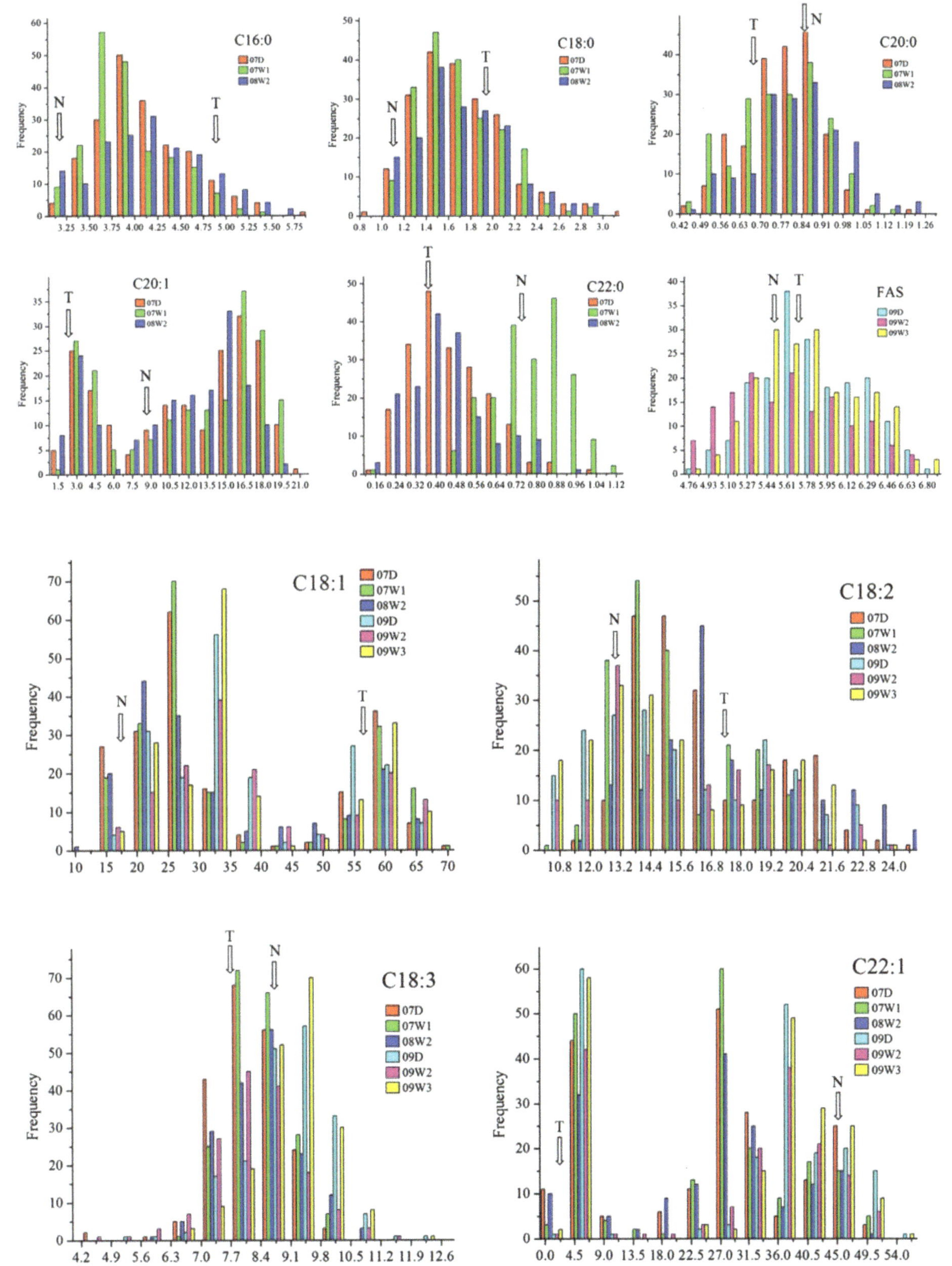

Figure 1 Distribution of fatty acid concentrations of TN DH population in multiple environments. The unit of x-axis means percentage of the specific fatty acid composition in the sum of all fatty acids. The unit of y-axis means the number of lines. N represents the parent "Ningyou7" and T the parent "Tapidor" of TN population.

Table 2 Pearson correlation coefficients for trait pairs affecting fatty acid compositions in the DH population

	C16:0	C18:0	C18:1	C18:2	C18:3	C20:0	C20:1	C22:0	C22:1
C16:0	1								
C18:0	0.62**	1							
C18:1	0.70**	0.75**	1						
C18:2	0.06	0.05	0.03	1					
C18:3	0.26**	0.06	0.03	−0.01	1				
C20:0	−0.23**	−0.12**	−0.33**	−0.03	−0.13**	1			
C20:1	−0.58**	−0.52**	−0.77**	−0.03	−0.12**	0.31**	1		
C22:0	−0.37**	−0.41**	−0.51**	−0.04	−0.16**	0.27**	0.11**	1	
C22:1	−0.77**	−0.78**	−0.92**	−0.04	−0.03	0.30**	0.63**	0.60**	1

**Significant at $P = 0.001$.

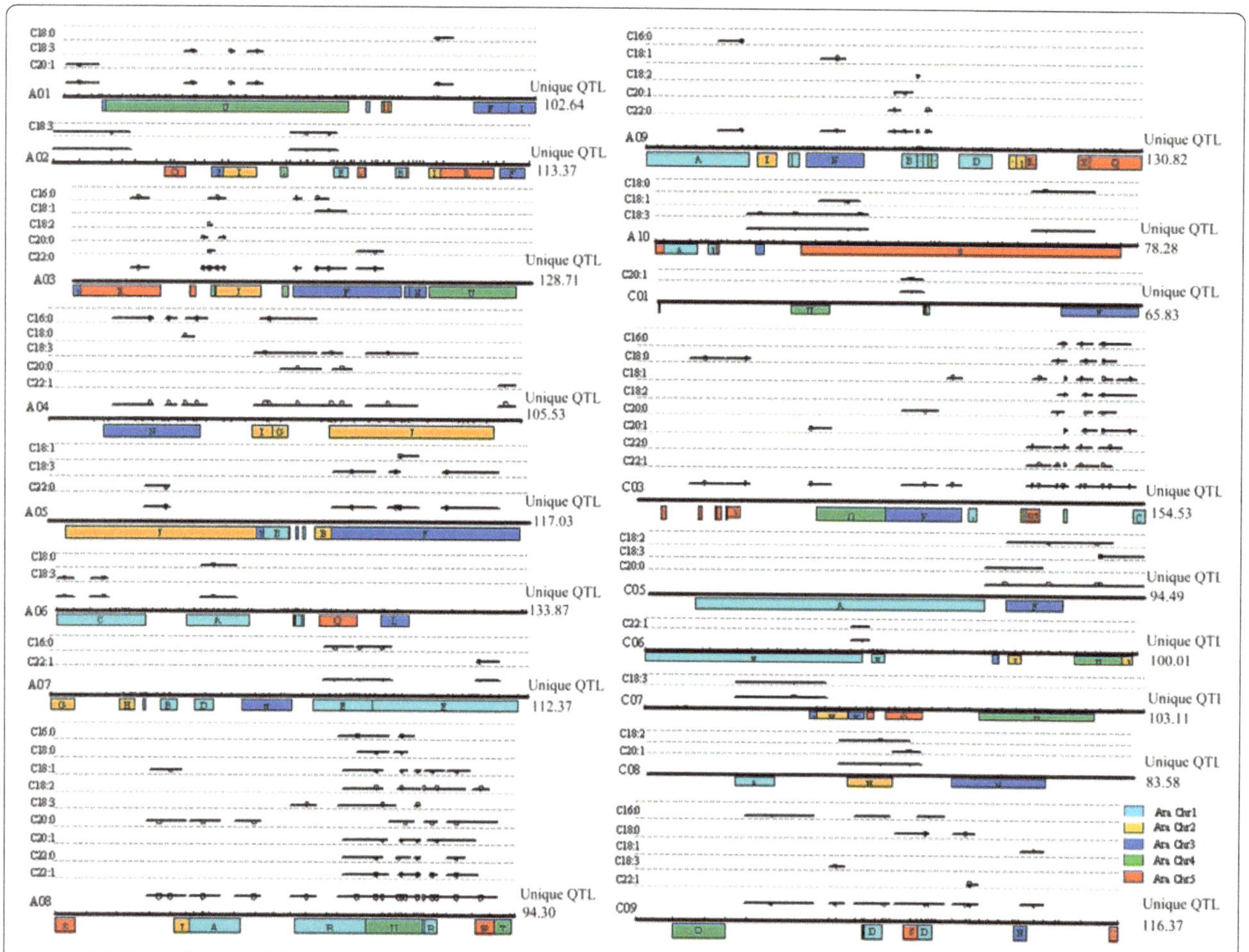

Figure 2 QTL distribution of fatty acid concentrations on linkage groups in *B. napus*. Whole linkage groups are shown with black lines labeled with molecular markers (short vertical bars) on the bottom, and the Arabic numerals listed on the right side show the length of linkage groups containing QTLs. The names of traits are listed on the left side of the linkage groups. The black lines on the linkage groups show the QTL confidence interval and the circles indicate the peak position. The pseudo-chromosomes of *Arabidopsis* are aligned under each linkage group of *B. napus*.

Table 3 The combined QTLs for fatty acid contents detected by WinQTLCart_2.5 and QTLNetwork_2.0

QTL	Chr.[a]	Marker interval	LOD	Additive	PV(%)	range	S[b]	Env.[c]	traits
qA1-1	**A1**	**znS13M26-100-CB10081**	**3.1**	**−1.28**	**4.87**	**0.3-7.2**	**W**	**07D**	**C20:1**
qA1-2	A1	CB10097-JICB0313	3.6	0.19	7.15	26.1-28.8	W	07 W1	C18:3
qA1-3	**A1**	**HBr006-ZAAS156a**	**7.6**	**0.29**	**14.64**	**35.9-37**	**W**	**07 W1**	**C18:3**
qA1-4	A1	JICB0455-znS08M15-320	4.8	0.24	11.42	39.8-43.1	W	07 W1	C18:3
qA1-5	**A1**	**em09me21-70-ZAAS165**	**4.0**	**0.07**	**3.43**	**80.4-84.6**	**W&Q**	**07 W1**	**C18:0**
qA2-1	**A2**	**CB10355-Ol10F04**	**3.1**	**0.23**	**6.64**	**0-18.2**	**W&Q**	**09 W3**	**C18:3**
qA2-2	**A2**	**HR-Sp1-210-pX154**	**3.3**	**0.31**	**7.85**	**56.2-61.5**	**W**	**09 W2**	**C18:3**
qA2-3	**A2**	**BRMS-082-HG-FT-A2**	**3.5**	**0.32**	**8.43**	**61.5-67.3**	**W**	**09 W2**	**C18:3**
qA3-1	A3	ZNS13M26-360-HBR160		0.05	2.02	15.7-21.4	Q		C16:0
qA3-2	**A3**	**RA2E11-BRMS-303**	**3.8**	**−0.03**	**4.02**	**35.9-37.7**	**W**	**07 W1**	**C20:0**
qA3-3	A3	CB10271-CNU250	3.3-4.0	−0.52-0.02	3.01-5.66	37.8-43.0	W&Q	07D/07 W1	C16:0/C18:2/C20:0/C22:0
qA3-4	A3	HBr124-KBrB043L02-12	3.1	−0.10	3.22	61.5-63.7	W	07D	C16:0
qA3-5	A3	HBr137-CNU270	3.0-3.5	−0.1-3.2	3.52-3.56	67.4-76.1	W&Q	07D/08 W2	C16:0/C18:1
qA3-6	A3	HR-S2-130-HR-Tp4-165	3.8	−0.03	4.78	79.0-86.5	W	07D	C22:0
qA4-1	**A4**	**OL11H02C-SR12307I**		**−1.02**	**0.39**	**100.1-104.1**	**Q**		**C22:1**
qA4-2	A4	znS13M26-220-sN3514f	4.4	0.18	8.86	14.0-23.2	W	08 W2	C16:0
qA4-3	A4	IGF3365B-SN13034		−0.09	5.43	25.7-28.5	Q		C16:0
qA4-4	**A4**	**BRMS-276-HR-C001-A4**	**4.6**	**0.06-0.17**	**1.04-8.02**	**29.5-35.2**	**W&Q**	**08 W2**	**C16:0/C18:0**
qA4-5	A4	niab048-JICB0134	3.4-5.4	−0.03-0.18	5.45-6.26	45.6-60.5	W	07D/07 W1	C16:0/C18:3/C20:0
qA4-6	A4	JICB0134-HBr091	5.22	−0.05_-0.04	0.37-5.44	60.5-67.2	W&Q	07 W1	C18:3/C20:0
qA4-7	A4	HBr202-HS-k02-2	3.2	0.24	6.98	70.4-82.3	W	07D	C18:3
qA5-1	**A5**	**BRMS-034-niab017**	**3.3**	**0.03**	**4.18**	**22.9-29.3**	**W**	**07D**	**C22:0**
qA5-2	A5	niab013-pX129a	3.8	0.09	4.40	57.9-59.9	W	09D	FAS
qA5-3	A5	CNU206-PIE1-6	3.2-3.7	0.09-0.26	4.38-8.17	65.3-79.0	W&Q	09 W3/09D	C18:3/FAS
qA5-4	A5	CNU268-HR-Tp4-200	3.1-3.3	−2.3-0.22	1.88-5.96	82.3-89.7	W	07D	C18:1/C18:3
qA5-5	A5	IGF3134a-CNU362	3.8	0.23	6.76	95.5-109.3	W	07D	C18:3
qA6-1	A6	Ol11F12a-ZAAS92a	4.8	0.29	10.56	0-4.6	W&Q	07D	C18:3
qA6-2	**A6**	**HBr201-BRMS-030**	**3.2**	**−0.07**	**3.05**	**40.7-51.1**	**W&Q**	**07D/07 W1**	**C18:0**
qA6-3	**A6**	**HR-Tp3-320-pW217**	**4.1**	**0.25**	**7.51**	**9.1-14.2**	**W**	**07D**	**C18:3**
qA7-1	**A7**	**IGF1226I-CNU053b**	**3**	**−1.70**	**1.27**	**100.2-105.5**	**W**	**07D**	**C22:1**
qA7-2	A7	HBr030-znS06M34-50	5.4	−0.10	4.36	64.0-70.7	W	07 W1	C16:0
qA7-3	**A7**	**CNU167-CNU044**	**4.4**	**−0.09**	**3.62**	**71.2-74.8**	**W**	**07 W1**	**C16:0**
qA7-4	**A7**	**BRMS-036-HBr021**	**4.5**	**−0.10**	**4.10**	**74.8-79.8**	**W**	**07 W1**	**C16:0**

Table 3 The combined QTLs for fatty acid contents detected by WinQTLCart_2.5 and QTLNetwork_2.0 *(Continued)*

qA8-1	A8	em10me26-120-JICB0335		−0.04-0.64	4.39-6.42	18.6-26.8	Q		C18:1/C20:0
qA8-2	A8	HBr107-HBr017	8.1	0.07	15.24	27.4-34.0	W	08 W2	C20:0
qA8-3	**A8**	**HBr104-HBr010**	**4.7**	**0.05**	**9.30**	**36.8-41.9**	**W**	**08 W2**	**C20:0**
qA8-4	A8	Na12B05a-HBr031	3.4	−0.22	6.16	48.0-53.3	W	07D	C18:3
qA8-5	**A8**	**IGF1108c-sR7178**	**3.1-59.9**	**−10.4-9.14**	**5.7-47.55**	**57.4-74.6**	**W&Q**	**07D/07 W1/08 W2/09D/ 09 W2/09 W3**	**C16:0/C18:0/C18:1/C18:2/C18:3/ C20:0/C20:1/C22:0/C22:1/FAS**
qA8-6	A8	HBr015-HBr026	3.9-33.2	−9.1-8.9	7.7-38.7	75.3-90.4	W	07D/07 W1/08 W2/09 W2	C18:1/C18:2/C20:0/C20:1/C22:0/ C22:1/FAS
qA9-1	A9	pW123aH-HBr178	3.6	0.08	2.74	18.3-25.3	W&Q	07 W1	C16:0
qA9-2	A9	CNU296-KBrB073D09-12	3.2	−3.20	4.64	45.5-52.0	W	09 W2	C18:1
qA9-3	**A9**	**HBr197-HBr205**	**4.5**	**0.04**	**5.59**	**63.5-66.9**	**W**	**07D**	**C20:1**
qA9-4	A9	HBr205-niab003	4.3	−1.50	6.47	66.9-70.0	W&Q	07 W1	C20:1
qA9-5	A9	IGF1087g-niab127	3.3	0.52	3.32	70.9-71.5	W	07D	C18:2
qA9-6	A9	em18me23-380-HBr196	3.9	0.03	4.97	73.2-75.1	W	07D	C22:0
qA9-7	A9	pW203b-HBr053	3.4	−0.08	3.96	86.1-95.5	W	09D	FAS
qA10-1	A10	RA2E03-HS-b14-1	3.5	0.24	6.29	14.6-18.9	W	07D	C18:3
qA10-2	A10	HS-j90-HG4-HG-CO-A10	3.3-4.6	−2.59-0.42	2.0-12.9	18.9-34.2	W&Q	07D/07 W1/09 W2	C18:0/C18:1/C18:3
qA10-3	A10	JICB0551-JICB0573		−0.05	0.28	60.8-71.3	Q		C18:0
qC1-1	**C1**	**EM18ME23-300-CB10258**		**0.58**	**3.85**	**33.0-35.9**	**Q**		**C20:1**
qC3-1	C3	IGF5376b-HBr014	3.4-15.1	−8.20-8.63	5.0-30.0	118.4-126.1	W	07D/09D	C18:1/C22:0/C22:1
qC3-2	C3	HBr014-Ol13C12	4.7-38.6	−9.83-8.86	6.3-37.7	126.1-131.5	W	07D/07 W1/08 W2/09D/ 09 W2/09 W3	C16:0/C18:0/C18:1/C18:2/C20:0/ C20:1/C22:0/C22:1/FAS
qC3-3	C3	IGF0235b-BRMS-093	3.8-64.9	−10.96-9.78	3.0-45.8	133.8-152.3	W&Q	07D/07 W1/08 W2/09D/ 09 W2/09 W3	C16:0/C18:0/C18:1/C18:2/C20:0/ C20:1/C22:0/C22:1/FAS
qC3-4	C3	pX141bE-HBr032	4.9	0.10	4.80	15.7-26.0	W&Q	07D	C18:0
qC3-5	C3	pW221-pX141aE	4.4	0.07	3.68	26.5-34.2	W	07D/07 W1	C18:0
qC3-6	C3	HBr211-HBr152	4.2	1.52	6.46	51.7-59.0	W	07 W1	C20:1
qC3-7	C3	Ol11G11b-pW143		−0.05	4.98	80.3-91.5	Q		C20:0
qC3-8	C3	HBr139-CNU099	4.7	−3.10	2.91	94.1-98.8	W	07D	C18:1
qC5-3	**C5**	**Ol10B02-JICB0509**	**3.2**	**0.10**	**5.01**	**30-62.6**	**W**	**09 W3**	**FAS**
qC5-1	C5	IGF3112a-em12me21-150	3.5-7.2	−0.06-0.55	3.7-13.9	63.8-82.9	W	07D/08 W2	C18:2/C20:0
qC5-2	**C5**	**em12me21-150-IGF0193C**	**4.2-4.6**	**0.28-0.56**	**3.9-9.8**	**82.9-88.3**	**W**	**07D/09 W3**	**C18:2/C18:3**
qC6-1	C6	HBr025-Na12E01a	5.5	0.24	11.4	74.3-80.8	W	09 W2	FAS
qC6-2	**C6**	**HBr057-HBr047**	**4.0**	**−0.21**	**8.64**	**91.0-95.4**	**W**	**09 W2**	**FAS**
qC6-3	**C6**	**SR12387-EM14ME28-200**		**−0.57**	**0.10**	**41.2-45.3**	**Q**		**C22:1**

Table 3 The combined QTLs for fatty acid contents detected by WinQTLCart_2.5 and QTLNetwork_2.0 *(Continued)*

qC7-1	C7	SNRH63-CNU400		−0.13	1.68	18.8-37.2	Q		C18:3
qC8-1	C8	CB10504-sN11670a	3.4	0.36-0.54	3.0-3.65	32.4-46.8	W&Q	07D/09 W3	C18:2/C20:1
qC9-1	**C9**	**CB10064-em20me27-230**	**6.1**	**−0.12**	**6.28**	**26.3-42.9**	**W**	**07 W1**	**C16:0**
qC9-2	**C9**	**EM20ME27-230-HR-TP3-360**		**0.13**	**3.83**	**46.2-50.3**	**Q**		**C18:3**
qC9-3	C9	BRMS-154-HBr144	5	−0.12	5.31	52.7-61.2	W&Q	07D/07 W1	C16:0
qC9-4	C9	HBr186-SA30	3.1-4.1	−0.14-0.08	3.97-5.4	62.5-74.2	W	07D/08 W2	C16:0/C18:0
qC9-5	**C9**	**S16071-3F3R-sNRG42**	**3.4**	**0.07-0.90**	**0.03-3.3**	**76.2-82.8**	**W&Q**	**07D**	**C18:0/C22:1**
qC9-6	C9	HR-Sp1-170-sR12384I	3.6	2.71	3.36	92.7-98.4	W	09 W3	C18:1

[a]Chromosome.
[b]The software used to detect QTL. W, WinQTLCart_2.5; Q, QTLNetwork_2.0.
[c]The environment in which the QTLs are detected.
The QTL with bold indicates that no candidate genes are located in the confidence interval of this QTL in the present study.

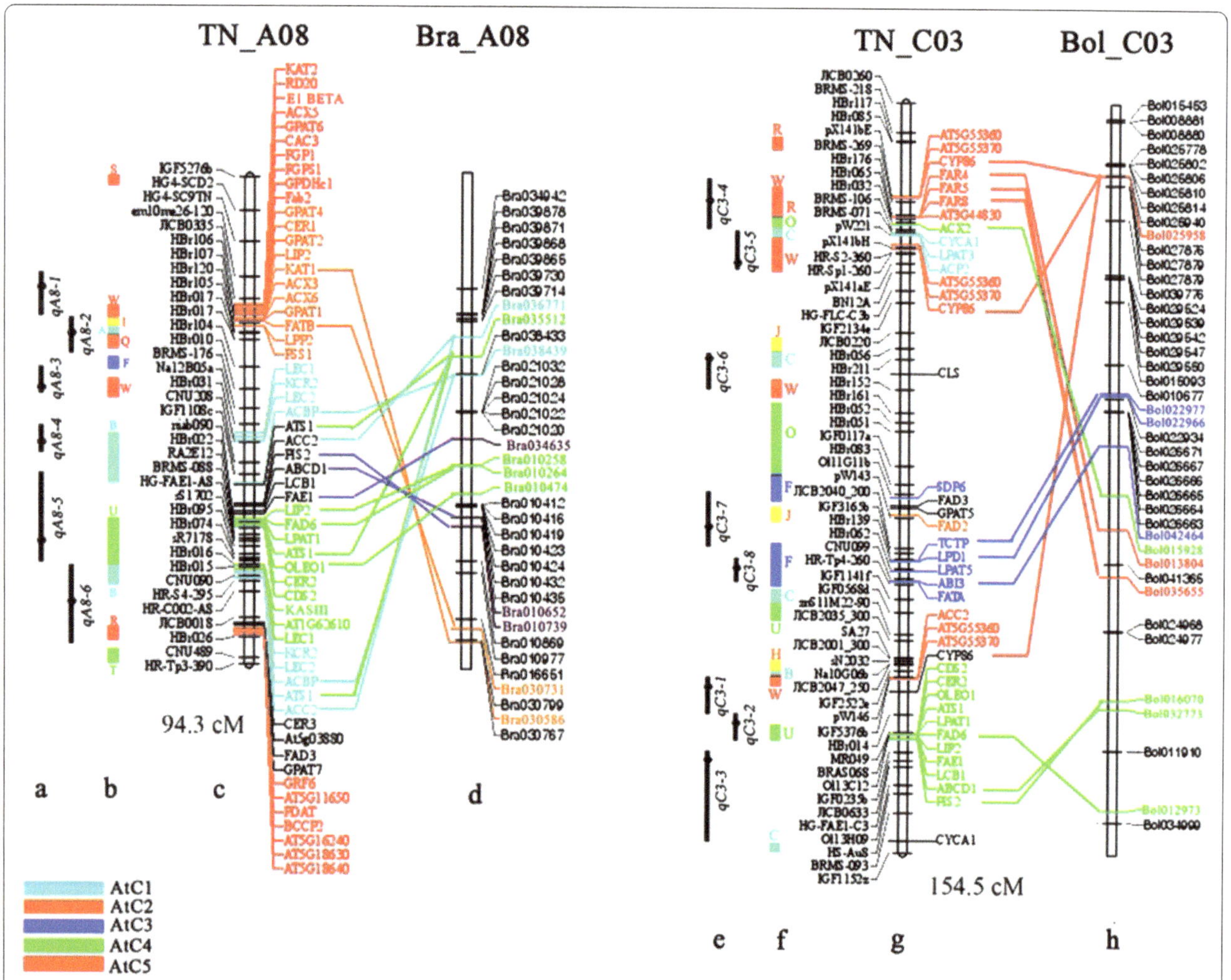

Figure 3 Comparative mapping of homologous linkage groups between *B. napus* and *B. rapa/B. oleracea*. (a, e) QTLs associated with 10 traits in multiple experiments, and the ring point in the bar indicates the QTL peak position. **(b, f)** The building blocks of *Brassica* genomes are identified on TN genetic maps. **(c, g)** On the left side are the markers on TN chromosome, and on the right side are candidate genes of *Arabidopsis* underlying the QTL confidence interval. **(d, h)** Genes with colored font are the candidate genes in *B. rapa/B. oleracea* of *Arabidopsis*, and genes with black font are the candidate genes in *B. rapa/B. oleracea* identified by each informative marker. The homologous genes in A/C genomes of *B. rapa/B. oleracea* and *B. napus* are connected with colored lines using the homologous genes as anchors. The genetic length of TN chromosome is annotated at the bottom of chromosome. The color-coding of the blocks is based on the report of Schranz *et al.* (2006) [66] and the candidate genes with the same color as the block.

interactions for fatty acid compositions. Twenty-four pairs of epistatic QTLs involving 28 loci were identified by QTLNetwork_2.0 for 10 measured traits (Additional files 5 and 6), with 1–5 epistatic QTL pairs for each trait. The proportion of total PV explained by all epistatic QTLs was 1.03–38.26% for each trait. Among 24 epistatic QTLs, two, three and 19 were NN, AN and AA interactions, respectively. A total of 395 loci interactions were identified by GMM, including 34 pairs of digenic and 361 pairs of trigenic interactions (Additional file 7). By comparing the epistatic interactions by GMM and single QTL based on common markers, 312 pairs of epistatic interactions were associated with QTLs. Some of the epistatic QTLs affected the level of more than one fatty acid composition. For example, the interactions of two loci associated with QTLs *qA8-5* and *qC3-3* were detected by both types of software, which controlled different fatty acid compositions (C16:0, C18:0, C18:1, C20:1, C22:0 and FAS by QTLNerwork_2.0 and nine traits except FAS by GMM; Additional files 6 and 7). This indicated that epistatic interactions were very important for fatty acid metabolism in *B. napus*.

For C16:0, four pairs of epistatic QTLs were detected by QTLNetwork_2.0, explaining 0.66–4.99% of PV. The interaction between QTL *qA8-5* and *qC9-3* explained 3.39% of PV (Additional file 6). The genes *OLEO1*, *ACC2*, *FAE1*, *LPAT1* and *ATS1* were underlying the QTL CI of *qA8-5*, while *LEC1*, *LEC2*, *ACC2*, *TAG1*, *KASIII* and *ATS1* were underlying the CI of *qC9-3* (Additional file 4). A total of 34 significant loci interactions were identified for C16:0 by GMM, containing one pair with digenic and 33 pairs with trigenic interactions, most of these trigenic interactions have loci underlying three QTLs of CIs, including QTLs *qA8-5*, *qA8-6* and *qC3-3* (Additional file 7). A number of homologous genes were mapped to the CI of *qA8-6*, including *LEC1*, *LEC2*, *ACC2*, *KASIII*, *FAD3*, *GPAT7*, *BCCP2*, *ACBP*, *PDAT*, and *ATS1*, but only one homologous gene *CYCA1* for *qC3-3* was found (Additional file 7). All the results above suggested that the potential of these gene interactions increased the level of C16:0. Similar to C16:0, a series of pairs of epistatic QTLs and loci interactions for another nine traits were identified by QTLNetwork_2.0 (Additional file 6) and GMM (Additional file 7), and numerous important genes underlying QTL CIs were identified (Additional file 4). For example, for C18:0, C18:1 and C18:2, besides the trigenic interaction of QTLs *qA8-5*, *qA8-6* and *qC3-3*, the interactions between QTLs *qC3-2* and *qA8-5* were also identified, and *FAE1* was underlying the CI of *qC3-2*.

Cytoscape_V2.6.3 software was used to investigate the interaction of candidate genes that were observed from the single and epistatic QTL results and the gene interaction network was constructed. The results revealed that the whole network incorporated 167 nodes and 416 edges, that could be divided into three sub-clusters: the five regulatory factors (*FUS3*, *ABI3*, *WRI1*, *LEC1* and *LEC2*) and the genes that were directly affected by only one regulatory factor (A cluster, Figure 4a), the genes affected by two or more regulatory factors (B cluster, Figure 4b) and the genes indirectly affected by regulatory factors (C cluster, Figure 4c). The A cluster consisted of 32 nodes and 105 edges in total. The five regulatory factors formed a tightly intra-linked group, with each regulatory factor under the influence of at least two other regulatory factors, and *LEC1* and *LEC2* were especially affected by all other four regulatory factors. In addition to the five regulatory factors, *BCCP2* and *CAC2* were underlying the QTL CIs. *BCCP2* was associated with QTLs *qA3-1* and *qA8-6*, which affected C16:0, C18:2 and C20:0; *CAC2* was associated with QTL *qC9-4* and affected C18:0. These two genes were both regulated by *WRI1*. The B cluster was composed of seven nodes and 40 edges, in which the genes were affected by at least two regulatory factors. *SEP2* and *LFY* in this group were directly affected

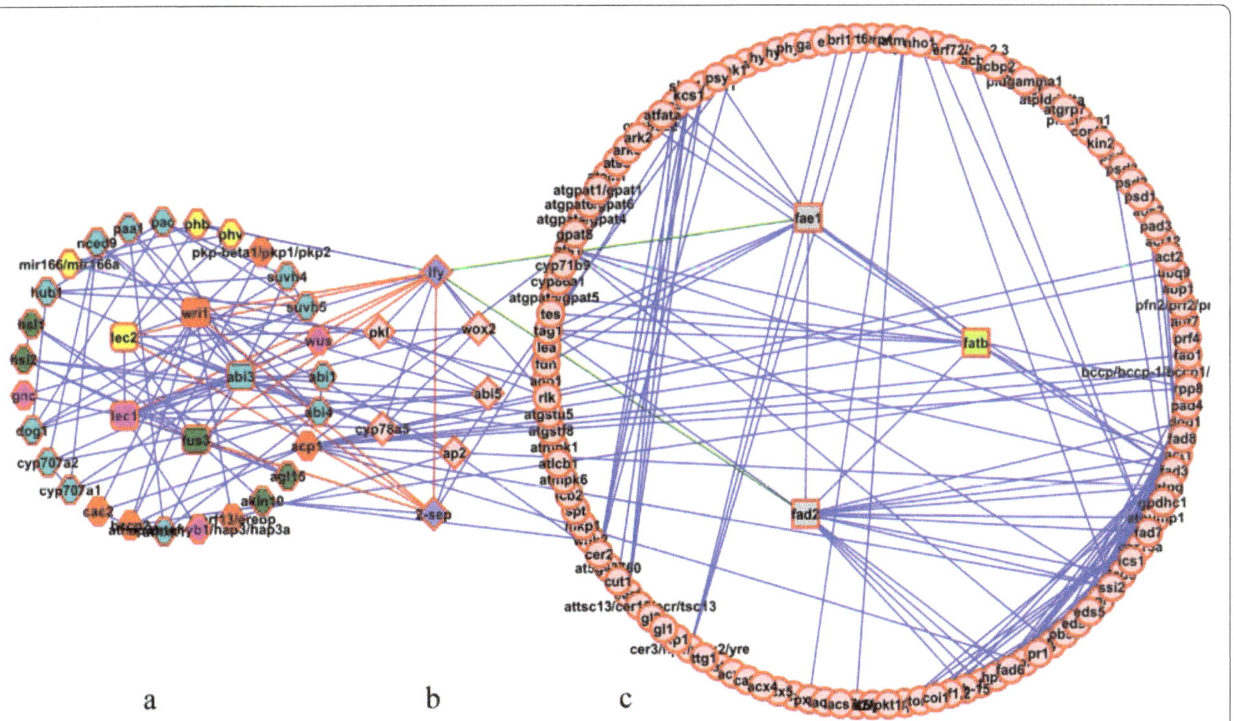

Figure 4 Gene interaction pathway. Network visualization for interaction of the 93 candidate genes and regulatory factors observed from the QTL results using Cytoscape_V2.6.3 software. Genes are presented as nodes and gene interactions are presented as edges. **(a)** The five regulatory factors and genes that are directly affected by only one regulatory factor: the five regulatory factors indicated by round rectangle with different color (yellow, *LEC2*; pink, *LEC1*; green, *FUS3*; light blue, *ABI3*; red, *WRI1*), while the genes are depicted as hexagon with the same color of the regulatory factors which linked to them. **(b)** The genes affected by two or more regulatory factors, denoting as diamond-shaped nodes, and the two diamond-shaped nodes with light blue color are *LFY* and *SEP2*, separately. **(c)** The genes indirectly affected by regulatory factors. In the middle of the cluster are three very important genes in fatty acid synthesis denoted as rectangle nodes with different color (pink, *FAD2*; green, *FAE1*; light yellow, *FATB*), and other genes are depicted as ellipse pink-filled nodes.

by all five regulatory factors. *LFY* was linked to other eight genes (Figure 4), and two of them (*FAD2* and *FAE1* in cluster C) were well-known important genes involved in fatty acid synthesis. The C cluster included 128 nodes and 319 edges, and 27 genes were detected to underline the QTL CIs. *FATB* associated with QTL *qA8-2* was notably affected by both *FAE1* and *FAD2*, and *FATB* also affected another eight genes (*ACP1, FAD3, FATA, ACT1, TES, TAG1, CYP86A2* and *GPAT5*) that were underlying the QTL CIs. The network mentioned above indicated that the genes involved in fatty acid synthesis were directly or indirectly affected by regulatory factors.

Construct of a potential controlling pathway of fatty acids in *B. napus*

Among the 97 genes of *Arabidopsis* located in the CIs of 47 QTLs in *B. napus*, 57 were shown to have roles in 32 different pathways (Additional file 8). Among them, 31 genes were involved in at least one of the five pathways, including biosynthesis of unsaturated fatty acids, fatty acid metabolism, glycerophospholipid metabolism, fatty acid biosynthesis and alpha-linolenic acid metabolism, with 13, nine, nine, eight and six candidate genes, respectively (Additional file 9). The potential pathway for fatty acid synthesis in *B. napus*, which included 50 genes and the five regulatory factors mentioned above, was constructed by combining the main pathways that involved interaction of candidate genes as well as knowledge of fatty acid regulatory pathways in *Arabidopsis* (Figure 5). The candidate genes in the deduced pathways could be divided into three types: genes associated with the fatty acid biosynthetic pathway (Figure 5a), the phospholipid synthesis and other pathways (Figure 5b) and the triacylglycerol and fatty acid degradation (Figure 5c).

The diagrams of the fatty acid biosynthetic pathway in the plastid and the cytosol/endoplasmic reticulum could be divided into four categories: genes participating in early stages of fatty acid synthesis, fatty acid elongation and desaturation, triacylglycerol biosynthesis and regulators of fatty acid biosynthesis. At least 24 enzymes were involved in the plastidial fatty acid synthetic pathway [38]. *ACCase* was a major control point of the pathway among these enzymes [39,40], and four subunits of *ACCase* were detected (Figure 5a) and assigned to 13 QTLs: *ACC2* was assigned to seven QTLs: *qA3-3, qA5-2, qA8-5, qA8-6, qA9-4, qC3-1* and *qC9-3*; *BCCP2* to QTLs *qA3-1* and *qA8-6*; *CAC2* to QTL *qC9-4*; and *CAC3* to QTLs *qA4-7, qA8-1* and *qC6-1* (Additional file 4). Downstream from *ACCase*, fatty acids were constructed and elongated by each cycle with four reactions (steps 3–5). In step 3, two genes *EMB3147* and *AT1G62610* were found underlying the CIs of one (*qA5-3*) and four QTLs (*qA8-6, qA9-7, qC9-3* and *qC9-4*), respectively. Candidate genes of *KASIII* (*qA8-6, qA9-7, qC9-3* and *qC9-4*) and *FAB1* (*qA7-2*) in step 4 and *FAB1* in step 5 were also observed. This suggested that all these candidate genes in QTL CIs could affect the traits through their action in early stages of fatty acid synthesis.

Most C18:0-ACP produced by elongation was desaturated by the stearoyl-ACP desaturase in step 6, and *FAB2* (*qA8-1* and *qC6-1*) and *AT5G16240* (*qA3-1, qA8-6* and *qA10-2*) were found. Some important candidate genes were found to be associated with fatty acid elongation and desaturation; for example, *FAE1* and *CER10* appeared in step 9. *FAE1* was associated with QTLs *qA1-4* detected by C18:3 and *qA8-5* with all 10 traits, and *qC3-2* with nine traits but not C18:3. *CER10* was assigned to *qA4-3, qA5-2, qA10-1* and *qC9-6*, which were detected with C16:0, FAS, C18:3 and C18:1, respectively.

Triacylglycerol synthesis could be divided into four main steps [38,41]. DHAP could be reduced to G3P catalyzed by GPDH, and *GPDHc1* (*qA8-1, qC6-1* and *qC9-4*) and *SDP6* (*qA3-5, qA5-4, qC3-7* and *qC5-1*) were revealed in this step. From G3P to PA of steps 13 and 14, fatty acids were sequentially transferred from CoA to positions 1 and 2 of G3P catalyzed by *GPAT* and *LPAT*, respectively. *GPAT* was composed of 10 subunits [38], seven of which except for *GPAT3, GPAT8* and *GPAT9* were observed in this step. Among them, *GPAT1, GPAT2* and *GPAT4* were associated with QTL *qA8-2*; *GPAT5* with QTLs *qA3-5, qC3-7* and *qC5-1*; *GPAT6* with QTLs *qA4-7, qA8-1* and *qC6-1*; *GPAT7* with QTLs *qA8-6* and *qA10-3*; and *ATS1* with QTLs *qA3-3, qA5-2, qA8-5, qA8-6, qA9-4* and *qC9-3* (Additional file 4). *LPAT* was a main enzyme in step 14 with five subunits [38], and *LPAT2, LPAT3, LPAT4* and *LPAT5* were associated with four (*qA4-2, qA9-2, qA10-1* and *qC9-6*), two (*qA5-2* and *qC3-5*), one (*qA9-6*) and four (*qA3-6, qA5-3, qC3-8* and *qC5-1*) QTLs, respectively. *LPP3* was associated with QTLs *qA3-4* and *qC5-1* that catalyze dephosphorylation of PA to release DAG in step 15. Step 16 was the final step of TAG synthesis and *TAG1* (*qC8-1* and *qC9-3*), *AT3G44830* (*qC3-4*) and *PDA* (*qA3-1* and *qA8-6*) were observed in this step. Many genes or QTLs involved in one step indicated that the fatty acid synthesis in *B. napus* was very complicated.

Besides those genes that had direct activities in catalyzing fatty acid synthesis, some regulators (*LEC1, LEC2, FUS3, WRI1* and *ABI3*) underlying QTL CIs were also found (Figure 5). The interaction analysis indicated that *WRI1* directly affected *BCCP2* and *CAC2*, and indirectly affected *FAB1* and *FATB* in early stages of fatty acid synthesis regulation, while *WRI1, LEC1/2, FUS3* and *ABI3* all indirectly affected *FAE1, FAD2* and *FAD3*, which played important roles in later modifying and transporting processes. *LEC1* and *LEC2* were underlying the CIs of five QTLs (*qA3-3, qA8-4, qA8-6, qA9-6* and *qC9-3*) and six QTLs (*qA3-3, qA5-2, qA8-4, qA8-6, qA9-4* and *qC9-3*), respectively. Both *LEC1* and *LEC2* affected nine

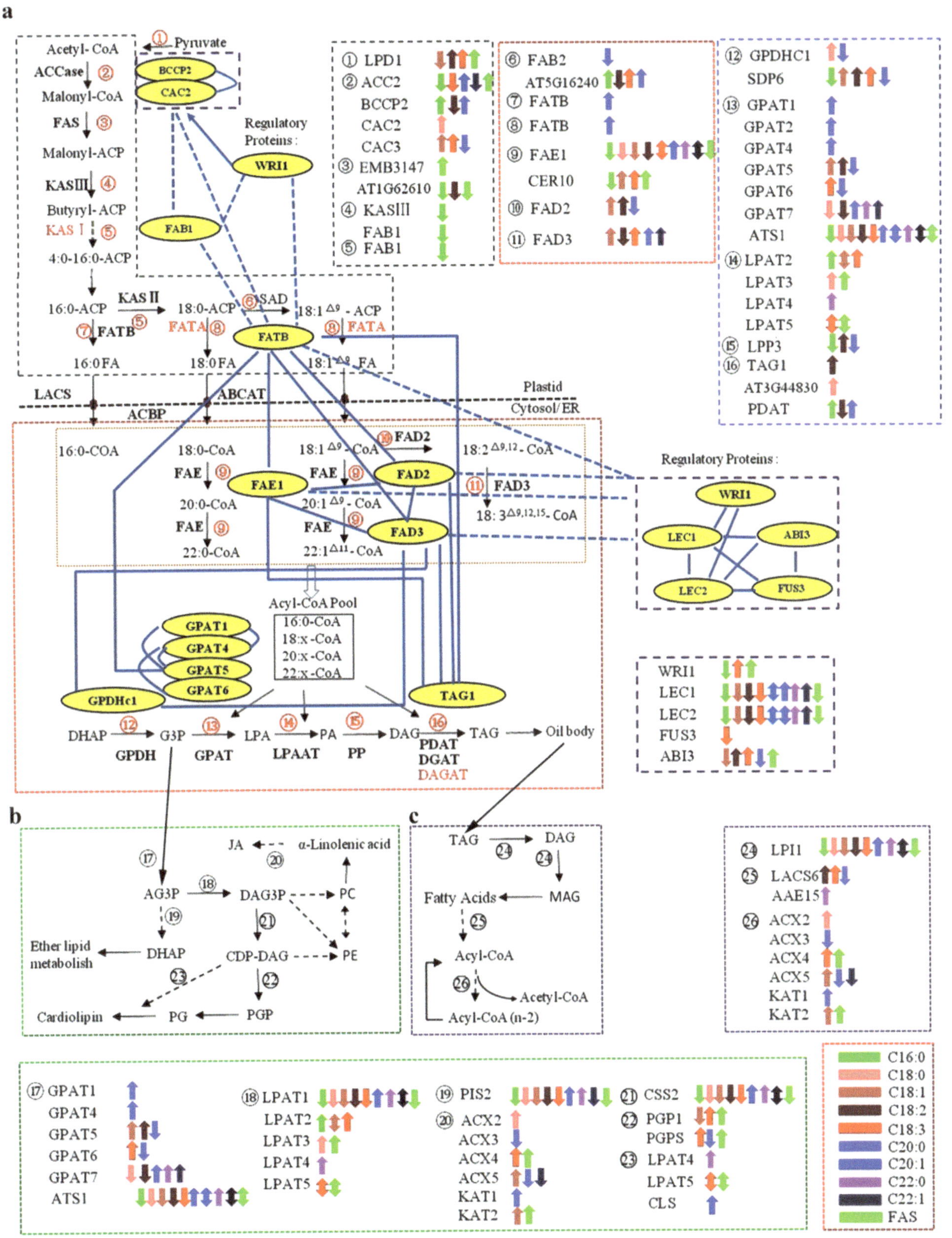

Figure 5 (See legend on next page.)

(See figure on previous page.)
Figure 5 Potential regulatory pathways and candidate genes associated with fatty acid synthesis in *B. napus*. **a**, the fatty acid biosynthetic pathway; **b**, the phospholipid synthesis and other pathways; **c**, the triacylglycerol and fatty acid degradation. The black characters indicate genes detected in previous studies and also in this study, the red characters indicate genes not found in this study but in previous studies. The colored arrows indicate genes underlying a QTL which affects traits with additive effect, up-arrows indicate QTLs with positive additive and down-arrows a negative additive effect. The double-headed arrows indicate genes underlying different QTLs which affect the same trait with opposite additive effect at the same time based on the WinQTLcart2.5 and QTLNetwork_2.0 results (not combined results of the two). Different colors denote different traits as the bar shown at the lower right corner of the picture. The abbreviations of gene name are same with the following reference papers [30,37,38].

fatty acid compositions, except for C18:0, in different environments. *FUS3* located in the CI of the QTL *qC7-1* that was only associated with C18:3. *WRI1* located in the CIs of QTLs *qA4-3*, *qA5-2* and *qA10-1* was associated with C16:0, FAS and C18:3, respectively. *ABI3* was associated with three QTLs (*qA5-3*, *qC3-8* and *qC5-1*) and affected the levels of C18:3, FAS, C18:1, C18:2 and C20:0 (Figure 5). Moreover, the epistatic effects of major gene–gene interactions were also identified in fatty acid synthesis (Figure 5). For example, the major genes *FATB*, *FAE1*, *FAD2* and *FAD3* were all under the influence of each other and all could affect the gene *TAG1* in the final step of TAG synthesis, and then affected the level of different fatty acids (Figure 5). All these findings indicated that the regulatory pathway of fatty acid synthesis was a complex network, controlled by a large number of genes and also affected by gene–gene interactions.

Potential phospholipid synthesis (Figure 5b), and triacylglycerol and fatty acid degradation (Figure 5c), were also predicted in *B. napus* based on knowledge of acyl-lipid metabolism in *Arabidopsis* [31]. G3P was converted to DAG3P by sequential acylation reactions in steps 17 and 18, which was regulated by *GPAT* and *LPAT*, and the homologous genes involved in the two steps were very similar to the genes in steps 13 and 14, respectively. Then DAG3P was converted to PC or PE by a series of enzyme reactions. In the pathway of triacylglycerol and fatty acid degradation, a number of homologous genes were also found, and the six genes in step 26 were the same as the genes in step 20. Moreover, gene *LPI1*, which functioned in degradation of TAG to MAG, affected the level of nine traits but not C20:0.

In addition to the 47 QTLs with candidate genes mapped in the CIs, the remaining 25 QTLs also affected the concentrations of one or two fatty acids. QTL *qA4-4* on chromosome A4 with additive effect on traits C16:0 and C18:0 might be assigned to step 5 (Figure 5) with potential ketoacyl ACP synthetase activity. QTL *qC5-2* might be assigned to step 11 with a similar function to *FAD3*, as the QTL had a consistent additive effect on traits C18:2 and C18:3. QTL *qC9-5* might be a pleiotropic locus with a similar function to *FAE1* as it had an additive effect on traits C18:0 and C22:1. The remaining 22 QTLs only had additive effects on one trait. Based on the premise that plastidic synthesis and modification should result in consistent QTL effects on products in any of the individual sets of saturated fatty acids [42], *qA7-3*, *qA7-4* and *qC9-1* affected *C16:0*, and *qA1-5* and *qA6-2* affected C18:0, which allows us to assign these QTLs tentatively to early stages of synthesis. The other 17 QTLs may be assigned to desaturation and elongase steps, as they affected C18:3, C20:0, C20:1, C22:0, C22:1 and FAS. Generally, it is important to take into account all the preceding synthetic and modification steps where a QTL could be identified for a particular fatty acid [42]; however, assigning a specific enzyme activity to these QTLs is relatively complex.

Discussion

A total of 72 QTLs and a large number of pairs of epistatic loci for 10 fatty acid compositions were identified using QTL analysis. Then 234 homologous genes were mapped in the CIs of 47 QTLs. QTL mapping and interaction analysis of candidate genes enabled the construction of a regulatory pathway that controlled the different concentrations of fatty acids in *B. napus*.

Single QTL mapping of each fatty acid concentration showed that QTLs on A8 and C3 were highly correlated. QTL *qA8-5* on A8 affected concentrations of all 10 types of fatty acids, and explained PV of individual fatty acids ranging from 15.89% for C20:1 to 47.55% for C20:0, and QTL *qA8-6* affected seven fatty acids; meanwhile, both QTLs *qC3-2* and *qC3-3* on C3 affected nine fatty acids but not C18:3. All these results revealed that these QTLs were major QTLs, consistent with previous studies [6,9,10]. For the two QTLs on C3, the alleles from 'Ningyou 7' conferred higher levels of C20:0, C20:1, C22:0 and C22:1, but had negative effects on levels of C16:0, C18:0, C18:1 and C18:2. The group C16:0, C18:0, C18:1 and C18:2 were positively correlated with each other, and likewise for the group C20:0, C20:1, C22:0 and C22:1; however, the two groups showed a negative correlation (Table 2). The fact that QTLs clustered in certain linkage groups and controlled different traits with opposite effects, corroborated strongly with the significant positive and negative correlations among the fatty acids analyzed. Besides the two major QTLs on A8 and C3, another 15 QTLs also affected more than one fatty acid. Moreover, the number of QTLs associated with a single fatty acid in the present study ranged from nine (C18:2, C20:0 and C22:1) to 21 (C18:3), and a total of 72 QTLs for all 10 traits detected were more than that from previous studies [6-10], with a number of new QTLs for different fatty acids. No QTLs for C16:0 were found on A4, A7 and C9 previously [6,8,10], suggesting that the 10 QTLs on these chromosomes in the TN population were all new QTLs. The unsaturated fatty acids were synthesized as a result of the fatty acid desaturation pathway, starting from C18:0; however, only a few QTLs for C18:0 were detected in previous studies. Zhao *et al.* identified two QTLs for C18:0 on A6 and A7 [8], Smooker *et al.* identified one QTL on A9 [9], and Burns *et al.* detected four QTLs respectively on A1, A6, A7 and A8 [6]. Herein, 12 QTLs across seven chromosomes were detected, and the QTLs on A4 (one QTL), A10 (two QTLs), C3 (four QTLs) and C9 (two QTLs) were new. QTLs for C18:1 were identified on 14 chromosomes except A4, A6, C2, C5 and C7 [6-10]. In the present study, there were 12 QTLs associated with C18:1 across seven chromosomes, nine of which also affected other fatty acids. QTLs for C18:2 were identified on 15 chromosomes except A4, A10, C5 and C9 [6,8-10], indicating that the two QTLs (*qC5-1* and *qC5-2*) for C18:2 on C5 were new. There were 21 QTLs for C18:3 identified here, and three QTLs (*qA2-1*, *qA2-2* and *qA2-3*) mapped on A2 and one QTL (*qC7-1*) on C7 were new QTLs, but the former 18 QTLs were distributed across 14 chromosomes excluding A2, A3, A9, C6 and C7 [6-10]. Many QTLs for C22:1 were identified on 14 chromosomes excluding A4, A5, A7, C1 and C5 [4-6,8-10,43]. Two major QTLs associated with C22:1 on A8 and C3 chromosomes were observed in the present study, which was consistent with previous studies [5,9], and two new QTLs (*qA4-1* and *qA7-1*) were mapped on A4 and A7 chromosomes (Table 3).

Generally, additive effects are considered as the main factors contributing to variations in quantitative traits [44,45]. However, epistatic effects can also play an important role in complex traits in *B. napus* [46,47]. Epistatic QTLs was found to explain PV ranging from 0.01% for C18:2 to 25.47% for C20:1 (Additional file 6), showing similar trends to previous studies [48]. Moreover, interactions among 395 loci were identified by GMM, including 34 pairs of digenic and 361 pairs of trigenic interactions (Additional file 7). A number of candidate genes were mapped to the CIs of these QTLs. These results showed that epistasis may substantially contribute to variation in oil concentration in different crops.

The regulatory pathway constructed in this study was based on QTL information and the interaction of candidate genes associated with the single and epistatic QTLs of 10 fatty acids, as well as previous knowledge concerning *Brassica* species and *Arabidopsis* [31,42]. A large number of important genes were found to be candidate genes for the processes of fatty acid metabolism. Among the 50 genes and five regulatory factors detected, 31 genes (62.0%) affected more than one trait, and *FAE1* and *ATS1* affected nine (but not C20:0) and 10 traits, respectively. Previous studies found that *FAE1* affected oil composition in several steps of seed fatty acid synthesis and modification pathways in *B. oleracea* [42] and *B. napus* [30,49]. In *B. napus*, the orthologs of *FAD2* and *FAD3* were mapped in different linkage groups of the genome. *BnaFAD2* loci were mapped in linkage groups A1, A5, C1 and C5 [28], and *BnaFAD2* located in different linkage groups was found to be responsible for the high C18:1 level [29] and erucic acid content [9]. It seemed clear that a major functional variant of *FAD2* orthologs in the A genome was located on linkage group A5, but unclear whether *FAD2* loci were present in the *Brassica* C genome [8,29]. The ortholog of *FAD3* was identified in the control of C18:3 level in *B. napus*, and a specific marker of *fad3* was located in the CI of the QTL for C18:3 in linkage group A1 [50]. In the present study, *FAD2* was located in three QTLs CIs: *qA3-5* by C16:0 and C18:1, *qC3-7* by C20:0, and *qC5-1* by C18:2 and C20:0 using *in silico* mapping. The homologous genes of *FAD3* were mapped in the CIs of seven QTLs: *qA3-5* by C16:0 and C18:1; *qA5-3* by C18:3 and FAS; *qA5-4* by C18:1 and C18:3; *qA8-6* by C18:1, C18:2, C20:0, C20:1, C22:0, C22:1 and FAS; *qA10-3* by C18:0; *qC3-7* by C20:0; and *qC5-1* by C18:2 and C20:0. Taken together, these findings increased the understanding of the important genes affecting the processes of fatty acid metabolism in *B. napus*.

Despite the multiple studies on the function of several important genes in fatty acid biosynthesis [23-26], the regulatory mechanism of the pathway is still not well understood in *B. napus*. Five regulatory factors (*LEC1*, *LEC2*, *WRI1*, *FUS3* and *ABI3*) were involved in the process concerning the fatty acid biosynthetic pathways, which might correspond to new important findings in this study. Both *LEC1* and *LEC2* could act as positive regulators upstream of *ABI3* and *FUS3* in the control of seed maturation [51-53], and several studies revealed that these four regulatory factors might have a function in regulation of fatty acid metabolism [37,51-54]. *WRI1*, which acted downstream of *LEC2*, controlled expression of a subset of genes involved in fatty acid biosynthesis [33], and overexpression of *WRI1* caused an increased TAG level in both seeds and leaves of *Arabidopsis* [34]. In this study, each of the five regulatory factors was under the influence of at least two other regulatory factors, and *LEC1* and *LEC2* were especially affected by all four other regulatory factors (Figures 4 and 5). Moreover, all five regulatory factors were associated with C18:3, but none affected C18:0. Conversion of acetyl-CoA to malonyl-CoA by acetyl carboxylase (ACCase) is one of the most committed steps in fatty acid biosynthesis [39,40], and reduction of ACCase activity caused a decrease of fatty acid content in the range of 1.5–16.0% in transgenic seeds of oilseed crops [55]. Most of the enzymes involved in *de novo* fatty acid synthesis seemed to be affected by *WRI1* [33], and only *WRI1* of the five regulatory factors had a direct impact on transcription levels of genes in the plastidial fatty acid synthetic pathway [38]. As a result of the interaction analysis of candidate genes, two genes *CAC2* and *BCCP2*, which encoded subunits of ACCase, were found to be under the influence of *WRI1*. In the process of fatty acid elongation and desaturation, no candidate genes were shown to be under the influence of the five regulatory factors in this study (Figure 5). Among the targets of the five regulatory factors, *LFY* might represent a key factor mediating the regulatory action of elongation and desaturation processes, as it was directly affected by all five regulatory factors and also regulated important downstream genes, including *FAD2*, *FAE1* and *FATB*, and these three genes regulated a series of downstream genes (Figure 4). *LFY* was arguably the most important floral meristem identity gene, which encoded a plant-specific transcription factor and controlled multiple aspects of floral morphogenesis [56,57]. It was striking that *LFY* played an important role in the regulatory pathway of fatty acid metabolism, and is worthy of future study.

Overall, these results provided important new insights into the regulatory model for the control of oil synthesis in *B. napus* and enhanced our understanding of the fatty acid synthesis pathways.

Conclusions

In this study, 72 individual QTLs and a large number pairs of epistatic interactions contributing to fatty acid biosynthesis were identified. By using *in silico* mapping analysis, 234 homologous genes of *A. thaliana* that could be involved in fatty acid metabolism within the CIs of 47 QTLs were found. After QTL mapping and candidate gene analysis, a potential regulatory network controlling fatty acid metabolism was revealed in *B. napus*. Our results provided new insights into the regulatory model for the control of oil synthesis in *B. napus*.

Methods

Plant materials and field experiments

The segregating DH population (the TN DH population), with 202 DH lines derived from the cross 'Tapidor' × 'Ningyou7' was constructed by Qiu *et al.* [5]. The DH lines as well as their parents were grown in six independent environments in China. The population was grown at Wuhan (W1) for one year (September 2007 to May 2008, 07 W1), Huanggang (W2) for two years (September 2008 to May 2009, 08 W2; September 2009 to May 2010, 09 W2) and Qichun (W3) for one year (September 2009 to May 2010, 09 W3) in Hubei Province; and Dali for two years (September 2007 to May 2008, 07D; September 2009 to May 2010, 09D) in Shaanxi Province. Wuhan, Qichun and Huanggang were the experiment bases of Huazhong University of Science and Technology, and Dali was the experiment base of Hybrid Rapeseed Research Center of Shaanxi Province. No specific permissions were required for the field trials. For each field trial, all lines were planted in a randomized complete-block designed with three replicates, and each plot contained 30 plants per genotype [58].

Measurements of fatty acid composition

Bulked seed samples (200 mg) from each of replicates in the three environments (07D, 07 W1 and 08 W2) were analyzed for their fatty acid compositions by gas liquid chromatography according to Rücker and Röbbelen [59]. In three other environments (09D, 09 W2 and 09 W3), fatty acid compositions were determined by near-infrared reflectance spectroscopy (NIRS) using standard methods [60] and all saturated fatty acid compositions in these three environments were considered as one trait named FAS.

Linkage map construction and map alignment with *Arabidopsis*, *B. rapa* and *B. oleracea*

The TN genetic linkage map of *B. napus*, which was introduced by Jiang *et al.* was updated in the present study [61]. A total of 932 markers were mapped to the new linkage map generated with the TN DH population using JoinMap_4.0 [62]. Detailed information of TN genetic linkage map was summarized in Additional file 3. Of the 932 linked markers, 429 with known sequence information or corresponding to *Arabidopsis* genes were used as anchored markers to carry out map alignment between *B. napus* and *Arabidopsis* according to the method of Long *et al.* [21]. If more than three homologous sequence-informative markers in the TN DH population were closely linked within one conserved block of *Arabidopsis* as described by Schranz *et al.* [17], a synteny block was considered to exist. If there were only one or two sequence-informative marker(s), it was recognized as an insertion segment. The homologous genes underlying the block or island were considered to be candidate genes if they were just underlying the QTL CI. There were 111 key genes (824 homologous genes) (Additional file 4) involved in fatty acid metabolism of *Arabidopsis* collected from the TAIR website [63] and *in silico* mapping in the genetic map as described by Long *et al.* [21]. The comparative mapping analysis between *B. napus* and *B. rapa/B. oleracea* was done using BLAST analysis on the *Brassica* Database [64] or *B. oleracea* Database [65]. The sequence of markers in the TN DH linkage map was used as a query sequence. All available sequence-informative markers on *B. napus* were subjected to BLASTn analysis to identify their physical positions on *Arabidopsis*, *B. rapa* and *B. oleracea* chromosomes. Then the Brassicaceae building blocks that corresponded to each informative marker were identified. The candidate gene(s) in *B. napus*, *B. rapa* and *B. oleracea* for each informative marker were also identified.

QTL and epistasis analysis

QTL analysis was done using the QTLNetwork_2.0 software described by Yang *et al.* [66,67] and WinQTLCart_2.5 [68], and epistasis QTL mapping analyses were performed using the QTLNetwork_2.0 and GMM [69]. For QTLNetwork_2.0, a mixed linear model was used to identify QTLs at 2-cM intervals with a window size of 10 cM. Two-dimensional (2D) genome scans were used to search for multiple interacting QTLs. For each trait, a genome-wide threshold value of the F-statistic ($P = 0.05$) for declaring the presence of a QTL was estimated by 1000 random permutations [70]. A Bayesian method with Gibbs sampling was used to estimate QTL effects [71]. The sum of individual PV explained by each QTL was calculated as the total PV explained by all QTLs for each trait. Based on genetic effects, mapped epistatic QTLs comprised three types: interactions between two QTLs with additive effects (AA), interactions between a QTL with additive effect and a locus without significant additive effect (AN or NA), and interactions between two loci within non-QTL (NN) [45]. For WinQTL-Cart_2.5, composite interval mapping (CIM) was used with a significance threshold of $P = 0.05$ using the permutation test method [68], based on 1000 runs of randomly shuffling the trait values [72]. QTL CIs were determined by 2-LOD intervals surrounding the QTL peak. QTLs for different traits that mapped to the same region with overlapping CIs were assumed to be the same [58,73]. For GMM, two and three loci interactions were detected with default parameters. For the QTL comparison between different populations, QTLs detected in this study were preliminarily compared with those in other publications, because few common markers were available between TN and other populations. If one or more QTLs were identified on a chromosome in this study but not in previous studies, these were considered as new QTLs.

The genetic interaction analysis candidate genes

The network was constructed and analyzed using Cytoscape_V2.6.3 with the Agilent Literature Search Plug-in [74]. The degree of connectivity, the clustering coefficient, the network density and the diameter were designed according to Toubiana *et al.* [75]. Search controls option by using the default parameter settings and extraction controls option with appropriate choice and choosing *Arabidopsis* in concept lexicon, the network was laid out by using group attributes layout.

Construction of potential controlling pathways of fatty acids in *B. napus*

To identify candidate genes underlying the QTL CIs involved in the fatty acid regulatory pathways, the Kyoto Encyclopedia of Genes and Genomes (KEGG) database was applied to identify pathways in which these genes were involved [76,77]. Potential controlling pathways of fatty acids in *B. napus* were inferred based on the pathways of fatty acid synthesis within *Arabidopsis* [31,38] and other plants [42,78] and also using gene interaction analysis and QTL epistasis analysis.

Additional files

Additional file 1: Single QTL detection by WinQTLCart_2.5 in TN DH population.

Additional file 2: Single QTL detection by QTLNetwork_2.0 in TN DH population.

Additional file 3: The TN DH genetic linkage map and its syntenic segmental alignment with the *Arabidopsis* genome.

Additional file 4: Positions of homologous genes of *Arabidopsis* mapped on TN DH linkage map by *in silico* mapping and orthologous genes between *B. napus* and *B. rapa/B. oleracea*.

Additional file 5: Demonstration of a complex epistatic network for fatty acids in *B. napus*.

Additional file 6: Epistatic QTLs for fatty acid compositions detected by QTLNetwork_2.0 in TN DH population.

Additional file 7: Digenic and trigenic interactions for fatty acid compositions detected by genotype matrix mapping software ver2.1 in TN DH population.

Additional file 8: The distribution of 57 candidate genes in 32 different pathways by the KEGG analysis.

Additional file 9: Five KEGG pathway maps of candidate genes in *Arabidopsis*.

Abbreviations

ABCAT: ABC acyl transporter; *ACBP*: Acyl CoA binding protein; *ACCase*: Acetyl-CoA carboxylase; CI: Confidence interval; DAG: Diacylglycerol; *DAGAT*: Diacylglycerol transacylase; *DGAT*: Acyl-CoA,diacylglycerol acyltransferase; DH: Doubled haploid; *FAE*: Fatty acid elongase; FAS: Fatty acid synthase; *FAT*: Fatty ACP thioesterase; G3P: Glycerol-3-phosphate; GMM: Genotype Matrix Mapping ver2.1software; *GPAT*: Glycerol-3-phosphate acyltransferase; *GPDH*: NAD-dependent glycerol-3-phosphate dehydrogenase; JA: Jasmonate; *KAS*: Ketoacyl-ACP synthase; *LACS*: Long-chain acyl-CoA synthetase; LPA (AG3P): Lysophosphatidic acid; *LPAAT*: 1-acylglycerol-3-phosphate acyltransferase; MAG: Monoacylglycerol; PA (DAG3P): Phosphatidic acid; PC: Phosphatidylcholine; *PDAT*: Phospholipid,diacylglycerol acyltransferase; PE: Phosphatidylethanolamine; PG: Phosphatidylglycerol; PGP: Phosphatidyglycerophosphate; *PP*: Phosphatidate phosphatase; PV: Phenotypic variance; QTL: Quantitative trait loci; *SAD*: Stearoyl-ACP desaturase; TAG: Triacylglycerol.

Competing interests

The authors declare that they have no competing interests.

Authors' contributions

XW and YL participated in field experiment and the fatty acid compositions measurement, carried out the QTL and epistasis analysis, inferred potential controlling pathway of fatty acids in *B. napus* and co-wrote the manuscript. YY, LG and CZ participated in the interaction analysis of candidate genes and mapped alignment with *A. thaliana*, *B. rapa* and *B. oleracea*. LL, LY and JM made helpful suggestions to the manuscript. ML designed, led and coordinated the overall study. All authors read and approved the final manuscript.

Acknowledgments

The authors are very grateful to Professor Xingwang Deng of Yale University for his critical reading of the manuscript. The work was supported by National Natural Science Foundation of China (31171582, 31071447), the High Technology Program of China (2011AA10A104) and the New Century Talents Support Program of the Ministry of Education of China (NCET110172).

Author details

[1]College of Life Science and Technology, Huazhong University of Science and Technology, Wuhan 430074, China. [2]National Key Lab of Crop Genetic Improvement, Huazhong Agricultural University, Wuhan 430070, China. [3]Institute of Biotechnology, Chinese Academy of Agricultural Sciences, Beijing 100081, China. [4]Key Laboratory of Cotton and Rapeseed, Ministry of Agriculture, Institute of Industrial Crops, Jiangsu Academy of Agricultural Sciences, Nanjing 210014, China. [5]College of Agronomy and Biotechnology, Southwest University, Chongqing 400716, China.

References

1. Hoekman B. Global Economic Prospects 2008: Technology Diffusion in the Developing World. Washington, DC: World Bank Publications; 2007. p. 40–1.
2. Sovero M. Rapeseed, a new oilseed crop for the United States. New York: Wiley; 1993. p. 302–7.
3. Velasco L, Becker HC. Estimating the fatty acid composition of the oil in intact-seed rapeseed (*Brassica napus* L.) by near-infrared reflectance spectroscopy. Euphytica. 1998;101(2):221–30.
4. Ecke W, Uzunova M, Weissleder K. Mapping the genome of rapeseed (*Brassica napus* L.). II. Localization of genes controlling erucic acid synthesis and seed oil content. Theor Appl Genet. 1995;91(6–7):972–7.
5. Qiu D, Morgan C, Shi J, Long Y, Liu J, Li R, et al. A comparative linkage map of oilseed rape and its use for QTL analysis of seed oil and erucic acid content. Theor Appl Genet. 2006;114(1):67–80.
6. Burns MJ, Barnes SR, Bowman JG, Clarke MH, Werner CP, Kearsey MJ. QTL analysis of an intervarietal set of substitution lines in *Brassica napus*: (i) Seed oil content and fatty acid composition. Heredity. 2003;90(1):39–48.
7. Hu X, Sullivan-Gilbert M, Gupta M, Thompson SA. Mapping of the loci controlling oleic and linolenic acid contents and development of *fad2* and *fad3* allele-specific markers in canola (*Brassica napus* L.). Theor Appl Genet. 2006;113(3):497–507.
8. Zhao J, Dimov Z, Becker HC, Ecke W, Möllers C. Mapping QTL controlling fatty acid composition in a doubled haploid rapeseed population segregating for oil content. Mol Breed. 2008;21(1):115–25.
9. Smooker AM, Wells R, Morgan C, Beaudoin F, Cho K, Fraser F, et al. The identification and mapping of candidate genes and QTL involved in the fatty acid desaturation pathway in *Brassica napus*. Theor Appl Genet. 2011;122(6):1075–90.
10. Yan XY, Li JN, Wang R, Jin MY, Chen L, Qian W, et al. Mapping of QTLs controlling content of fatty acid composition in rapeseed (*Brassica napus*). Genes Genomics. 2011;33(4):365–71.
11. Koch MA, Haubold B, Mitchell-Olds T. Comparative evolutionary analysis of chalcone synthase and alcohol dehydrogenase loci in *Arabidopsis*, *Arabis*, and related genera (*Brassicaceae*). Mol Biol Evol. 2000;17(10):1483–98.
12. Yang YW, Lai KN, Tai PY, Li WH. Rates of nucleotide substitution in angiosperm mitochondrial DNA sequences and dates of divergence between *Brassica* and other angiosperm lineages. J Mol Evol. 1999;48(5):597–604.
13. Lan TH, DelMonte TA, Reischmann KP, Hyman J, Kowalski SP, McFerson J, et al. An EST-enriched comparative map of *Brassica oleracea* and *Arabidopsis thaliana*. Genome Res. 2000;10:776–88.
14. Parkin I, Sharpe AG, Lydiate DJ. Patterns of genome duplication within the *Brassica napus* genome. Genome. 2003;46:291–303.
15. Babula D, Kaczmarek M, Barakat A, Delseny M, Quiros CF, Sadowski J. Chromosomal mapping of *Brassica oleracea* based on ESTs from *Arabidopsis thaliana*: complexity of the comparative map. Mol Genet Genomics. 2003;268(5):656–65.
16. Parkin IA, Gulden SM, Sharpe AG, Lukens L, Trick M, Osborn TC, et al. Segmental structure of the *Brassica napus* genome based on comparative analysis with *Arabidopsis thaliana*. Genetics. 2005;171(2):765–81.
17. Schranz ME, Lysak MA, Mitchell-Olds T. The ABC's of comparative genomics in the *Brassicaceae*: building blocks of crucifer genomes. Trends Plant Sci. 2006;11(11):535–42.
18. Wang X, Wang H, Wang J, Sun R, Wu J, Liu S, et al. The genome of the mesopolyploid crop species *Brassica rapa*. Nat Genet. 2011;43(10):1035–9.
19. Chalhoub B, Denoeud F, Liu S, Parkin IA, Tang H, Wang X, et al. Plant genetics. Early allopolyploid evolution in the post-Neolithic *Brassica napus* oilseed genome. Science. 2014;345(6199):950–3.
20. Liu S, Liu Y, Yang X, Tong C, Edwards D, Parkin IA, et al. The *Brassica oleracea* genome reveals the asymmetrical evolution of polyploid genomes. Nat Commun. 2014;5:3930.
21. Long Y, Shi J, Qiu D, Li R, Zhang C, Wang J, et al. Flowering time quantitative trait loci analysis of oilseed *Brassica* in multiple environments and genomewide alignment with *Arabidopsis*. Genetics. 2007;177(4):2433–44.
22. Zhao JY, Huang JX, Chen F, Xu F, Ni XY, Xu HM, et al. Molecular mapping of *Arabidopsis thaliana* lipid-related orthologous genes in *Brassica napus*. Theor Appl Genet. 2012;124(2):407–21.
23. Kachroo A, Shanklin J, Whittle E, Lapchyk L, Hildebrand D, Kachroo P. The *Arabidopsis* stearoyl-acyl carrier protein-desaturase family and the contribution of leaf isoforms to oleic acid synthesis. Plant Mol Biol. 2007;63(2):257–71.
24. Browse J, McConn M, James DJ, Miquel M. Mutants of *Arabidopsis* deficient in the synthesis of alpha-linolenate. Biochemical and genetic characterization of the endoplasmic reticulum linoleoyl desaturase J Biol Chem. 1993;268(22):16345–51.
25. Miquel M, Browse J. *Arabidopsis* mutants deficient in polyunsaturated fatty acid synthesis. Biochemical and genetic characterization of a plant oleoyl-phosphatidylcholine desaturase J Biol Chem. 1992;267(3):1502–09.
26. Kunst L, Taylor DC, Underhill EW. Fatty acid elongation in developing seeds of *Arabidopsis thaliana*. Plant Physiol Biochem. 1992;30(4):425–34.
27. Yang Q, Fan C, Guo Z, Qin J, Wu J, Li Q, et al. Identification of *FAD2* and *FAD3* genes in *Brassica napus* genome and development of allele-specific markers for high oleic and low linolenic acid contents. Theor Appl Genet. 2012;125(4):715–29.
28. Scheffler JA, Sharpe AG, Schmidt H, Sperling P, Parkin I, Lühs W, et al. Desaturase multigene families of *Brassica napus* arose through genome duplication. Theor Appl Genet. 1997;94(5):583–91.
29. Schierholt A, Becker HC, Ecke W. Mapping a high oleic acid mutation in winter oilseed rape (*Brassica napus* L.). Theor Appl Genet. 2000;101(5–6):897–901.
30. Fourmann M, Barret P, Renard M, Pelletier G, Delourme R, Brunel D. The two genes homologous to *Arabidopsis FAE1* co-segregate with the two loci governing erucic acid content in *Brassica napus*. Theor Appl Genet. 1998;96(6–7):852–8.
31. Li-Beisson Y, Shorrosh B, Beisson F, Andersson MX, Arondel V, Bates PD, et al. Acyl-lipid metabolism. The Arabidopsis book/American Society of Plant Biologists. 2010;8:e0133.
32. Lysak MA, Koch MA, Pecinka A, Schubert I. Chromosome triplication found across the tribe *Brassiceae*. Genome Res. 2005;15(4):516–25.
33. Baud S, Mendoza MS, To A, Harscoet E, Lepiniec L, Dubreucq B. *WRINKLED1* specifies the regulatory action of *LEAFY COTYLEDON2* towards fatty acid metabolism during seed maturation in *Arabidopsis*. Plant J. 2007;50(5):825–38.
34. Cernac A, Benning C. *WRINKLED1* encodes an AP2/EREB domain protein involved in the control of storage compound biosynthesis in *Arabidopsis*. Plant J. 2004;40(4):575–85.
35. Stone SL, Kwong LW, Yee KM, Pelletier J, Lepiniec L, Fischer RL, et al. *LEAFY COTYLEDON2* encodes a B3 domain transcription factor that induces embryo development. Proc Natl Acad Sci U S A. 2001;98(20):11806–11.
36. Luerßen H, Kirik V, Herrmann P, Miséra S. *FUSCA3* encodes a protein with a conserved VP1/ABI3-like B3 domain which is of functional importance for the regulation of seed maturation in *Arabidopsis thaliana*. Plant J. 1998;15(6):755–64.
37. Lotan T, Ohto M, Yee KM, West MA, Lo R, Kwong RW, et al. *Arabidopsis LEAFY COTYLEDON1* is sufficient to induce embryo development in vegetative cells. Cell. 1998;93(7):1195–205.
38. Beisson F, Koo AJ, Ruuska S, Schwender J, Pollard M, Thelen JJ, et al. *Arabidopsis* genes involved in acyl lipid metabolism. A 2003 census of the candidates, a study of the distribution of expressed sequence tags in organs, and a web-based database. Plant Physiol. 2003;132(2):681–97.
39. Sasaki Y, Nagano Y. Plant acetyl-CoA carboxylase: structure, biosynthesis, regulation, and gene manipulation for plant breeding. Biosci Biotechnol Biochem. 2004;68(6):1175–84.

40. Baud S, Guyon V, Kronenberger J, Wuilleme S, Miquel M, Caboche M, et al. Multifunctional acetyl-CoA carboxylase 1 is essential for very long chain fatty acid elongation and embryo development in *Arabidopsis*. Plant J. 2003;33(1):75–86.
41. Ohlrogge J, Browse J. Lipid biosynthesis. Plant Cell. 1995;7(7):957–70.
42. Barker GC, Larson TR, Graham IA, Lynn JR, King GJ. Novel insights into seed fatty acid synthesis and modification pathways from genetic diversity and quantitative trait loci analysis of the *Brassica* C genome. Plant Physiol. 2007;144(4):1827–42.
43. Jourdren C, Barret P, Horvais R, Foisset N, Delourme R, Renard M. Identification of RAPD markers linked to the loci controlling erucic acid level in rapeseed. Mol Breed. 1996;2(1):61–71.
44. Delourme R, Falentin C, Huteau V, Clouet V, Horvais R, Gandon B, et al. Genetic control of oil content in oilseed rape (*Brassica napus* L.). Theor Appl Genet. 2006;113(7):1331–45.
45. Zhao JY, Becker HC, Zhang DQ, Zhang YF, Ecke W. Oil content in a European × Chinese rapeseed population: QTL with additive and epistatic effects and their genotype-environment interactions. Crop Sci. 2005;45(1):51–9.
46. Carlborg OR, Haley CS. Epistasis: too often neglected in complex trait studies? Nat Rev Genet. 2004;5(8):618–25.
47. Lark KG, Chase K, Adler F, Mansur LM, Orf JH. Interactions between quantitative trait loci in soybean in which trait variation at one locus is conditional upon a specific allele at another. Proc Natl Acad Sci U S A. 1995;92(10):4656–60.
48. Zhao J, Becker HC, Zhang D, Zhang Y, Ecke W. Conditional QTL mapping of oil content in rapeseed with respect to protein content and traits related to plant development and grain yield. Theor Appl Genet. 2006;113(1):33–8.
49. Barret P, Delourme R, Renard M, Domergue F, Lessire R, Delseny M, et al. A rapeseed *FAE1* gene is linked to the E1 locus associated with variation in the content of erucic acid. Theor Appl Genet. 1998;96(2):177–86.
50. Jourdren C, Barret P, Brunel D, Delourme R, Renard M. Specific molecular marker of the genes controlling linolenic acid content in rapeseed. Theor Appl Genet. 1996;93(4):512–8.
51. To A, Valon C, Savino G, Guilleminot J, Devic M, Giraudat J, et al. A network of local and redundant gene regulation governs *Arabidopsis* seed maturation. Plant Cell. 2006;18(7):1642–51.
52. Kagaya Y, Okuda R, Ban A, Toyoshima R, Tsutsumida K, Usui H, et al. Indirect ABA-dependent regulation of seed storage protein genes by *FUSCA3* transcription factor in *Arabidopsis*. Plant Cell Physiol. 2005;46(2):300–11.
53. Kroj T, Savino G, Valon C, Giraudat J, Parcy F. Regulation of storage protein gene expression in *Arabidopsis*. Development. 2003;130(24):6065–73.
54. Mu J, Tan H, Zheng Q, Fu F, Liang Y, Zhang J, et al. *LEAFY COTYLEDON1* is a key regulator of fatty acid biosynthesis in *Arabidopsis*. Plant Physiol. 2008;148(2):1042–54.
55. Thelen JJ, Ohlrogge JB. Metabolic engineering of fatty acid biosynthesis in plants. Metab Eng. 2002;4(1):12–21.
56. Siriwardana NS, Lamb RS. The poetry of reproduction: the role of *LEAFY* in *Arabidopsis thaliana* flower formation. Int J Dev Biol. 2012;56(4):207–21.
57. Grandi V, Gregis V, Kater MM. Uncovering genetic and molecular interactions among floral meristem identity genes in *Arabidopsis thaliana*. Plant J. 2012;69(5):881–93.
58. Shi J, Li R, Qiu D, Jiang C, Long Y, Morgan C, et al. Unraveling the complex trait of crop yield with quantitative trait loci mapping in *Brassica napus*. Genetics. 2009;182(3):851–61.
59. Rücker B, Röbbelen G. Impact of low linolenic acid content on seed yield of winter oilseed rape (*Brassica napus* L.). Plant Breed. 1996;115(4):226–30.
60. Mika V, Nerusil P, Koprna R, Kucera V. Fast prediction of quality parameters in whole seeds of oilseed rape (*Brassica napus*). Plant Soil Environ. 2003;49(4):141–5.
61. Jiang C, Ramchiary N, Ma Y, Jin M, Feng J, Li R, et al. Structural and functional comparative mapping between the *Brassica* A genomes in allotetraploid *Brassica napus* and diploid *Brassica rapa*. Theor Appl Genet. 2011;123(6):927–41.
62. Van Ooijen JW. JoinMap 4. Software for the calculation of genetic linkage maps in experimental populations. Wageningen, Netherlands: Kyazma BV; 2006.
63. The Arabidopsis Information Resource. [http://www.arabidopsis.org/]
64. Brassica Database. [http://brassicadb.org/brad/]
65. Brassica oleracea Database. [http://www.ocri-genomics.org/bolbase/]
66. Yang J, Zhu J, Williams RW. Mapping the genetic architecture of complex traits in experimental populations. Bioinformatics. 2007;23(12):1527–36.
67. Yang J, Hu C, Hu H, Yu R, Xia Z, Ye X, et al. QTLNetwork: mapping and visualizing genetic architecture of complex traits in experimental populations. Bioinformatics. 2008;24(5):721–3.
68. Wang S, Basten CJ, Zeng ZB. Windows QTL cartographer 2.5. Raleigh, NC: Department of Statistics, North Carolina State University; 2007.
69. Isobe S, Nakaya A, Tabata S. Genotype matrix mapping: searching for quantitative trait loci interactions in genetic variation in complex traits. DNA Res. 2007;14(5):217–25.
70. Doerge RW, Churchill GA. Permutation tests for multiple loci affecting a quantitative character. Genetics. 1996;142(1):285–94.
71. Wang CS, Rutledge JJ, Gianola D. Bayesian analysis of mixed linear models *via* Gibbs sampling with an application to litter size in Iberian pigs. Genet Sel Evol. 1994;26(2):91–115.
72. Churchill GA, Doerge RW. Empirical threshold values for quantitative trait mapping. Genetics. 1994;138(3):963–71.
73. Arcade A, Labourdette A, Falque M, Mangin B, Chardon F, Charcosset A, et al. BioMercator: integrating genetic maps and QTL towards discovery of candidate genes. Bioinformatics. 2004;20(14):2324–6.
74. Shannon P, Markiel A, Ozier O, Baliga NS, Wang JT, Ramage D, et al. Cytoscape: A software environment for integrated models of biomolecular interaction networks. Genome Res. 2003;13:2498–504.
75. Toubiana D, Semel Y, Tohge T, Beleggia R, Cattivelli L, Rosental L, et al. Metabolic profiling of a mapping population exposes new insights in the regulation of seed metabolism and seed, fruit, and plant relations. PLoS Genet. 2012;8(3):e1002612.
76. Ogata H, Goto S, Fujibuchi W, Kanehisa M. Computation with the KEGG pathway database. Biosystems. 1998;47(1–2):119–28.
77. Kanehisa M, Goto S. KEGG: kyoto encyclopedia of genes and genomes. Nucleic Acids Res. 2000;28(1):27–30.
78. Rawsthorne S. Carbon flux and fatty acid synthesis in plants. Prog Lipid Res. 2002;41(2):182–96.

16

Columnar apple primary roots share some features of the columnar-specific gene expression profile of aerial plant parts as evidenced by RNA-Seq analysis

Romina Petersen*, Haris Djozgic, Benjamin Rieger, Steffen Rapp and Erwin Robert Schmidt

Abstract

Background: Primary roots (radicles) represent the first visible developmental stages of the plant and are crucial for nutrient supply and the integration of environmental signals. Few studies have analyzed primary roots at a molecular level, and were mostly limited to *Arabidopsis*. Here we study the primary root transcriptomes of standard type, heterozygous columnar and homozygous columnar apple (*Malus* x *domestica*) by RNA-Seq and quantitative real-time PCR. The columnar growth habit is characterized by a stunted main axis and the development of short fruit spurs instead of long lateral branches. This compact growth possesses economic potential because it allows high density planting and mechanical harvesting of the trees. Its molecular basis has been identified as a nested Gypsy-44 retrotransposon insertion; however the link between the insertion and the phenotype as well as the timing of the phenotype emergence are as yet unclear. We extend the transcriptomic studies of columnar tissues to the radicles, which are the earliest developmental stage and investigate whether homozygous columnar seedlings are viable.

Results: Radicles mainly express genes associated with primary metabolism, growth and development. About 200 genes show differential regulation in a comparison of heterozygous columnar radicles with non-columnar radicles, whereas the comparison of homozygous columnar radicles with non-columnar radicles yields about 300 differentially regulated genes. Genes involved in cellulose and phenylpropanoid biosynthesis, cell wall modification, transcription and translation, ethylene and jasmonate biosynthesis are upregulated in columnar radicles. Genes in the vicinity of the columnar-specific Gypsy-44 insertion experience an especially strong differential regulation: the direct downstream neighbor, *dmr6-like*, is downregulated in heterozygous columnar radicles, but strongly upregulated in columnar shoot apical meristems.

Conclusions: The transcriptomic profile of primary roots reflects their pivotal role in growth and development. Homozygous columnar embryos are viable and form normal radicles under natural conditions, and selection towards heterozygous plants most likely occurs due to breeders' preferences. Cell wall and phytohormone biosynthesis and metabolism experience differential regulation in columnar radicles. Presumably the first step of the differential regulation most likely happens within the region of the retrotransposon insertion and its tissue-specificity suggests involvement of one (or several) tissue-specific regulator(s).

Keywords: Primary root, Homozygous, Columnar, Apple, Gypsy-44, RNA-Seq, DESeq, MapMan, qRT-PCR

* Correspondence: peterser@uni-mainz.de
Department of Molecular Genetics, Johannes Gutenberg-University, Mainz D-55128, Germany

Background

Columnar apple trees show a characteristic pillar-like growth habit with a thick, stunted main axis and short lateral fruit spurs [1,2]. This growth habit could be of potential benefit to apple growers because columnar trees can be planted closer together and require less pruning than standard tree types [2,3]. However, none of the columnar cultivars available to date can compete with commercially successful cultivars in terms of fruit quality and disease resistance [2-7]. Columnar growth arose as a spontaneous somaclonal mutation of a McIntosh tree in Canada in 1961 [8,9]. With one exception [10], all columnar cultivars that have been described so far are heterozygous (hemizygous) for the columnar mutation (*Co/-*) [11]. Whether the lack of homozygous individuals (*Co/Co*) is due to a decreased viability of homozygous columnar seeds/seedlings or just the decision of apple growers to preferentially choose non-columnar breeding partners is unclear.

The molecular cause of the columnar phenotype has recently been identified as the insertion of a Gypsy-44 long terminal repeat (LTR) retrotransposon into the LTR of another retrotransposon on chromosome 10 [10]. Wolters et al. [12] detected a smaller columnar-specific insertion at the same position, which is the solo-LTR of Gypsy-44 and thus most likely represents an artefact (unpublished data). The Gypsy-44 insertion is probably responsible for the upregulation of a nearby gene encoding a 2OG-Fe(II) oxygenase (also called *downy mildew resistance 6-like* (*dmr6-like*)) of unknown function in apical meristems and axillary buds of columnar trees, whereas leaves do not show any expression of *dmr6-like* [10,12]. Overexpression of *dmr6-like* in *Arabidopsis thaliana* led to a columnar-like phenotype [12]. In addition to the upregulation of *dmr6-like*, a significant increase of the expression level of other genes within the retrotransposon vicinity has been shown [10,13]. The consequence is a change in the overall gene expression pattern of the columnar plants. The shoot apical meristems and leaves of columnar apple trees show a differential regulation of defense-associated genes, genes involved in secondary metabolism such as terpene and phenylpropanoid synthesis, as well as genes related to auxin and jasmonate synthesis and signaling [10,13,14]. Since a reliable detection of the columnar growth habit is only possible after about two to three years, it is as yet unclear at which developmental time point the gene expression patterns leading to the formation of the columnar habit are established. Up to now, it is also not known whether and how the gene expression pattern of roots is affected by the *Co* mutation. Even the phenotype of own roots of columnar apple trees has never been analyzed, which is probably due to the fact that the vast majority of columnar trees are grown as scions on non-columnar rootstocks.

Germination and radicle emergence are the first developmental steps towards the formation of a new plant. The radicle is important for anchorage, nutrient and water supply of the plantlet as well as for the perception and integration of a multitude of environmental signals such as gravity or pest attacks. Its tip has even been described as the "brain" of the plant by Charles Darwin [15] (cited in [16]). Despite their crucial regulatory and pioneering role in development, little research has been conducted on primary roots at the molecular level (e.g. deciphering their transcriptome profiles) and those were mainly limited to the model plants *Arabidopsis thaliana* and *Zea mays* (for example [17-21]). Only one publication has dealt explicitly with the transcriptome of adult roots of poplar [22]. While apple seeds have been the subject of several studies mostly owing to their deep and well-pronounced dormancy [23-28], research interest seems to fade significantly when they finish germination. Apple seed dormancy can be overcome by cold treatment (stratification) for 60 – 90 days depending on the cultivar and environmental conditions [23,28]. After this time, the radicle protrudes the testa as the primary root. Primary roots of the dicotyledonous model plant *Arabidopsis* consist of four longitudinal sections: the root cap (columella) at the tip, followed by the zone of division, the elongation zone and the differentiation zone [29]. A cross section of the root reveals a radial organization of different cell layers: epidermis, cortex, endodermis and the vascular cylinder encompassing the pericycle, protoxylem, protophloem and procambium [30,31]. This radial symmetry is established at the root apical meristem, a small set of cells near the root tip surrounding the mitotically less active quiescent center [32].

In this study we analyzed and compared the transcriptomes of heterozygous columnar, homozygous columnar and non-columnar primary apple roots. Our aims were 1) to gather general information about the gene expression profile of this poorly studied plant tissue, 2) to analyze whether homozygous columnar seedlings exist and are viable, 3) to determine whether the columnar-specific gene expression profile observed in aerial plant parts can already be observed at the earliest stages of development in the root and 4) to further investigate how the Gypsy-44 insertion might be linked to the formation of the growth phenotype. For this purpose, apple seeds were subjected to stratification and radicles were harvested when they had reached a length of about 3 cm. Half of the radicle was used for DNA isolation and subsequent genotyping, while the remaining half containing the root tip was used for RNA isolation followed by Illumina sequencing. The transcriptomic reads were assembled and contigs were subjected to Basic Local Alignment Search Tool (BLAST) searches [33] as well as Blast2GO analyses [34,35] to gain a comprehensive view of the genes expressed in primary apple roots in general. Furthermore, the reads were mapped to the apple draft genome [36] and individual gene expression levels (normalized read counts) were compared between columnar and non-columnar radicles, with a special focus on the genes in the vicinity of Gypsy-44, the *Co* mutation. Gene expression patterns in the *Co* target region were compared across primary roots, leaves and shoot apical meristems by additional quantitative real-time PCRs (qRT-PCRs). Similarities and differences in the gene expression patterns of the different tissues were found. Our data make a substantial contribution to the understanding of the development of the columnar growth habit and primary root function in general.

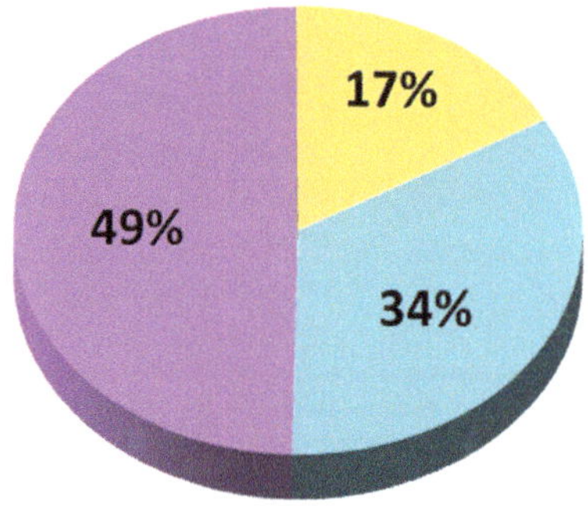

Figure 1 Genotypes of primary roots. The genotype of 119 apple primary roots was determined by marker PCRs with regard to the presence of the Gypsy-44 retrotransposon insertion.

Results

Homozygous columnar apple seedlings are viable

To investigate whether homozygous columnar apple seedlings show reduced viability or phenotypic effects compared to standard-type seedlings in early developmental stages, seeds obtained from apples of the heterozygous columnar cultivar 'Procats 28' (P28) that had been subjected to open pollination were germinated. As the trees were grown surrounded by other columnar apple varieties, the chance of pollination by a columnar father was high. After about 12 weeks of incubation at 4°C, the germination rate of the seeds was approximately 80%. No obvious phenotypic differences could be observed between individual radicles. The radicle genotype with regard to the presence of the columnar-specific Gypsy-44 transposable element (TE), which most likely represents the original *Co* mutation [10], was determined via PCRs. The diagnostic PCR assays as established by Otto et al. discriminate unambiguously between the non-columnar, the heterozygous columnar and the homozygous columnar genotype [10]. Of 119 seedlings subjected to genotyping, 40 seedlings were detected to be non-columnar, 59 showed a heterozygous columnar genotype, while the remaining 20 seedlings carried the columnar-specific Gypsy-44 insertion homozygously. These results could be confirmed by PCR assays using our indel-based markers I2_3_M1 and H1_M1 that are tightly linked to the *Co* mutation [37]. In total, a genotype ratio of non-columnar : heterozygous columnar : homozygous columnar seedlings of 2 : 3 : 1 (Figure 1) was detected. This suggests that homozygous columnar apple embryos are viable and most likely germinate at normal ratios.

Primary roots mainly express genes for growth, development and signaling

We sequenced RNA extracted from one primary root of each genotype and obtained about 118 million, 104 million and 126 million reads for the non-columnar, the heterozygous columnar and the homozygous columnar sample, respectively [EMBL: PRJEB6212] (Table 1). In order to investigate biological replicates, in a second approach, three radicles of each genotype were pooled prior to RNA isolation and were subjected to Illumina sequencing, yielding about 40 million, 28 million and 67 million reads for the non-columnar, the heterozygous columnar and the homozygous columnar sample, respectively [EMBL: PRJEB6212]. Illumina reads were assembled in the CLC Assembly Cell using different k-mer sizes, and assemblies yielding the highest N50 value were used for downstream analysis. For the first datasets comprising more than 100 million sequences, the highest N50 values were obtained for k = 17, for the smaller replicate datasets for k = 18. Table 1 summarizes the assembly results. In the larger datasets, about 44,000 – 49,000 contigs were produced, whereas assemblies of the smaller datasets yielded about 28,000 – 38,000 contigs. N50 values were in the range of 1,200 – 1,500 bp. With no mismatches allowed, more than 40% of the trimmed reads of each dataset could be mapped back to the corresponding contigs.

In order to find out how many and which genes were represented in the datasets, BLAST searches were conducted with all contigs against the annotated apple genes (MDPs), *Malus* x *domestica* expressed sequence tags (ESTs), *Malus* x *domestica* unigenes and the SwissProt/UniProtKB database (Table 2 and Table 3). A maximum of 89% of the sequences matched to a homolog in the MDP list, whereas only up to 60% of the contigs had a match against the SwissProt/UniProtKB database. Only up to 19 contigs had the same hit in the MDP database (meaning they most likely represent fragments of the same gene), indicating that about 89% of the contigs represent different genes. By contrast, up to 266 contigs yielded the same SwissProt/UniProtKB hit, thereby reducing the number of individual genes detected to about 11,500 (44% of the contigs created).

Replicate datasets were subjected to Blast2GO analysis. Of the sequences yielding BLAST hits against SwissProt/UniProtKB, 51% – 63% were successfully annotated.

Table 1 Results of assemblies

	# raw seqs	k-mer size	# contigs	GC content (%)	Longest contig	Shortest contig	N50	Reads mapped back (%)
PR (−/−)	118400555	17	48023	43	9994	200	1484	63
PR (Co/-)	104464241	17	44197	44	10381	200	1455	59
PR (Co/Co)	125731385	17	48932	43	16563	200	1487	60
PRs (−/−)	40261320	18	32471	45	11741	200	1288	43
PRs (Co/-)	27708416	18	28493	45	13248	200	1242	40
PRs (Co/Co)	67322394	18	37527	44	8970	200	1329	52

Illumina sequences were assembled and key values for contigs are indicated. The percentage of reads that was successfully mapped back to the contigs as a reference is shown in the last column.

About 86,000 – 100,000 Level 2 gene ontology (GO) terms were assigned (see Additional file 1), half of which were associated with growth and development (categories "metabolic process", "growth", "developmental process" and "cellular component organization or biogenesis") or with the integration and reaction to environmental signals (categories "response to stimulus", "biological regulation" and "signaling"). This is in line with the major roles of primary roots.

Differential gene expression in columnar versus non-columnar radicles is widespread

For evaluation of differential gene expression we chose a mapping approach rather than an assembly approach. For each sample, reads were mapped against the annotated Golden Delicious genome [36], total read counts for each MDP were extracted, and differential gene expression across the three genotypes was statistically analyzed in DESeq [38]. Highly expressed (base mean > 100), significantly up- or downregulated genes (fold change > 2 or < 0.5 and p-value < 0.05) were visualized in MapMan [39]. A Venn diagram showed high overlap of active genes (represented as MDPs to which at least one read was mapped) between the three genotypes (see Additional file 2). Furthermore, all libraries showed a high degree of correlation, with Pearson correlation coefficients ranging from 0.80 – 0.98, indicating that overall gene expression was similar across genotypes and that replicate datasets showed consistency for gene expression values (Table 4).

The comparison of non-columnar versus heterozygous columnar primary roots identified 194 significantly differentially expressed genes (see Additional file 3). Visualization in MapMan (Figure 2) showed that many genes were upregulated in heterozygous columnar roots when compared with non-columnar roots. This applied to genes involved in cellulose synthesis, cell wall modification and degradation, glycolysis, the oxidative pentose phosphate pathway, the biosynthesis of phenylpropanoids, starch and fatty acids and the mitochondrial electron transport. On the other hand, single genes involved in lipid and starch degradation and photorespiration were downregulated in the heterozygous columnar radicles (Figure 2A). Some genes associated with stress reactions such as ethylene biosynthesis and signaling (Figure 2B) were also more highly expressed in heterozygous columnar than in non-columnar primary roots. This also holds true for genes encoding the plastidic ribosomal proteins (Figure 2C).

The comparison of non-columnar versus homozygous columnar primary roots yielded 269 significantly differentially expressed genes (see Additional file 4) and most of them were induced in homozygous columnar radicles (Figure 3). Single genes encoding components of the carbohydrate metabolism, cellulose synthesis, cell wall modification and degradation, the oxidative pentose phosphate pathway, fermentation, starch synthesis and ascorbate/

Table 2 Results of BLASTx searches against MDPs and Malus ESTs

	MDPs		Malus ESTs	
	# contigs with hits (% of total contigs)	# different hits (% of contigs with hits)	# contigs with hits (% of total contigs)	# different hits (% of contigs with hits)
PR (−/−)	40010 (83)	25197 (63)	38094 (79)	31484 (83)
PR (Co/-)	37666 (85)	23966 (64)	35816 (81)	30183 (84)
PR (Co/Co)	40110 (82)	25100 (63)	38208 (78)	31583 (83)
PRs (−/−)	28377 (87)	19456 (69)	26746 (82)	23621 (88)
PRs (Co/-)	25278 (89)	17481 (69)	23550 (83)	21262 (90)
PRs (Co/Co)	32067 (85)	21350 (67)	30071 (80)	26012 (87)

The number and percentage of contigs yielding a hit and the number of different hits are indicated. The percentage of different hits refers to the total number of contigs and thus represents an estimate for the percentage of contigs representing individual genes.

Table 3 Results of BLASTx searches against Malus Unigene and SwissPro/UniProtKB

	Malus Unigene		SwissProt/UniProt KB	
	# contigs with hits (% of total contigs)	# different hits (% of contigs with hits)	# contigs with hits (% of total contigs)	# different hits (% of contigs with hits)
PR (–/–)	31593 (66)	17379 (55)	26221 (55)	11467 (44)
PR (Co/-)	29806 (67)	16883 (57)	24992 (57)	11238 (45)
PR (Co/Co)	31636 (65)	17288 (55)	26044 (53)	11404 (44)
PRs (–/–)	22545 (69)	14437 (64)	19102 (59)	9821 (51)
PRs (Co/-)	19923 (70)	13398 (67)	17111 (60)	9253 (54)
PRs (Co/Co)	25278 (67)	15373 (61)	21343 (57)	10331 (48)

The number and percentage of contigs yielding a hit and the number of different hits are indicated. The percentage of different hits refers to the total number of contigs and thus represents an estimate for the percentage of contigs representing individual genes.

glutathione metabolism were more highly expressed in the homozygous columnar radicles than in the non-columnar radicles (Figure 3A). On the other hand, single genes involved in fatty acid synthesis and lipid degradation, glycolysis, the tricarboxylic acid cycle, the mitochondrial electron transport chain and the Calvin cycle showed lower expression in the homozygous columnar roots than in the non-columnar primary roots. With regard to genes involved in stress reactions (Figure 3B), abscisic acid, ethylene and jasmonate-associated genes, genes encoding Myb transcription factors and peroxidases as well as genes involved in the maintenance of the redox state were upregulated. Genes linked to transcription and translation were also upregulated, and this holds true for nuclear as well as plastidic genes (Figure 3C).

When heterozygous and homozygous primary root samples were compared with each other, a similar picture emerged as for the analysis of non-columnar versus homozygous columnar radicles (see Additional file 5). 264 genes showed significant differential regulation (see Additional file 6) and of these, most genes were upregulated in the homozygous columnar primary roots. These genes are linked to primary and secondary metabolism (see Additional file 5A), stress reactions (see Additional file 5B) as well as transcription and translation (see Additional file 5C). The only exceptions were some genes involved in glycolysis and mitochondrial electron transport, which showed a lower expression in homozygous columnar radicles than in heterozygous columnar radicles.

The differentially expressed genes identified in each of the three binary analyses were compared among each other (Figure 4). No genes were identified as differentially expressed in all three comparisons. 80 of the genes showing differences in gene expression between heterozygous and non-columnar radicles also show differential expression in homozygous columnar compared with non-columnar radicles. 30 of the genes identified as differentially expressed in the heterozygous columnar versus the non-columnar radicles were also differentially expressed in the heterozygous versus the homozygous columnar radicles. Another 39 genes differentially expressed in the heterozygous versus the homozygous columnar radicles were also differentially expressed in the homozygous columnar versus non-columnar radicles.

Genes downstream of Gypsy-44 are downregulated in radicles but upregulated in aerial organs of columnar plants

In order to identify the link between the columnar-specific Gypsy-44 insertion and the columnar growth habit (most likely corresponding to the first, causal level of differential gene expression) we performed detailed expression analyses of all genes in the vicinity of the Gypsy-44 insertion. Eight genes that had previously been found to be transcribed and were annotated on our BAC metacontig based on the transcriptomic data [10] as well as Gypsy-44 itself were used for in-depth RNA-Seq and qRT-PCR analyses (Table 5, Figure 5). Fold changes were calculated for the primary root datasets as well as for three RNA-Seq Illumina datasets

Table 4 Pearson correlation coefficients of RNA-Seq libraries

	PR (–/–)	PRs (–/–)	PR (Co/-)	PRs (Co/-)	PR (Co/Co)	PRs (Co/Co)
PR (–/–)	1.00	0.84	0.80	0.83	0.85	0.86
PRs (–/–)		1.00	0.91	0.98	0.95	0.96
PR (Co/-)			1.00	0.93	0.93	0.89
PRs (Co/-)				1.00	0.95	0.96
PR (Co/Co)					1.00	0.92
PRs (Co/Co)						1.00

Correlation coefficients were calculated based on total read counts obtained for individual MDPs.

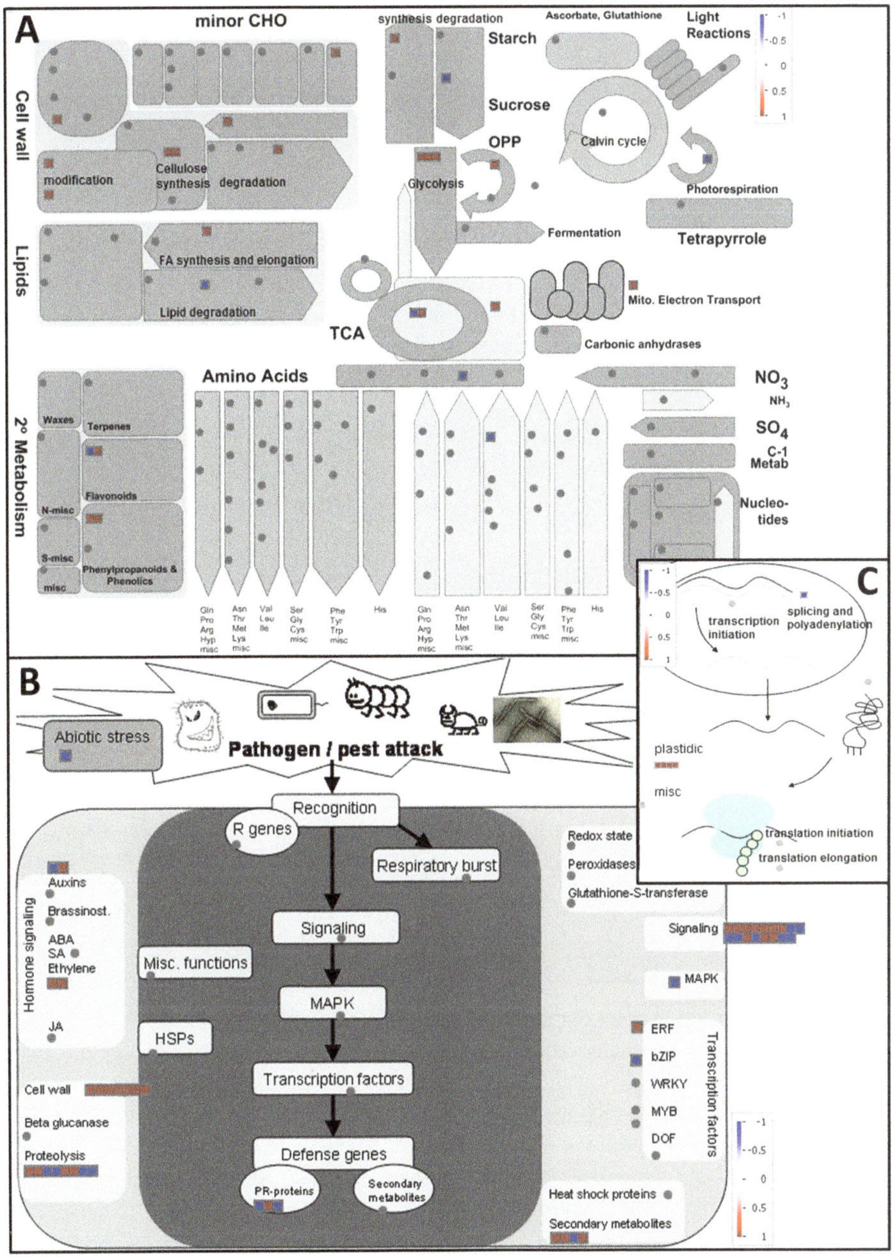

Figure 2 Differential gene expression in heterozygous compared with non-columnar primary roots. LOG2 fold changes of significantly differentially expressed genes (normalized to the non-columnar sample) as listed in Additional file 3 were imported and visualized in MapMan for the heterozygous columnar sample with regard to a metabolism overview **(A)**, pathogen/pest attack **(B)** and transcription and translation **(C)**. Genes upregulated in heterozygous columnar radicles are shown as red boxes, downregulated genes are shown as blue boxes.

each, generated from shoot apical meristems of A14 and P28 ([13,14], [EMBL: PRJEB2506]) and two transcriptomic datasets each, obtained from the total RNA of a McIntosh and a Wijcik leaf ([10], [EMBL: PRJEB1902]). If all genotypes of one tissue displayed read counts of 0 or 1 for a specific gene, this gene was considered not to be expressed (numeric read counts and fold changes can be found in Additional file 7). The results are shown in the upper part of Figure 5. *At1g08530-like* and *MDP0000934866* (*At1g06150-like*) showed similar expression levels for columnar and non-columnar varieties in all samples analyzed. For *MDP0000927091* (*Autophagy9-like*) and *5NG4-like*, similar levels of transcription were reached in all but the SAM3 dataset, in which they were induced in the columnar sample when compared with the non-columnar sample. Fold changes were less consistent between different tissues and across biological replicates for *MDP0000912172* (*PP2C15-like*), *MDP0000163720* (*ACC1-like*) and Gypsy-44

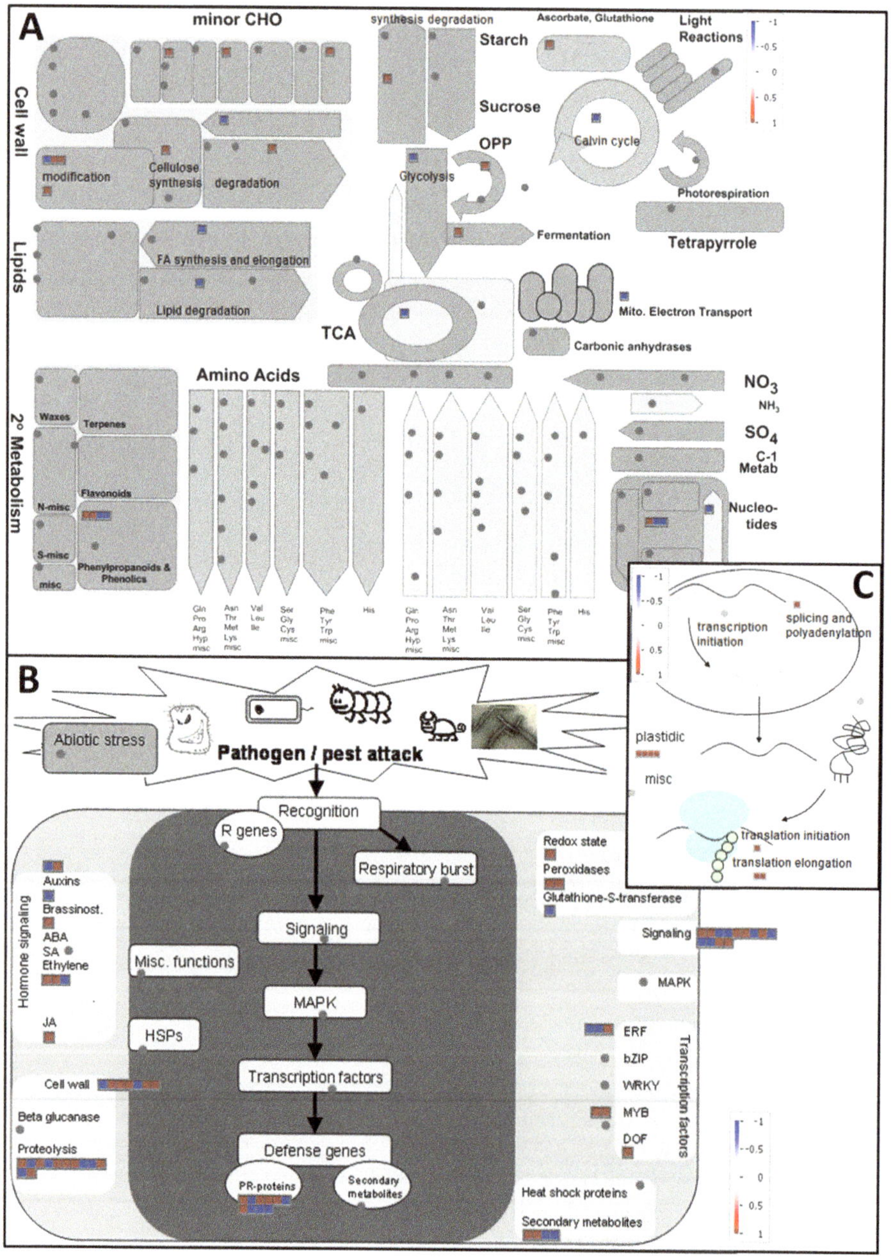

Figure 3 Differential gene expression in homozygous compared with non-columnar primary roots. LOG2 fold changes of significantly differentially expressed genes (normalized to the non-columnar sample) for the homozygous columnar sample as listed in Additional file 4 were imported and visualized in MapMan with regard to a metabolism overview **(A)**, pathogen/pest attack **(B)** and transcription and translation **(C)**. Genes upregulated in homozygous columnar radicles are shown as red boxes, downregulated genes are shown as blue boxes.

itself. For the former two genes, this might be caused by the generally lower expression level making fold changes prone to fluctuation. The most striking results were obtained for *dmr6-like* and *MDP0000927098* (*ATL5K-like*), the first two protein coding genes that follow downstream of the Gypsy-44 insertion: they were downregulated in the primary root samples and strongly upregulated in the shoot apical meristem of columnar varieties. These effects were most pronounced for *dmr6-like* in the shoot apical meristem, where no or only basal transcription (0 reads or 1 read) occurred in non-columnar A14, while expression was 30-, 53- and 56-times higher in the three biological replicates of P28.

Gypsy-44 itself, its direct neighboring genes and all other genes within the *Co* target region showing interesting differential gene expression were subjected to qRT-PCRs in order to verify the RNA-Seq results (Figure 5, lower part). Overall, the qRT-PCR results were mostly consistent with

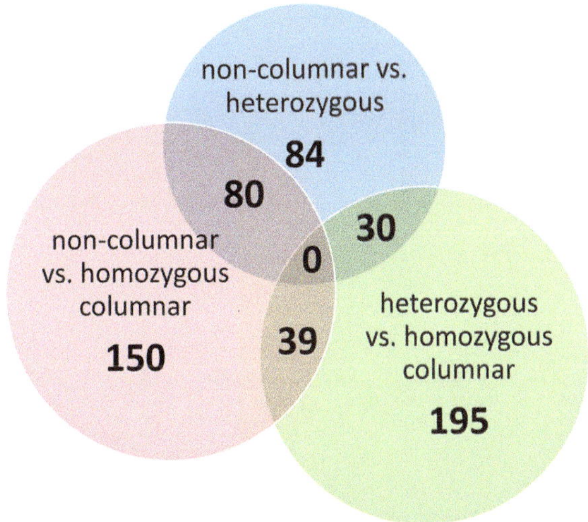

Figure 4 Venn diagram summarizing the differentially expressed genes identified in the three binary comparative gene expression analyses. While no genes were found to be differentially expressed across all three analyses, there was a high degree of overlap of the genes detected as differentially expressed in either columnar genotype versus the non-columnar genotype.

the RNA-Seq results. However, fold changes showed less variation across biological replicates in the qRT-PCRs than in the RNA-Seq data. Fold changes of around 1 were obtained for *At1g08530-like* and *MDP0000934866* (*At1g06150-like*) for all tissues with the exception of a 2.5-fold induction of *MDP0000934866* in one of the shoot apical meristem samples. Gypsy-44 was upregulated in the shoot apical meristem, but downregulated in leaves and heterozygous primary roots. *MDP0000163720* (*ACC1-like*) was induced in shoot apical meristems of P28 and slightly induced in heterozygous and homozygous primary roots of one replicate dataset. With regard to the two genes downstream of Gypsy-44, downregulation of *MDP0000927098* (*ATL5K-like*) in columnar primary roots was corroborated, whereas its upregulation in the shoot apical meristem was less pronounced in the qRT-PCR than in the RNA-Seq data. In the qRT-PCR results for *dmr6-like*, downregulation was only detected when heterozygous radicles were compared with non-columnar radicles, but not when homozygous columnar radicles were compared with non-columnar radicles. Strong upregulation was detected in the Wijcik leaf when compared with the McIntosh leaf. The very strong induction of *dmr6-like* in the shoot apical meristem samples was validated by a 41-fold and 202-fold induction in the columnar shoot apical meristem replicates.

In order to identify any conserved cis-regulatory sequences of *dmr6-like* whose effect might be impaired by the insertion of Gypsy-44, we conducted comparative sequence analyses among the Rosaceae species apple, pear (*Pyrus communis*), peach (*Prunus persica*), Chinese plum (*Prunus mume*) and strawberry (*Fragaria vesca*). In these species, the genes flanking the Gypsy-44 insertion are microcolinear (with an inverse orientation on linkage group 2 of *Fragaria*), enabling the analysis of conserved non-coding sequences (CNSs) in the intergenic region. Remarkably, the intergenic region between *At1g08530-like* and *dmr6-like* or their orthologs has a size of about 33 kb in *Malus* (without the additional 8.2 kb of Gypsy-44), 6.8 kb in *Pyrus*, 1.7 kb in *Fragaria* and 1.4 kb in *Prunus*, suggesting that it has served as a popular target of TE insertions in the Pyreae and especially in the *Malus* lineage. Within this region, one CNS of about 400 bp showing two peaks in the identity plot could be observed about 31 kb upstream of *dmr6-like* in *Malus* (Figure 6). The sequence of the second peak region yielded a hit against a class II TE of the Mariner group in a CENSOR BLAST search [40], whereas no information could be obtained from database searches for the first part, suggesting that it might contain sequences conserved owing to their importance for gene regulation.

In conclusion, there is evidence that Gypsy-44 influences the expression level of its direct downstream gene, *dmr6-like*, and most likely also some genes located further downstream, possibly via impairment of the function of a CNS.

Table 5 Genes annotated on the BAC metacontig

Gene Name	Position on BAC Metacontig (strand)	Position on Chr 10 of GD Genome	Probable Function
At1g08530-like	49164 – 52233 (+)	18743425 – 18746497	unknown
dmr6-like	93949 – 96036 (+)	18814292 – 18815452	defense reaction
MDP0000927098 (ATL5K-like)	103945 – 104770 (+)	18853768 – 18854596	ubiquitination
MDP0000927091 (Autophagy9-like)	119780 – 126200 (–)	18832988 – 18836788	recycling of cell components
5NG4-like	140776 – 142998 (–)	18862782 – 18864480	auxin-induced transporter
MDP0000912172 (PP2C15-like)	145267 – 146261 (+)	18866768 – 18867762	serine/threonine phosphatase
MDP0000934866 (At1g06150-like)	169780 – 175840 (+)	18884175 – 18886818	bHLH transcription factor
MDP0000163720 (ACC1-like)	184078 – 185923 (–)	18905698 – 18907541	ethylene biosynthesis

Eight genes were found to be expressed in at least one of the tissues investigated and were annotated on the BAC metacontig. Their name, position on the BAC metacontig and on chromosome 10 of the Golden Delicious (GD) draft genome sequence [36] as well as the possible function (according to BLAST searches) are indicated. bHLH, basic helix-loop-helix.

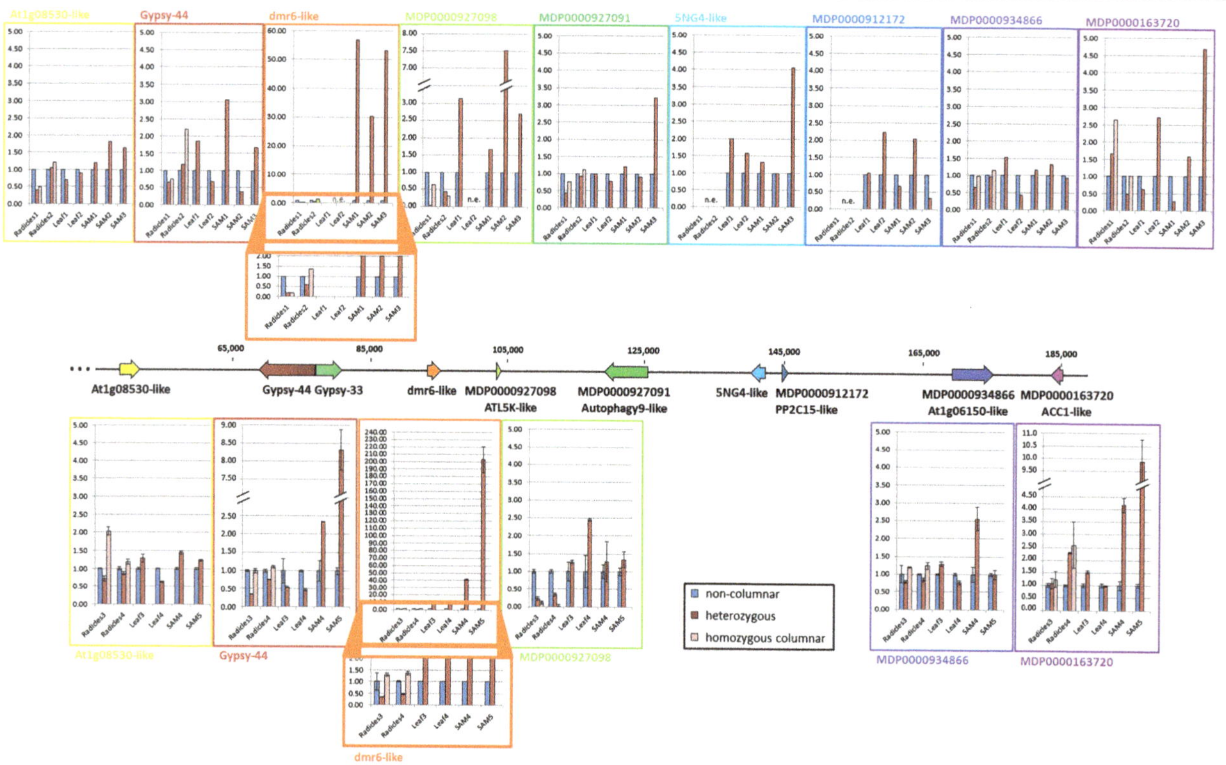

Figure 5 Differential gene expression in the *Co* target region. Fold changes of RNA-Seq samples (upper panel) and n-fold expression of qRT-PCR experiments (lower panel) in different tissues were calculated for up to eight genes located in the vicinity of Gypsy-44 as well as for Gypsy-44 itself in at least two biological replicates. Positions on the metacontig are given. Blue bars indicate fold changes for the non-columnar sample (normalized to 1), red and light pink bars indicate fold changes in the heterozygous and homozygous columnar sample, respectively, for a comparison with the non-columnar sample. Two parallel lines on a bar and the y axis represent a broken axis. Absolute read counts of 0 or 1 in both genotypes of a tissue are considered no expression (n.e.). Error bars in qRT-PCR bar chart represent standard deviations of three technical replicates. The two orange boxes below the two panels of graphs signify zoom-ins on the *dmr6-like* graphs within the region 0 – 2.

Discussion

Genotyping, sequencing and assemblies

So far, only heterozygous columnar cultivars have been described in the available literature, with one exception of a homozygous cultivar at Geisenheim University [10]. Therefore, Meulenbroek et al. [4] suggested that *Co* or a linked gene might negatively influence the fitness of pollen, seeds or early seedlings. However, crosses between two columnar apple trees have been shown to yield up to 75% columnar progeny [4,41], which is in accordance with the result of dominant Mendelian inheritance comprising 25% homozygous and 50% heterozygous columnar plants. In our genotyping experiments with the progeny of a heterozygous columnar cultivar we detected 17% homozygous columnar radicles. Hence, homozygous columnar primary roots are clearly viable. They did not show any deviating phenotypic

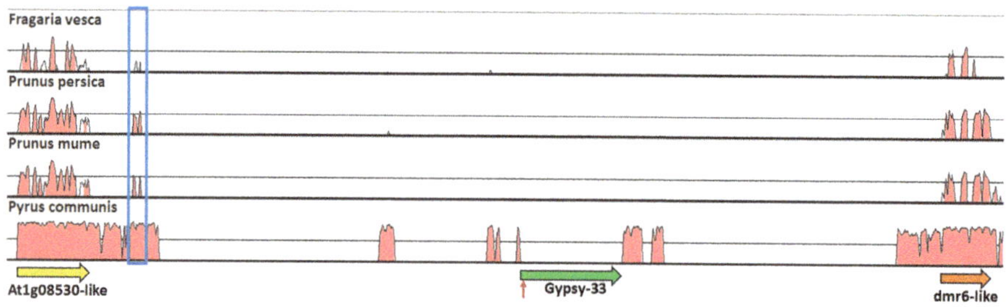

Figure 6 Conserved regions within the Gypsy-44 region. An mVISTA plot of sequence identity between the *Malus* x *domestica* sequence spanning the region from *At1g08530-like* to *dmr6-like* (x axis) and different Rosaceae species indicates sequence conservation in exon regions and one CNS present in all species investigated (blue box). Other conserved regions in Pyrus correspond to TEs that probably inserted before the divergence of pear and apple. The red arrow marks the Gypsy-44 insertion site.

differences either. The genotype ratio of non-columnar : heterozygous columnar : homozygous columnar of 2 : 3 : 1 is not really surprising because seeds were obtained by open pollination.

In this study, we detected that apple primary roots express between 9,000 and 32,000 distinct genes (represented as contigs yielding different BLAST hits). The number of genes identified as expressed is dependent on the dataset and the criteria used to distinguish individual genes. Assembling the smaller replicate datasets produced less contigs than assembling the larger datasets because increased sequencing depth facilitates detection of a higher number of transcripts. Regarding the total number of active genes in primary roots, we consider the numbers obtained from BLAST searches against the Malus Unigene dataset, around 17,000, to be the best estimate for those genes that are assembled. MDP and EST most likely show redundancy, listing alleles or fragments of one gene as individual genes, and the SwissProt/UniProtKB probably does not contain all the apple-specific genes. However, considering that not all contigs matched to a Malus Unigene entry and that not all reads were assembled into contigs, the actual number of active genes is probably significantly higher than estimated. There is still a lot of discussion about the real number of genes in apple. Within its genomic sequence, almost 58,000 genes have been anchored, which is the highest gene number reported for plants so far, and this has even been considered an underestimate [36]. However, pear only has about 42,000 genes, and when the apple genome is re-assembled filtering out overlapping genes in apple chromosomes that might correspond to alleles instead of individual genes, the gene number drops down to 45,293 [42]. This would be more consistent with gene numbers of other close relatives such as peach (27,852) [43] and strawberry (34,809) [44]. Newcomb et al. [45] conducted one of the first exhaustive EST analyses in apple and found about 43,000 non-redundant sequences, which they considered to be approximately half of the apple genes. By contrast, Allan et al. [46] assumed the apple EST dataset of 68,599 sequences from databases to be an overestimate. On the other hand, based on EST analyses by Sanzol [47], 68% of the apple genes fall into families with a mean copy number of 4.6 owing to several genome duplications, and the members of one family can have high sequence similarity but still represent different genes rather than alleles. Hence, the exact gene number remains debatable.

With regard to gene function, the well-known non-model-organism problem of an unsatisfying proportion of genes being successfully annotated is encountered [48]. We were able to assign GO terms to about 35% of contigs. This is slightly lower than the percentage of contigs annotated in RNA-seq studies of most other non-model organisms such as 50% in ten different invertebrates [49], 47 and 48% in the non-model plants *Streptocarpus rexii* and olive, respectively [50,51], and 40% in the bloom-forming alga *Emiliania huxleyi* [52]. However, it is much higher than the 15 – 19% of contigs contigs that could be successfully annotated in the non-model gastropod *Nerita melanotragus* [53]. Further efforts would most likely enable the assignment of possible functions to a higher number of contigs, similar to the techniques applied by [50]. Of those genes that were annotated, the majority had the expected function in growth, development and signaling, which is in agreement with the results obtained from transcriptome analyses of the poplar root and the *Arabidopsis* pericycle, where the most highly expressed genes were found to be involved in protein synthesis, metabolism, cellular communication and signal transduction [20,22].

Evaluation of differential gene expression

DESeq was used to evaluate differential gene expression since it has been found to perform well in a comparison of DESeq, edgeR, baySeq and a method employing a two-stage Poisson model [54]. The comparison of gene expression in columnar and non-columnar radicles yields a lower number of significantly differentially expressed genes than the comparison of gene expression in columnar and non-columnar shoot apical meristems [13]. Differential expression was detected for 200 – 300 and more than 600 genes in primary roots and shoot apical meristems, respectively, despite a more stringent definition of significant differential expression in the shoot apical meristem study of Krost et al. [13]. This might indicate that gene expression of columnar and non-columnar varieties is more similar in underground than in aerial organs. Alternatively, the difference in gene activity might be smaller in early developmental stages than at a higher age of the plants, which would explain the fact that the columnar growth habit can only be reliably detected after about 2 – 3 years [55,56]. However, interpretation of the differential expression data is hampered by the fact that the genetic background of the individual radicles used for RNA extraction is unclear due to the open pollination of the flowers. While the mother plant is P28, the father could be any of the dozens of different varieties grown on the field. Additionally we have no knowledge about recombination that might have occurred within the gametes. We chose this material because we wanted to find out whether homozygous columnar individuals occur under natural field conditions. However, for gene expression analyses seeds obtained from targeted crossings would be much more favorable. Moreover, radicles consist of distinct developmental regions in the longitudinal direction and radicles were halved without detailed prior investigation so that different developmental zones most likely harboring different gene expression patterns might have been grouped to each RNA isolation sample. In future studies, the radicles should be subdivided into their individual zones by microscopic control (possibly after fluorescence *in situ* hybridization against marker genes of the distinct zones, similarly to [17]) and the gene expression should be investigated separately for each zone. Nevertheless, due to the high degree of correlation across samples with regard to active genes (see Additional file 2) and read counts for each gene (Table 4), we are confident that we analyzed samples of similar developmental zones and stages.

The pattern of differential gene expression in primary roots is in concordance with that of aerial plant parts. The differential expression of genes involved in cell wall biosynthesis and modification observed for heterozygous and homozygous columnar primary roots has already been described for shoot apical meristems of columnar trees [14]. In addition, an increased transcription of genes associated with jasmonate signaling and biosynthesis has been found for shoot apical meristems and leaves of columnar varieties [10,13,14], and elevated ethylene signaling and biosynthesis seem to occur not only in columnar radicles, but also in leaves of columnar Wijcik compared with non-columnar McIntosh [10]. Therefore, alterations in cell boundary components and an increased level of jasmonates and ethylene might represent features of columnar apple trees that are already established in the early developmental stages and across different organs of the plantlet.

The first level of differential gene expression occurs in the Gypsy-44 vicinity

In a previous study [10] we found that columnar growth is most likely caused by the insertion of the Gypsy-44 LTR retrotransposon on chromosome 10 at 18.8 Mb, in accordance with the target region determined for the *Co* gene by Moriya et al. [57] and Baldi et al. [56]. Since it is a nested retrotransposon insertion, the link between its presence and the formation of the columnar phenotype remains to be unraveled. In this study, gene expression of Gypsy-44 itself and its neighboring genes was found to differ in columnar compared with non-columnar varieties in the individual plant organs. The expression of Gypsy-44 is induced in the shoot apical meristem. Its nearest downstream neighbor gene, *dmr6-like*, shows very strong (up to 200-fold) upregulation in shoot apical meristems and downregulation in the primary root samples of columnar individuals. The next gene downstream of *dmr6-like*, *MDP0000927098* (*ATL5K-like*), and one other downstream gene, *MDP0000163720* (*ACC1-like*), are also induced in shoot apical meristems of columnar P28. In contrast, expression of the upstream neighbor of Gypsy-44, *At1g08530-like*, remains unchanged. This indicates that Gypsy-44 influences the gene expression pattern of the downstream genes, especially *dmr6-like*, which has already been discussed as a possible candidate gene for the mediation of the columnar phenotype [10,12]. In addition to its induction in axillary meristems [12] and shoot apical meristems [10], *dmr6-like* is induced in our Wijcik leaf qRT-PCR samples, which were generated from RNA of the newly developing leaves near the top of the tree, but it shows no expression in the Wijcik leaf RNA-Seq data, which were obtained from RNA of young but fully developed leaves. We therefore conclude that *dmr6-like* is upregulated in all aerial organs of columnar trees that are important for the development of the plant organs showing differences between columnar and non-columnar plants: the shoot apical meristem, which is responsible for the formation of the short internodes and short lateral branches, the axillary meristem, which does not develop into lateral branches but instead produces short spurs, and the young leaves, which form a thicker palisade parenchyma with a high chlorophyll content [58]. In contrast, this upregulation does not occur in organs that have finished their development, such as older leaves, as well as in primary roots, which are less significant for the growth habit of the aerial organs; instead, a reverse regulation seems to take place on the below-ground organs. Wolters et al. [12] have already overexpressed *dmr6-like* in *Arabidopsis* and found the plants to have shorter internodes and smaller branches, strongly suggesting that *dmr6-like* is involved in the formation of the columnar growth habit. Our results support this hypothesis, but also raise further questions, e.g. how the tissue specificity of this induction is achieved.

Possible effects of the Gypsy-44 insertion like alterations in the methylation pattern or the allocation of enhancers and/or silencers have already been discussed [10]. The tissue specificity of this effect suggests that the Gypsy-44 insertion with a size of 8.2 kb might move a tissue-specific silencer of *dmr6-like* that is normally active in apical and axillary meristems and young developing leaves to a distance from the gene that abolishes silencer function, resulting in upregulation of *dmr6-like*. Alternatively, an enhancer being active in those tissues crucial for development might be moved into spatial proximity of *dmr6-like* due to an altered chromosome architecture (e.g. looping of the Gypsy-44 containing region) and thereby cause induction in tissues where *dmr6-like* is normally not expressed. We identified a candidate for a CNS that might have a role in gene regulation at 31.4 kb distance upstream of *dmr6-like*. This is reminiscent of the situation for the *vgt1* locus in maize, a CNS positioned 70 kb downstream of *ZmRap2.7*. *ZmRap2.7* is a repressor of flowering and its expression is downregulated in varieties that carry a miniature inverted repeat TE (MITE) insertion within *vgt1*, resulting in early flowering [59]. However, in our case the TE insertion does not occur at the CNS site, but downstream of it. This might still influence its function due to the increased distance between the CNS and *dmr6-like*. The fact that the region crucial for the formation of columnar growth, including *dmr6-like*, is conserved across several Rosaceae species might provide an opportunity for the creation not only of new columnar apple cultivars, but also of columnar pear and plum varieties by genetic engineering in future.

In *Arabidopsis*, *dmr6* loss-of-function mutants show enhanced expression of a subset of defense-associated genes, suggesting a control different from that in apple [60]. Interestingly, the truncated protein of the *dmr6-1* mutant is constitutively expressed in the shoot apical meristem and leaf primordia of *Arabidopsis* [60], in line with the expression profile of *dmr6-like* in columnar apple trees. The regulatory pathways influenced by columnar growth downstream of the *dmr6-like* upregulation still await elucidation. Since *dmr6-like* is associated with downy mildew resistance in *Arabidopsis* [60] it might trigger a defense response leading to a diseased-looking growth habit. On the other hand, members of the 2OG Fe(II) oxygenase family are involved in the biosynthesis of ethylene, gibberellins and flavonoids, the latter of which modulate polar auxin transport, so that upregulation of *dmr6-like* might directly influence phytohormone concentrations [12,61-64]. In apple, silencing of the chalcone synthase, the first committed enzyme in flavonoid biosynthesis, leads to decreased concentrations of flavonoids and causes altered cellular organization and cell wall composition, increased auxin transport and a reduction of plant size due to shortened internodes [65]. Specifically some genes encoding indole-3-acetic acid (IAA) transporters show upregulation in shoot apical meristem of P28 compared with A14, in line with an altered auxin transport capacity [13].

Furthermore, besides the induction of *dmr6-like*, other gene expression changes in the Gypsy-44 region or the upregulation of Gypsy-44 itself in shoot apical meristems might play a role in phenotype generation. Induction of *MDP0000163720* in columnar shoot apical meristems, which encodes a homolog of Aminocyclopropane-1-carboxylate oxidase, might lead to increased levels of ethylene. Since ethylene crosstalks with many other phytohormones such as auxin, cytokinin, abscisic acid, jasmonates and salicylic acid (for example [66-70]), this could contribute to the altered hormonal balance in columnar trees.

Conclusions

Open pollination of heterozygous columnar apple trees yield non-columnar, heterozygous columnar and homozygous columnar seedlings in a genotype ratio of 3 : 2 : 1. This suggests that homozygous columnar plants are viable within their early life stages and the lack of described homozygous columnar varieties is probably due to breeders' selection criteria. Apple primary roots display a transcriptome profile typical for a developing and actively growing tissue. About 200 – 300 genes are differentially regulated in columnar compared with non-columnar radicles, and these are involved in cell wall synthesis and modification, stress reactions and phytohormone signaling and biosynthesis. The molecular cause of columnar growth, the Gypsy-44 insertion, leads to upregulation of its nearest downstream gene, *dmr6-like*, in all tissues that are crucial to the development of the visible plant growth habit. The role of *dmr6-like* in pathogen resistance possibly results in upregulation of defense reactions in aerial plant organs leading to a phenotype reminiscent of a diseased plant. Alternatively or additionally, the upregulation of *dmr6-like* might lead to alterations in flavonoid content causing altered auxin transport, which would be in line with changes in auxin signaling and metabolism that have been detected in previous studies. The Gypsy-44 insertion is the causative event, then leading to upregulation of *dmr6-like* as a second step. In columnar primary roots, however, *dmr6-like* is downregulated, so that the activity of a tissue-specific transcriptional regulator might play a pivotal role. We are confident that our data will contribute to the elucidation of the whole signaling cascade causing columnar growth in apples.

Methods

Plant material

For primary root transcriptome studies, apples from the heterozygous columnar cultivar P28 obtained by open pollination on a field with many columnar apple trees were collected in August 2011 at Geisenheim University. Seeds were extracted and dried at room temperature for about 1 – 2 weeks.

For qRT-PCRs, shoot tips of columnar P28 and non-columnar 'A14-190-93K' (A14) sampled on September 29, 2009 and very young apical leaves of 'McIntosh' and 'McIntosh Wijcik' sampled on July 19, 2013 were snap frozen in dry ice or liquid nitrogen, respectively, and stored at −80°C until further use.

Seed germination

Dried seeds were swabbed with 70% ethanol to roughly remove contamination and were placed in Petri dishes equipped with towel-paper that had been moistened with tap water. The petri dishes were wrapped in tinfoil and were stored in the dark at 4°C for about 12 weeks. When the radicles reached a length of about 2 – 3 cm they were snap frozen in liquid nitrogen, cut off the seed, cut in half and stored at −80°C until further use.

DNA isolation

The upper half of each radicle was subjected to DNA extraction according to a manual protocol. One radicle was transferred into 500 μL Cetyltrimethyl ammonium bromide (CTAB) extraction buffer, to which 168 μL 8 M urea and a spatula tip of polyvinylpyrrolidone (PVPP) were added. The radicle was thoroughly homogenized with a micropestle followed by incubation at 65°C for

30 min. After centrifugation (10 min at 13,000 rpm and room temperature) the supernatant was purified by three chloroform isoamyl alcohol (24:1) extractions and subsequent ethanol precipitation.

Primary root genotyping

In order to investigate the genotype of each radicle with regard to the presence of the Gypsy-44 retrotransposon associated with the columnar growth habit, PCRs were performed with primers spanning the left border, the right border or the whole TE insertion as described by Otto et al. [10]. Results were double checked by PCRs with our indel-based markers I2_3_M1 and H1_M1 [37]. All primers were purchased from Invitrogen^{TM} (Invitrogen, Darmstadt, Germany). PCRs were carried out in 50 µL reaction volume containing at least 10 ng of radicle DNA and 1 U Go-Taq polymerase (Promega, Madison, United States) as follows: initial denaturation at 94°C for 5 min followed by 40 amplification cycles of denaturation at 94°C for 30 sec, annealing at 58.4°C or 60°C for 30 sec and elongation at 72°C for 30 sec. A final elongation step at 72°C for 10 min was included. PCR products were analyzed on horizontal 2% agarose gels.

RNA extraction

The distal halves (tips) of radicles were used for RNA extraction. For the initial Illumina RNA-Seq, one radicle of each genotype was subjected to total RNA isolation. For replicate datasets, three frozen radicles of one genotype were pooled prior to RNA extraction. Total RNA was isolated using the innuPREP Plant RNA Kit (Analytik Jena, Jena, Germany) following the manufacturer's instructions. A DNase I (Fermentas, St. Leon-Roth, Germany) digestion step on the column was included to remove genomic DNA. RNA concentration and integrity were assessed with Bioanalyzer RNA Nano or Pico chips (Agilent, Santa Clara, USA).

For qRT-PCRs with leaf and shoot tip material, samples were transferred to liquid nitrogen, disrupted with a mortar and pestle and RNA was isolated using the innuPREP Plant RNA kit (Analytik Jena, Jena, Germany) following the manufacturer's instructions.

Illumina sequencing

For RNA-Seq library construction 500 ng of total RNA of each primary root sample were sent to GENterprise Genomics (Mainz, Germany). RNA Integrity Number (RIN) values were 7.6, 7.8 and 8.3 for the total RNA obtained from the single non-columnar, heterozygous columnar and homozygous columnar radicle, repectively. The RNA of the replicate samples had RIN values of 8.2, 7.0 and 7.0 for the non-columnar, the heterozygous columnar and the homozygous columnar genotype, respectively. 100 bp paired-end runs were conducted on the Illumina HiSeq 2000 or Illumina HiSeq 2500 (Illumina, San Diego, USA) of the IMSB Mainz.

Assembly of Illumina data

Prior to assembly, Illumina raw data were trimmed with an in-house Perl script to remove low quality stretches. 6 bp at the 5' end and 5 bp at the 3' end were cut off, and bases with a Phred score < 20 or called as "N" were removed. After trimming, only sequences with a minimum length of 30 bp were maintained. All remaining sequences were assembled with the CLC Assembly Cell (CLC bio, Aarhus, Denmark) using different k-mer sizes and a bubble size of 300. The assemblies yielding the highest N50 value were used for BLAST searches. To evaluate these assemblies, trimmed reads were mapped back to contigs with Bowtie [71], allowing no mismatches.

BLAST searches

Contigs generated by assembling were subjected to BLAST searches [33] against 1) MDPs representing the genes annotated in the Golden Delicious genome sequence [36] as extracted from the Genome Database of Rosaceae (http://www.rosaceae.org/), 2) the NCBI Malus expressed sequence tag (EST) dataset, 3) the Malus Unigene v5.0 dataset representing filtered ESTs assembled with CAP3 [72], available at the Genome Database of Rosaceae (http://www.rosaceae.org/), and 4) the NCBI SwissProt/UniProtKB database, all databases downloaded on February 3, 2014. BLAST results were filtered and analyzed at an E-Value cutoff of 10^{-5}.

Blast2GO

Results of BLAST searches with the replicate datasets against SwissProt/UniProtKB were imported into CLC Genomics Workbench v6.5 (CLC bio, Aaarhus, Denmark) and analyzed with the Blast2GO plugin using default parameters [34,35]. Following GO Slim reduction, combined graphs of the GO term biological process were created and were converted to pie charts of the second-level terms.

RNA-Seq analysis

Illumina raw sequencing data were imported into CLC Genomics Workbench v6.5 (CLC bio, Aarhus, Denmark) as paired-end reads and were subjected to trimming and RNA-Seq mappings against our BAC metacontig of the transposon region or the Golden Delicious reference genome as previously described [10]. A similarity of at least 98% was demanded for at least 90% of the read length and using the "include broken pairs" counting scheme. For visualization of differential gene expression between two datasets each, total gene read counts for each MDP were extracted from RNA-Seq mapping results and analyzed in DESeq [38] with default parameters and including the biological replicates. Only genes with base mean > 100, fold change > 2 (or < 0.5) and p-value < 0.05 were maintained, and genes were ranked according to their p-values. Subsequently, the LOG2 fold changes were loaded into MapMan [39] and mapped against the *Malus* x *domestica* mapping file.

qRT-PCRs

Genes showing interesting differential regulation were subjected to qRT-PCR analysis in order to verify the bioinformatics results. Primers were designed to span the last intron of each gene and were purchased from Invitrogen (Darmstadt, Germany). A list of primers can be found in Additional file 8. 800 ng of total RNA extracted from three pooled primary roots of each genotype (RIN > 7) and 200 ng of human MDA-MB468 RNA serving as exogenous control were converted to cDNA using SuperScript III Reverse Transcriptase (Invitrogen, Darmstadt, Germany) and 100 μM oligo$(dT)_{18}$VN primer according to the manufacturer's instructions. In order to compare results of different tissues, the same procedure was applied to RNA extracted from shoot tips (shoot apical meristems) of A14 and P28 and to RNA isolated from young leaves of McIntosh and McIntosh Wijcik. All qRT-PCR experiments were carried out in a 7500 Fast Real-Time PCR System using Power SYBR Green PCR Master Mix (Applied Biosystems, Carlsbad, USA). Reactions were conducted in triplicate and two biological replicates were analyzed.

Conserved sequence analysis

Orthologs were retrieved from pear, peach, Chinese plum (Mei) and strawberry based on BLASTx searches yielding PCP006734.1 and PCP039177.1 (pear), ppa009607m and ppa014841m (peach), Pm020608 and Pm020609 (Chinese plum) and mrna08066.1-v1.0-hybrid and mrna08065.1-v1.0-hybrid (strawberry) as the best hits for At1g08530-like and dmr6-like, respectively. The corresponding genomic sequences were extracted from GDR and were aligned in mVISTA (http://genome.lbl.gov/vista/index.html) [73] with parameters "calculated window" of 100 bp, "minimum consensus width" 100 bp and "consensus identity" 70%.

Availability of supporting data

Transcriptomic Illumina data of primary roots, SAMs and leaves can be found in the EMBL EBI Short Read Archive under study accession numbers PRJEB6212 (http://www.ebi.ac.uk/ena/data/view/PRJEB6212), PRJEB2506 (http://www.ebi.ac.uk/ena/data/view/PRJEB2506) and PRJEB1902 (http://www.ebi.ac.uk/ena/data/view/PRJEB1902), respectively. The annotated genomic contig of the Gypsy-44 region including all annotated genes can be retrieved from the EMBL EBI GenBank repository under accession number HF968765 (http://www.ebi.ac.uk/ena/data/view/HF968765). All other supporting data can be found in the supplementary online material of this article.

Additional files

Additional file 1: Pie charts of level 2 GO terms assigned to contigs of primary root transcriptome assemblies. Assemblies of primary root transcriptome Illumina reads were subjected to BLAST searches against SwissProt/UniProtKB and subsequent Blast2GO analysis. Pie charts of level 2 GO terms are shown for data obtained from three pooled non-columnar **(A)**, heterozygous columnar **(B)** and homozygous columnar **(C)** primary roots.

Additional file 2: Venn diagram of active genes in the datasets of the non-colmnar, heterozygous and homozygous columnar primary roots. Annotated MDPs to which at least one read of the original or replicate RNA-Seq dataset of primary roots could be mapped were considered active. Active genes were compared across genotypes. 46,417 genes are active in all three genotypes, 1385 genes are active in the non-columnar and the heterozygous columnar samples, but not in the homozygous colmnar samples. 1005 genes are expressed in the non-clumnar and the homozygous columnar roots, but not in the heterozygous columnar roots. 655 active genes are shared between heterozygous and homozygous columnar primary roots and are not expressed in non-columnar radicles. 630, 430 and 2845 genes are expressed only in the non-columnar, the hetrozygous and the homozygous columnar radicles, respectively.

Additional file 3: Table of significantly differentially expressed genes in heterozygous compared with non-columnar primary roots. Total gene reads of the two biological replicate data were imported into DESeq and analysed for differential expression. Only genes with base mean > 100, fold change > 2 and p-value < 0.05 are displayed. The number of the MapMan BIN each gene was assigned to as well as the gene name, identifier (PAC-ID) and description are given as in the MapMan mapping file for apple.

Additional file 4: Table of significantly differentially expressed genes in homozygous columnar compared with non-columnar primary roots. Total gene reads of the two biological replicate data were imported into DESeq and analyzed for differential expression. Only genes with base mean > 100, fold change > 2 and p-value < 0.05 are displayed. The number of the MapMan BIN each gene was assigned to as well as the gene name, identifier (PAC-ID) and description are given as in the MapMan mapping file for apple.

Additional file 5: Figure of differentially expressed genes in homozygous columnar compared with heterozygous primary roots. LOG2 fold changes of significantly differentially expressed genes (normalized to the heterozygous sample) as listed in Additional File 6 were imported and were visualized in MapMan for the homozygous columnar sample with regard to a metabolism overview (A), pathogen/pest attack (B) and transcription and translation (C). Genes upregulated in homozygous columnar radicles are shown as red boxes, downregulated genes are shown as blue boxes.

Additional file 6: Table of significantly differentially expressed genes in homozygous columnar compared with heterozygous primary roots. Total gene reads of the two biological replicate data were imported into DESeq and analyzed for differential expression. Only genes with base mean > 100, fold change > 2 and p-value < 0.05 are displayed. The number of the MapMan BIN each gene was assigned to as well as the gene name, identifier (PAC-ID) and description are given as in the MapMan mapping file for apple.

Additional file 7: Table of total read counts and fold changes of genes and Gypsy-44 within the *Co* target region. Total Read counts of RNA-Seq data are displayed.

Additional file 8: Primer List. Primers used for marker PCRs and qRT-PCR.

Abbreviations

ACC1: 1-aminocyclopropane-1-carboxylate synthase 1; BLAST: Basic Local Alignment Search Tool; *Co*: *Columnar*; CNS: conserved non-coding sequence; CTAB: cetyltrimethyl ammonium bromide; *dmr6*: *Downy mildew resistant 6*; EST: Expressed sequence tag; GO: Gene Ontology; IAA: Indole-3 acetic acid; LTR: Long terminal repeat; MDP: *Malus* x *domestica* protein; PP2C: Protein phosphatase 2C; PVPP: Polyvinylpyrrolidone; qRT-PCR: Quantitative real-time polymerase chain reaction.

Competing interests
The authors declare that they have no competing interests.

Authors' contributions
RP carried out differential gene expression analysis and qRT-PCRs and drafted the manuscript. HD participated in RNA-Seq analysis. BR conducted assemblies and BLAST searches. BR and SR analyzed assemblies and provided bioinformatics support. ERS initiated the project and supervised the project throughout. All authors read and approved the final manuscript.

Acknowledgements
We would like to thank the Institutes of Botany and Pomology of the Geisenheim University for providing Procats 28 apples, Clemens Krost for advice on DESeq and qRT-PCR analysis and Ramona Petersen for linguistic proofreading. We are grateful to Oliver Dautermann, Inessa Penner and Nina Seiwert for help with genotyping and RNA extraction. This work was supported by grants of the Federal Ministry of Agriculture and Nutrition (no. 511–06.01-28-1-43.042-07 and no. 313–06.01-28-1-43.042-07)

References
1. Lapins KO. Segregation of compact growth types in certain apple seedling progenies. Can J Plant Sci. 1969;49:765–8.
2. Tobutt KR. Breeding columnar apple trees at east malling. Acta Hortic. 1985;159:63–8.
3. Gelvonauskienë D, Gelvonauskis B, Sasnauskas A. Impact of rootstock on columnar apple tree growth in a nursery. Sci Work Lith Inst Hortic Lith Univ Agric. 2006;25:51–6.
4. Meulenbroek J, Verhaegh J, Janse J. Inheritance studies with columnar type trees. Acta Hortic. 1998;484:255–8.
5. Moriya S, Iwanami H, Kotoda N, Takahashi S, Yamamoto T, Abe K. Development of a marker-assisted selection system for columnar growth habit in apple breeding. J Jpn Soc Hortic Sci. 2009;78:279–87.
6. Lauri PE, Lespinasse JM. The Relationship between cultivar fruiting-type and fruiting branch characteristics in apple trees. Acta Hortic. 1993;349:259–63.
7. Petersen R, Krost C. Tracing a key player in the regulation of plant architecture: the columnar growth habit of apple trees (*Malus × domestica*). Planta. 2013;238:1–22.
8. Fisher DV. Spur-type strains of mcintosh for high density plantings. Br Columbia Fruit Grow Assoc Q Rep. 1969;14:3–10.
9. Fisher DV. The "Wijcik Spur McIntosh.". Fruit Var J. 1995;49:212–3.
10. Otto D, Petersen R, Brauksiepe B, Braun P, Schmidt ER. The columnar mutation ("*Co* gene") of apple (*Malus × domestica*) is associated with an integration of a *Gypsy*-like retrotransposon. Mol Breed. 2014;33:863–80.
11. Tian Y-K, Wang C-H, Zhang J-S, James C, Dai H-Y. Mapping *Co*, a gene controlling the columnar phenotype of apple, with molecular markers. Euphytica. 2005;145:181–8.
12. Wolters PJ, Schouten HJ, Velasco R, Si-Ammour A, Baldi P. Evidence for regulation of columnar habit in apple by a putative 2OG-Fe(II) oxygenase. New Phytol. 2013;200:993–9.
13. Krost C, Petersen R, Lokan S, Brauksiepe B, Braun P, Schmidt ER. Evaluation of the hormonal state of columnar apple trees (*Malus × domestica*) based on high throughput gene expression studies. Plant Mol Biol. 2013;81:211–20.
14. Krost C, Petersen R, Schmidt ER. The Transcriptomes of Columnar and Standard Type Apple Trees (*Malus × domestica*) – a Comparative Study. Gene. 2012;498:223–30.
15. Darwin CR, assisted by DF. The Power of Movement in Plants. London: John Murray; 1880.
16. Baluska F, Mancuso S, Volkmann D, Barlow PW. The "root-brain" hypothesis of Charles and Francis Darwin: Revival after more than 125 years. Plant Signal Behav. 2009;4:1121–7.
17. Birnbaum K, Shasha DE, Wang JY, Jung JW, Lambert GM, Galbraith DW, et al. A gene expression map of the *Arabidopsis* root. Science. 2003;302:1956–60.
18. Nawy T, Lee J-Y, Colinas J, Wang JY, Thongrod SC, Malamy JE, et al. Transcriptional profile of the *Arabidopsis* root quiescent center. Plant Cell. 2005;17:1908–25.
19. Brady SM, Orlando DA, Lee J-Y, Wang JY, Koch J, Dinneny JR, et al. A high-resolution root spatiotemporal map reveals dominant expression patterns. Science. 2007;318:801–6.
20. Dembinsky D, Woll K, Saleem M, Liu Y, Fu Y, Borsuk LA, et al. Transcriptomic and proteomic analyses of pericycle cells of the maize primary root. Plant Physiol. 2007;145:575–88.
21. Hoecker N, Lamkemeyer T, Sarholz B, Paschold A, Fladerer C, Madlung J, et al. Analysis of nonadditive protein accumulation in young primary roots of a maize (Zea mays L.) F1-hybrid compared to its parental inbred lines. Proteomics. 2008;8:3882–94.
22. Kohler A, Dalaruelle C, Martin D, Encelot N, Martin F. The poplar root transcriptome: analysis of 7000 expressed sequence tags. FEBS Lett. 2003;542:37–41.
23. Eichholtz DA, Robitaille HA, Herrmann KM. Protein Changes during the Stratification of *Malus domestica* Borkh. Seed Plant Physiol. 1983;72:750–3.
24. Bogatek R, Dziewanowska K, Lewak S. Hydrogen cyanide and embryonal dormancy in apple seeds. Physiol Plant. 1991;83:417–21.
25. Gniazdowska A, Dobrzyńska U, Babańczyk T, Bogatek R. Breaking the apple embryo dormancy by nitric oxide involves the stimulation of ethylene production. Planta. 2007;225:1051–7.
26. Gniazdowska A, Krasuska U, Debska K, Andryka P, Bogatek R. The beneficial effect of small toxic molecules on dormancy alleviation and germination of apple embryos is due to NO formation. Planta. 2010;232:999–1005.
27. Gniazdowska A, Krasuska U, Bogatek R. Dormancy removal in apple embryos by nitric oxide or cyanide involves modifications in ethylene biosynthetic pathway. Planta. 2010;232:1397–407.
28. Dębska K, Krasuska U, Budnicka K, Bogatek R, Gniazdowska A. Dormancy removal of apple seeds by cold stratification is associated with fluctuation in H_2O_2, NO production and protein carbonylation level. J Plant Physiol. 2013;170:480–8.
29. Petricka JJ, Winter CM, Benfey PN. Control of *Arabidopsis* root development. Annu Rev Plant Biol. 2012;63:563–90.
30. Dolan L, Janmaat K, Willemsen V, Linstead P, Poethig S, Roberts K, et al. Cellular organisation of the *Arabidopsis thaliana* root. Development. 1993;119:71–84.
31. Van den Berg C, Weisbeek P, Scheres B. Cell fate and cell differentiation status in the *Arabidopsis* root. Planta. 1998;205:483–91.
32. Van den Berg C, Willemsen V, Hendriks G, Weisbeek P, Scheres B. Short-range control of cell differentiation in the *Arabidopsis* root meristem. Nature. 1997;390:287–9.
33. Altschul SF, Gish W, Miller W, Myers EW, Lipman DJ. Basic local alignment search tool. J Mol Biol. 1990;215:403–10.
34. Conesa A, Götz S, García-Gómez JM, Terol J, Talón M, Robles M. Blast2GO: a universal tool for annotation, visualization and analysis in functional genomics research. Bioinformatics. 2005;21:3674–6.
35. Conesa A, Götz S. Blast2GO: a comprehensive suite for functional analysis in plant genomics. Int J Plant Genomics. 2008;2008:619832.
36. Velasco R, Zharkikh A, Affourtit J, Dhingra A, Cestaro A, Kalyanaraman A, et al. The genome of the domesticated apple (*Malus × domestica* Borkh). Nat Genet. 2010;42:833–9.
37. Otto D, Petersen R, Krost C, Schmidt ER, Brand R, Brauksiepe B, Braun P: Molecular Characterization of the *Co* Gene Region in *Malus × domestica*. *Acta Hortic*, under review.
38. Anders S, Huber W. Differential expression analysis for sequence count data. Genome Biol. 2010;11:R106.
39. Thimm O, Bläsing O, Gibon Y, Nagel A, Meyer S, Krüger P, et al. Mapman: a user-driven tool to display genomics data sets onto diagrams of metabolic pathways and other biological processes. Plant J. 2004;37:914–39.
40. Kohany O, Gentles AJ, Hankus L, Jurka J. Annotation, submission and screening of repetitive elements in Repbase: RepbaseSubmitter and Censor. BMC Bioinformatics. 2006;7:474.
41. Lapins KO. Inheritance of compact growth type in apple. J Am Soc Hortic Sci. 1976;101:133–5.
42. Wu J, Wang Z, Shi Z, Zhang S, Ming R, Zhu S, et al. The genome of the pear (*Pyrus bretschneideri* Rehd.). Genome Res. 2013;23:396–408.
43. Initiative TIPG. The high-quality draft genome of peach (*Prunus persica*) identifies unique patterns of genetic diversity, domestication and genome evolution. Nat Genet. 2013;45:487–94.
44. Shulaev V, Sargent DJ, Crowhurst RN, Mockler TC, Folkerts O, Delcher AL, et al. The genome of woodland strawberry (*Fragaria vesca*). Nat Genet. 2011;43:109–16.

45. Newcomb RD, Crowhurst RN, Gleave AP, Rikkerink EHA, Allan AC, Beuning LL, et al. Analyses of expressed sequence tags from apple. Plant Physiol. 2006;141:147–66.
46. Allan AC, Crowhurst R, Gleave A, Newcomb R, Schaffer R. Apple Functional Genomics. In: Folta KM, Gardiner SE, editors. Genet Genomics Rosaceae. 2009. p. 121–42.
47. Sanzol J. Dating and functional characterization of duplicated genes in the apple (*Malus domestica* Borkh.) by analyzing EST data. BMC Plant Biol. 2010;10:87.
48. Horan K, Jang C, Bailey-Serres J, Mittler R, Shelton C, Harper JF, et al. Annotating genes of known and unknown function by large-scale coexpression analysis. Plant Physiol. 2008;147:41–57.
49. Riesgo A, Andrade SCS, Sharma PP, Novo M, Pérez-Porro AR, Vahtera V, et al. Comparative description of ten transcriptomes of newly sequenced invertebrates and efficiency estimation of genomic sampling in non-moel taxa. Front Zool. 2012;9:33.
50. Chiara M, Horner DS, Spada A. *De novo* assembly of the transcriptome of the non-model plant *streptocarpus rexii* employing a novel heuristic to recover locus-specific transcript clusters. PLoS One. 2013;8:e80961.
51. Alagna F, D'Agostino N, Torchia L, Servili M, Rao R, Pietrella M, et al. Comparative 454 pyrosequencing of transcripts from two olive genotypes during fruit development. BMC Genomics. 2009;10:399.
52. Feldmesser E, Rosenwasser S, Vardi A, Ben-Dor S. Improving transcriptome construction in non-model organisms: integrating manual and automated gene definition in *Emiliana huxleyi*. BMC Genomics. 2014;15:148.
53. Amin S, Prentis PJ, Gilding EK, Pavasovic A. Assembly and annotation of a non-model gastropod (*Nerita melanotragus*) transcriptome: a comparison of *De novo* Assemblers. BMC Res Notes. 2014;7:488.
54. Kvam VM, Liu P, Si Y. A comparison of statistical methods for detecting differentially expressed genes from RNA-seq data. Am J Bot. 2012;99:248–56.
55. Blazek J. Segregation and general evaluation of spur type or compact growth habits in apples. Acta Hortic. 1992;317:71–9.
56. Baldi P, Wolters PJ, Komjanc M, Viola R, Velasco R, Salvi S. Genetic and physical characterisation of the locus controlling columnar habit in apple (*Malus × domestica* Borkh.). Mol Breed. 2012;31:429–40.
57. Moriya S, Okada K, Haji T, Yamamoto T, Abe K. Fine mapping of *Co*, a gene controlling columnar growth habit located on apple (*Malus × domestica* Borkh.) Linkage Group 10. Plant Breed. 2012;131:641–7.
58. Zhang Y, Dai HY. Comparison of photosynthetic and morphological characteristics, and micro-structure of roots and shoots, between columnar apple and standard apple trees of hybrid seedlings. Phyton. 2011;80:119–25.
59. Salvi S, Sponza G, Morgante M, Tomes D, Niu X, Fengler KA, et al. Conserved noncoding genomic sequences associated with a flowering-time quantitative trait locus in maize. Proc Natl Acad Sci U S A. 2007;104:11376–81.
60. Van Damme M, Huibers RP, Elberse J, Van den Ackerveken G. *Arabidopsis DMR6* encodes a putative 2OG-Fe(II) oxygenase that is defense-associated but required for susceptibility to downy mildew. Plant J. 2008;54:785–93.
61. Prescott AG, John P. DIOXYGENASES: molecular structure and role in plant metabolism. Annu Rev Plant Physiol Plant Mol Biol. 1996;47:245–71.
62. Brown DE, Rashotte AM, Murphy AS, Normanly J, Tague BW, Peer WA, et al. Flavonoids act as negative regulators of auxin transport in vivo in *Arabidopsis*. Plant Physiol. 2001;126:524–35.
63. Peer WA, Murphy AS. Flavonoids and auxin transport: modulators or regulators? Trends Plant Sci. 2007;12:556–63.
64. Buer CS, Imin N, Djordjevic MA. Flavonoids: new roles for old molecules. J Integr Plant Biol. 2010;52:98–111.
65. Dare AP, Tomes S, Jones M, McGhie TK, Stevenson DE, Johnson RA, et al. Phenotypic changes associated with RNA interference silencing of chalcone synthase in apple (*Malus × domestica*). Plant J. 2013;74:398–410.
66. Chae HS, Cho YG, Park MY, Lee MC, Eun MY, Kang BG, et al. Hormonal cross-talk between auxin and ethylene differentially regulates the expression of two members of the 1-aminocyclopropane-1-carboxylate oxidase gene family in rice (*Oryza sativa* L.). Plant Cell Physiol. 2000;41:354–62.
67. Stepanova AN, Yun J, Likhacheva AV, Alonso JM. Multilevel interactions between ethylene and auxin in *Arabidopsis* roots. Plant Cell. 2007;19:2169–85.
68. El-Showk S, Ruonala R, Helariutta Y. Crossing paths: cytokinin signalling and crosstalk. Development. 2013;140:1373–83.
69. Leon-Reyes A, Spoel SH, De Lange ES, Abe H, Kobayashi M, Tsuda S, et al. Ethylene modulates the role of NONEXPRESSOR OF PATHOGENESIS-RELATED GENES1 in cross talk between salicylate and jasmonate signaling. Plant Physiol. 2009;149:1797–809.
70. Ghassemian M, Nambara E, Cutler S, Kawaide H, Kamiya Y, McCourt P. Regulation of abscisic acid signaling by the ethylene response pathway in *Arabidopsis*. Plant Cell. 2000;12:1117–26.
71. Langmead B, Trapnell C, Pop M, Salzberg SL. Ultrafast and memory-efficient alignment of short DNA sequences to the human genome. Genome Biol. 2009;10:R25.
72. Huang X, Madan A. CAP3: A DNA sequence assembly program. Genome Res. 1999;9:868–77.
73. Frazer KA, Pachter L, Poliakov A, Rubin EM, Dubchak I. VISTA: computational tools for comparative genomics. Nucleic Acids Res. 2004;32(Web Server issue): W273–9.

17

Transcriptional regulation of the paper mulberry under cold stress as revealed by a comprehensive analysis of transcription factors

Xianjun Peng[1], Qingqing Wu[1,2], Linhong Teng[1], Feng Tang[1], Zhi Pi[1] and Shihua Shen[1*]

Abstract

Background: Several studies have focused on cold tolerance in multiple regulated levels. However, a genome-scale molecular analysis of the regulated network under the control of transcription factors (TFs) is still lacking, especially for trees. To comprehensively identify the TFs that regulate cold stress response in the paper mulberry and understand their regulatory interactions, transcriptomic data was used to assess changes in gene expression induced by exposure to cold.

Results: Results indicated that 794 TFs, belonging to 47 families and comprising more than 59% of the total TFs of this plant, were involved in the cold stress response. They were clustered into three groups, namely early, intermediate and late responsive groups which contained 95, 550 and 149 TFs, respectively. Among of these differentially expressed TFs, one bHLH, two ERFs and three CAMTAs were considered to be the key TFs functioning in the primary signal transduction. After that, at the intermediate stage of cold stress, there were mainly two biological processes that were regulated by TFs, namely cold stress resistance (including 5 bHLH, 14 ERFs, one HSF, 4 MYBs, 3 NACs, 11 WRKYs and so on) and growth and development of lateral organ or apical meristem (including ARR-B, B3, 5 bHLHs, 2 C2H2, 4 CO-like, 2 ERF, 3 HD-ZIP, 3 YABBYs, G2-like, GATA, GRAS and TCP). In late responsive group, 3 ARR-B, C3H, 6 CO-like, 2 G2-like, 2 HSFs, 2 NACs and TCP. Most of them presented the up-regulated expression at 12 or 24 hours after cold stress implied their important roles for the new growth homeostasis under cold stress.

Conclusions: Our study identified the key TFs that function in the regulatory cascades mediating the activation of downstream genes during cold tress tolerance in the paper mulberry. Based on the analysis, we found that the AP2/ERF, bHLH, MYB, NAC and WRKY families might play the central and significant roles during cold stress response in the paper mulberry just as in other species. Meanwhile, many other TF families previously reported as involving in regulation of growth and development, including ARF, DBB, G2-like, GRF, GRAS, LBD, WOX and YAABY exhibited their important potential function in growth regulation under cold stress.

Keywords: Cold stress, Plant development, Signaling crosstalk, Transcription factor

Background

The paper mulberry (*Broussonetia papyrifera*) belongs to the family of Moraceae and is naturally distributed in Eastern Asia and pacific countries. The paper mulberry has the shallow roots morphology with advanced lateral roots but without an obvious taproot. The paper mulberry is one of the multifunctional tree species in agroforestry systems [1], as well as being one of the traditional forages [2] and Chinese medicines in many countries of Asia [3]. It is the ideal tree species for ecological and gardening purposes [4]. Due to its fast growth and adaptability, the paper mulberry is commonly used for the ecological afforestation and landscape in both sides of highway, mined areas and on barren land [5]. However, the molecular mechanism of strong adaptability and tolerance to biotic or abiotic stress of the paper mulberry has not been studied, which limits the exploitation of the paper mulberry.

In recent years, many reports have provided new and exciting information that has allowed us to better understand the genes involved in cold adaptation and freezing

* Correspondence: shshen@ibcas.ac.cn
[1]Key Laboratory of Plant Resources, Institute of Botany, the Chinese Academy of Sciences, Beijing 100093, People's Republic of China
Full list of author information is available at the end of the article

response [6]. One of the major advances in the past decade of cold tolerance research is the discovery of the cold stress related TFs. Many TFs, including DREB [7], MYB [8], NAC [9] and WRKY [10], have been found to be involved in the cold response of plants. Recently, more and more researches concentrate on TF families involved in abiotic stress including cold stress response based on genome wide analyses, such as WRKY [11], NAC [12,13], MYB [14] and AP2/ERF [15]. However, there are few reports on the role in cold responses of other TF families, such as ARF, E2F, GRF and GRAS, though their functions in development and growth have been characterized. Moreover, most studies about cold stress in plants concentrate on a few species, such as Arabidopsis [16,17] and some crops [18]. The universality of the mechanism under TF control is not explicit, especially for tree species. Because of low domestication, open-pollinated native populations and high levels of genetic variation, trees are ideal organisms to unveil the molecular basis of population adaptive divergence in nature and have gained much attention in recent years as non-classical model plants for environment adaptation, evolutionary and genomic studies [19]. A substantial number of ESTs (expressed sequence tags) encoding putative transcription factors, including CBFs (C-repeat binding factor), are observed in *Eucalyptus* under low temperature [20]. In subgroup III of the *PtWRKY* genes in *Populus*, eight were induced expression under cold stress [21]. A comprehensive analysis of the NAC gene family in *Populus* revealed the functional divergence among members in NAC family [22]. Over-expression of *JcERF* or *JcDREB*, isolated from *Jatropha curcas*, a woody oil plant, in transgenic Arabidopsis exhibited enhanced salt and freezing stress responses [23,24]. *BpDREB2*, cloned from the paper mulberry, could enhanced the freezing tolerance of Arabidopsis significantly without causing growth retardation [25]. Genetic diversity revealed by SRAP (Sequence-related amplified polymorphism) marker and cluster analysis show that there is a relationship between the genetic variation and geographical distribution [26]. These results provide reference for making genetic map and guiding the breeding of the paper mulberry. In this study, we discern the potential transcriptional regulatory network regulated by the TFs in the seedlings of the paper mulberry from transcriptomic data under cold stress.

Results and discussion

Identification and classification of TFs in the paper mulberry under cold stress

TFs play the significant roles in plant development and stress tolerance. To identify the TFs involved in the cold stress response, we surveyed the biological functions of putative TFs that were differentially expressed in the paper mulberry under cold treatment. After retrieving annotation results for every unigene, there were total of 1,337 TFs in the paper mulberry and classified into 55 families based on their DNA-binding domains and other conserved motifs [27]. The expression level of all TFs was normalized by using their RPKM values. Among of these, total of 1,180 TFs, belonging to 52 families, were identified as being expressed in paper mulberry during cold stress (Additional files 1 and 2A). Of these, 441 TFs had the complete ORF (opening read frame) representing 37.02% of the total (Additional file 3).

Under non-stress conditions, 841 TFs belonging to 52 families were expressed in the leaves. This number reached 978 after 2 hours of cold stress and 1,009 after 6 hours. When 12 hours passed under low temperature, the number of expressed TFs was reduced to 986. After 24 hours of exposure to cold, the number of expressed TFs was 993 (Figure 1A). A total of 739 TFs were expressed in all five samples, representing 62.04% of the total TFs. A relatively small number of TFs were uniquely expressed in each sample: 3, 19, 73, 17 and 24, respectively.

Differentially expressed TFs the paper mulberry under cold stress

Temperature is an important environmental factor that affects plant growth and development. Temperature stresses include cold, freezing and heat. Expression data from different plants species have indicated that the members of the TFs participate in plant responses to cold stresses. After four hours of cold treatment, a total of 70 up-regulated and 18 down-regulated TFs were identified in the well-developed leaf of *Vitis amurensis* while 68 up-regulated and 43 down-regulated TFs were identified in *Vitis vinifera* [28]. The transcripts of 35 TFs from 6 TF families were responsive to cold stress in *Populus simonii* [29]. When *Ammopiptanthus mongolicus*, a desert shrub, was cultured at 4°C for 14 days, a total of 720 TFs were identified as DEGs, 209 of which showed significant up-regulation and 511 down-regulation [30]. In our study, a total of 794 TFs were responsive to cold stress, representing 59.38% of the total number of TFs expressed in the paper mulberry (Figure 1B and Additional file 4). 491 TFs were expressed in all five samples, representing 61.84% of the total differentially expressed TFs. The number of specifically expressed TFs was 3, 3, 62, 3 and 8, respectively. These results show that the number of TFs involved in cold response varies considerably among different tissues of different species under different conditions. This implies that the individual plant species have distinct cold response.

After 2 hours of cold stress treatment, 175 TFs showed differential expression characteristics, of which 69 TFs were up-regulated and 106 were down-regulated (Figure 2). The number of repressed TFs was significantly more than the number of induced expression TFs. Compared with

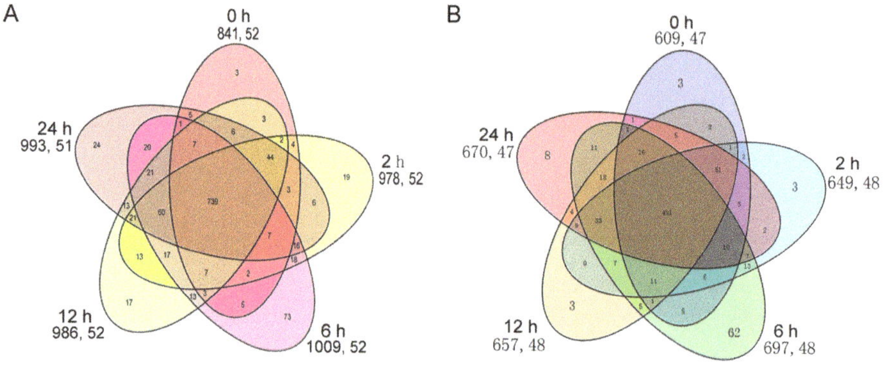

Figure 1 No. of expressed and differentially expressed TFs in each sample. **A** Venn diagram of expressed TFs distributed in five samples **B** Venn diagram of differentially expressed TFs in every sample. The time was represented the sample that were treated under cold stress. The former and latter number presented below the treat time, such as (841, 52), was the number of TFs and families presented in this sample, respectively.

the untreated leaves, a total of 564 TFs were differentially expressed after 6 hours of cold stress treatment, of which 356 were up-regulated and 198 down-regulated TFs. Meanwhile, there were 566 differentially expressed TFs between 6 hours and 2 hours after cold stress treatment, with 347 up-regulated TFs and 219 down-regulated. After 12 hours of cold stress, the number of differentially expressed TFs (273) decreased dramatically compared with the untreated leaves, with 169 up-regulated and 114 down-regulated TFs. However, when comparing gene expression at 12 hours of cold exposure to 6 hours, there is a significant increase in the number of down-regulated TFs (320), which is more than the number of up-regulated TFs (214). This suggests that most of the TFs up-regulated at 6 hours were reduced upon another 6 hours of exposure to cold, and thus are only transiently activated in response to this stress. After 24 hours of cold stress treatment, there were 261 differential expression TFs, including 167 up-regulated and 94 down-regulated TFs when compared with 12 hours of treatment.

To validate the transcriptomic data, we selected 10 TFs for confirmation by qPCR (Additional file 5). The qPCR results suggested that the expression profile of the TFs were consistent with the transcriptomic data, though there was no biological replicate in the transcriptomic experiment. This was mainly benefit from using the clonal seedlings in this study.

Family distribution of the differentially expressed TFs

A total of 794 differentially expressed TFs could be classified in to 487 families (Figure 3 and Additional file 2B). The bHLH family contained the most differentially expressed TFs (89), followed by WRKY (71), ERF (66), MYB (45), C2H2 (41) and NAC (39). Regarding the CPP (7), DBB (5), HRT (1), NF-YB (5), RAV (2), SRS (1), VOZ (2), Whirly (2), YABBY (3) families, all known members were responsive to cold stress. Meanwhile, all the members of BTF3, CAMTA, LSD, NF-X1 and S1Fa family did not change in expression during cold stress treatment. However, according to the results of qPCR (Additional file 5), three members of the CAMTA family showed a significant expression difference in the early stage (15 min or 30 min) of cold stress treatment though this was not detected in the transcriptomics data.

There were 175 TFs belonging to 37 families (Figure 4A) responded to cold stress after 2 hours of treatment; most

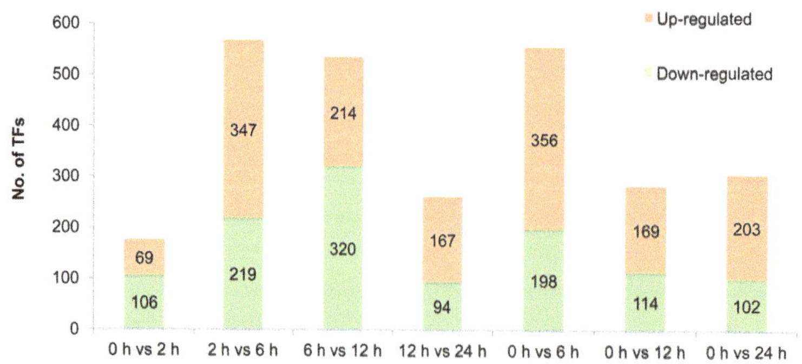

Figure 2 Cold stress responsive TFs between every two samples. Taken pairwise comparison of 0 h vs 2 h as an example, up-regulated was referred to the expression was higher than that after 2 hours of cold stress treatment while down-regulated referred to the expression was lower than that after 2 hours of cold stress treatment.

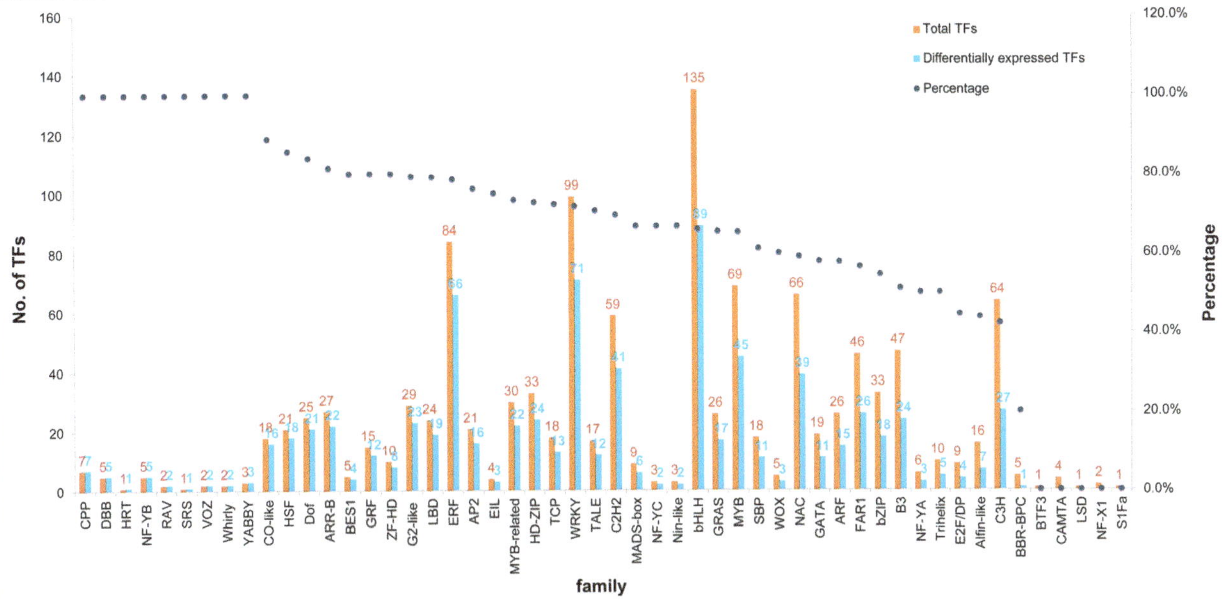

Figure 3 Total and cold stress responsive TFs distributed in every family. "Total TFs" referred to the total number of TFs in each family expressed in the leaf of the paper mulberry. "Differentially expressed TFs" referred to the TFs with dramatic change in the leaf of the paper mulberry under cold stress. "Percentage" referred to the percentage of the "Differentially expressed TFs" number accounting for the "Total TFs" number of each family.

of these TFs showed a decrease in expression. For example, there were 12 down-regulated ERF family TFs compared with 7 up-regulated; 6 down-regulated NAC family members, and only one up-regulated; and 18 down-regulated and 3 up-regulated WRKY family TFs. Total 43 families are involved in the cold stress response at 6 hours of treatment (Figure 4B). Whether it is compared with the untreated sample or with the 2 hours treated sample, the up-regulated TFs in most families increased significantly, while there was little to no change in the number of down-regulated TFs. For example, the number of up-regulated TFs in the WRKY family increased to 57 and the number down-regulated reduced to 3 (Figure 4C). After 12 hours of cold stress, there were 44 TF families that exhibited differential expression. For most families, when compared with the untreated leaf sample, the number of up-regulated TFs was clearly more than down-regulated TFs (Figure 4D). However, the numbers of down-regulated TFs were significantly more than the up-regulated in the bHLH, WRKY, ERF, MYB, C2H2, NAC and LBD families when compared with 6 hours cold stress treated leaf (Figure 4E). The Whirly (1 up-regulated and 1 down-regulated), RAV (1 up-regulated), and YABBY (3 up-regulated) TF families emerged as having members with differential expression at this time, while the NF-YC and VOZ families were no longer expressed.

Compared with the untreated leaf, a total of 305 TFs belonging to 38 families were differentially expressed in the leaf treated under cold stress for 24 hours. The numbers of up-regulated TFs was significantly more than that of down-regulated TFs in most families, except the bHLH family with 9 up-regulated and 19 down-regulated TFs (Figure 4F). When compared with the leaf treated with cold stress after 12 hours, only 261 TFs belonging to 37 families showed the differential expression traits at 24 hours of cold stress. The B3 family changed greatly with 15 TFs down-regulated and 4 up-regulated. Regarding the ERF, GRF, HSF, NAC and WRKY families, the number of up-regulated TFs was obviously higher than that of down-regulated TFs. The TF family HRT (1 up-regulated) and NF-YC (1 down-regulated) emerged as having differentially regulated members after 24 hours of cold exposure, while the BBR-BPC, BES1, EIL, NF-YB, MADS-box, RAV, Whirly, WOX and YABBY families were no longer expressed (Figure 4G).

According above analysis, from 0 to 24 hours of cold stress treatment (Figure 4A-G), the expression pattern of TFs in the AP2/ERF, bHLH, MYB, NAC, and especially WRKY families experienced a dramatic change. Genes encoding 82 AP2/ERF, 89 bHLH, 41 C2H2, 45 MYB, 39 NAC and 71WRKY family members were responsive to cold stress in the paper mulberry, which comprise 46.22% of the total cold-regulated TFs in this species (Figure 3), and the number of expressed TFs from these families obviously increased after the cold treatment while the number of the remaining families was almost same within each treatment sample (Additional file 2A). This result was similar with previous studies. For instance,

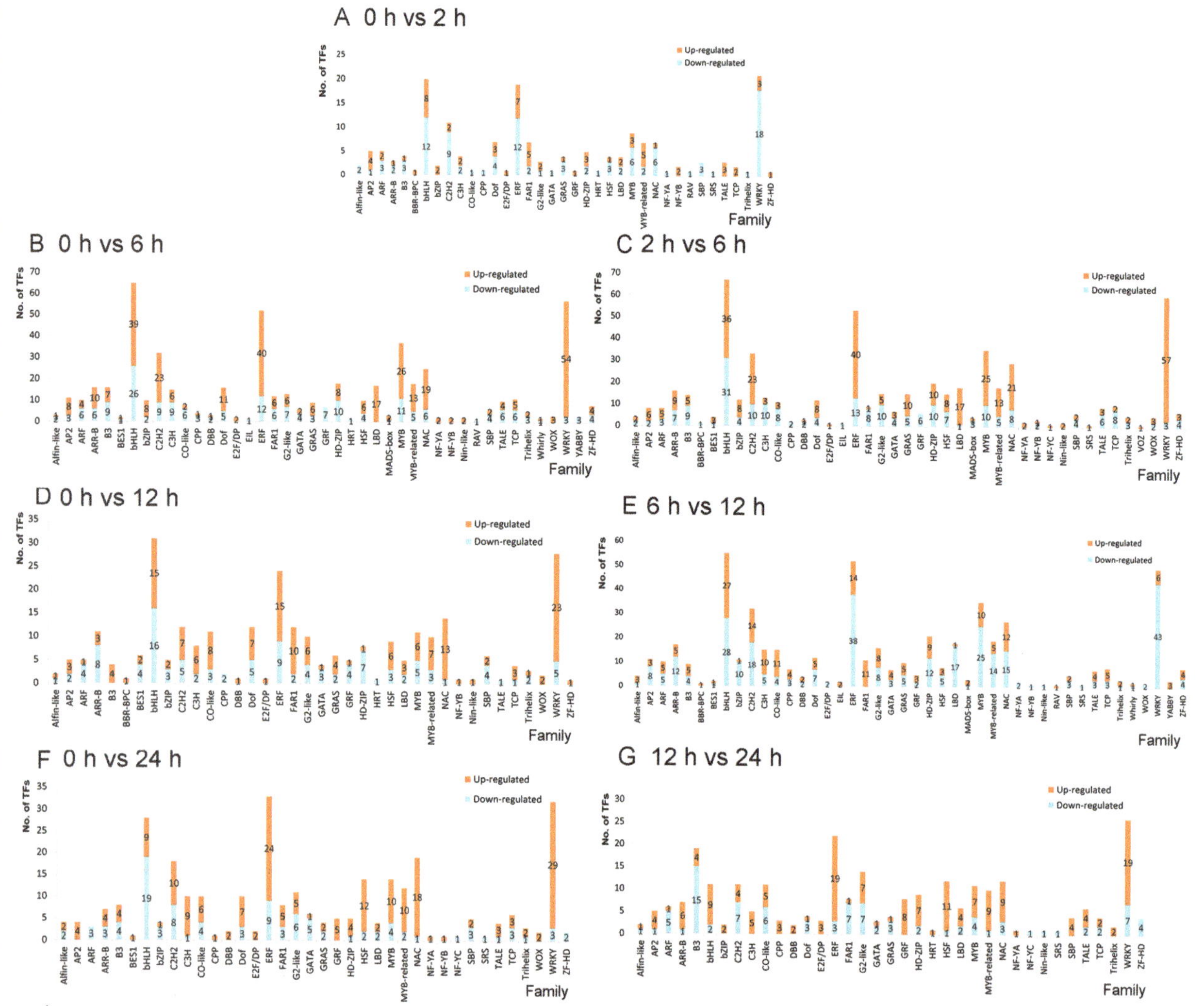

Figure 4 Cold stress responsive TFs distributed in each family of every pairwise comparison. **A-G** Family distribution in each pairwise comparison. Numbers on the column represent the amount of up-and –down regulated TFs distributed in the family.

in the mature pollen of Arabidopsis, a total 93 TFs from eight families, including AP2/ERF, bZIP, bHLH, HD-ZF MYB, MADS, NAC, and WRKY, have been reported to be involved in the cold stress response process [31]. In apical shoots of cassava subjected to cold, a total of 32 genes were identified as early cold-responsive TFs; AP2/ERF, MYB, and GRAS were the three major TF families involved in this process, with 6, 5, and 5 TFs represented, respectively [32]. The WRKY [11], NAC [12,13], MYB [14] and AP2/ERF [15] TF families have been characterized as being involved in abiotic stress, including cold stress response, based on genome-wide analyses and many intensive researches have revealed their significant roles in the cold stress response.

Cold-inducible DREB1/CBFs, one of the subfamily in AP2/ERF superfamily, have been identified from numerous plant species, such as Arabidopsis [33], sheep grass [34], barley [35], *Prunus mume* [36], rice [18] and maize [37]. The DREB1/CBF cold response pathway play a central role in cold acclimation. DREB1/CBF proteins bind to DRE/CRT elements in enhancers and activate transcription of target cold-inducible genes, including other transcription factors, and thus bring about transcriptomic and metabolomics changes, which eventually cause cold stress responses [15,38]. The DREB1/CBF function in cold tolerance by ABA-dependent and ABA independent pathway. Meanwhile, under cold stress, an acute rise in cellular ABA levels induces the expression of downstream genes which contain ABA-responsive *cis* elements in their promoters, represented by various types of TFs, such as bZIP, MYB, and MYC. MYB proteins are a superfamily of TFs that have known regulatory roles in developmental processes and defense responses in plants [39]. The *HOS10* gene encodes an R2R3-type MYB and works as an important coordinating factor for responses to cold stress [17]. Expression of *OsMYB2* was up-regulated by cold and the *OsMYB2* overexpressing plants were more tolerant against cold stress than wild type plants [8]. Cold stimulates the activation of NTL6, a NAC family TF, which induces a subset of *PR* genes and the transgenic exhibited enhanced disease resistance [40]. This indicates that the mechanism by which these paralogous TFs mediate stress tolerance is complex in plants, though these TFs can bind to the same core recognition sequence [12,41].

Transgenic Arabidopsis plants overexpressing *GmWRKY21* shows increased tolerance to cold stress when compared with wild-type plants [42]. Recent research suggest that the three structurally related WRKY proteins, AtWRKY18, AtWRKY40 and AtWRKY60, participate in at least three phytohormone-mediated signaling pathways, including SA (Salicylic acid), JA (Jasmonic acid) and ABA (Abscisic acid) [43], Two closely related WRKY transcription factors (At*WRKY25* and *AtWRKY33*) respond to both biotic and abiotic stresses, e.g. *P. syringae*, NaCl, cold and heat [11].

Cross-talk and mutual regulations exist among these TFs. CBF2/DREB1C negatively regulates *CBF1/DREB1B* and *CBF3/DREB1A*, ensuring that their expression is tightly controlled, which, in turn, guarantees the proper induction of downstream genes and the accurate development of Arabidopsis tolerance to freezing and related stresses [44]. The cold regulation of *CBF3* involves upstream bHLH [45] and MYB [46] factors. Cold stress induces simulation of ICE1 at K393, which is critical for ICE1-mediated activation of transcription of *CBF*s and repression of *MYB15*. CBFs regulate the expression of *COR* genes that confer freezing tolerance. The expression of *CBF*s is negatively regulated by MYB15 and ZAT12 [47]. The MYB15 protein interacts with ICE1 and binds to MYB recognition sequences in the promoters of *CBF* genes, which suggest MYB15 is part of a complex network controlling the expression of *CBFs* in response to cold stress [46]. CAMTA3, one member of the calmodulin binding transcription activator (CAMTA) family, binds to the CM2 motif and works as a positive regulator of *CBF2* [48]. AtWRKY34 negatively mediates cold sensitivity of mature Arabidopsis pollen through regulating the expression of genes encoding CBF family transcriptional activators [10]. Therefore, we inferred that the AP2/ERF, bHLH, C2H2, MYB, NAC and WRKY family members interacted with ABA, JA and other of phytohormone might play the central and significant roles during cold stress response in the paper mulberry.

In addition, there were still many TFs distributed in other small TF families, such as ARF, DBB, G2-like, LBD, WOX the G2-like, GRF, LBD, WOX and YABBY (Figure 3 and Additional file 2B), even though they have been characterized primarily for their participation in the regulation of plant photo-morphogenesis, development and growth. Most members of these families also showed changes in expression upon exposure to cold, suggesting they also have important roles in cold stress tolerance in the paper mulberry.

The ARF TFs play important roles in regulating diverse biological processes, including development, growth, cell division and responses to environmental stimuli [49,50]. *AtGRF5* is required for the development of appropriate leaf size and shape through the promotion and/or maintenance of cell proliferation activity in leaf primordial [51]. Cold stress inhibits shoot-ward Auxin transport and alters the intracellular Auxin gradient [52]. However, knowledge about its role under cold stress is limited. A total of 11 up-regulated and 4 down-regulated ARFs might function as the positive and negative regulator by sensing the Auxin signal in the paper mulberry under cold stress.

BBX (another abbreviation form of DBB) proteins are key factors in regulatory networks controlling growth and developmental processes that include seedling photomorphogenesis, photoperiodic regulation of flowering, shade avoidance, and responses to biotic and abiotic stresses [53]. Their functions are not entirely redundant, as judged by the fact that some DBBs were apparently implicated in light signal transduction in a negative manner, whereas another was implicated in a positive manner with regard to light-induced inhibition of elongation of hypocotyls [54]. For instance, BBX25 and BBX24 function as transcriptional co-repressors, forming inactive heterodimers with HY5 (bZIP) that down-regulate *BBX22* expression for fine-tuning of light-mediated seedling development [55]. Therefore, five *DBB*s are thought to play regulated roles, respectively in the photomorphogenesis of the paper mulberry under cold stress.

G2-like (GOLDEN2-LIKE) TFs are required for chloroplast development and have been reported to co-regulate and synchronize the expression of a suite of nuclear photosynthetic genes and thus act to optimize photosynthetic capacity in varying environmental and developmental conditions [56]. Although no study has reported that G2-like are related to cold stress, fifteen *G2-like* TFs were induced and eight had repressed expression exposure to cold, which suggests they may function to regulate chloroplast development in the paper mulberry under cold stress.

The LBD family is of plant-specific and implicated in plant development. Two members of the Arabidopsis LBD family, *LBD30* and *LBD18* are expressed in immature tracheary elements (TEs), and their expression is dependent on VND6 and VND7, which are NAC family TFs required for TE differentiation. *ASL20* appears to be involved in a positive feedback loop for *VND7* expression that regulates TE differentiation-related gene [57]. Four *LBD* genes downstream of ARFs, *LBD16, LBD17, LBD18* and *LBD29*, are rapidly and dramatically induced by callus-inducing medium; LBD as key regulators in the callus induction process, thereby establishing a molecular link between Auxin signaling and the plant regeneration program [58]. In addition, bHLH048 post-translated regulates the function of LOB (LBD) at lateral organ boundaries [59]. However, TCP TFs play a pivotal role in the control of morphogenesis of shoot organs by negatively regulating the expression of boundary specific genes, including *LBD*s [60]. There are 17 LBDs highly expressed in the leaf of the paper mulberry under cold stress and this might account for the advanced lateral organ development in the paper mulberry even in the cold or other hostile environment.

The WOX TFs have been identified to function in SAM (shoot apical meristem) and RAM (root apical meristem) by a dynamic feedback loop involving the CLAVATA3 (CLV3) peptide ligand and the CLV1 receptor in SAM [61]. The *dwt1* tillers have shorter internodes with fewer and unelongated cells compared with the wild type; further study showed that the DWT1 (homologous to the Arabidopsis WOX8 and WOX9) activity in the internode elongation is directly or indirectly associated with GA (Gibberellin) signaling, suggesting a new function of WOX genes in balancing branch growth in rice [62]. Overexpression of *WOX3* induced ectopic expression of *Knotted 1-like homeobox1* (*KNOX1*) genes in leaves and consequently produced a phenotype similar to plants ectopically expressing *KNOXI* genes [63]. However, *HOS9*, encoding the WOX6, is important for growth and development, and for a part of freezing tolerance, by affecting the activity of genes independent of the CBF pathway in Arabidopsis [64]. This is the only report that WOX TFs are involved in the cold response. In our study, three of five genes encoding WOX TFs were induced by cold exposure. Up-expression of these genes implied that they might have important roles in cambial meristems during leaf growth and development of the paper mulberry under cold stress.

Thus, TFs of the ARF, DBB, G2-like, LBD, WOX, YABBY and other families, interacted with Auxin, GA and other various hormones, were thought to be essential for the regulation of photo-morphogenesis, growth and development of the paper mulberry under cold stress.

Clustering of the differentially expressed TFs and the identification of key TFs

Discerning the expression pattern and the regulatory cascades of the TFs will be benefit for the identification of key TFs that mediating the activation of downstream genes during cold stress from 0 to 24 hours in the paper mulberry. So, according to the expression profiles, the 794 differentially expressed TFs were clustered into three groups, namely early, intermediate and late responsive groups (Figure 5). Besides, the key TFs was screened from these three groups following the threshold: fold change was more than ten or the difference value of RPKM was more than 300.

TFs in group 1 was early responsive to cold stress at 2 hour, including 55 up-regulated (Subgroup 1) and 40 down-regulated (Subgroup 2). Subgroup 1 is characterized by a rapid induction of expression at 2 hours of cold treatment, followed by a slight decrease in expression for the duration of the experiment. This group contained 55 TF accounting for 6.9% of the total. A total of 40 TFs in subgroup 2 clearly underwent transcriptional repression as the cold stress began (Figure 6). Among of these 95 TFs, one up-regulated *bHLH* (T4-18787), and two down-regulated *ERF*s (T6-23630 and T7-28635) were considered to playing key role in the early activation of downstream genes. In addition, according to the results of qPCR, 3 *CAMTA*s were highly induced to express under cold stress at 15 min or 30 min (Additional file 5), though they were not detected in the transcriptomic data. Even though, this was consistent with the fact that Ca^{2+} is the second message of cold stress and CAMTA is one of the pathways that senses cold stress downstream of Ca^{2+} signal transduction. Because the transduction of Ca^{2+} signal is rapid and the balance of intracellular calcium concentration is critical, the induced expression of *CAMTA* is always fast and transient [65]. Thus, three *CAMTA*s were also considered to be the primary key TFs that sensed and transduced the cold stress signal.

The intermediate group (group 2) contained 550 differentially expressed TFs that obviously began in response to cold stress since 6 hours. According to the expression profile, group 2 has been divided into 6 subgroup, of which group 3, 4 and 5 was up-regulated while group 6, 7 and 8 was down-regulated (Figure 6). A total of 38 TFs were categorized as subgroup 3. These genes remained relatively stable through 2 hours after induction of cold stress, but increased suddenly at 6 hours and remained at heightened expression until 12 hours of cold stress. Then the expression gradually decreased to the level of untreated leaves after 24 hours treatment. Subgroup 4 contained a total of 47 TFs that were induced at 6 hours treatment of cold exposure, but the expression of these genes was decreased again by 12 hours of treatment, though their expression levels were relatively high again at 24 hours of treatment. The members of subgroup 5 were up-regulated rapidly at 6 hours treatment and then sharply declined. It contained 282 TF and was the largest subgroup, representing 35.5% of the total.

A total of 65 TFs are contained in subgroup 6. These genes have a gradual decrease in expression until 6 hours of exposure to cold, at which point they reached their lowest level of expression. At 12 hours of treatment, they reached their highest expression levels and then

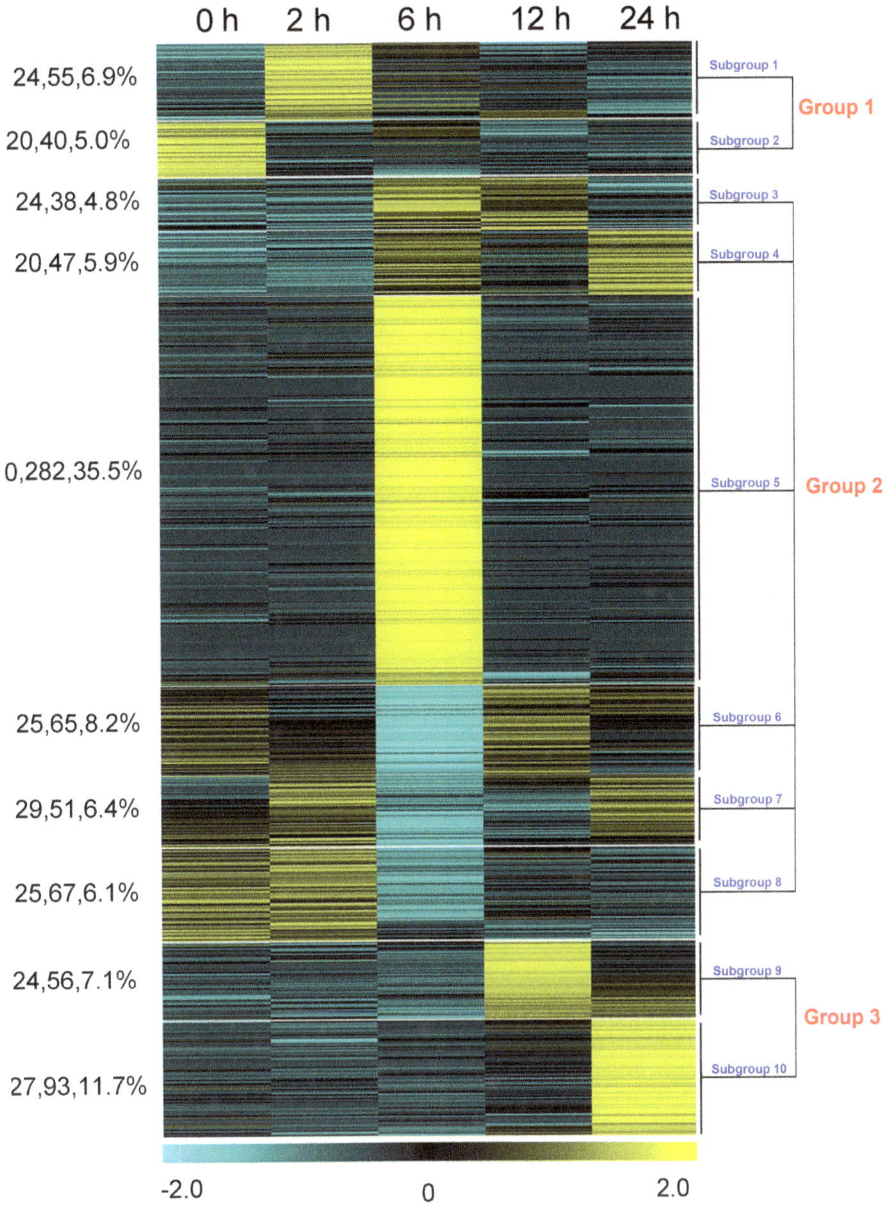

Figure 5 Heat map of differentially expressed TFs under cold stress. To be considered differentially expressed, the transcript must have RPKM ≥ 2 in at least one tissue, 2-fold or greater change, and $P \leq 0.05$. According to their expression pattern under cold stress, the differentially expressed TFs were categorized in to 3 groups and 10 subgroups. The number at the left of the heat map (such as 24, 55, 6.9%) represented the amount of the families, differentially expressed TFs and the percentage of the total differential expressed TFs in each group, respectively. Yellow indicates high expression, black indicates intermediate expression, and blue indicates low expression.

gradually returned to the pretreatment or slightly lower than pre-treatment levels by 24 hours. In subgroup 7, which contains 51 TFs, expression levels began to rise or slightly fluctuate from 0 to 2 hours cold stress treatment. They were then reduced to the lowest levels at 6 hours of treatment. After that, expression gradually increased and returned to relatively high levels by 24 hours of cold exposure. Unlike the TFs in subgroup 6 and 7, those showing subgroup 8 expression profiles remained relatively stable after suppressed expression at 6 hours of cold stress. This subgroup includes 67 TF belonging to 25 families.

In group 2, total 83 TFs were considered to be the key TFs mediating downstream genes in the intermediate stage of cold stress (Additional file 6). Among of them, there were 52 TFs were induced expression, including 5 bHLH, 14 ERFs, one HSF, 4 MYBs, 3 NACs and 11 WRKYs, which mainly functioned in abiotic stress

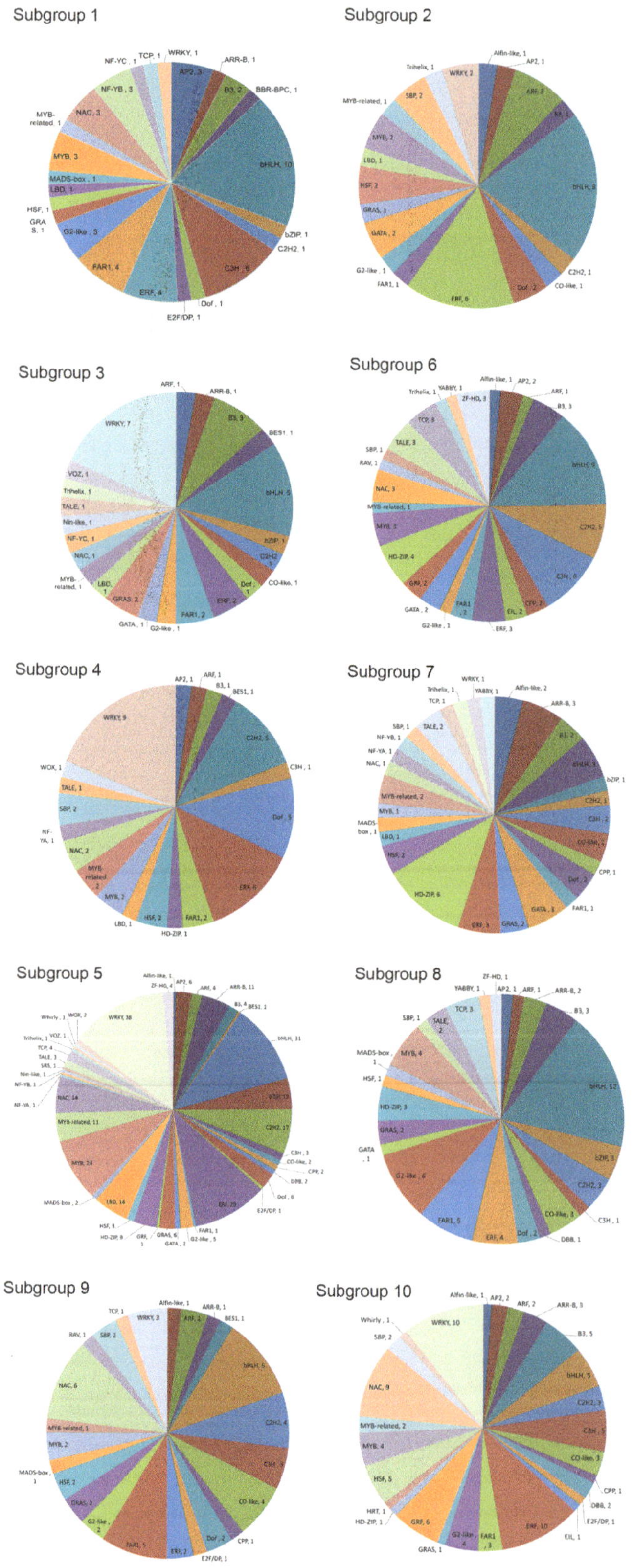

Figure 6 The family composition of each subgroup in the heat map. The number represented the amount of the differentially expressed TFs in each family.

response [12,22,38,43]. Meanwhile, total 31 TFs were significantly suppressed expression, such as ARR-B, B3, 5 bHLHs, 2 C2H2, 4 CO-like, 2 ERF, 3 HD-ZIP, 3 YABBYs, G2-like, GATA, GRAS and TCP. Many of these TFs have been reported play important roles in growth and development of lateral organ or apical meristem [66-68] as well as the photosynthesis [56]. These result suggested that, at the intermediate stage under cold stress, low temperature began to repress the development and growth of vegetative organ in the paper mulberry. Simultaneously, the paper mulberry initiated more stress related genes to enhance the cold tolerance.

Group 3, namely late responsive group, contained 149 TFs that were clustered into subgroup 9 and 10. These two subgroups were significantly up-regulated at the 12 and 24 hours of treatment, respectively. A total of 56 TFs were in subgroup 9 and belonged to 24 TF families accounting for 7.1% of the total. There were 93 TFs in subgroup 10, belonging to 27 families and accounting for 11.7% of the total. In late responsive group, total 18 TFs were defined as the key TFs functioning in the terminal stage because of their great changes of RPKM value just beginning from 12 or 24 hours cold stress (Additional file 6), including 3 ARR-B, C3H, 6 CO-like, 2 G2-like, 2 HSFs, 2 NACs and TCP. Most of them presented the up-regulated expression, which implied that the expression level of a large number of TFs went back to the originally expressed status and only a few of them were highly expressed to be responsible for the recovery growth and continuous cold stress.

Based on the above differential expression analysis, the proposed cold response regulatory cascades under the regulation of TFs in the paper mulberry was illustrated in Figure 7. After signal reception, stress-activated Ca^{2+} signaling and other secondary signal modulate the expression of early cold stress responsive TFs. One bHLH, two ERFs and three CAMTAs played the key roles in the primary signal transduction. After that, at the intermediate stage of cold stress, there were mainly two biological processes that were regulated by TFs, namely cold stress resistance and the growth homeostasis. On the one hand, much cold stress related genes could be repressed or induced by TFs, such as WRKY, HSF, ERF, NAC, MYB, etc., which played the central roles in endowing paper mulberry with the cold tolerance by regulating the content of osmoprotectant, including soluble sugar, proline and so on. On the other hand, the major growth physical process, e.g. photosynthesis, chloroplast malfunction, cell division in cambial meristem and lateral organ development, were mainly regulated by YABBY, WOX, LBD, DBB, GRF, etc., which made

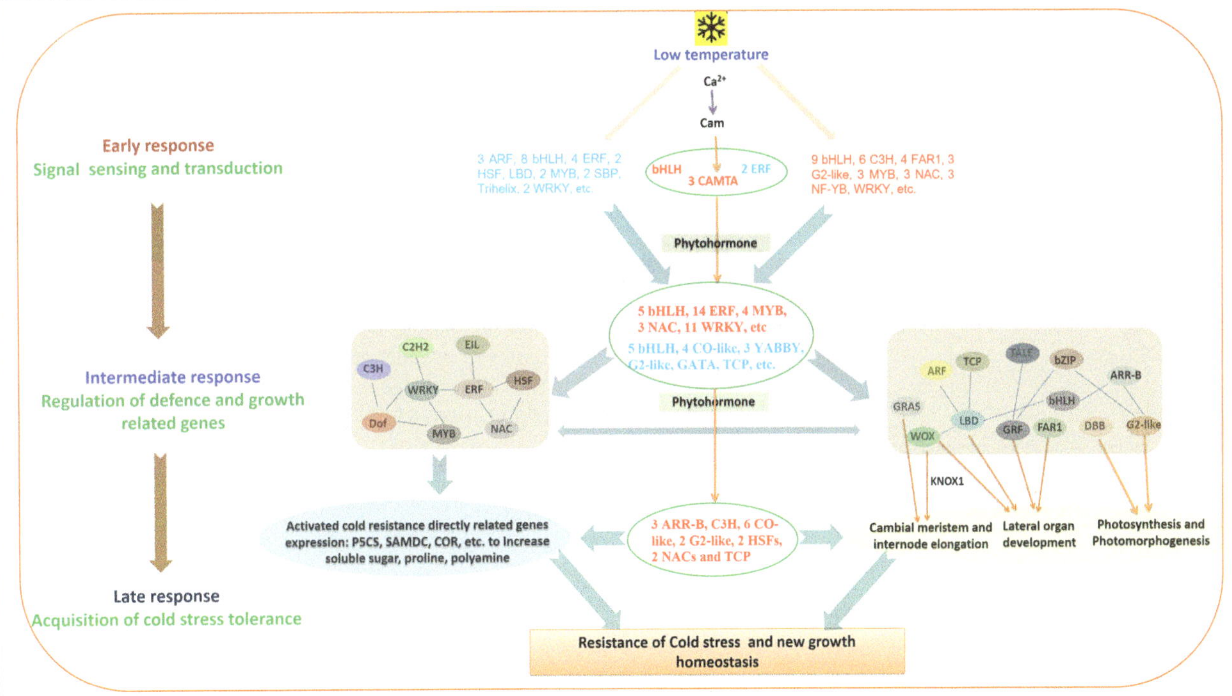

Figure 7 The supposed regulated network under the control of TFs of the paper mulberry exposed to cold stress. After signal reception, stress-activated Ca^{2+} signaling and other signaling modulate the expression of stress-responsive transcription factors. The thick arrows show different biological processes in the stress response under the control of TFs. The key TFs was included in the ellipse drawing by green line. Blue front TFs represented the repressed expressed TFs and the red front TFs represented the induced expression TFs. The blue line between TFs showed in the intermediate response group presented the existed the interaction among these TFs.

the growth state of paper mulberry reach to a new homeostasis. Of cause, the crosstalk and the intricate regulation network could exist in these processes. At the late stage of cold stress, the expression of most TFs restored to the level before cold stress. A few of TFs, mainly concentrated in the ARR-B, CO-like, G2-like and TCP families, were drastically induced expression to respond to the new growth homeostasis under low temperature stress.

Conclusions

Our study is the comprehensive transcriptomic-wide identification of cold stress-responsive TFs in the paper mulberry. Total of 794 TFs belonging to 47 families were responded to cold stress and they were clustered into three groups, namely early, intermediate and late responsive groups. The key TFs were also screened as playing the important roles in three stages. In addition, based on the analysis, the AP2/ERF, bHLH, MYB, NAC and WRKY families play central and significant roles during cold stress response in the paper mulberry just as in other species, as many members of these families are up-regulated during exposure to cold. The finding indicates a substantial conservation of TF families that regulate cold stress response across different species of plant. In addition, we identified many other TF families previously associated with growth and development, including ARF, DBB, G2-like, GRF, GRAS, LBD, WOX and YAABY, which exhibit obviously altered expression under cold stress in the paper mulberry. Although it seemed that their important roles in the cold stress response have been neglected, these proteins might contribute to the regulation of cold response in this species. This study provides us the profile of the TFs involved in the cold stress adaptation of the paper mulberry and also highlights that many plant growth and development related TFs are affected by cold stress, suggesting the potential of using these TFs to improve the plant growth under cold conditions instead of simply enhancing the cold stress tolerance.

Methods

Plant material and RNA extraction

Plantlets were cultured on MS culture media in an artificial climatic chamber kept at 26°C with a 14/10 h photoperiod (day/night). For low temperature treatment, plantlets were grown as above, and transferred to 4°C. Leaves were sampled at different time points (0, 2, 6, 12 and 24 h). In this study, a mixed sampling strategy was adopted to eliminate differences between individuals.

Total RNA was isolated with TRIzol® Reagent (Life Technologies, Shanghai, China) from each sample according to the manufacturer's instructions. It was treated with RNase-free DNase I (Takara, Dalian, China) to remove the residual DNA. RNA quality and purity were assessed with OD260/230 ratio and RNA integrity number (RIN) by using the NanoDrop 2000 (Thermo Fisher, Waltham, USA) and the Agilent 2100 Bioanalyzer (Agilent Technologies, Santa Clara, USA), respectively.

cDNA library preparation, sequencing (RNA-seq) and transcriptome assembly

The libraries were prepared according to former study [69]. After clustering on a flow cell using the cBOT, the cDNA libraries were loaded on the Illumina Genome Analyzer ΠX platform and sequenced. Raw sequence data were generated by Illumina pipeline and were available in NCBI's Short Read Archive (SRA) database (http://www.ncbi.nlm.nih.gov/Traces/sra/sra.cgi) under accession number SRP029966. All of the Illumina reads generated from cDNA libraries were pooled together and *de novo* assembled with the Trinity program to form the global transcriptome of the paper mulberry.

Annotation, classification and expression analysis of TFs

For functional annotation, unigenes were firstly aligned by Blastx to protein databases NCBI nr, Swiss-Prot, TrEMBL, KEGG and COG. A significance cut off of $E \leq$ 1e-5 was used to ensure that only the proteins with the highest sequence similarity to the given unigenes were retrieved, along with their functional annotations. After getting annotation result for every unigene, all of the TFs of the paper mulberry were identified and classified into different families based on their DNA-binding domains and other conserved motifs [70,71]. In addition, the TF families' abbreviations were referenced to PlantTFDB 3.0 (http://planttfdb.cbi.pku.edu.cn/index.php) and PlnTFDB (3.0) (http://plntfdb.bio.uni-potsdam.de/v3.0/). All of these TF sequences information were provided in Additional file 1.

For gene expression analysis, the expression level of each TF in each sample was calculated by quantifying the number of Illumina reads that mapped to transcriptome of the paper mulberry with default parameters. The raw gene expression counts were normalized using the RPKM method (Reads per kb per million reads).

Identification of differentially expressed TFs in response to cold stress

For screening of differentially expressed TFs, p value corresponds to differentially expressed genes (DEGs) was obtained via a general Chi squared test that was performed by using IDEG6 (http://telethon.bio.unipd.it/bioinfo/IDEG6/). The threshold of p value in multiple tests was checked through manipulating the false discovery rate (FDR) value. The TFs with a ratio of RPKM between samples of more than 2 (Fold change ≥ 2 or ≤ 0.5) and an FDR ≤ 0.01 were considered to have significant changes in expression in response to cold stress. The Multiexperiment

Viewer (v4.9) was used to make the heat map and expression classification.

Validation by qPCR

The qPCR was adopted to validate the DEGs identified in analysis of the RNA-seq data. Ten TFs were chosen for verification. RNA used for validation was the same as that isolated for RNA-seq. In addition, two samples (samples were kept under cold stress for 15 min and 30 min) were added to detect the early expression profile of CAMTA TFs. First-strand cDNA for each sample was made from 1 μg of total RNA using SuperScript II reverse transcriptase (Takara, Dalian, China) following the manufacturer's recommendations and diluted 3 times before use in PCR. Gene-specific primers based on the selected considerate unigenes were subsequently designed using the Primer Premier 5 program and are listed in Additional file 7. QPCR reaction conditions and volume was performed as described by our former study [69]. Relative transcript levels for each sample were obtained using the comparative cycle threshold method using the cycle threshold value of the *actin* gene for each sample as a control.

Availability of supporting data

The data set supporting the results of this article is available in the NCBI's Short Read Archive (SRA) database (http://www.ncbi.nlm.nih.gov/Traces/sra/sra.cgi) under accession number SRP029966.

Additional files

Additional files 1: The sequence information and expression values of the TFs expressed in these five samples.

Additional files 2: The expressed and the differently expressed TFs distributed in every family in each sample. A The expressed TFs statistic in every family in each sample B The differentially expressed TFs statistic in every family in each sample.

Additional files 3: The differentially expressed TFs with complete ORF distributed in every family.

Additional files 4: The expression values of the differentially expressed TFs in these five samples.

Additional files 5: The expression profile of ten selected differentially expressed TFs validated by qPCR. The left axis represents the results of transcriptomics analysis while the right axis represents relative expression detected by qPCR.

Additional files 6: The key TFs identified in the paper mulberry under cold stress.

Additional files 7: The selected TFs and their primer for qPCR.

Abbreviations

ABA: Abscisic acid; CBF: C-repeat binding factor; ESTs: Expressed sequence tags; GA: Gibberellin; JA: Jasmonic acid; ORF: Opening read frame; RAM: Root apical meristem; RPKM: Reads per kb per million reads; SA: Salicylic acid; SAM: Shoot apical meristem; SRAP: Sequence-related amplified polymorphism; TF: Transcription factors.

Competing interests

The authors declare that they have no competing interests.

Authors' contributions

PXJ performed the RNA-seq experiments including all statistical and bioinformatics analyses. WQQ and TLH performed the expression profile cluster analyses. TF and PZ contributed to the transcriptomic and qPCR analyses. SSH was responsible for the overall concept and experimental designs, data integration, analysis and interpretation, and manuscript preparation. All authors read and approved the final manuscript.

Acknowledgement

This work was supported by the the National Natural Science Foundation of China (31270653), Beijing Municipal Science and Technology Project, Science and Technology Service Network Initiative (STS) (KFJ-EW-STS-061-5), and Thanks Dr. Brenna from Baylor College of Medicine for her kindly help of the English proofing.

Author details

[1]Key Laboratory of Plant Resources, Institute of Botany, the Chinese Academy of Sciences, Beijing 100093, People's Republic of China. [2]University of the Chinese Academy of Sciences, Beijing, People's Republic of China.

References

1. Saito K, Linquist B, Keobualapha B, Shiraiwa T, Horie T. *Broussonetia papyrifera* (paper mulberry): its growth, yield and potential as a fallow crop in slash-and-burn upland rice system of northern Laos. Agrofor Syst. 2009;76(3):525–32.
2. Zhai X, Zeng H, Liu Y, Liu F. Change of nutrients and shape of *Broussonetia papyrifera* leaves from different clones. J Northeast Forest Univ. 2012;11:011.
3. Ko HJ, Jin JH, Kim HP. Inhibition of experimental systemic inflammation (septic inflammation) and chronic bronchitis by new phytoformula BL containing *broussonetia papyrifera* and *lonicera japonica*. Biomol Ther. 2013;21(1):66–71.
4. Wang J, Liu J, Peng X, Ni Z, Wang G, Shen S. Applied hybrid paper mulberry in ecological virescence of the Coastal Saline. Tianjin Agric Sci. 2014;20(02):95–101.
5. Yan J, Wu P, Du H, Zhang Q. First report of black spot caused by Colletotrichum gloeosporioides on paper Mulberry in China. Plant Dis. 2011;95(7):880.
6. Knight MR, Knight H. Low-temperature perception leading to gene expression and cold tolerance in higher plants. New Phytol. 2012;195(4):737–51.
7. Peng X, Zhang L, Zhang L, Liu Z, Cheng L, Yang Y, et al. The transcriptional factor LcDREB2 cooperates with LcSAMDC2 to contribute to salt tolerance in *Leymus chinensis*. Plant Cell Tiss Org Cult. 2013;113:245–56.
8. Yang A, Dai X, Zhang W-H. A R2R3-type MYB gene, *OsMYB2*, is involved in salt, cold, and dehydration tolerance in rice. J Exp Bot. 2012;63(7):2541–56.
9. Hu H, You J, Fang Y, Zhu X, Qi Z, Xiong L. Characterization of transcription factor gene *SNAC2* conferring cold and salt tolerance in rice. Plant Mol Biol. 2008;67(1–2):169–81.
10. Zou CS, Jiang WB, Yu DQ. Male gametophyte-specific WRKY34 transcription factor mediates cold sensitivity of mature pollen in Arabidopsis. J Exp Bot. 2010;61(14):3901–14.
11. Chen L, Song Y, Li S, Zhang L, Zou C, Yu D. The role of WRKY transcription factors in plant abiotic stresses. Biochim Biophys Acta, Gene Regul Mech. 2012;1819(2):120–8.
12. Puranik S, Sahu PP, Srivastava PS, Prasad M. NAC proteins: regulation and role in stress toleranc. Trends Plant Sci. 2012;17(6):369–81.
13. Distelfeld A, Pearce SP, Avni R, Scherer B, Uauy C, Piston F, et al. Divergent functions of orthologous NAC transcription factors in wheat and rice. Plant Mol Biol. 2012;78(4–5):515–24.
14. Du H, Yang S, Liang Z, Feng B, Liu L, Huang Y, et al. Genome-wide analysis of the MYB transcription factor superfamily in soybean. BMC Plant Biol. 2012;12(1):106.
15. Licausi F, Ohme-Takagi M, Perata P. APETALA2/Ethylene Responsive Factor (AP2/ERF) transcription factors: mediators of stress responses and developmental programs. New Phytol. 2013;199(3):639–49.
16. Dubos C, Stracke R, Grotewold E, Weisshaar B, Martin C, Lepiniec L. MYB transcription factors in Arabidopsis. Trends Plant Sci. 2010;15(10):573–81.

17. Zhu J, Verslues PE, Zheng X, Lee B-h, Zhan X, Manabe Y, et al. *HOS10* encodes an R2R3-type MYB transcription factor essential for cold acclimation in plants. Proc Natl Acad Sci U S A. 2005;102(28):9966–71.
18. Wang Q, Guan Y, Wu Y, Chen H, Chen F, Chu C. Overexpression of a rice *OsDREB1F* gene increases salt, drought, and low temperature tolerance in both Arabidopsis and rice. Plant Mol Biol. 2008;67(6):589–602.
19. González-Martínez SC, Krutovsky KV, Neale DB. Forest-tree population genomics and adaptive evolution. New Phytol. 2006;170(2):227–38.
20. Keller G, Marchal T, SanClemente H, Navarro M, Ladouce N, Wincker P, et al. Development and functional annotation of an 11,303-EST collection from Eucalyptus for studies of cold tolerance. Tree Genet Genom. 2009;5(2):317–27.
21. He H, Dong Q, Shao Y, Jiang H, Zhu S, Cheng B, et al. Genome-wide survey and characterization of the *WRKY* gene family in *Populus trichocarpa*. Plant Cell Rep. 2012;31(7):1199–217.
22. Hu R, Qi G, Kong Y, Kong D, Gao Q, Zhou G. Comprehensive analysis of NAC domain transcription factor gene family in *Populus trichocarpa*. Plant Mol Biol. 2010;10(1):145.
23. Tang M, Sun J, Liu Y, Chen F, Shen S. Isolation and functional characterization of the *JcERF* gene, a putative AP2/EREBP domain-containing transcription factor, in the woody oil plant *Jatropha curcas*. Plant Mol Biol. 2007;63(3):419–28.
24. Tang M, Liu X, Deng H, Shen S. Over-expression of JcDREB, a putative AP2/EREBP domain-containing transcription factor gene in woody biodiesel plant Jatropha curcas, enhances salt and freezing tolerance in transgenic Arabidopsis thaliana. Plant Sci. 2011;181(6):623–31.
25. Sun J, Peng X, Fan W, Tang M, Liu J, Shen S. Functional analysis of *BpDREB2* gene involved in salt and drought response from a woody plant *Broussonetia papyrifera*. Gene. 2013;535(2):140–9.
26. Liu Z, Fan W, Shen S. SRAP marker in *Broussonetia papyrifera*. Sci Silvae Sin. 2009;45(12):54–8.
27. Xianjun P, Yucheng W, Ruiping H, Meiling Z, Shihua S. Global transcriptomics identification and analysis of transcriptional factors in different tissues of the paper mulberry. Plant Mol Biol. 2014;14(1):194.
28. Xin H, Zhu W, Wang L, Xiang Y, Fang L, Li J, et al. Genome wide transcriptional profile analysis of *Vitis amurensis* and *Vitis vinifera* in response to cold stress. PLoS One. 2013;8(3), e58740.
29. Song Y, Chen Q, Ci D, Zhang D. Transcriptome profiling reveals differential transcript abundance in response to chilling stress in *Populus simonii*. Plant Cell Reports. 2013;32(9):1–19.
30. Pang T, Ye C-Y, Xia X, Yin W. De novo sequencing and transcriptome analysis of the desert shrub, *Ammopiptanthus mongolicus*, during cold acclimation using Illumina/Solexa. BMC Genomics. 2013;14(1):488.
31. Changsong Z, Diqiu Y. Analysis of the cold-responsive transcriptome in the mature pollen of Arabidopsis. Int J Plant Biol. 2010;53(6):400–16.
32. An D, Yang J, Zhang P. Transcriptome profiling of low temperature-treated cassava apical shoots showed dynamic responses of tropical plant to cold stress. BMC Genomics. 2012;13(1):64.
33. Liu Q, Kasuga M, Sakuma Y, Abe H, Miura S, Yamaguchi-Shinozaki K, et al. Two transcription factors, DREB1 and DREB2, with an EREBP/AP2 DNA binding domain separate two cellular signal transduction pathways in drought-and low-temperature-responsive gene expression, respectively, in Arabidopsis. Plant Cell. 1998;10(8):1391.
34. Peng X, Ma X, Fan W, Su M, Cheng L, Alam I, et al. Improved drought and salt tolerance of *Arabidopsis thaliana* by transgenic expression of a novel DREB gene from *Leymus chinensis*. Plant Cell Rep. 2011;30:1493–502.
35. Choi DW, Rodriguez EM, Close TJ. Barley *Cbf3* gene identification, expression pattern, and map location. Plant Physiol. 2002;129(4):1781.
36. Zhang J, Yang W, Cheng T, Pan H, Zhang Q. Functional and evolutionary analysis of two *CBF* genes in *Prunus mume*. Can J Plant Sci. 2013;93:1–10.
37. Wang C, Yang Q, Yang Y. Characterization of the *ZmDBP4* gene encoding a CRT/DRE-binding protein responsive to drought and cold stress in maize. Acta Physiol Plant. 2011;33(2):575–83.
38. Mizoi J, Shinozaki K, Yamaguchi-Shinozaki K. AP2/ERF family transcription factors in plant abiotic stress response. Biochim Biophys Acta, Gene Regul Mech. 2012;1819((2):86–96.
39. Chen Y, Yang X, He K, Liu M, Li J, Gao Z, et al. The MYB transcription factor superfamily of Arabidopsis: expression analysis and phylogenetic comparison with the rice MYB family. Plant Mol Biol. 2006;60(1):107–24.
40. Seo PJ, Kim MJ, Park JY, Kim SY, Jeon J, Lee YH, et al. Cold activation of a plasma membrane-tethered NAC transcription factor induces a pathogen resistance response in Arabidopsis. Plant J. 2010;61(4):661–71.
41. Nuruzzaman M, Sharoni AM, Satoh K, Kondoh H, Hosaka A, Kikuchi S. A genome-wide survey of the NAC transcription factor family in monocots and eudicots. In: Introduction to Genetics-DNA Methylation, Histone Modification and Gene Regulation. Hong Kong: iConcept Press; 2012. ISBN, 978-14775549-4-4.
42. Zhou QY, Tian AG, Zou HF, Xie ZM, Lei G, Huang J, et al. Soybean WRKY-type transcription factor genes, *GmWRKY13*, *GmWRKY21*, and *GmWRKY54*, confer differential tolerance to abiotic stresses in transgenic Arabidopsis plants. Plant Biotechnol J. 2008;6(5):486–503.
43. Chen H, Lai Z, Shi J, Xiao Y, Chen Z, Xu X. Roles of Arabidopsis WRKY18, WRKY40 and WRKY60 transcription factors in plant responses to abscisic acid and abiotic stress. BMC Plant Biol. 2010;10(1):281.
44. Novillo F, Alonso JM, Ecker JR, Salinas J. CBF2/DREB1C is a negative regulator of CBF1/DREB1B and CBF3/DREB1A expression and plays a central role in stress tolerance in Arabidopsis. Proc Natl Acad Sci U S A. 2004;101(11):3985–90.
45. Zarka DG, Vogel JT, Cook D, Thomashow MF. Cold induction of Arabidopsis CBF genes involves multiple ICE (inducer of CBF expression) promoter elements and a cold-regulatory circuit that is desensitized by low temperature. Plant Physiol. 2003;133(2):910–8.
46. Agarwal M, Hao Y, Kapoor A, Dong C, Fujii H, Zheng X, et al. A R2R3 type MYB transcription factor is involved in the cold regulation of CBF genes and in acquired freezing tolerance. J Biol Chem. 2006;281(49):37636–45.
47. Chinnusamy V, Zhu J, Zhu J-K. Cold stress regulation of gene expression in plants. Trends Plant Sci. 2007;12(10):444–51.
48. Doherty CJ, Van Buskirk HA, Myers SJ, Thomashow MF. Roles for Arabidopsis CAMTA transcription factors in cold-regulated gene expression and freezing tolerance. Plant Cell. 2009;21(3):972–84.
49. Van Ha C, Le DT, Nishiyama R, Watanabe Y, Sulieman S, Tran UT, et al. The auxin response factor transcription factor family in soybean: genome-wide identification and expression analyses during development and water stress. DNA Res. 2013;20(5):511–24.
50. Ubeda-Tomás S, Beemster GT, Bennett MJ. Hormonal regulation of root growth: integrating local activities into global behaviour. Trends Plant Sci. 2012;17(6):326–31.
51. Horiguchi G, Kim GT, Tsukaya H. The transcription factor AtGRF5 and the transcription coactivator AN3 regulate cell proliferation in leaf primordia of *Arabidopsis thaliana*. Plant J. 2005;43(1):68–78.
52. Rahman A. Auxin: a regulator of cold stress response. Physiol Plant. 2013;147(1):28–35.
53. Gangappa SN, Botto JF. The BBX family of plant transcription factors. Trends Plant Sci. 2014;19(7):460–70.
54. Kumagai T, Ito S, Nakamichi N, Niwa Y, Murakami M, Yamashino T, et al. The common function of a novel subfamily of B-Box zinc finger proteins with reference to circadian-associated events in *Arabidopsis thaliana*. Biosci Biotechnol Biochem. 2008;72(6):1539–49.
55. Gangappa SN, Crocco CD, Johansson H, Datta S, Hettiarachchi C, Holm M, et al. The Arabidopsis B-box protein BBX25 interacts with HY5, negatively regulating *BBX22* expression to suppress seedling photomorphogenesis. Plant Cell. 2013;25(4):1243–57.
56. Waters MT, Wang P, Korkaric M, Capper RG, Saunders NJ, Langdale JA. GLK transcription factors coordinate expression of the photosynthetic apparatus in Arabidopsis. Plant Cell. 2009;21(4):1109–28.
57. Soyano T, Thitamadee S, Machida Y, Chua N-H. Asymmetric leaves2-like19 lateral organ boundaries domain30 and Asl20/Lbd18 regulate tracheary element differentiation in Arabidopsis. Plant Cell. 2008;20(12):3359–73.
58. Fan M, Xu C, Xu K, Hu Y. LATERAL ORGAN BOUNDARIES DOMAIN transcription factors direct callus formation in Arabidopsis regeneration. Cell Res. 2012;22(7):1169–80.
59. Husbands A, Bell EM, Shuai B, Smith HM, Springer PS. LATERAL ORGAN BOUNDARIES defines a new family of DNA-binding transcription factors and can interact with specific bHLH proteins. Nucleic Acids Res. 2007;35(19):6663–71.
60. Koyama T, Furutani M, Tasaka M, Ohme-Takagi M. TCP transcription factors control the morphology of shoot lateral organs via negative regulation of the expression of boundary-specific genes in Arabidopsis. Plant Cell. 2007;19(2):473–84.
61. Van der Graaff E, Laux T, Rensing SA. The WUS homeobox-containing (WOX) protein family. Genome Biol. 2009;10(12):248.
62. Wang W, Li G, Zhao J, Chu H, Lin W, Zhang D, et al. DWARF TILLER1, a WUSCHEL-Related homeobox transcription factor, is required for tiller growth in rice. PLOS Gen. 2014. DOI: 10.1371/journal.pgen.1004154.

63. Dai M, Hu Y, Zhao Y, Zhou D-X. Regulatory networks involving YABBY genes in rice shoot development. Plant Physiol. 2007;144:380–90.
64. Zhu J, Shi H, Lee B-h, Damsz B, Cheng S, Stirm V, et al. An Arabidopsis homeodomain transcription factor gene, *HOS9*, mediates cold tolerance through a CBF-independent pathway. Proc Natl Acad Sci U S A. 2004;101(26):9873–8.
65. Miura K, Furumoto T. Cold signaling and cold response in plants. Int J Mol Sci. 2013;14(3):5312–37.
66. Tanaka W, Toriba T, Ohmori Y, Yoshida A, Kawai A, Mayama-Tsuchida T, et al. The YABBY gene TONGARI-BOUSHI1 is involved in lateral organ development and maintenance of meristem organization in the rice spikelet. Plant Cell. 2012;24(1):80–95.
67. Wang H, Zhang Z, Li H, Zhao X, Liu X, Ortiz M, et al. CONSTANS-LIKE 7 regulates branching and shade avoidance response in Arabidopsis. J Exp Bot. 2013;64(4):1017–24.
68. Kim JH, Choi D, Kende H. The AtGRF family of putative transcription factors is involved in leaf and cotyledon growth in Arabidopsis. Plant J. 2003;36(1):94–104.
69. Peng X, Teng L, Wang X, Wang Y, Shihua S. De Novo assembly of expressed transcripts and global transcriptomic analysis from seedlings of the paper Mulberry (*Broussonetia kazinoki* x *Broussonetia papyrifera*). PLoS One. 2014;9(5), e97487.
70. Jin J, Zhang H, Kong L, Gao G, Luo J. PlantTFDB 3.0: a portal for the functional and evolutionary study of plant transcription factors. Nucleic Acids Res. 2013;42(D1):D1182–7.
71. Pérez-Rodríguez P, Riaño-Pachón DM, Corrêa LGG, Rensing SA, Kersten B, Mueller-Roeber B. PlnTFDB: updated content and new features of the plant transcription factor database. Nucleic Acids Res. 2010;38 Suppl 1:D822–7.

Genetic analysis of inflorescence and plant height components in sorghum (Panicoidae) and comparative genetics with rice (Oryzoidae)

Dong Zhang[1,2], Wenqian Kong[1,3], Jon Robertson[1], Valorie H Goff[1], Ethan Epps[1], Alexandra Kerr[1], Gabriel Mills[1], Jay Cromwell[1], Yelena Lugin[1], Christine Phillips[1] and Andrew H Paterson[1,2,3,4,5*]

Abstract

Background: Domestication has played an important role in shaping characteristics of the inflorescence and plant height in cultivated cereals. Taking advantage of meta-analysis of QTLs, phylogenetic analyses in 502 diverse sorghum accessions, GWAS in a sorghum association panel (n = 354) and comparative data, we provide insight into the genetic basis of the domestication traits in sorghum and rice.

Results: We performed genome-wide association studies (GWAS) on 6 traits related to inflorescence morphology and 6 traits related to plant height in sorghum, comparing the genomic regions implicated in these traits by GWAS and QTL mapping, respectively. In a search for signatures of selection, we identify genomic regions that may contribute to sorghum domestication regarding plant height, flowering time and pericarp color. Comparative studies across taxa show functionally conserved 'hotspots' in sorghum and rice for awn presence and pericarp color that do not appear to reflect corresponding single genes but may indicate co-regulated clusters of genes. We also reveal homoeologous regions retaining similar functions for plant height and flowering time since genome duplication an estimated 70 million years ago or more in a common ancestor of cereals. In most such homoeologous QTL pairs, only one QTL interval exhibits strong selection signals in modern sorghum.

Conclusions: Intersections among QTL, GWAS and comparative data advance knowledge of genetic determinants of inflorescence and plant height components in sorghum, and add new dimensions to comparisons between sorghum and rice.

Keywords: Sorghum, GWAS, Biparental QTL mapping, Inflorescence, Flowering time, Plant height, Domestication, Genetic correspondence

Background

The Sorghum genus has recently become an important botanical model for Andropogoneae grasses, by virtue of its relatively small and largely-sequenced genome, a minimum of gene duplication thanks to 70 million years of abstinence from polyploidy, and its close relationship to grasses such as maize, sugarcane and Miscanthus that have much more complex genomes [1]. Cultivated sorghum (*Sorghum bicolor*) ranks fifth in importance among the world's grain crops, is a versatile source of food, fodder, and fuel, and possesses a great diversity of cultivated forms that may reflect its wide range of adaptation [2-4].

The ~30 year history of using linked molecular markers to dissect complex traits in plants has broadly used two complementary approaches. Conventional biparental QTL mapping [5] has been widely used and has provided foundational information that led to some successes in the identification of causal genes in many organisms. However, biparental QTL mapping generally offers relatively coarse resolution that is not sufficient to determine causative genes. Highly saturated recombination maps, multiparent advanced generation intercrosses (MAGIC) [6] or nested association mapping (NAM) [7], offer options to enhance mapping resolution of QTLs. Dramatic increases

* Correspondence: paterson@uga.edu
[1]Plant Genome Mapping Laboratory, University of Georgia, Athens, GA 30602, USA
[2]Institute of Bioinformatics, University of Georgia, Athens, GA 30602, USA
Full list of author information is available at the end of the article

in genomic data provide rich resources with which to investigate genes and gene functions on a much finer scale than QTL mapping by taking advantage of historical accumulation of recombination events in a gene pool using 'association genetics' [8]. However, association mapping can require extremely high DNA marker densities to thoroughly scan a genome for genes influencing a trait, and complex measures to distinguish between artifacts such as relatedness among genotypes (especially in improved germplasm) and true evidence of functional association between a mutation and a phenotype [8-11]. Although GWAS is able to explore for causative loci on a genome-wide scale, population structure and genetic relatedness may confound associations at causative loci. Some GWAS conducted in rice and *A. thaliana* have suggested that known causative loci showed weaker signals than nearby markers [12]. In contrast, carefully designed crossing schemes in QTL mapping may be more targeted to locate relevant QTLs. Identifying intersections between results from biparental QTL mapping and association genetic data is a potentially powerful means to mitigate constraints associated with each approach, accelerating progress toward identifying specific genes that function in biological processes of relevance to agriculture.

The grass inflorescence, the primary food source for humanity, has been repeatedly selected during domestication [13]. Some well-characterized domestication traits related to the inflorescence include pericarp color of seed, seed shattering, awn length/presence and seed size/yield. Of similarly high and recurring importance in plant domestication and crop improvement are flowering time and plant height, which often show significant genetic correlation with one another [13,14].

In this study, we use GWAS to investigate 6 components of sorghum inflorescence morphology and 6 traits related to plant height, then compare GWAS-based associations to positional evidence from meta-analysis of QTL likelihood intervals. QTL meta-analysis, the comparison of multiple independent QTL studies in different germplasm and environments, is used here to provide a more comprehensive picture of the true genetic control of a trait than analysis of any single population [15], for example revealing plant height and flowering of sorghum to be genetically more complex than had been realized after more than 70 years of investigation [16]. We note that a few classically-identified loci (*dw1-dw4)* had already been compared to GWAS data [4], one of which (*dw3*) had been cloned [17], however here we address additional loci identified only by meta-analysis of QTL likelihood intervals.

Inbreeding organisms have limits to the precision in the association mapping studies [12]. Sorghum is largely inbreeding, which can result in strong LD patterns, and may lead to low genetic resolution in specific local regions along the genome. Hence, we reported the hotspots underlying 12 traits in sorghum, instead of gene candidates. The identified 'hotspots' provide a valuable advance toward the goal of uncovering causative variants for the trait of interest.

It is accepted widely that QTL intervals controlling common traits have non-random correspondence across and within cereals, and GWAS adds a new dimension to the ability to compare the genetic control of common traits in different cereals (and other taxa). For example, in an early comparison, the observed probabilities that seed mass (size) QTLs in sorghum, rice, and maize would correspond so frequently by chance was conservatively estimated as 0.1 to 0.8% [18]. Another early study [14] indicated that 8 of 25 regions affecting flowering of maize fall into 4 homoelogous regions. Numerous studies have shown that some orthologs across taxa have similar functions underlying common phenotypes, but other causative genes have no obvious counterparts that contribute to similar traits even in their close relatives. Thus, genetic correspondence may reflect functionally conserved 'hotspots' existing across taxa, but may or may not be conserved 'genes'. Genes in a pathway exhibit significantly higher genomic clustering than expected by chance in eukaryotes [19]; for example, co-regulated clusters of genes have been implicated in QTLs affecting cotton fiber traits [20].

The common ancestor of rice, maize and sorghum experienced a whole-genome duplication (WGD; named rho) that is still readily discernible in their genomes [21], making it possible to test the hypothesis of convergent evolution across an estimated 70 million years, with the possibility of subfunctionalization among homoeologous regions. A hypothesis worthy of further exploration is that a co-regulated cluster of genes in the cereal common ancestor may have experienced gain/loss and functional divergence of some members in the subsequent 70 my of divergence, with independent domestications conferring additional functional changes in similar locations of different taxa but which are not strictly orthologous.

Methods

Genotype

The imputed version of ~265,000 published SNPs characterized in 971 worldwide accessions based on genotyping-by-sequencing (GBS) was employed [4]. About 72% of annotated genes contain ≥1 SNP site. A total of 228 of 354 accessions from a US sorghum association panel (SAP) [22] are converted tropical lines that are photoperiod insensitive, early maturing, and short stature phenotypes, produced via crossing exotic lines and modern U.S. cultivars. It has been demonstrated [4] that the population has sufficient power to dissect a trait, such as inflorescence architecture, that was not a target of selection in the sorghum conversion program. We used the 354 accessions from the SAP to perform GWAS. A total of 502 accessions, each characterized to a unique morphological type, were used to analyze the population structure. Wild sorghums [*Sorghum bicolor* ssp. *verticilliflorum* (L.) Moench] (n = 31) of four races, namely *aethiopicum, arundinaceum, verticilliflorum* and *virgatum*, were used to calculate expected heterozygosity for the wild population, and to compare with the heterozygosity in the 502 accessions of cultivated sorghum.

Phenotype for GWAS

Phenotypic data for the 354 accessions in the SAP from three different growouts was utilized. We used a completely randomized design for all the 354 accessions (unreplicated trial with two observations per plot). On seed grown during 2008 in Lubbock, TX, the average RGB values for pericarp color for each genotype were determined from images of the surfaces of 5 seeds. Using the conversion formula of RGB→CIE-L*ab, RGB was transformed to CIE-L*ab color space. RGB is summarized by three values: R, G and B, and CIE-L*ab is summarized by: L, a and b. From growouts in 2009 (seeds sowed on May 19th) and 2010 (seeds sowed on May 26th), near Watkinsville GA, we took two representative samples (from each plot) per genotype in each year for the following measurements. Awn presence: presence or absence of awn, with 2 for abundant, 1 for occasional, and 0 for absent; base-flag length: measured in cm, from the plant base to the flag leaf; flag-rachis length: measured in cm, from the flag leaf to the top node, with positive values indicating the flag leaf below the rachis and negative values indicating the flag leaf above the rachis; inflorescence length: measured in cm, from the rachis to the top of inflorescence; inflorescence width: measured in cm, at the widest point; nodes: the number of nodes; whorls: the number of whorls; dry inflorescence weight; dry stalk weight; total plant height; and flowering time: the average number of days for the first five heads to flower, from the planting date. For traits measured on two samples, we used the average values per genotype to assess heritability and phenotypic correlations. Since the raw data in all the quantitative traits in our study is distributed in near-normal fashion, we conducted GWAS by using the data without transformation. Considering that data combined across years may weaken association levels for those traits with low heritability, we performed GWAS on data for each year individually. The phenotypic correlations determined by Pearson correlation are available from Additional file 1: Table S1.

QTL mapping

QTL confidence intervals were from two resources. (1) We compiled 1-LOD likelihood intervals, which have been identified to underlie any of the 12 traits from published literature [13,14,18,23-29]. (2) We used a recombinant inbred line (RIL) from an interspecific cross between *S. bicolor* and *S. propinquum* (BTxSP), the widest cross that can be made with *S. bicolor* using conventional techniques, containing 141 loci on 10 linkage groups collectively spanning 773.1 cM to map the confidence intervals for base-flag length, flowering time, nodes and total plant height [29] for 2010 and 2011 data. Phenotypic data were collected with two replications per genotype in both 2010 and 2011. A LOD score of 2.5 was used for QTL detection. The methods for anchoring QTL intervals to the reference genome have been discussed in Zhang et al., 2013 [16]. Briefly, based on colinearity between genetic and physical positions of markers, a QTL region is delineated by two flanking markers nearest to the likelihood peak that have alignment information (BLASTN hits). Genomic positions for intervals are available from Additonal file 2: Table S2.

GWAS

The Compressed Mixed Linear Model (CMLM) involves genetic marker-based kinship matrix modeling of random effects, used jointly with population structure estimated by principal components analysis (PCA) to model fixed effects [30-32]. Since more extensive genetic variations may confer a phylogeny closer to the true one, the total of 265,487 SNPs in the SAP were used to analyze population structure. The compression level and optimal number of principal components that adequately explain population structure were previously determined by the Genomic Association and Prediction Integrated Tool [30-32]. Log quantile–quantile (QQ) P-value plots for 265,487 single-SNP tests of association (Additional file 2: Figure S1-S3) implied that there were few systematic sources of spurious association using CMLM, noting the close adherence of P values to the null hypothesis over the most of the range. Genomic positions for identified hotspots are available from Additional file 1: Table S3.

Significance threshold

We performed Bonferroni-like multiple testing correction [33] to determine significance thresholds for GWAS. Instead of 265,487 independent tests assumed in the Bonferroni method, the total number of tests was estimated by using the average extent of LD across the genome. On average, LD decays to background levels ($r^2 < 0.1$) within 150 kb in the current GBS data [4]. The effective number of independent tests was defined as LD bins [reference genome size (730 Mb)/average LD extent (150 kb)]. Given 0.05 as the desired experiment wide probability of type I error, a significance cutoff within about an order of magnitude of 10^{-5} was estimated.

Overlap between QTL and heterozygosity reduction

Since the hypergeometric probability distribution (sampling without replacement) $p = \frac{\binom{1}{m}\binom{n-1}{s-m}}{\binom{n}{s}}$ can assess the correspondence between QTLs [14,23], we used the hypergeometric probability distribution to evaluate genetic overlap between plant height/flowering time QTLs and significant heterozygosity reduction in wild sorghum. n is the total number of intervals (defined as 30 cM, approximating a QTL likelihood interval) along the whole genome; l is the number of intervals having significant heterozygosity reduction; s is the number of intervals having plant height/flowering time QTL; m is the number of intervals having both features (overlapping intervals). The purpose of this test is to show that biparental QTL mapping may capture genetic variations for domestication traits that were evolved from wild sorghum and of importance in the history of sorghum selection. Thus, we used QTL intervals for plant height/flowering time that were only determined by biparental QTL mapping. The regions with the largest 1% of heterozygosity reduction values were selected for testing. We utilized alignment between a high-density genetic recombination map [34] and the sorghum reference genome [1] to unify the genetic positions of QTLs and heterozygosity reduction regions (based on windows of 500 consecutive SNPs).

Reference genomes

The gene annotations refer to JGI annotation release Sbi1.4 [1] and Michigan State University Rice Genome Annotation Project (MSU-RGAP release 7) [35].

Results and discussion

Phylogenetic relationships of five main sorghum races

Morris et al used a genome-wide SNP map to explore the population structure of 971 sorghum accessions, and illustrated the differentiation of their geographic origins [4]. However, several intriguing questions remain, for example: (1) what is the most primitive sorghum type?, and (2) how many independent domestications has sorghum experienced?

It is generally accepted that the domestication of sorghum started in Africa. Bicolor, guinea, caudatum, durra and kafir are five main morphological types that are well recognized to represent genetic diversity in the cultivated sorghum. Wild sorghum [*Sorghum bicolor* ssp. *verticilliflorum* (L.) Moench] including four races (namely *aethiopicum*, *arundinaceum*, *verticilliflorum* and *virgatum*), is the progenitor of the cultivated sorghum [*Sorghum bicolor* (L.) Moench] [36,37]. The phylogenetic relatedness in sorghum races has been discussed in several studies [38-41], but ambiguous clustering patterns have often been found, part of which may be attributable to the limitations of either low-density markers or small population size. To re-investigate inferences drawn before, we focused on a subset of 502 accessions, in which 471 are each characterized uniquely to one and only one of the five primary cultivated races and 31 are wild types. Both the phylogenetic tree and the PCA plots indicate that bicolor is the most primitive race, based on having close phylogenetic relationship with wild types (Figure 1a and b). The level of population differentiation, fixation index (F_{ST}), was measured between wild types and each of five primary types. Race bicolor ($F_{ST} = 0.04$) exhibits closer genetic relationship with wild sorghums than any of the other 4 primary races (F_{ST} (guinea-wild) = 0.11, (durra-wild) = 0.20, (kafir-wild) = 0.33, (caudatum-wild) = 0.14), with guinea and caudatum apparently representing early derivatives. A large block of bicolor accessions are intermediate among durra types, potentially consistent with an ancestral relationship (Figure 1b), noting that F_{ST} (0.14) supports small population differentiation between bicolor and durra. Likewise, a large block of guinea accessions clustering within caudatum may suggest another derivation in the history of sorghum selection, also supported by evidence of minimal population differentiation ($F_{ST} = 0.17$). Races caudatum, durra and kafir show clustering patterns that are substantially distinct from one another, save for occasional single accessions that could be misclassifications (Figure 1b). Pairs of these 3 sorghum races show relatively high levels of population differentiation [F_{ST} (durra-caudatum) = 0.26, (durra-kafir) = 0.46, (caudatum-kafir) = 0.33]. Although the first three components of PCA are able to explain 89.1% of the total genetic variance, ambiguous clustering patterns are still occasionally observed, especially for the two primitive races bicolor and guinea. Both the PCA plot and neighbor-joining tree show quite similar clustering patterns by using the complete SNP set (265,487 SNPs) (Figure 1) and a set including more informative SNPs (missing rate ≤ 50% and MAF ≥ 0.02) (Additional file 2: Figure S4). The complex history of diffusion and selection in sorghum may confound phylogenetic inference using the clustering patterns. The phylogenetic relationships of sorghum races are still open to questions that may be more accurately addressed with growing genotype data.

Using representative SNPs across the *Shattering1* (*Sh1*) gene, a key domestication locus [2], a recent discovery revealed four haplotypes, in which three represent non-shattering forms. Specifically, guinea and durra share a common haplotype (SC265); a second haplotype (Tx623) is prevalent in kafir; and a third haplotype (Tx430) is dominant in caudatum. Three non-shattering haplotypes strongly suggest at least three domestication episodes for

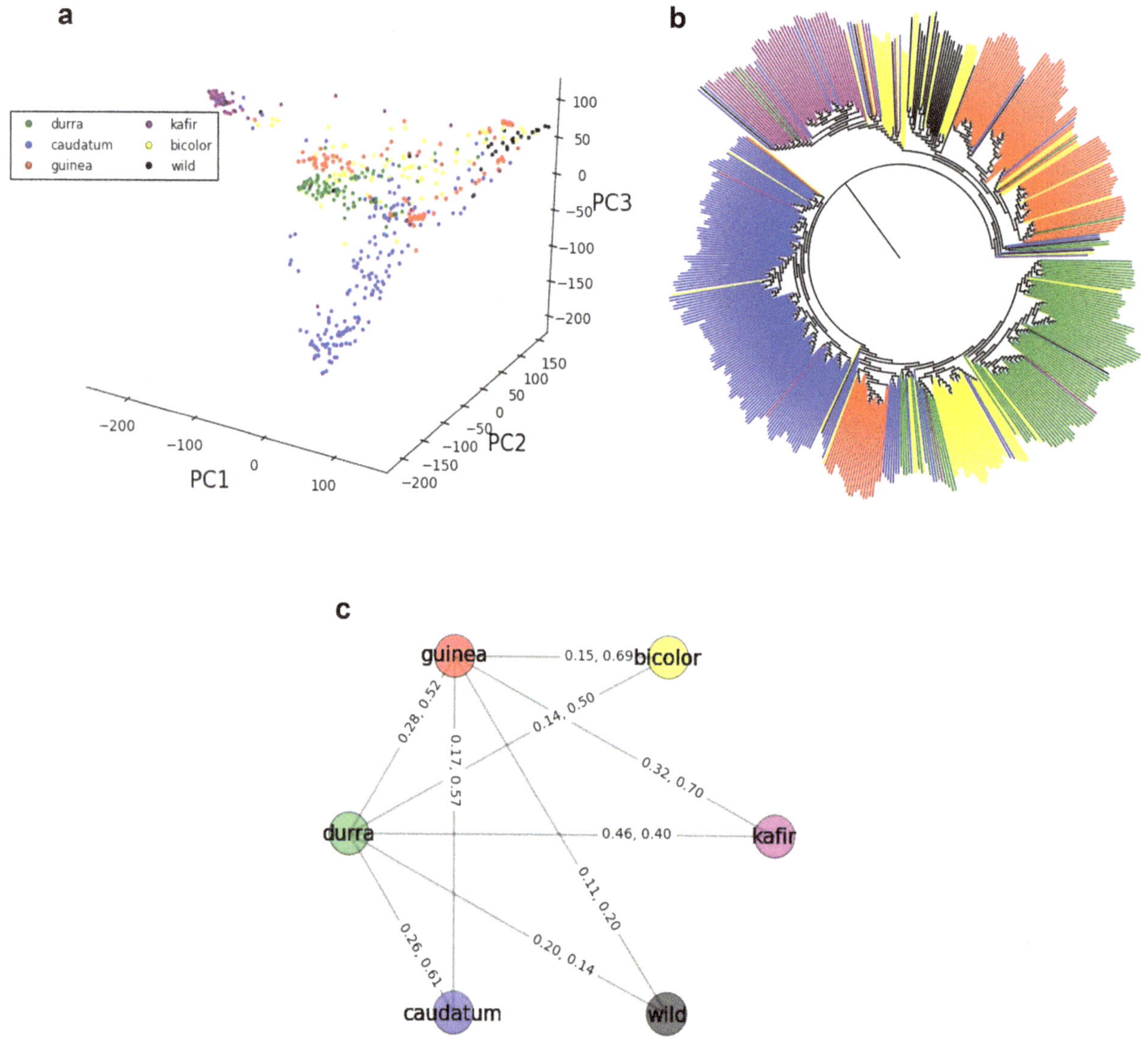

Figure 1 Population structure of 502 worldwide sorghum accessions. 476 belong to the five main cultivated races and 26 are wild types. **(a)** PCA plots of the first three components for 265,487 SNPs. The five main cultivated races and the wild type are color-coded. **(b)** Neighbor-joining tree of 502 sorghum accessions. **(c)** Population differentiations and frequencies of common two-locus haplotypes for 100 SNPs adjacent to the *Sh1* gene [2] for pairs of populations. All the connections for guinea and durra are shown. The F_{ST} values paired with frequencies of common two-locus haplotypes are indicated.

reduced shattering in the five main sorghum races [2,3]. In general, domestication loci can exhibit very strong LD, because of reduction in genetic divergence. Unusually high genetic divergence at *Sh1* in cultivated sorghum races may explain why the surrounding region does not show stronger LD than the background level. The clustering pattern in the neighbor-joining tree and F_{ST} values support the inference that kafir and caudatum have experienced two independent domestication events for non-shattering. However, is the non-shattering allele derived from a common ancestor of guinea and durra? Generally, we did not observe a close phylogenetic relationship between guinea and durra on the basis of the clustering patterns of the genetic tree (Figure 1b), and the large value of F_{ST} (0.28). This relatively high level of genetic differentiation implies that guinea and durra may have experienced convergent domestication events. The types of haplotypes resulting from selection could be restricted by the limited number of representative SNPs detected across *Sh1*. To overcome this limitation, we examined the region adjacent to *Sh1* to provide a pool enriched by SNPs that are tightly linked to the shattering locus. Since *Sh1* was not genotyped in the current data set, it would be challenging to know the exact number of SNP sites that are linked to *Sh1*. Instead, using 100 SNPs in a 200 kb region adjacent to *Sh1*, we examined haplotypes in consecutive pairs of loci for each sorghum population. The ratio of common haplotypes/ total haplotypes was calculated for each pair of sorghum types. Among the 9 possible pairwise comparisons of populations (Figure 1c), guinea-durra only shows a modest frequency (0.52) of common two-locus haplotypes, that is lower than guinea-bicolor (0.69), guinea-caudatum (0.57), guinea-kafir (0.70) and durra-caudatum (0.61). Both guinea and durra have extremely low frequencies (0.20 and 0.14) of common haplotypes with wild sorghum. Below, using two additional domestication genes mapped herein, we further investigate the hypothesis that guinea and durra may have experienced independent domestication events but achieved non-shattering by convergence.

Meta-analysis of sorghum QTLs

Biparental QTL mapping is based on the principle that genes and linked DNA markers largely co-segregate during meiosis save for occasional recombination events, thus allowing their analysis in the progeny. The limited number of recombination events captured in progeny of recent crosses may result in QTL likelihood intervals that contain dozens or even hundreds of genes. Further, the environment and parental lines used in a cross can limit the power to accurately estimate the number of QTLs and magnitude of their effects. Using a database that we have recently described [16], we compiled 1-LOD QTL likelihood intervals resulting from 11 independent biparental QTL mapping studies to yield a more complete picture of the genetic control of a trait than could be obtained from any individual study.

The pericentromeric region of sorghum chromosome 6 (Sb06) has repeatedly shown evidence of genetic control of plant height (Additional file 2: Figure S11), and was thought to harbor two classic dwarfing genes (*dw2* and *dw4*) [4]. The general lack of recombination in this region allowed QTL confidence intervals to cross centromeres and cover broad genomic areas. Two other regions repeatedly associated with plant height are in the euchromatin of Sb07 and Sb09 (Additional file 2: Figure S11), which were considered to contain *dw3* and *dw1* respectively [4]. An additional 9 nonoverlapping regions in the sorghum genome containing height QTLs (Additional file 2: Figure S11) show that genetic control of sorghum plant height involves substantially more than the four genes reported in classical studies [42]. Likewise, 14 flowering QTL likelihood intervals published in six studies fall into at least 11 non-overlapping regions (Figure 2 and Additional file 2: Figure S12), strongly indicating far more than the classically suggested six genes, *Maturity1* (*Ma1*) to *Ma6*, in genetic control of sorghum flowering time [16,42,43]. Flowering time and plant height show significant genetic correlations on chromosomes Sb01, Sb03, Sb04, Sb06, and Sb09, indicating that their inheritance is linked either functionally (pleiotropy) or physically (linkage disequilibrium) [13].

GWAS for sorghum

To dissect the genetic basis of 12 traits in sorghum by GWAS, we used a compressed mixed linear model (CMLM) [30,32] to assess evidence of phenotype-genotype association. Three steps were taken into consideration: (1) determination of significance thresholds for association, (2) identification of linkage disequilibrium (LD) regions for significant association signal and (3) positive control of association.

A major issue with genome scans, which involve many thousands of independent statistical tests, is multiple testing. The Bonferroni method approximates the significance cutoff for an overall (i.e. genome-wide) 5% probability of type I error as $0.05/265{,}487 = 1.89 \times 10^{-7}$ in our studies. However, this method has been criticized for its stringency [44] owing to the fact that genotype at some SNP loci are correlated thus are not independent hypotheses. Sorghum is largely inbreeding, which can result in strong LD patterns along the genome, so that an appropriate significance threshold may be larger than 1.89×10^{-7}. Here, we used the quantified average LD information in sorghum to adjust the Bonferroni correction [33] (See details in Methods). A significance cutoff within about an order of magnitude of 10^{-5} is inferred to balance an acceptable false positive rate with sufficient power to detect true associations.

It is also important to determine LD with single SNP association, especially when causative variants are not genotyped (or at least not known). On the basis of pairwise measures of LD (r^2), 'block-like' structures can be visually apparent. It is now well understood that the extent of LD in the pericentromeric region, which experiences relatively little recombination, is greater than in the euchromatin, which experiences more frequent recombination. A long LD block with association signals is most likely to contribute striking features to the 'skyline' of a genome-wide Manhattan plot.

Known genes and biparental QTL intervals that have been identified previously are useful to assess association validity. If knowledge of such candidate genes/intervals is limited in the species of interest, information from closely related species might be utilized, using synteny-based approaches to deduce orthology. In addition to the compilation of QTL likelihood intervals and *Dwarf1* (*Dw1*)-*Dw4* loci [4] in sorghum, sorghum maturity (*Ma*) genes *Ma1* [14] and *Ma6* [45], and genes *yellow seed1* (*y1*) [46,47] and *Tannin1* (*Tan1*) [48] for sorghum pericarp color provide positive controls for GWAS.

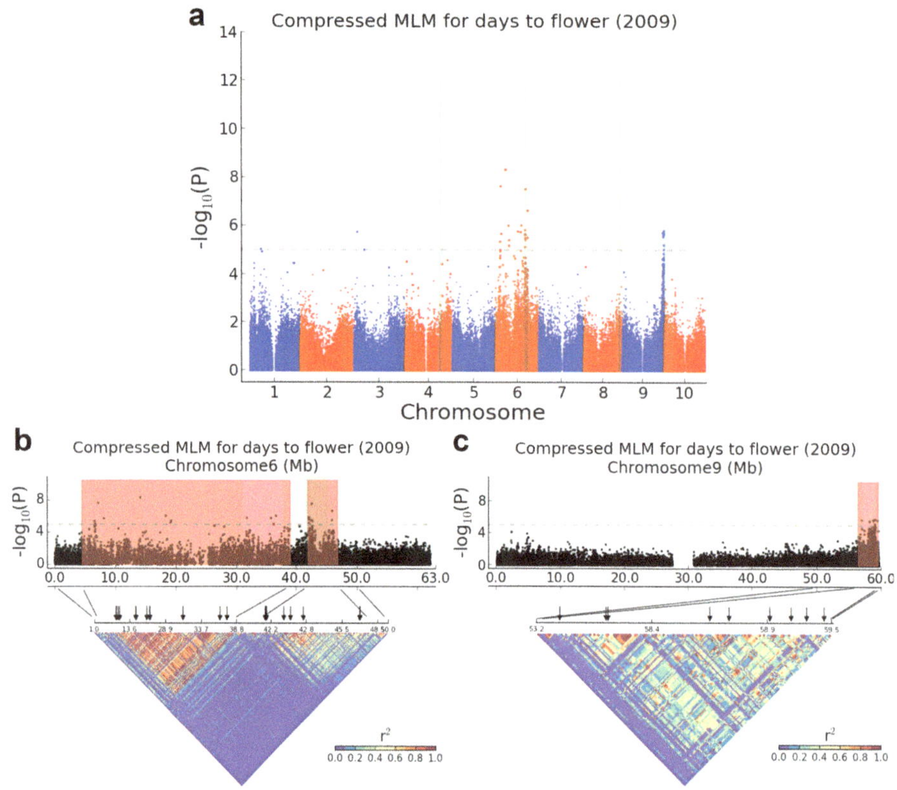

Figure 2 GWAS for flowering time in 2009. **(a)** Genome-wide Manhattan plot of CMLM. Significance threshold is denoted by the gray dashed line. The 10 sorghum chromosomes are plotted against the negative base-10 logarithm of the association P value. Areas highlighted in green indicate confidence intervals for flowering time determined by QTL mapping. **(b)** Chromosome Sb06 Manhattan plot of CMLM (top). Red areas show hotspots for 2009 flowering time identified by association mapping. Linkage disequilibrium (r^2 were calculated from SNPs with association $p \leq 0.05$ and missing data ≤ 50%) matrices (bottom) are plotted for regions denoted by anchoring lines. Regions of strong LD are shown in red. Significant association markers are denoted by black arrows. **(c)** Chromosome Sb09 Manhattan plot of CMLM.

Traits related to the sorghum inflorescence

We investigated 6 properties of the sorghum inflorescence, including awn presence, pericarp color, dry inflorescence weight, inflorescence length and width, and whorl number.

Using a *S. bicolor* intraspecific map (BTxIS), Hart et al., [24] identified an interval controlling the presence of awns in euchromatin near the 3′ end of chromosome Sb03. The most striking association based on GWAS of awn presence for two years was in the genetically mapped interval (Figure 3a, Additional file 2: Figure S5 and Figure 4b). Both mapping strategies achieve similar genetic resolution, with intervals spanning ~4.7 megabases (Mb). We found 10 additional significant association hotspots in 2009 and 7 in 2010, none of which are consistent in the two years, which could represent modifiers affected by environment, or false positive associations.

Inheritance studies of endosperm color in sorghum proposed that the trait was oligogenic [49,50]. To date, a single gene (*Rc*) has been verified to be responsible for pericarp pigmentation of rice grains (Figure 4a), while gene *Rd* is assumed to play a role in spreading pigment [51,52]. In view of the similar levels of gene duplication in rice and sorghum, it is plausible that pericarp color is also an oligogenic trait in sorghum. In sorghum, it is known that the gene *yellow seed1* (*y1*) is required for the production of red phlobaphene pigments in the grain pericarp [46,47], and the *Tan1* gene controls tannin biosynthesis to affect pericarp color [48]. In order to minimize the limitations of artificial descriptions of colors, we applied two commonly used color models, RGB and CIE-L*ab (See the Methods also for the details), to phenotype pericarp color of sorghum seeds. On chromosome Sb01, an association peak (~61.45 Mb) in a hotspot (58 Mb-67 Mb) for red, green and blue in RGB measurement, and the values of 'L' and 'a' in the Lab model (Additional file 2: Figure S6), is close to *y1* (~61 Mb). An additional association peak (~61.86 Mb) near *Tannin1* (*Tan1*) (~61.6 Mb), is localized in a hotspot 57.5 Mb-62 Mb on chromosome Sb04 to be associated with red and green in the RGB and 'L' in the Lab.

Dry inflorescence weight, inflorescence length/width, and whorl number, were each very sensitive to the environment, with no colocalized hotspots between the

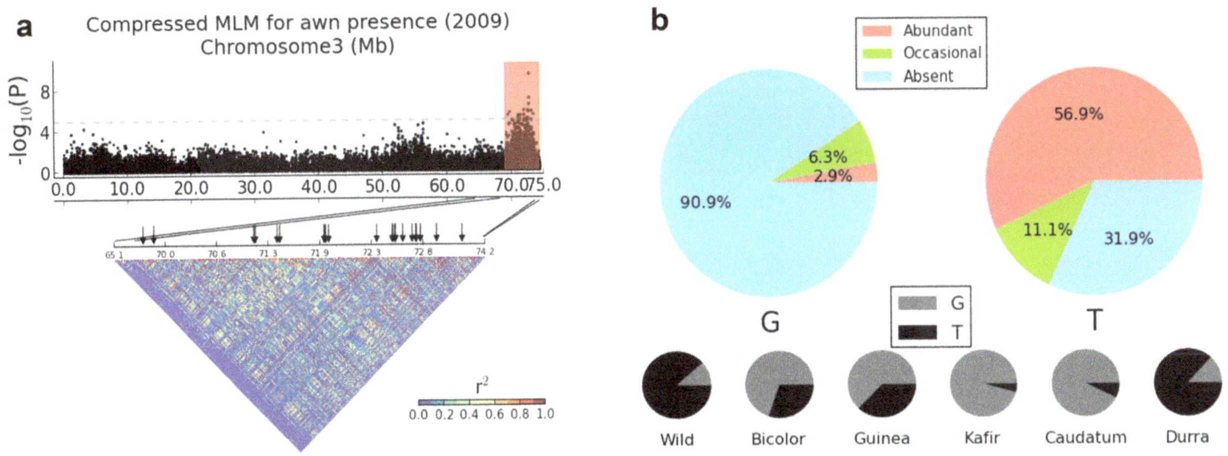

Figure 3 The spectrum of awn presence variation and allele frequencies at locus "S3_72702502" on chromosome Sb03. **(a)** Chromosome Sb03 Manhattan plot for 2009 awn presence in sorghum is plotted with the hotspot (red area) identified by GWAS, with the prior interval (green area) determined by QTL mapping [24], and with the LD pattern determined by r^2. Significant association markers are denoted by black arrows. **(b)** Three types of awn classification, which are "abundant", "occasional" and "absent" are color-coded in their frequencies plots for alleles "G" and "T". The allele frequencies are plotted for the five main cultivated races and wild sorghum.

two years for which we have data. The most striking 'skyline' determined by GWAS for dry inflorescence weight is centered at ~57 Mb on chromosome Sb09 (Additional file 2: Figure S7), which is also associated with plant height and flowering time. For inflorescence length the most striking association hotspot centered at ~45 Mb on chromosome Sb06 (Additional file 2: Figure S8), a location also connected with plant height and flowering.

To further investigate the inference that guinea and durra may have experienced independent domestication events but achieved non-shattering by convergence, we examined two-locus haplotypes for loci associated with pericarp color and awn presence respectively, both considered to be traits that may be subject to selection. For pericarp color, the association peak (S1_61453639) and 7 linked ($r^2 \geq 0.6$) SNPs in the hotspot (chromosome Sb01: 57,017,032 bp-68,413,103 bp) are used to calculate the ratio of common haplotypes/total haplotypes for each pair of sorghum types (See the "Phylogenetic relationships of five main sorghum races" also for the details). Similarly, we used an association peak (S3_72702502) and 18 linked SNPs in the hotspot (chromosome Sb03: 68,912,931 bp-74,241,979 bp) to calculate two-locus haplotype ratios for awn presence. Compared to other sorghum race pairs, guinea-durra only shows modest frequencies of common two-locus haplotypes for pericarp color (freq = 0.22) (Additional file 2: Figure S18a) and awn presence (freq = 0.26) (Additional file 2: Figure S18b). Our findings further support the hypothesis that independent domestication of guinea and durra involved convergent selection for non-shattering.

Race-specific patterns in awn presence variation

Geographic origins and domestication history can result in patterns of phenotypic variation among genotypes within a gene pool. We investigated whether awn presence exhibits variation patterns correlated with race-specific alleles for sorghum. The hotspot (chromosome Sb03: 68,912,931 bp-74,241,979 bp) for sorghum awn presence, with the association peak (S3_72702502), is detected by both GWAS and QTL mapping (See "Traits related to the sorghum inflorescence" also). The association peak is located at an intergenic locus, which is flanked by two annotated genes [*Sb03g045420* (chromosome Sb03: 72,681,274 bp-72,687,688 bp) (similar to Hexokinase-3) and *Sb03g045430* (chromosome Sb03: 72,703,668 bp-72,704,913 bp) (similar to Putative uncharacterized protein)], based on published reduced representation sequence [4]. In this scenario, it is probable that the causative genes/loci were not genotyped, and that S3_72702502 and the causative loci shared history of mutation and recombination.

An early study [53] described morphology of panicles for two of the five main sorghum races. We (Figure 3b) show allele distribution at the locus S3_72702502, reflecting alleles 'T' and 'G' associated with awned and awnless sorghum panicles individually, finding correlation between the alleles and the race-specific morphology of awns. Allele 'G' with the dominant frequency in race kafir corresponds to awnless cylindrical-shaped panicles of kafir [53]. Indeed, we found 84% of kafir accessions to have awnless panicles in our phenotypic data. Similarly, allele 'T' with the dominant frequency in race durra, corresponds to bearded and hairy panicles of durra [53]. The major allele

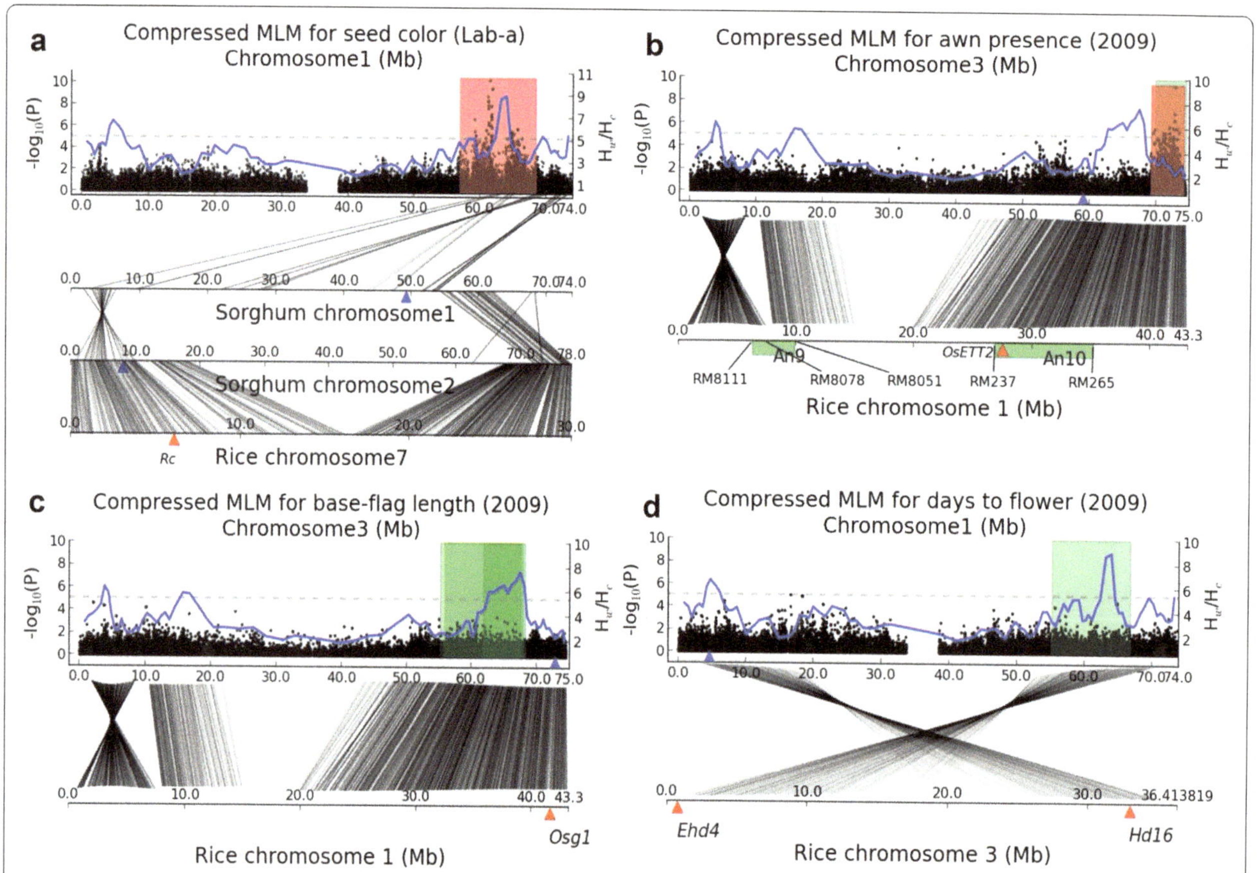

Figure 4 Genetic correspondence across taxa. **(a)** Chromosome Sb01 Manhattan plot for pericarp color (a value of L*ab model) of sorghum is plotted with the hotspot (red area) identified by GWAS, and with the curve of H_w/H_c ratios. The *Rc* [52] gene is denoted by a red triangle. Two sorghum homologs of *Rc* on chromosomes Sb01 and Sb02 are denoted by blue triangles. Gray connecting lines indicate pairs of duplicated genes. **(b)** Chromosome Sb03 Manhattan plot for 2010 awn presence in sorghum is plotted with the hotspot (red area) identified by GWAS, and with the prior interval (green area) determined by QTL mapping [24]. Rice awns co-segregated with SSR marker RM8078 tightly linked to An9 on chromosome Os01 [63]. The interval An10 for rice awn [63] was associated with SSR markers RM265 and RM237 on chromosome Os01. The *OsETT2* [64] gene is denoted by a red triangle. Sorghum ortholog of *OsETT2* is indicated by a blue triangle. **(c)** The genomic interval (~55 Mb-67 Mb) on chromosome Sb03 is implicated by three linkage studies [14,26,28] to affect plant height and flowering time, but doesn't harbor association signal in our GWAS. The *Osg1* [66] gene and its sorghum ortholog are indicated by a red and a blue triangle individually. **(d)** The genomic interval (~58 Mb-64 Mb) on chromosome Sb01 is implicated by two linkage studies [24,26] to control plant height, but doesn't harbor association signal in our GWAS. The genes *Ehd4* [67] and *Hd16* [68] are indicated by red triangles, and sorghum ortholog of *Hd16* is denoted by a blue triangle.

in race caudatum is 'G', consistent with the finding that 92% of caudatum accessions for which we have data are awnless. Our findings also suggest that the two most primitive sorghum types, bicolor and guinea, derived more allele 'G' from wild sorghum which is dominated by more ancestral allele 'T'.

Traits related to plant height

We conducted GWAS for 6 traits related to sorghum plant height, including total plant height, distance from base to flag leaf (base-flag), distance from rachis to flag leaf (rachis-flag), number of nodes on the main stalk (nodes), dry stalk weight and days to flowering. Strong positive correlation (Additional file 1: Table S1) are observed among the phenotypic data of total plant height, base-flag, nodes, dry stalk weight and days to flowering. It is important to remember that 228 accessions in the sorghum association panel (SAP), are converted tropical lines that are photoperiod insensitive, early maturing, and short statured phenotypes, developed by crossing exotic lines and U.S. cultivars [14,22]. Three independent studies [4,14,54] revealed three consistent genomic regions in sorghum that contribute to the introgression of cultivar-specific alleles to exotic sorghum lines. It is also clear that *Dwarf* (*Dw*)*4* and *Dw2* are associated with the introgression region in the heterochromatin of chromosome Sb06, and *Dw3* and *Dw1* are associated with the introgression regions in the 3′ terminal euchromatin of chromosome Sb07 and Sb09 respectively.

In our studies, the *Dw1-Dw4* regions reported by Morris et al. [4] show the most striking signals associated with base-flag leaf length for two years (Additional file 2: Figure S12).

Likewise, the association results of total plant height (Additional file 2: Figure S11) are consistent with that of Morris et al [4]. This suggests that environmental variables have relatively little effect on *Dw1-Dw4*.

In addition to *Dw1-Dw4*, we found noteworthy hotspot (s) for plant height located on chromosome Sb04. Our BTxSP mapping population [29] and Shiringani et al., [28] each suggested two overlapping QTL likelihood intervals on Sb04, 57.98 Mb-64.93 Mb and 48.80 Mb-58.58 Mb, to contribute to plant height (Additional file 2: Figure S11). Within both intervals, we found multiple SNPs significantly associated with base-flag and dry stalk weight (Additional file 2: Figure S12 and Figure S15). Because the terminal region of Sb04 holding these QTLs exhibited weak LD, we could not set clear boundaries for the hotspot(s). GWAS of another plant height related trait, dry stalk weight in 2009 data, also shows a clear association 'skyline' in the region of 48.80 Mb-64.93 Mb on Sb04 (Additional file 2: Figure S15a).

Plant height and flowering time show significant genetic correlation. In further agreement, three association loci for flowering time show strong LD with dwarfing genes, and are distributed in the introgression regions. There is some confusion concerning the identities of *Ma1* and *Ma6*, with one group suggesting that *Ma1* is a sorghum ortholog of a Triticeae flowering gene, *PRR37* [55], but which is located very near the published position of *Ma6* [45]. Many of the same authors who report that *PRR37* is *Ma1* have stated under separate cover that *PRR37* is *Ma6* [56]. The LD pattern suggests two haplotype blocks in the region of 6 Mb-46 Mb on chromosome Sb06 (Figure 2b), suggesting that *Ma1* (or *Ma6*) and *Dw4* [4] are tightly linked in the region of 6 Mb-39 Mb, and *Dw2* [4] is linked with *Ma6* (or *Ma1*) in the region of 40 Mb-46 Mb on Sb06. We also identified a haplotype block of 56 Mb-59.5 Mb on Sb09 (Figure 2c), which is strongly associated with plant height and flowering time. Greater mapping resolution is required to pinpoint the causative genetic mutation(s) that affect each trait.

Selective signatures and phenotype-genotype association

A phenotype-genotype association is no guarantee that the trait or its candidate gene has been historically important or is an adaptation [57]. Reduction in genetic diversity, which can be assessed by heterozygosity, can bear the signature of domestication [4,58,59]. Using quantified genome-wide heterozygosity, low heterozygosity regions in sorghum have been reported [4] but have not shown significant correspondence with identified loci associated with plant height, one of the key domestication traits. Here, we refined analysis of selective signatures from domestication and explored the genomic regions that have been important in sorghum adaptation for inflorescence morphology and plant height. Genetic diversity of the sorghum population was assessed by the ratio of expected heterozygosity in wild sorghum to that in cultivated types (H_w/H_c) across the sorghum genome (Figure 5). Overall, the refined pattern of reduction in genetic diversity is

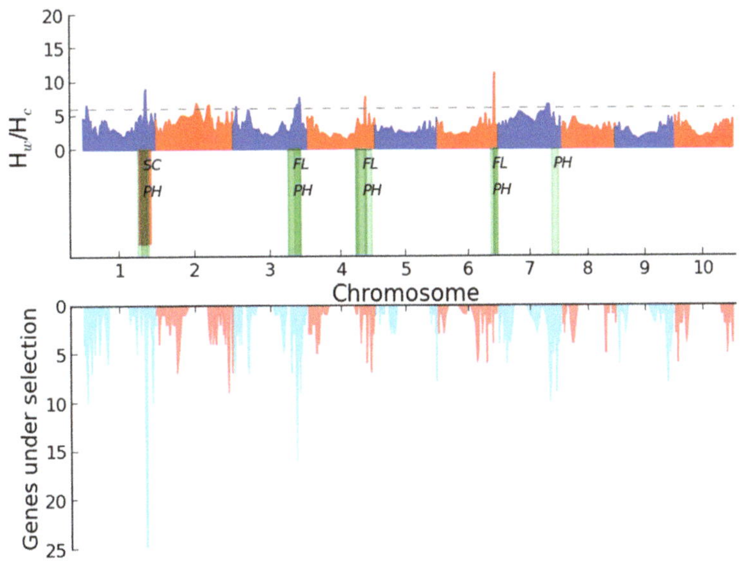

Figure 5 Genome-wide selection signatures. We calculated average ratios (expected heterozygosity of wild type sorghum (H_w) / expected heterozygosity of cultivated sorghum (H_c)) for windows of 500 consecutive SNPs throughout the genome. The gray dashed line is the cutoff for the top 5% of heterozygosity ratios. Green areas indicate confidence intervals determined by QTL mapping to contribute to either flowering time (FL) or plant height (PH), that also have strong selection signals. The hotspot on chromosome Sb01 identified by GWAS for pericarp color (SC) is highlighted by red. The distribution of 725 candidate genes implicated in domestication and/or improvement via gene-based population summary statistics [41] is shown below the heterozygosity ratios.

consistent with that measured previously [4]. Among the hotspots determined by association mapping for the 12 traits, the genomic region 58 Mb-64 Mb on chromosome Sb01 for pericarp color shows a strong selection signal based on reduction in heterozygosity ratio (Figure 4a). Such correspondence was also discovered in rice [60], consistent with the observation that pericarp color has profoundly influenced the popularity of cultivated cereal varieties [61].

Some confidence intervals identified by QTL mapping for plant height and flowering time [14,24-28] colocalized non-randomly (p = 0.231×10^{-6}) (See the overlapping testing in the "Methods") with regions having striking reductions in heterozygosity ratio (Figure 5). For example, Hart et al., [24] and Ritter et al., [26] each identified a region (55.1 Mb-66.7 Mb) for plant height on chromosome Sb01 which showed strong selection signals. A region of 55 Mb-62 Mb on chromosome Sb06, showing strong selection signals, was identified for plant height by Kebede et al., [25], Srinivas et al., [27] and Shiringani et al., [28], and for flowering time by Shiringani et al., [28]. Additional heterozygosity ratio peaks overlapping with plant height and flowering time intervals are found on Sb03 and Sb04. Most such colocalizing intervals affect plant height and flowering time simultaneously, and do not colocalize with hotspots determined by association mapping.

Using 8 M SNPs characterized in 44 sorghum lines, a recent study [41] implicated 725 candidate genes in sorghum domestication and/or improvement via gene-based population summary statistics. We show (Figure 5) at least two regions having striking reductions in heterozygosity ratio, to also be enriched for candidate genes under selection. One striking case is the 58 Mb-64 Mb interval on chromosome Sb01, in genetic control of pericarp color and plant height [24,26]. The other case is the 62 Mb-64 Mb interval on chromosome Sb03, in genetic control of plant height and flowering time [14,26,28].

Much of the 'missing heritability' from GWAS has been speculated to be attributed to the low/rare frequency of causative mutations, which are unlikely to be detected by most association studies [62]. Some genetic variations with large effect may be hidden by neutral genetic diversity in a natural population. By contrast, biparental QTL mapping uses genetic markers (e.g. SSR and RFLP), and is often derived from parental lines that differ strongly in phenotype such as our *S. bicolor* × *S. propinquum* cross between parents which are separated by 1-2 million years. Sometimes, biparental QTL mapping can be more efficient than GWAS to detect alleles that are rare in populations. QTL mapping and GWAS are thus two complementary approaches to determine "saturation" in terms of QTL discovery. Our findings lead to the prediction that the likelihood intervals on chromosome Sb01, Sb03, Sb04 and Sb06 identified by QTL mapping should bear the signature of selection, and may encode important variants with low/rare frequency for plant height and flowering time.

Genetic correspondence between sorghum and rice

Synteny and colinearity have been well conserved between grass species such as rice, maize, and sorghum since their divergence about 50 Mya [21], enabling us to compare causal loci in corresponding regions across taxa. SSR mapping in rice revealed that two intervals affecting awns are located on chromosome Os01, and designated as *An9* and *An10*, respectively (Figure 4b) [63]. *An10* produced shorter and sparser awns than *An9*, and is thought to be more important for awn presence in rice. Recently, a rice gene, *OsETT2*, which colocalizes with *An10*, has been identified using genetic analyses and RNA-silencing experiments, to be involved in awn formation [64]. Chromosome Sb03 harbors the most significant hotspot for awn presence in sorghum, bounded in a region sharing large-scale homoeology with a region on Os01 containing *An10* and *OsETT2* (Figure 4b). The two hotspots do not appear to have direct correspondence across taxa, making it unclear whether there has been some localized rearrangement, functional divergence of different members of a co-regulated cluster of genes, or some other explanation.

Duplication and subsequent rearrangements can result in chromosomes exhibiting a complex pattern of homoeology across taxa. The *Rc* gene is responsible for conditioning red pericarp in rice [52]. Association mapping shows no noteworthy signal around the sorghum ortholog (*Sb02g006380*) of *Rc*, or for another sorghum gene copy (*Sb01g028230*). Rearrangement following polyploidization can lead to a scrambled patchwork in genome organization [65]. In the case of pericarp color, the correspondence of gene arrangement between the most significant hotspot in sorghum and the rice *Rc* region is still discernible through multiple alignment (Figure 4a). The hotspot on sorghum chromosome Sb01 shares homoeology with three small genomic segments on Sb01, one of which has correspondence with the region enclosing *Sb02g006380* on sorghum chromosome Sb02. Thus, despite discernible parallels in genome organization, corresponding genes do not appear to be responsible for phenotypic variation in seed color of sorghum and rice. This may simply reflect that 40-50 My of rice-sorghum divergence, did involve many genetic changes that were different, as well as some that may have been convergent [18].

Both linkage studies and reductions in heterozygosity ratios (Figure 4c and d) implicate two genomic regions in genetic control of plant height and flowering time, which are located in the 58 Mb-64 Mb on chromosome Sb01 [24,26] and in the 62 Mb-64 Mb on chromosome Sb03 [14,26,28] individually. In rice, gene *Osg1* [66], which is located on chromosome Os01 (Figure 4c) orthologus to sorghum chromosome Sb03, has been known to affect plant height via reduction in its expression; genes *Ehd4* [67] and *Hd16* [68] on chromosome Os03 orthologous to sorghum chromosome Sb01, are characterized by their function in flowering time of rice. We observed that sorghum has experienced gene loss for *Ehd4*, and none of three rice genes shows direct correspondence with the two genomic regions in sorghum. These findings indicate that different genes, rather than orthologs of *Osg1*, *Ehd4* and *Hd16*, were identified by QTL mapping on Sb03 and Sb01.

QTL correspondence across homoeologous regions within sorghum

QTLs may share homoeology not only across taxa, but also within taxa. Maize, in which many homoeologous chromosome segments have been identified as a result of lineage-specific genome duplication, is a particularly favorable organism in which to show such QTL correspondence [14]. Only occasional cases have been found in sorghum [14], probably due to much greater antiquity of genome duplication [1].

A total of 4 pairs of homoeologous QTLs for plant height/flowering time are found in our study. For flowering time, we observed two instances of homoeologous QTLs in sorghum (Figure 6). The confidence interval mapped on chromosome Sb03 [28] shares homoeology with the hotspot of 54 Mb-59.6 Mb on chromosome Sb09 which has been identified by QTL mapping [23] and our GWAS. The other correspondence was determined between 48.6 Mb-59.3 Mb on chromosome Sb04 [26,28] and 38 Mb-50.7 Mb on chromosome Sb06 [14,18]. Likewise, we found homoeologous QTLs for plant height on chromosome pairs of Sb03-Sb09 and Sb04-Sb06 (Additional file 2: Figure S17). The additional cases for flowering time are shown on chromosome pairs of Sb02-Sb07 and Sb05-Sb08 (Additional file 2: Figure S17). Such genetic correspondence within sorghum implies that some homoeologous regions created in a genome duplication 70 million years ago [21] may still retain some genes with similar functions. In most instances, we found that one QTL hotspot exhibits much stronger selection signal than its homoeologous counterpart on the basis of genetic diversity reduction in the wild population. Independent episodes of domestication and crop improvement may preferentially select one counterpart, but may or may not select the other.

Conclusions

Our studies illustrate how GWAS may be used to complement the genetic resolution of causal elements (genes) of quantitative phenotypes that is typically attained from conventional likelihood intervals determined by QTL mapping. The degree of improvement in resolution by GWAS over QTL mapping is related to the nature of the 'genomic environment' surrounding a gene – with substantial improvement in recombinationally-active euchromatin but much less improvement in recombinationally recalcitrant heterochromatin with long LD blocks.

Understanding and utilizing the relative strengths and weaknesses of QTL mapping and GWAS can aid in dissecting the genetic basis of a complex trait. A carefully chosen cross can allow QTL mapping to have better statistical power to detect variants with low/rare frequency in a natural population. For example, the classical maturity locus *Ma3/phyB* was initially identified with a map-based strategy (QTL mapping) [69], but association mapping is

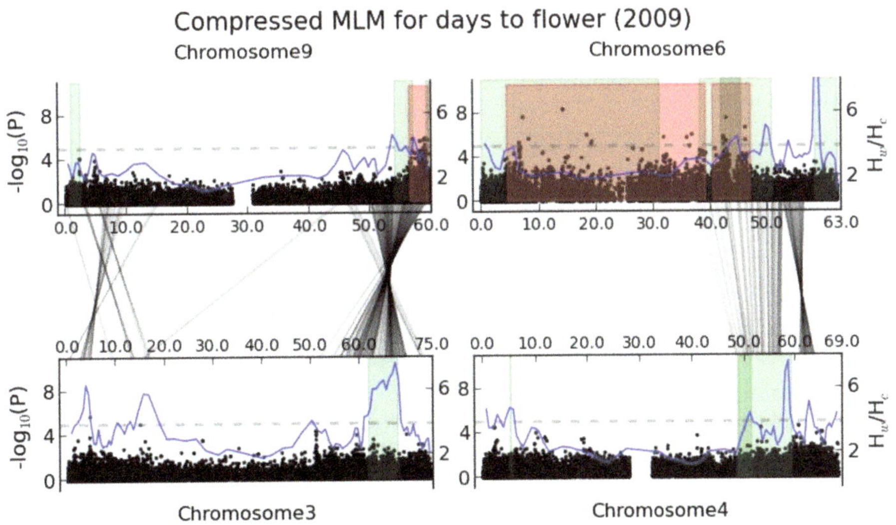

Figure 6 Genetic correspondence within taxa. Intra-genomic genetic correspondence for QTL intervals for flowering time on chromosome pairs of Sb03-Sb09 and Sb04-Sb06.

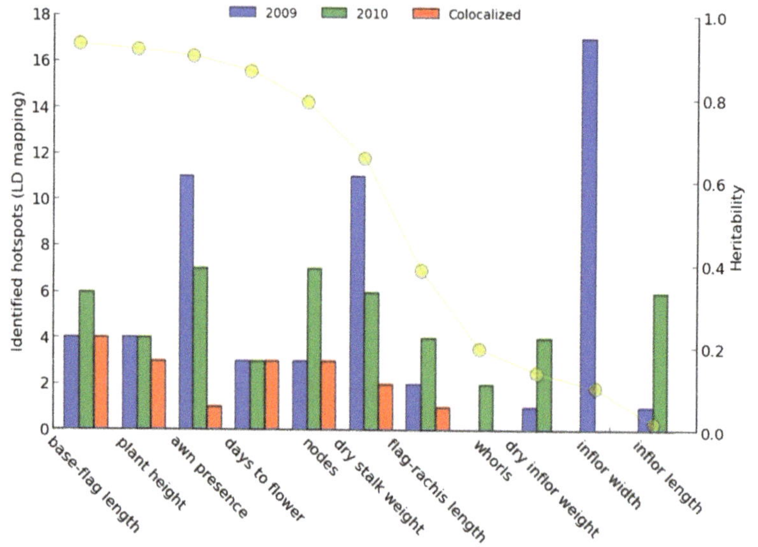

Figure 7 Summary of hotspots identified by GWAS for multi-year trait data. For each trait, numbers of hotspots for 2009, 2010 and both years are indicated with bars. The values of heritability for the traits are indicated with the curve.

unable to detect striking signal near *Ma3* because virtually all members of the panel are wild type (i.e. the mutation is rare). Another case is that a map-based (QTL mapping) strategy was used to determine one major-effect QTL (*sh1*) controlling shattering in sorghum, and revealed three different non-shattering haplotypes widely existing in five major cultivated races that may reflect low/rare frequency of causative polymorphism [2,3]. Hence, association mapping may have reduced power to find phenotype-genotype association for *sh1*. Additionally, complex population structure in unrelated individuals may compromise genetic variation at the true causative loci, resulting in cases of true negative detection in GWAS. More generally, multiple independent domestications may create a scenario under which many QTLs cannot be verified with the current sorghum association panel.

An important fundamental question in gene mapping is why the GWAS approach has revealed so much less variation than anticipated [62]. Even given 'perfect' (i.e. 100% accurate) information about phenotype and genotype, some associations may not be repeatable due to interaction between genotype and environment [57]. We observed that traits related to the inflorescence were more greatly affected by environment than traits related to plant height (Figure 7). The interaction between genotype and environment plays an important role in the variation between individuals in their observable characteristics, even for traits with high heritability.

Identification of intersections among QTL, GWAS and comparative data advance knowledge of the genetic determinants of variation in sorghum, and provide a finer-scale comparison than was previously possible of the genetics of important traits between sorghum and rice. Non-random correspondence of sorghum GWAS signals to rice QTLs, i.e. between divergent panicoid and oryzoid grasses, adds a new dimension to evidence of the ability to leverage genetic data about the important traits across divergent plants. More variants can be gained through re-sequencing gene candidates. Joint evidence from hotspots and re-sequencing data are likely to be extremely powerful to answer the question of whether such genetic correspondence implicates 'conserved' single genes or co-regulated clusters of genes that exist for the trait of interest after cereal divergence.

Rich SNP data also clarifies the phylogenetic relatedness of sorghum races, improving on several prior studies [38-41]. Based on the high-density genotype in the large population, we showed joint evidence from the striking clustering patterns in population structure and fixation index values, suggesting several hypotheses that might shed light on the unknown ancestry of sorghum races.

Availability of supporting data

In addition to the supplementary files listed below, Newick files for all the phylogenetic trees are available for download at Dryad (http://datadryad.org/).

Additional files

Additional file 1: The file contains supplementary Tables S1, S2 and S3. (**Table S1.** Phenotypic correlation determined by Pearson correlation coefficient). (**Table S2.** 1-LOD likelihood intervals determined by QTL mapping. Trait names, genomic positions and references are indicated). (**Table S3.** Hotspots identified by GWAS. Trait names, years, genomic positions are indicated).

Additional file 2: The file contains supplementary Figures S1-S18.

Abbreviations

QTL: Quantitative trait locus; GWAS: Genome-wide association study; SNP: Single nucleotide polymorphism; SAP: Sorghum association panel; PCA: Principal component analysis; LD: Linkage disequilibrium; MAF: Minor allele frequency; F_{ST}: Fixation index; WGD: Whole-genome duplication; GBS: Genotyping-by-sequencing; CMLM: Compressed mixed linear model.

Competing interests

The authors declare that they have no competing interests.

Authors' contributions

AHP conceived of the project and guided the process of analyses. DZ and WK conducted the genome data analyses. AHP, DZ, WK, JR, VHG, EE, AK, GM, JC, YL and CP collected phenotypic data. JR built and curated phenotypic databases. AHP and DZ wrote the paper. All authors read and approved the final manuscript.

Acknowledgements

This work was supported by the United Sorghum Checkoff Program, and the US Department of Energy-US Department of Agriculture Plant Feedstock Program (project grant numbers: 2011-02786, 2012-03304).

Author details

[1]Plant Genome Mapping Laboratory, University of Georgia, Athens, GA 30602, USA. [2]Institute of Bioinformatics, University of Georgia, Athens, GA 30602, USA. [3]Department of Crop and Soil Sciences, University of Georgia, Athens, GA 30602, USA. [4]Department of Plant Biology, University of Georgia, Athens, GA 30602, USA. [5]Department of Genetics, University of Georgia, Athens, GA 30602, USA.

References

1. Paterson AH, Bowers JE, Bruggmann R, Dubchak I, Grimwood J, Gundlach H, et al. The Sorghum bicolor genome and the diversification of grasses. Nature. 2009;457:551–6.
2. Lin Z, Li X, Shannon LM, Yeh C-T, Wang ML, Bai G, et al. Parallel domestication of the Shattering1 genes in cereals. Nat Genet. 2012;44:720–4.
3. Olsen KM. One gene' s shattering effects. Nat Genet. 2012;44:616–8.
4. Morris GP, Ramu P, Deshpande SP, Hash CT, Shah T, Upadhyaya HD, et al. Population genomic and genome-wide association studies of agroclimatic traits in sorghum. Proc Natl Acad Sci U S A. 2012;110:453–8.
5. Paterson AH, Lander ES, Hewitt JD, Peterson S, Lincoln SE, Tanksley SD. Resolution of quantitative traits into Mendelian factors by using a complete RFLP linkage map. Nature. 1988;335:721–6.
6. Kover PX, Valdar W, Trakalo J, Scarcelli N, Ehrenreich IM, Purugganan MD, et al. A multiparent advanced generation inter-cross to fine-map quantitative traits in Arabidopsis thaliana. PLoS Genet. 2009;5.
7. Buckler ES, Holland JB, Bradbury PJ, Acharya CB, Brown PJ, Browne C, et al. The genetic architecture of maize flowering time. Science. 2009;325:714–8.
8. Yu J, Pressoir G, Briggs WH, Vroh Bi I, Yamasaki M, Doebley JF, et al. A unified mixed-model method for association mapping that accounts for multiple levels of relatedness. Nat Genet. 2006;38:203–8.
9. Thornsberry JM, Goodman MM, Doebley J, Kresovich S, Nielsen D, Buckler ES. Dwarf8 polymorphisms associate with variation in flowering time. Nat Genet. 2001;28:286–9.
10. Yu J, Holland JB, McMullen MD, Buckler ES. Genetic design and statistical power of nested association mapping in maize. Genetics. 2008;178:539–51.
11. Myles S, Peiffer J, Brown PJ, Ersoz ES, Zhang Z, Costich DE, et al. Association mapping: critical considerations shift from genotyping to experimental design. Plant Cell. 2009;21:2194–202.
12. Morrell PL, Buckler ES, Ross-Ibarra J. Crop genomics: advances and applications. Nat Rev Genet. 2011;13:85–96.
13. Brown PJ, Klein PE, Bortiri E, Acharya CB, Rooney WL, Kresovich S. Inheritance of inflorescence architecture in sorghum. Theor Appl Genet. 2006;113:931–42.
14. Lin Y, Keith F, Paterson AH. Comparative analysis of QTLs affecting plant height and maturity across the poaceae, in reference to an interspecific sorghum population. Genetics. 1995;141:391–411.
15. Rong J, Feltus FA, Waghmare VN, Pierce GJ, Chee PW, Draye X, et al. Meta-analysis of polyploid cotton QTL shows unequal contributions of subgenomes to a complex network of genes and gene clusters implicated in lint fiber development. Genetics. 2007;176:2577–88.
16. Zhang D, Guo H, Kim C, Lee T-H, Li J, Robertson J, et al. CSGRqtl, a comparative quantitative trait locus database for Saccharinae grasses. Plant Physiol. 2013;161:594–9.
17. Multani DS, Briggs SP, Chamberlin MA, Blakeslee JJ, Murphy AS, Johal GS. Loss of an MDR transporter in compact stalks of maize br2 and sorghum dw3 mutants. Science. 2003;302:81–4.
18. Paterson AH, Lin YR, Li Z, Schertz KF, Doebley JF, Pinson SR, et al. Convergent domestication of cereal crops by independent mutations at corresponding genetic Loci. Science. 1995;269:1714–8.
19. Lee JM, Sonnhammer ELL. Genomic gene clustering analysis of pathways in eukaryotes. Genome Res. 2003;13:875–82.
20. Paterson AH, Wendel JF, Gundlach H, Guo H, Jenkins J, Jin D, et al. Repeated polyploidization of Gossypium genomes and the evolution of spinnable cotton fibres. Nature. 2012;492:423–7.
21. Paterson AH, Bowers JE, Chapman BA. Ancient polyploidization predating divergence of the cereals, and its consequences for comparative genomics. Proc Natl Acad Sci U S A. 2004;101:9903–8.
22. Casa AM, Pressoir G, Brown PJ, Mitchell SE, Rooney WL, Tuinstra MR, et al. Community resources and strategies for association mapping in sorghum. Crop Sci. 2008;48:30–40.
23. Feltus FA, Hart GE, Schertz KF, Casa AM, Casa AM, Kresovich S, et al. Alignment of genetic maps and QTLs between inter- and intra-specific sorghum populations. Theor Appl Genet. 2006;112:1295–305.
24. Hart GE, Schertz KF, Peng Y, Syed NH. Genetic mapping of sorghum bicolor (L.) Moench QTLs that control variation in tillering and other morphological characters. TAG Theor Appl Genet. 2001;103:1232–42.
25. Kebede H, Subudhi PK, Rosenow DT, Nguyen HT. Quantitative trait loci influencing drought tolerance in grain sorghum (Sorghum bicolor L. Moench). TAG Theor Appl Genet. 2001;103:266–76.
26. Ritter KB, Jordan DR, Chapman SC, Godwin ID, Mace ES, Lynne McIntyre C. Identification of QTL for sugar-related traits in a sweet × grain sorghum (Sorghum bicolor L. Moench) recombinant inbred population. Mol Breed. 2008;22:367–84.
27. Srinivas G, Satish K, Madhusudhana R, Reddy RN, Mohan SM, Seetharama N. Identification of quantitative trait loci for agronomically important traits and their association with genic-microsatellite markers in sorghum. Theor Appl Genet. 2009;118:1439–54.
28. Shiringani AL, Frisch M, Friedt W. Genetic mapping of QTLs for sugar-related traits in a RIL population of Sorghum bicolor L. Moench. Theor Appl Genet. 2010;121:323–36.
29. Kong W, Jin H, Franks CD, Kim C, Bandopadhyay R, Rana MK, et al. Genetic analysis of recombinant inbred lines for Sorghum bicolor × Sorghum propinquum. G3 (Bethesda). 2013;3:101–8.
30. Bradbury PJ, Zhang Z, Kroon DE, Casstevens TM, Ramdoss Y, Buckler ES. TASSEL: software for association mapping of complex traits in diverse samples. Bioinformatics. 2007;23:2633–5.
31. Zhang Z, Ersoz E, Lai C-Q, Todhunter RJ, Tiwari HK, Gore MA, et al. Mixed linear model approach adapted for genome-wide association studies. Nat Genet. 2010;42:355–60.
32. Lipka AE, Tian F, Wang Q, Peiffer J, Li M, Bradbury PJ, et al. GAPIT: genome association and prediction integrated tool. Bioinformatics. 2012;28:2397–9.
33. Matthies I, Malosetti M, Röder MS, Eeuwijk FV. Genome-Wide association mapping for kernel and malting quality traits using historical European barley records. PLoS One. 2014;9(11):e110046.
34. Bowers JE, Abbey C, Anderson S, Chang C, Draye X, Hoppe AH, et al. A High-density genetic recombination map of sequence-tagged sites for sorghum, as a framework for comparative structural and evolutionary genomics of tropical grains and grasses. Genetics. 2003;386:367–86.
35. Kawahara Y, de la Bastide M, Hamilton JP, Kanamori H, McCombie WR, Ouyang S, et al. Improvement of the Oryza sativa Nipponbare reference genome using next generation sequence and optical map data. Rice (N Y). 2013;6:4.
36. Doggett H. Sorghum, 2nd edition. UK: Longman; 1988.
37. Wiersema JH, Leon B. World Economic Plants: A standard Reference. New York: CRC press; 1999.
38. Brown PJ, Myles S, Kresovich S. Genetic support for phenotype-based racial classification in sorghum. Crop Sci. 2011;51:224.

39. Bouchet S, Pot D, Deu M, Rami J-F, Billot C, Perrier X, et al. Genetic structure, linkage disequilibrium and signature of selection in Sorghum: lessons from physically anchored DArT markers. PLoS One. 2012;7:e33470.
40. Billot C, Ramu P, Bouchet S, Chantereau J, Deu M, Gardes L, et al. Massive sorghum collection genotyped with SSR markers to enhance use of global genetic resources. PLoS One. 2013;8:e59714.
41. Mace ES, Tai S, Gilding EK, Li Y, Prentis PJ, Bian L, et al. Whole-genome sequencing reveals untapped genetic potential in Africa's indigenous cereal crop sorghum. Nat Commun. 2013;4.
42. Quinby JR. Sorghum Improvement and the Genetics of Growth. College Station: Texas A&M University Press; 1974.
43. Rooney WL, Aydin S. Genetic control of a photoperiod-sensitive response in Sorghum bicolor (L.) Moench. Crop Sci. 1999;39:397–400.
44. The Wellcome Trust Case Control Consortium. Genome-wide association study of 14,000 cases of seven common diseases and 3,000 shared controls. Nature. 2007;447:661–78.
45. Brady JA. Sorghum Ma5 and Ma6 Maturity Genes. Ph.D. dissertation. Texas A&M University; 2006.
46. Chopra S, Gevens A, Svabek C, Wood KV, Peterson T, Nicholson RL. Excision of the Candystripe1 transposon from a hyper-mutable Y1-cs allele shows that the sorghumY1 gene controls the biosynthesis of both 3-deoxyanthocyanidin phytoalexins and phlobaphene pigments. Physiol Mol Plant Pathol. 2002;60:321–30.
47. Ibraheem F, Gaffoor I, Chopra S. Flavonoid phytoalexin-dependent resistance to anthracnose leaf blight requires a functional yellow seed1 in Sorghum bicolor. Genetics. 2010;184:915–26.
48. Wu Y, Li X, Xiang W, Zhu C, Lin Z, Wu Y, et al. Presence of tannins in sorghum grains is conditioned by different natural alleles of Tannin1. Proc Natl Acad Sci U S A. 2012;109:10281–6.
49. Worzella WW, Khalidy R, Badawi Y, Daghir S. Inheritance of Beta-carotene in grain sorghum hybrids. Crop Sci. 1965;5:591–2.
50. Gorbet DW, Weibel DE. Inheritance and genetic relationships of six endosperm types in sorghum. Crop Sci. 1972;12:378–82.
51. Kinoshita T. Gene analysis and Linkage in Biology of Rice. In: Tsunoda S, Takahashi N, editors. JSSP. Tokyo: Elsevier; 1984. p. 187–274.
52. Sweeney MT, Thomson MJ, Pfeil BE, Mccouch S. Caught red-handed: Rc encodes a basic helix-loop-helix protein conditioning red pericarp in rice. Plant Cell. 2006;18:283–94.
53. Magness JR, Markle GM, Compton CC. Food and feed crops of the United States. Interregional Research Project IR-4, IR Bul. 1 (Bul. 828 New Jersey Agr. Expt. Sta.); 1971.
54. Thurber CS, Ma JM, Higgins RH, Brown PJ. Retrospective genomic analysis of sorghum adaptation to temperate-zone grain production. Genome Biol. 2013;14:R68.
55. Murphy RL, Klein RR, Morishige DT, Brady JA, Rooney WL, Miller FR, et al. Coincident light and clock regulation of pseudoresponse regulator protein 37 (PRR37) controls photoperiodic flowering in sorghum. Proc Natl Acad Sci U S A. 2011;108:16469–74.
56. Mullet JE, Rooney WL, Klein PE, Morishige DT, Murphy R, Brady JA. Discovery and utilization of sorghum genes (MA5/MA6). Office USP ed., vol. US 8,309793 B2. USA: The Texas A&M University System; 2012.
57. Ross-Ibarra J, Morrell PL, Gaut BS. Plant domestication, a unique opportunity to identify the genetic basis of adaptation. Proc Natl Acad Sci U S A. 2007;104 Suppl:8641–8.
58. Doebley JF, Gaut BS, Smith BD. The molecular genetics of crop domestication. Cell. 2006;127:1309–21.
59. Xu X, Liu X, Ge S, Jensen JD, Hu F, Li X, et al. Resequencing 50 accessions of cultivated and wild rice yields markers for identifying agronomically important genes. Nat Biotechnol. 2012;30:105–11.
60. Huang X, Kurata N, Wei X, Wang Z-X, Wang A, Zhao Q, et al. A map of rice genome variation reveals the origin of cultivated rice. Nature. 2012;490:497–501.
61. Sieglinger JB. Seed-color inheritance in certain grain-sorghum crosses. J Agric Res. 1934;XXVII(1).
62. Manolio TA, Collins FS, Cox NJ, Goldstein DB, Hindorff LA, Hunter DJ, et al. Finding the missing heritability of complex diseases. Nature. 2009;461:747–53.
63. Matsushita S, Kurakazu T, Sobrizal Doi K, Yoshimura A. Mapping of genes for awn in rice using Oryza meridionalis introgression lines. Rice Genet Newsl. 2003;20:17.
64. Toriba T, Hirano H-Y. The DROOPING LEAF and OsETTIN2 genes promote awn development in rice. Plant J. 2014;77:616–26.
65. Shoemaker RC, Polzin K, Labate J, Specht J, Brummer EC, Olson T, et al. Genome duplication in soybean (Glycine subgenus soja). Genetics. 1996;144:329–38.
66. Wan L, Zha W, Cheng X, Liu C, Lv L, Liu C, et al. A rice β-1,3-glucanase gene Osg1 is required for callose degradation in pollen development. Planta. 2011;233:309–23.
67. Gao H, Zheng X-M, Fei G, Chen J, Jin M, Ren Y, et al. Ehd4 encodes a novel and Oryza-genus-specific regulator of photoperiodic flowering in rice. PLoS Genet. 2013;9:e1003281.
68. Hori K, Ogiso-Tanaka E, Matsubara K, Yamanouchi U, Ebana K, Yano M. Hd16, a gene for casein kinase I, is involved in the control of rice flowering time by modulating the day-length response. Plant J. 2013;76:36–46.
69. Childs KL, Miller FR, Cordonnier-Pratt MM, Pratt LH, Morgan PW, Mullet JE. The sorghum photoperiod sensitivity gene, Ma3, encodes a phytochrome B. Plant Physiol. 1997;113:611–9.

19

Genetic mapping of centromeres in the nine *Citrus clementina* chromosomes using half-tetrad analysis and recombination patterns in unreduced and haploid gametes

Pablo Aleza[1†], José Cuenca[1†], María Hernández[1], José Juárez[1], Luis Navarro[1*] and Patrick Ollitrault[1,2*]

Abstract

Background: Mapping centromere locations in plant species provides essential information for the analysis of genetic structures and population dynamics. The centromere's position affects the distribution of crossovers along a chromosome and the parental heterozygosity restitution by 2n gametes is a direct function of the genetic distance to the centromere. Sexual polyploidisation is relatively frequent in *Citrus* species and is widely used to develop new seedless triploid cultivars. The study's objectives were to (i) map the positions of the centromeres of the nine *Citrus clementina* chromosomes; (ii) analyse the crossover interference in unreduced gametes; and (iii) establish the pattern of genetic recombination in haploid clementine gametes along each chromosome and its relationship with the centromere location and distribution of genic sequences.

Results: Triploid progenies were derived from unreduced megagametophytes produced by second-division restitution. Centromere positions were mapped genetically for all linkage groups using half-tetrad analysis. Inference of the physical locations of centromeres revealed one acrocentric, four metacentric and four submetacentric chromosomes. Crossover interference was observed in unreduced gametes, with variation seen between chromosome arms. For haploid gametes, a strong decrease in the recombination rate occurred in centromeric and pericentromeric regions, which contained a low density of genic sequences. In chromosomes VIII and IX, these low recombination rates extended beyond the pericentromeric regions. The genomic region corresponding to a genetic distance < 5cM from a centromere represented 47% of the genome and 23% of the genic sequences.

Conclusions: The centromere positions of the nine citrus chromosomes were genetically mapped. Their physical locations, inferred from the genetic ones, were consistent with the sequence constitution and recombination pattern along each chromosome. However, regions with low recombination rates extended beyond the pericentromeric regions of some chromosomes into areas richer in genic sequences. The persistence of strong linkage disequilibrium between large numbers of genes promotes the stability of epistatic interactions and multilocus-controlled traits over successive generations but also maintains multi-trait associations. Identification of the centromere positions will allow the development of simple methods to analyse unreduced gamete formation mechanisms in a large range of genotypes and further modelling of genetic inheritance in sexual polyploidisation breeding schemes.

Keywords: Clementine, Triploid, Second-division restitution, Chromosome interference, Physical location, Genetic recombination

* Correspondence: lnavarro@ivia.es; patrick.ollitrault@cirad.fr
[†]Equal contributors
[1]Centro de Protección Vegetal y Biotecnología, Instituto Valenciano de Investigaciones Agrarias (IVIA), Moncada, Valencia, Spain
[2]CIRAD, UMR AGAP, Avenue Agropolis - TA A-75/02 F-34398, Montpellier, France

Background

The centromere is a specialised structure within a chromosome, recognisable morphologically as the primary constriction. Centromeres mediate chromosome segregation at mitosis and meiosis, provide the proteinaceous kinetochore, promote sister chromatid cohesion and suppress recombination. Centromeric regions include large arrays of satellite DNA flanked by middle repetitive DNA rich in repetitive elements including transposons, retroelements and pseudogenes [1]. Although the structural characteristics of plant centromeres have been well defined, there is no conservation of centromeric sequences and these differ both from chromosome to chromosome and between species [1-3], highlighting the rapid rate of centromere evolution [1,3-5]. Centromere mapping allows the development of improved linkage maps, deciphering of chromosome arms, investigation of crossover events and understanding of crossover interference during meiosis. It is thus essential for analysis of the genetic structures of animal and plant species [6-9].

In some plant species, which keep their meiotic products together in tetrads, centromere mapping can be performed via tetrad analysis [10]. However, this mechanism is limited to a few species [9,10]. In many more species, the centromeres can be localised using half-tetrad analysis (HTA) of unreduced (2n) gametes. Unreduced gametes are the main cause of polyploidisation in plant species [11-14]. Their presence has been described in several crop species, including alfalfa [15], maize [16], *Solanum* [9,17-20] and *Citrus* species [21,22]. Several meiotic aberrations related to spindle formation, spindle function and cytokinesis result in 2n gamete formation in plants. The type of 2n gamete produced, however, depends essentially upon which of two basic processes, first-division restitution (FDR) and second-division restitution (SDR), has occurred. FDR and SDR depend upon the mode of nuclear restitution [12,15] and result from the omission of the first or the second meiotic division, respectively. A FDR 2n gamete contains non-sister chromatids, whereas an SDR 2n gamete contains two sister chromatids [23]. This implies that the patterns of parental heterozygosity restitution (HR) seen along the chromosome with respect to the genetic distance from the centromere will be completely opposite in the two types of diploid gamete [9,15,17,24]. Molecular marker analysis is therefore a powerful means of identifying the mechanism underlying unreduced gamete formation [15,25-27] and of locating centromeres genetically [6,9,17,24,28,29]. Tavoletti *et al.* [15] developed a multilocus maximum likelihood method of HTA assuming complete chromosome interference. Cuenca *et al.* [24] proposed an alternative approach based on HR functions along a chromosome and their relationship to locus-centromere genetic distance, allowing different chromosome interference models to locate the positions of the centromeres in linkage groups (LGs).

Citrus species make up the world's leading fruit crop with 131.3 million tons produced in 2012 [30]. International efforts have led to an increase in genomic and genetic resources, allowing better preservation of biodiversity and improvements in breeding strategies and their efficiency. Ollitrault *et al.* [31] established the current reference clementine genetic map (*Citrus clementina* Hort. ex Tan). This includes 961 co-dominant markers spread across nine LGs and spanning 1084.1 cM, with an average marker spacing of 1.13 cM. Recently, a reference whole clementine genome sequence, anchored on the clementine genetic map, was obtained by the International Citrus Genome Consortium [32] from a haploid plant of 'Clemenules' clementine [33].

Diploidy is the general rule in *Citrus* and related genera, and the basic chromosome number is $x = 9$. However, sexual polyploidisation is relatively frequent and is central to the current approach of developing triploid citrus-breeding programmes with an aim of producing new seedless mandarin cultivars [34,35]. Esen and Soost [21] proposed that, in *Citrus*, unreduced ovules arose from the abortion of the second meiotic division. This hypothesis was corroborated by molecular marker analyses of clementine [26] and 'Fortune' mandarin (*C. clementina* × *C. tangerina*) [24]. However, Chen *et al.* [36] proposed that female 2n gametes produced by the sweet orange (*C. sinensis* (L.) Osb.) were generated by FDR. This paved the way for HTA mapping of centromere locations in *Citrus*, although only the centromere of LG 2 has been mapped to date [24].

The major objective of this work was to establish the genetic location of centromeres within the nine LGs of the clementine genetic map [31] anchored on the sequences of the corresponding nine chromosome of the citrus haploid set [32]. Centromere positions were located using HTA by determining the genotypes of 87 triploid hybrids, recovered from 2x × 2x sexual hybridisation, at 104 co-dominant molecular markers, including simple sequence repeats (SSRs), insertion-deletions (InDels) and single nucleotide polymorphisms (SNPs). This information was used to determine the distributions of crossovers and reveal the variation in interference levels across the different chromosome arms in clementine 2n gametes. Finally, we analysed the pattern of genetic recombination along the physical sequences of haploid gametes with respect to the centromere location and frequency of genic sequences.

Methods

Plant material

The mechanism of 2n gamete formation was investigated in the progeny of 87 triploid hybrids recovered from a cross between diploid 'Fina' clementine (female parent) and 'Nadorcott' tangor (male parent). Practical details of recovery of the triploid hybrids from 2× × 2× hybridisation, using embryo rescue followed by flow cytometry to select triploid hybrids, may be found in Aleza *et al.* [35]. No selection was made between the triploid hybrids and all genotypes were grown at the Instituto Valenciano de Investigaciones Agrarias (IVIA, Moncada, Valencia, Spain). Genomic DNA was isolated from triploid hybrids and their parents using the Plant DNAeasy kit (Qiagen, Madrid, Spain), following the manufacturer's protocol.

Genotyping of triploid progeny using molecular marker analysis

The male and female parents and 87 triploid hybrids were genotyped using a total of 104 molecular markers (48 SSRs, 11 InDels and 45 SNPs). Genotyping revealed heterozygosity of the 'Fina' clementine female parent and polymorphism within the 'Nadorcott' tangor male parent. SNP markers were selected using previous genotyping data obtained from the Illumina Golden Gate™ platform [37]. These markers are widely distributed across the current genetic map of Clementine [31].

PCR amplifications of genomic DNA with the 59 SSR and InDel markers were performed using a Thermocyclerep gradient S (Eppendorf®). Each reaction contained 0.8 U Taq DNA polymerase (Fermentas®), 2 ng/mL *Citrus* DNA, 0.2 mM wellRED (Sigma®) dye-labelled forward primer, 0.2 mM non-dye-labelled reverse primer, 0.2 mM each dNTP, 10× PCR buffer and 1.5 mM$MgCl_2$in a final volume of 10 mL. The PCR protocol was as follows: an initial denaturation at 94°C for five minutes followed by 40 cycles of 30 seconds at 94°C, one minute at 50°C or 55°C, 45 seconds at 72°C, and a final elongation step of four minutes at 72°C. Capillary electrophoresis was performed using a CEQ™ 8000 Genetic Analysis System (Beckman Coulter Inc., Fullerton, CA, USA). Data were collected and analysed using GenomeLab® GeXP (Beckman Coulter Inc.) version 10.0 software. Allele dosage was calculated using the MAC-PR (microsatellite DNA allele counting-peak ratio) method [38], validated in *Citrus* by Cuenca *et al.* [24].

The genotypes of triploid progenies at 45 SNP markers were determined using KASPar technology by Kbioscience® services (now LGC Genomics; http://www.lgcgenomics.com). The KASPar™ Genotyping System is a competitive, allele-specific dual Förster Resonance Energy Transfer (FRET) based assay for SNP genotyping. Primers were designed by LGC Genomics based on the SNP locus flanking sequence (approximately 50 nucleotides each side of the SNP). A detailed explanation of the specific conditions and reactions may be found in Cuppen [39]. Allele doses in the heterozygous triploid hybrids were determined using the relative allele signals of the SNP markers, based on competitive allele-specific PCR, as described by Cuenca *et al.* [40]. Detailed information on all the markers used in this study is given in Additional file 1.

Identification of the parent producing the unreduced gamete and inference of the unreduced gamete genotype

The use of markers which differentiated between the alleles from the female and male parents allowed the unequivocal identification of the parent which produced the 2n gamete for each triploid hybrid, based on HR or allele dosage estimation. Once the origin of the 2n gamete had been identified, the allelic configurations of the unreduced gametes were inferred using the data obtained from genotyping the triploid hybrids, as previously described in Cuenca *et al.* [24].

The origin of the 2n gamete for each hybrid was determined by identifying which parent had passed on a double dose of genetic information. For markers scored in the parents (female × male) as $A_1A_2 \times A_1A_1$ or $A_1A_2 \times A_1A_3$, identification of the triploid hybrids as $A_1A_2A_2$ or $A_2A_2A_3$ (i.e., possessing a double dosage of A_2, the allele specific to the female parent) revealed the 2n gamete had originated from the female parent but the observation of $A_1A_3A_3$ or $A_2A_3A_3$ (i.e., with a double dosage of A_3, the allele specific to the male parent) indicated a male origin for the 2n gamete. For markers scored as $A_1A_2 \times A_3A_3$ in the parents, the triploid hybrids $A_1A_2A_3$, $A_1A_1A_3$ or $A_2A_2A_3$ indicated a maternal origin for the unreduced gamete, whilst $A_1A_3A_3$ or $A_2A_3A_3$ indicated a paternal origin. For markers scored as $A_1A_2 \times A_3A_4$ in the parents, the triploid hybrids $A_1A_1A_3$, $A_1A_1A_4$, $A_1A_2A_3$, $A_1A_2A_4$, $A_2A_2A_3$ and $A_2A_2A_4$ indicated a female origin for the 2n gamete and the hybrids $A_1A_3A_3$, $A_2A_3A_3$, $A_1A_3A_4$, $A_2A_3A_4$, $A_1A_4A_4$ and $A_2A_4A_4$ a male origin.

Once the parent producing the 2n gamete had been identified, the allelic configurations of the unreduced gametes were inferred from triploid hybrid genotyping, as previously described by Cuenca *et al.* [24]. For loci where the parental alleles were completely different (for example, $A_1A_2 \times A_3A_4$), the genotype of the 2n gamete was directly read from the triploid hybrid structure. If both parents shared one allele (for example, $A_1A_2 \times A_2A_2$ or $A_1A_2 \times A_2A_3$), the inference of the 2n female gamete structure was carried out using the estimated allele dosage for those triploid hybrids that inherited the common allele from the male parent. For each locus, the parental heterozygosity restitution (HR) was calculated as the percentage of individuals with the same heterozygous allelic configuration as that of the female parent.

Identification of the mechanism of unreduced gamete formation by single-locus analysis

The parental HR for 2n gametes arising from FDR or SDR directly relates to the genetic distance of a given locus from the centromere, but the two types of diploid gamete produce a completely opposite pattern of HR [24]. In the absence of crossovers, all loci heterozygous in the parent will be heterozygous in FDR gametes but, in SDR gametes, all the loci situated between the centromere and the first crossover will be homozygous. As the genetic distance from the centromere increases, HR decreases in the case of FDR; it remains, however, at over 50% regardless of the chromosome interference model. In the case of SDR, HR increases with the genetic distance until it reaches 100% under the total chromosome interference model [9,15,17,24]. In this study, the identification of mechanism was based on an analysis of HR at each locus across the entire inferred 2n gamete population. In the absence of previous knowledge of the relative positions of markers to the centromere, the observation that HR is greater than 50% at a single locus is not informative, since it could result from either FDR or SDR. Theoretical HR values below 50%, however, are only found in the case of SDR [9]. When such low values of HR were observed at a marker, a LOD score test was calculated to compare the probabilities of the observed level of HR occurring, under both the SDR and FDR hypotheses. In the case of SDR, the highest probability of the observed HR occurs at the centromere position, leading to a match between the theoretical and observed proportions of heterozygous gametes. In the case of FDR, the best fit between the theoretical proportion of heterozygous gametes and observed data is obtained for a theoretical proportion of 50%, because 50% is the minimum value of HR expected under the hypothesis of FDR. Thus, logarithm of the odds ratios (LOD) were estimated as follows:

$$\mathrm{LOD} = \log_{10}\left[\frac{p(SDR)}{p(FDR)}\right] = \log_{10}\left[\frac{h^{nh} \cdot (1-h)^{(1-h)\cdot n}}{0.5^{nh} \cdot (0.5)^{(1-h)\cdot n}}\right]$$

With h being the HR observed for the marker and n the number of genotypes analysed at this marker. LOD = 3 (*i.e.*, the probability of the SDR hypothesis being more than 1000-fold that of FDR) was considered the significance threshold for concluding in favour of SDR rather than FDR.

Identifying the preliminary locations of centromeres using HR functions under no interference and partial interference chromosome models

The methodology of Cuenca *et al.* [24] was used to identify the preliminary locations of the centromeres. This method is based on comparing the observed HR values along each LG with the theoretical restitution functions under the SDR mechanism of 2n gamete formation for both the no interference and partial interference models on a chromosome arm ($Cx(Co)^4$). These functions were derived from those developed by Zhao and Speed [41] for ordered tetrads, based on the random spindle–centromere attachment hypothesis [42], and extended by the same authors to HTA [43]. Discrepancies between the different models (no interference and partial interference coupled with the different location of the centromere) and the observed data were estimated by the sum of the squared differences between the observed and theoretical values at the marker map positions. Let Fit(c) be the value of the sum of the squared distance for each position of the centromere for one interference model; the best theoretical centromere position under this model is deduced by searching c, which minimises Fit(c). The confidence interval (95%) for the centromere position was estimated by bootstrap on the loci (500 bootstraps).

The locations of genetic markers were obtained from the reference clementine map [31] established using Kosambi's map function and, for the no interference model, the genetic positions were established from the same genotypic data [31] but using Haldane's map function.

Centromere mapping using multilocus half-tetrad structure analysis

Multilocus analyses were performed on the 87 hybrid triploid genotypes at four loci in each LG. These four loci were selected following the preliminary localisation of the centromeres, as described above, and consisted of two flanking loci on each side of the preliminary location. The analyses were conducted according to Tavoletti *et al.* [15] and assumed multiple crossovers did not occur between contiguous markers. The position of the centromere was moved virtually stepwise at intervals of 0.01 cM along the LG from before the first selected marker (M1) to after the last selected marker (M4), and the probability of the observed populations occurring was estimated at each position. The best estimate for the centromere location was that producing the highest probability. The confidence interval was calculated using the LOD drop-off method [44].

Crossover interference analysis

After determination of the centromere position, three-point linkage mapping was used to estimate the level of crossover interference for each chromosome arm. The centromere (considered to be homozygous) was used as the first point and two markers were selected in each arm. The chromosome interference coefficient (IC) was defined as follows by Griffiths *et al.* [42]:

$$\mathrm{IC} = 1 - \left[\frac{rd}{r_{CM_1} \cdot r_{M_1M_2}}\right]$$

When r_{CM1} denotes the observed recombination rate (heterozygous to homozygous and *vice versa*) between the centromere and locus 1, r_{M1M2} is the observed recombination rate between locus 1 and 2 and rd is the observed rate of double recombination between the centromere and locus 2.

Relationship between genetic and physical locations

The physical locations of the genetic markers used to establish the clementine genetic map were identified by searches, using the flanking sequences of the markers [31,38], of the clementine reference genome [32] using the blast N option [45] of the online Galaxy tool (http://gohelle.cirad.fr/galaxy/).

The cM/Mb rates between genetic and physical positions of markers for each LG were estimated using local linear regression. Each LG was divided in intervals of two or three Mb and the slope between the values of the genetic and physical marker positions was calculated for each interval.

Results and discussion

Mechanism of 2n gamete formation

Parental origin of the 2n gamete producing triploid hybrids in 2x × 2x crosses

Of the 104 markers analysed, six loci (mCrCIR01C06, mCrCIR07B05, Cx6F03, CID0591, CID2493 and MEST473) were able to differentiate completely between the female and male parents and thus enabled unequivocal identification of the parent producing the 2n gamete for each triploid hybrid (Additional file 1). The diploid gamete was of maternal origin in all the triploids analysed and it was therefore possible to infer the 2n gamete structure of each hybrid. Our results were in agreement with previous, pioneering works, which concluded citrus triploid hybrids arose from 2n megagametophytes [21,46,47]. Cytogenetic studies [48,49] showed triploid embryos are associated with pentaploid endosperm, a strong indication that they result from the fertilisation of unreduced ovules by normal haploid pollen [48]. Depending on the genotype, the frequency of duplication among female gametes varied from between less than 1% to more than 20% [50].

Mechanism of 2n megagametophyte formation in Clementine

Potential allelic segregation distortion in the 2n gamete population was tested at each marker using χ^2-analysis (0.05 probability threshold) with the Bonferroni correction for multiple testing applied (Additional file 2). None of the 104 markers tested displayed significant segregation distortion.

Maternal HR in each 2n gamete varied between 27.18 and 61.39% across the loci analysed, with a mean value of 42.45%. The unimodal distribution of HR in the 2n megaspores suggested all the 2n gametes arose from the same mechanism (Figure 1a). Analyses of 2n gamete origin and the centromere location were conducted under this hypothesis and it was confirmed *a posteriori* by a LOD score analysis conducted at the individual level (Additional file 3).

The rate of HR at a population level was calculated for each of the 104 loci (Figure 1b). The average rate across all loci was 41.73% but values ranged from 0% for the marker CiC6278-01, to 82.76% for the marker mCrCIR03G05. HR values were lower than 50% at 57 of the markers analysed. LODs of SDR/FDR probabilities were calculated for these markers, and found to vary between 0 and 23.8 (Additional file 2). When a LOD = 3 was taken as the significance threshold for accepting SDR rather than FDR, the corresponding%HR threshold was 30.35%. Forty-two markers displayed an LOD > 3 (*i.e.*,%HR < 30.35%; Figure 1b), supporting the SDR hypothesis and ruling out the possibility of FDR in this population.

These results are consistent with earlier studies which indicated an SDR origin for 2n megagametophytes in clementine [24,26] and 'Fortune' mandarin [51]. FDR, however, may be the mechanism producing unreduced female gametes in sweet orange [36] and in lemon [51], although these conclusions were based on HR values greater than 50% in a small number of markers without any knowledge of the distances between markers and centromeres. They are therefore questionable, since such values could also result from SDR if there was a large distance between the markers analysed and the centromeres. In the absence of additional information on centromere position, drawing a definitive conclusion of FDR would require the analysis of a considerable number of genotypes with a large number of markers well distributed across the different chromosomes.

SDR is frequently observed in plants in the formation of 2n ovules whereas, even within the same species, 2n pollen may be produced by both FDR and SDR [12]. The mechanism of 2n gamete formation is a strong determinant of the genetic and phenotypic structures of polyploid hybrid populations, because of its effect on parental HR. Knowledge of the particular meiotic nuclear restitution mechanism producing the unreduced gametes is crucial, therefore, for the optimisation of plant breeding strategies based on sexual hybridisation [52,53]. Several studies of genetic markers indicate that gametes formed by FDR transmit 70–80% of parental heterozygosity to progeny but those gametes formed by SDR transmit only 30–40% [54-56]. These values are in agreement with our estimation of 41.7% transmission in clementine.

In progenies produced by 2n gametes derived from FDR, parental heterozygosity and epistatic interactions are maintained across a higher number of individuals and loci than in SDR-derived progenies. Moreover, because a greater percentage of the parental genome is transferred intact following FDR than SDR, FDR produces a more uniform population of 2n gametes and thus a lower range of variation between individuals is expected in populations derived from FDR [57]. For this reason, gametes derived from FDR are considered better than those from SDR for

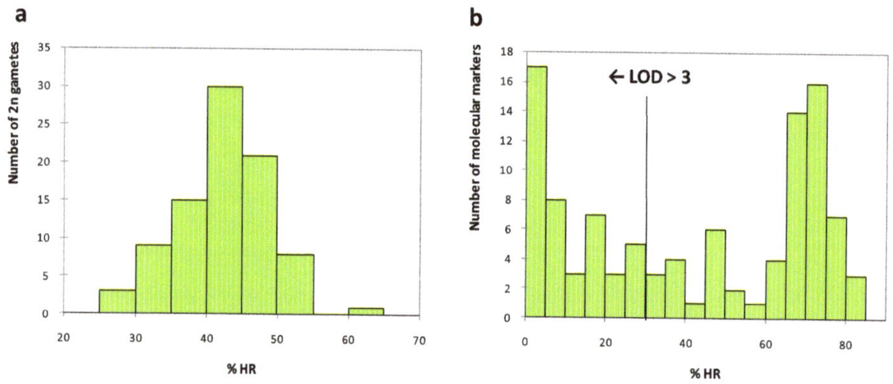

Figure 1 Heterozygosity restitution percentages (%HR) for all 2n gametes and molecular markers used in this study. a: Distribution of %HR estimated for each 2n gamete. **b**: Distribution of %HR estimated for each molecular marker and %HR corresponding to LOD scores greater than three with significance for SDR mechanism of 2n gamete formation.

breeding purposes, as they create offspring similar to the female parent and thus allow transmission of the genetic gain at the maternal level into progeny and optimisation of heterotic responses [17]. The superiority of progeny resulting from sexual polyploidisation involving 2n gametes produced by FDR has been demonstrated, for instance, in potato [58,59]. In contrast, 2n gametes generated by SDR produce more variable offspring and thus create a greater number of new genetic combinations, increasing the likelihood of obtaining novel phenotypes [60].

For specific characters controlled by a major gene, the allelic effect (dominance; recessivity; heterosis) and genetic distance to the centromere are both crucial in determining the proportion of the progeny that will show the favourable trait. With respect to tuber yield in potato, for example, increased yield due to heterosis in FDR-derived progeny is associated with the location of genes with major effects on tuber yield between the centromeres and proximal crossovers [61]. In *Citrus*, the recessive resistance gene for Alternaria Brown Spot fungus disease is located close to the centromere. This situation favours the operation of SDR and allows for up to 40% of the progeny showing disease resistance when the heterozygous 'Fortune' mandarin is used as the female parent [62]. Therefore, being able to locate the position of centromeres accurately on a genetic map anchored on the annotated whole genome sequence is a critical step for further modelling of the inheritance of traits in breeding schemes utilising sexual polyploidisation.

Centromere mapping

Centromere mapping by HTA is used in plants and animals to integrate centromeres with linkage maps [6,9,15,24,28,62]. In this study, mapping of the centromere positions was done in two steps: initially, we identified a preliminary position for each centromere by comparing the observed and theoretical patterns of HR along the LG under the SDR mechanism of female 2n gamete formation and the no interference and partial chromosome interference models, according to the methodology of Cuenca *et al.* [24]. Once the preliminary location of the centromere had been established, we selected markers flanking this position and used multilocus HTA to locate the centromeres more accurately [15].

Preliminary location of centromeres

Between nine and fourteen molecular markers per LG were used to locate the position of the centromere by comparing the observed and theoretical patterns of HR rate within each LG. Earlier results had allowed us to discard the FDR mechanism and so we only tested the two interference models under the SDR hypothesis; Figure 2 displays the pattern of HR for each model of interference on LG 2. As we moved from one end of LG 2 to the other, HR decreased from 70.11% at the SSR marker mCrCIR02D09 to 0% at the SNP marker CiC6278-01, and then increased again to 75.86% at the SSR marker JK-TAA41. A better adjustment between the theoretical curves and the observed pattern of HR was obtained using the partial interference model.

The patterns of HR observed across all LGs provided a picture of a typical 2n gamete population resulting from SDR (Additional file 4). For each LG, the best fit value (the lowest value of the sum of the squared distance between theoretical and observed HR for each marker for the best identified location of the centromere) was obtained using the partial interference model. The 'no interference/partial interference' ratios ranged between 1.4 for LG 8 and 8.5 for LG 7 (Table 1). The maximum value of theoretical HR for the no interference model was 66.67% [63,64] and 38 markers displayed a HR value greater than 66.67%. The statistical significance was tested using χ^2-analysis and seven markers displayed

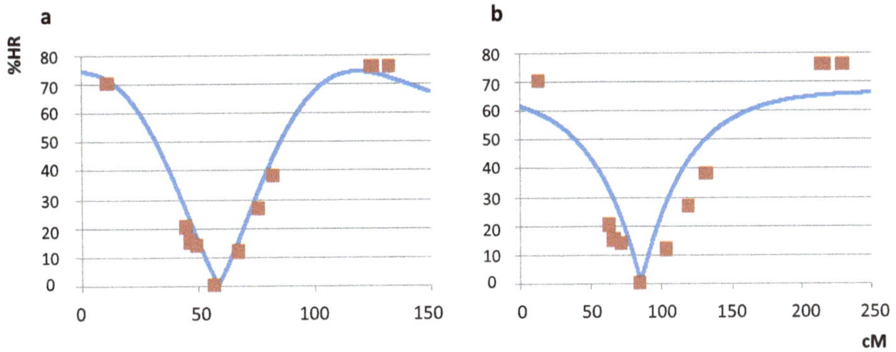

Figure 2 Observed heterozygosity restitution percentages (%HR) for markers on LG 2 (squares) and theoretical HR (line) for the best-fitting centromere position. a: under SDR and the Cx(Co)[4] model of partial interference (Kosambi's map function) and **b**: under the SDR model without crossover interference (Haldane's map function). Markers on the x-axis are shown according to their position on the clementine genetic map [31].

significant p values ($p < 0.05$) (Additional file 2). The better fit of values to the partial interference model than the no interference model and the observation that HR values at several markers significantly exceeded 66.67% suggested the presence of crossover interference, which is in agreement with previous conclusions for 'Fortune' mandarin chromosome II [24].

Centromere mapping using multilocus half-tetrad structure analysis

Having obtained a preliminary location for the centromere in each LG, we selected four flanking markers (two markers on the right and two on the left side of this location) to perform HTA. Four markers were predicted to produce 16 different multilocus profiles, considering homozygosity and heterozygosity at each locus. The number of 2n gametes that corresponded to each of these profiles (Additional file 5) was used to locate the centromere of each LG (Table 1; Figure 3). The 95% CI was calculated using the LOD drop-off method [44].

The centromeres of LGs 1, 3, and 8 were located approximately in the middle of the LG whereas those of LGs 2, 8 and 9 were positioned off-centre, being slightly closer to one end of the LG than the other. In LGs 4, 5, 6 and 7, the centromeres were located very close to one of the LG ends (Figure 3). This genetic mapping of the position of the centromere in each linkage group paves the way for accurate genetic analysis [9,29] and will, in particular, greatly simplify analysis of the mechanisms of 2n gamete formation in different *Citrus* varieties. Very simple, routine tests that identify the mechanism of 2n gamete formation can be performed at the individual level using co-dominant centromeric markers, as previously demonstrated using the *Pgm-2* locus in potatoes [25,65]. In this paper, however, knowledge of the centromere locations was used to analyse the recombination patterns in 2n and haploid clementine gametes, as described below.

Table 1 Localisation of centromere positions using multilocus half-tetrad analysis (HTA)

LG	N	Fit ratio NI/PI	Centromere location (cM)	
			Cent. Position	Conf. Interval[a]
1	13	1.5	60.66	58.82 - 61.64
2	11	6.9	56.87	55.25 - 58.74
3	11	1.8	90.59	89.08 - 93.92
4	13	5.0	16.14	15.12 - 16.82
5	14	1.7	23.12	22.32 - 25.63
6	12	2.4	6.20	4.79 - 6.80
7	11	8.5	96.43	95.02 - 97.56
8	10	1.4	54.21	50.92 - 56.72
9	9	6.9	52.16	50.29 - 53.77

cM: Centimorgans.
N: Number of used markers.
PI: Partial interference model ($m = 4$), $Cx(Co)^4$.
NI: No Interference model.
a. Confidence interval calculated using drop off method [44].

Crossovers and interference analysis in 2n gametes

'Crossover interference' refers to the observation that the occurrence of one crossover affects the likelihood of occurrence and/or the location of other crossovers in its neighbourhood [66]. Partial crossover interference in 'Fortune' mandarin was proposed by Cuenca *et al.* [24]. In the current study, theoretical partial crossover interference models were a better fit to the patterns of HR observed along all the clementine LGs, as some markers displayed HR values greater than the 66.6% threshold for the no interference model. Using the genetic location of the centromere estimated by HTA, we studied the crossovers occurring in each of the chromosome arms (Table 2). The percentage of 2n gametes that showed multiple crossovers (MCO) in a particular region ranged from 0% (arm 1 of chromosomes IV and V) to 44.16% (arm 2 of chromosome VI). A maximum of four crossovers (CO) was observed;

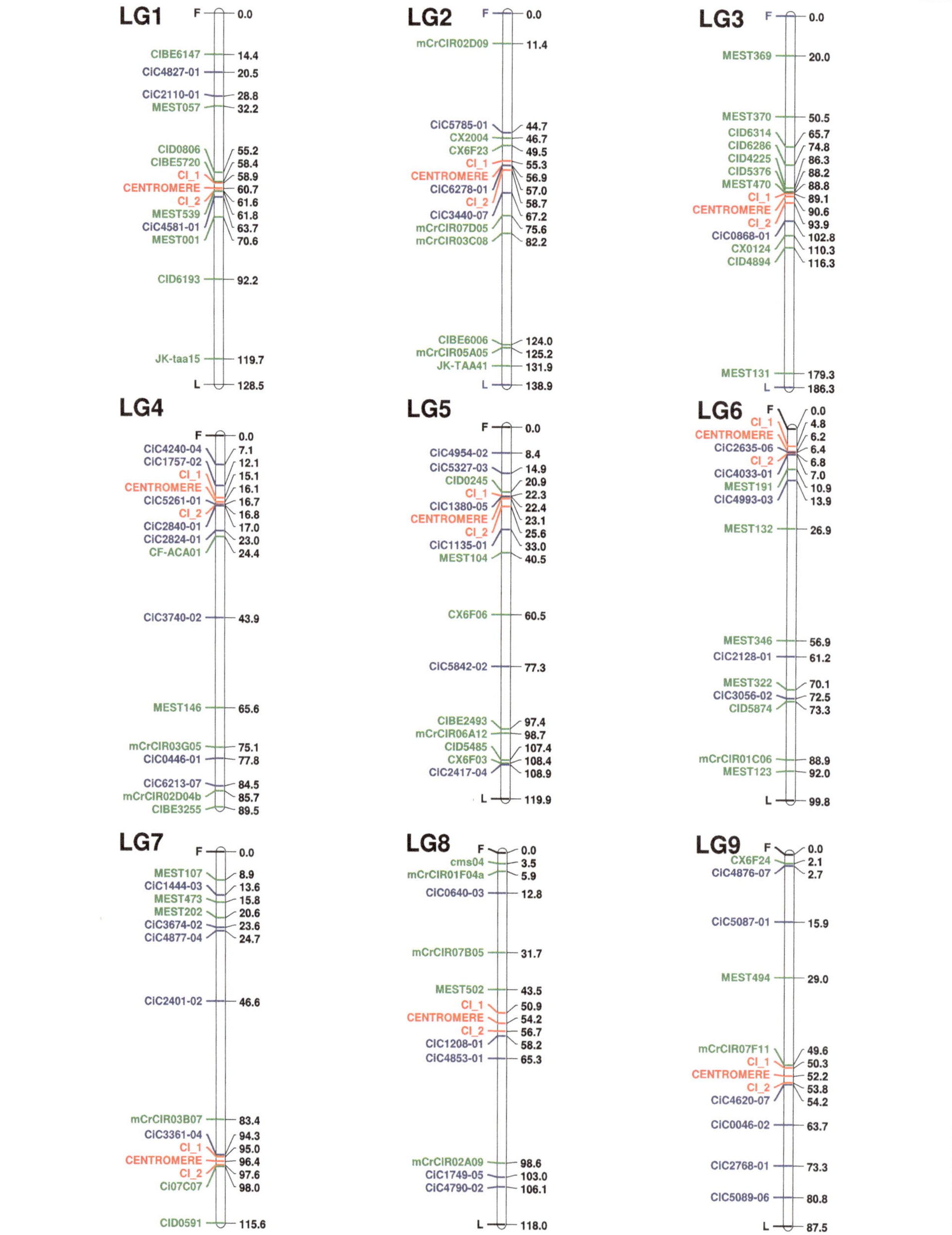

Figure 3 Location of centromeres on the current clementine genetic map using HTA and drop-off calculations of the confidence interval. Markers F and L correspond to the first and the last marker for each LG. Green and blue indicate SSR and SNP markers, respectively. Locations of centromeres and CI are highlighted in red.

Table 2 Multiple and complementary crossovers observed in 2n gametes, estimation of the interference coefficient (IC) for each chromosome arm and test for interference between arms (TI)

LG	Arm 1			Arm 2			Ratio IC1/IC2	TI	
	% MCO	% CCO	IC 1	% MCO	% CCO	IC 2		X^2	p-value
1	6.17	6.17	0.82	22.22	6.17	0.48	1.71	12.36	0.194
2	4.76	0.00	0.60	21.43	15.48	0.53	1.13	5.82	0.667
3	15.48	10.71	0.59	13.10	4.76	0.53	1.11	5.55	0.937
4	0.00	0.00	1.00	22.78	10.13	0.56	1.79	2.73	0.603
5	0.00	0.00	1.00	30.56	19.44	0.38	2.63	1.39	0.708
6	-	-	-	44.16	19.48	0.19		-	-
7	33.77	19.48	0.39	1.30	1.30	1.00	0.39	14.33	0.074
8	8.86	3.80	0.75	16.46	3.80	0.43	1.74	2.49	0.869
9	7.50	2.50	0.62	0.00	0.00	1.00	0.62	2.64	0.268

% MCO: Percentage of 2n gametes with multiple crossovers.
% CCO: Percentage of 2n gametes with complementary crossovers (phase change).
IC: Interference coefficient.

this was seen in arm 1 of chromosome VII and in arm 2 of chromosomes II, IV and VI (Additional file 6). Complementary crossovers (double crossovers implying the presence of four chromatids; CCO) were identified by the occurrence of an allelic phase change in homozygosity between markers (Additional file 7), according to Cuenca *et al.* (24). The maximum percentage of CCO was 19.48%, which was observed both in arm 2 of chromosome VI and in arm 1 of chromosome VII (Table 2).

The interference coefficient (IC) was estimated for each chromosome arm. Values of IC ranged considerably from 0.19 (arm 1 of chromosome VI) to 1.0 (arm 1 of chromosomes IV and V and arm 2 of chromosomes VII and IX). Three of the four chromosome arms displaying total interference were very short genetic arms; partial interference was observed in the longer arm of chromosomes V and VII. Different levels of interference between arms of the same LG have been previously reported for chromosome II of 'Fortune' mandarin [24]. Potential interference between arms was tested for each chromosome by χ^2-analysis, based on the contingency table of the number of crossovers in each chromosome arm (Additional file 7), but no interference between the different arms of the same chromosome was observed (Table 2).

Crossover interference plays an important role in determining the number of crossovers per chromosome, but little is known about the mechanisms controlling interference [67]. In clementine, we found no evidence for interference between different arms of a chromosome, a maximum number of four crossovers per chromosome (chromosomes II, IV, VI and VII) and observed total interference in one arm of four different chromosomes (IV, V, VII and IX). Variation in the level of interference between different parts of the genome has been observed in Arabidopsis [68], in humans [69] and in mice [70]. The last work also suggested that levels of interference are also higher in the smaller chromosomes of mice.

Pattern of recombination in clementine haploid gametes ad its relation with centromere location and genic sequences distribution

The data from the clementine genetic map [31] were anchored on the whole genome sequence of clementine [32] and used to analyse the pattern of recombination along each chromosome and its relationship with centromere location (Figure 4 and Additional file 8). The cM/Mb rates were estimated using local linear regression between the genetic and physical positions of markers in the region under consideration. The physical locations of the confidence intervals around genetic positions of centromeres were also inferred using local regression between the genetic and physical positions of the flanking markers. From this inference of the physical location of their centromeres, the nine clementine chromosomes should be classified as metacentric (I, II, V and VIII), submetacentric (III, IV, VII and IX) and acrocentric chromosomes (VI), according to the criteria of Levan *et al.* [71].

The average recombination rate over the entire *Citrus* genome was around 3.0 cM/Mb, but there was large variation across the genome. The higher local rates of recombination were less than 14 cM/Mb in most chromosomes, with only chromosomes VI and VIII displaying regions with elevated recombination rates of 26 cM/Mb and 33 cM/Mb, respectively. Centromeric areas displayed very low levels of recombination (<1.0 cM/Mb), but the distance over which this reduction in the rate recombination extended varied considerably between chromosomes. The genomic region with a genetic distance from the centromere of under 5.0 cM was less than 13 Mb for chromosomes I, II, IV, VI and VII, but extended to 23 Mb for chromosome IX and to 30 Mb for chromosome III

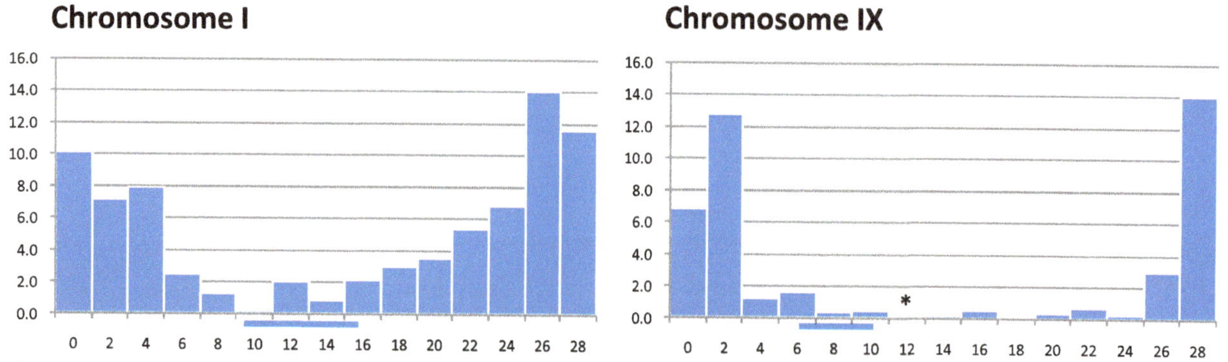

Figure 4 Variation in recombination rates along chromosomes I and IX of clementine. The x-axis displays the physical position in megabases (Mb) along each chromosome, and the y-axis represents the ratio of genetic distance to physical distance (cM/Mb). The bars beneath the x-axis indicate the approximate locations of the centromeres (CI). *These data have been calculated from up and down intervals (no marker on the genetic map on the considered genomic segment).

(Figure 5 and Additional file 9). The fraction of the entire genome within 5.0 cM of a centromere was around 47% but, again, this value varied considerably between chromosomes, being lowest (24%) for chromosome VII and highest (74%) for chromosome IX. Low values of recombination close to the centromeres were first reported by Dobzhansky [72] in *Drosophila melanogaster*. Suppression of crossovers in centromeric and pericentromeric regions has been observed in many plant species, including tomato [73], wheat [74], Arabidopsis [75], rice [76], maize [77] and soybean [71], and this reduction ranges from 5-fold to > 200-fold, depending on the organism [78].

Most chromosomes showed a strong positive correlation between the genetic distance from the centromere and the frequency of genic sequences (Figures 4 and 5 and Additional files 8 and 9). As a consequence, across the whole genome, the 47% of the genome situated within 5 cM of a centromere contained only 23% of the genic sequences. On chromosome IX, however, because the recombination rate was suppressed over a much larger region (Figure 5), 48% of genic sequences were located within 5 cM of the centromere. Chromosome VII displayed a unique pattern, as both the rate of recombination and the frequency of genic sequences were relatively homogeneous across its length. A low frequency of genes in centromeric and pericentromeric regions has been described previously for many plant species [79]. In wheat, for example, most genes are clustered in the distal regions of chromosomes and the large centromeric regions (as large as 100 Mb) are gene-poor and recombinationally inactive [74,80,81]. Outside the centromeric and pericentromeric regions, the relationship between gene density and recombination rate may

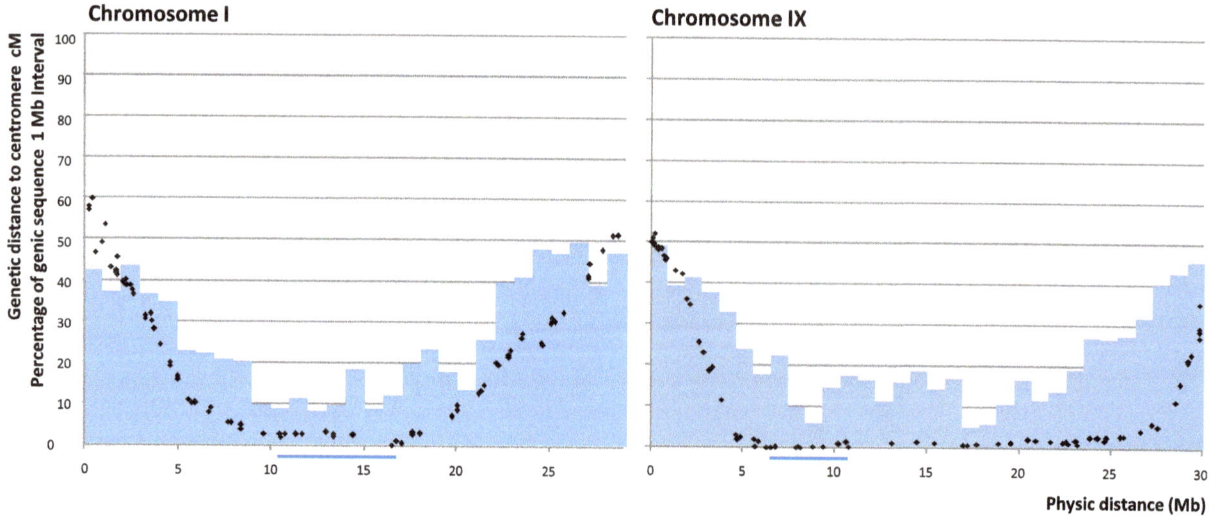

Figure 5 Relationship between physical location (x-axis), genetic distances between the centromere and markers on the clementine genetic map (black dots) and proportion of genic sequences (blue bars) along chromosomes I and IX. Bars beneath the x-axis indicate the approximate location of the centromeres (CI).

vary. In rice, Tian *et al.* [82] observed a positive correlation between genomic recombination rates and gene densities, but such a relationship was not found in wheat [81]. In *Citrus*, in addition to containing low densities of genes, the centromeric and pericentromeric areas are characterised by a high frequency of retrotransposon DNA [32]; a similar pattern has been observed in wheat [83]. In the current study, the suppression of crossovers extended beyond the centromeric and pericentromeric regions in chromosome IX and, to a lesser extent, in chromosome VIII.

The interspecific heterozygosity of the clementine genome [32] is likely to be a contributing factor to the level of crossover suppression beyond the centromeric and pericentromeric regions. As a consequence, high levels of linkage disequilibrium will be maintained across large regions of each of these chromosomes, affecting numerous genes. Linkage disequilibrium across extended portions of the genome serves to maintain epistatic interactions and retain phenotypic traits determined by the interactions of specific alleles at multiple loci but, simultaneously, strongly limits the possibility of generating recombined lines and hence new, potentially more favourable, genotypes. Moreover, as 2n gametes of clementine are produced by SDR, many genes on these chromosomes will be transmitted in a homozygous state.

Conclusions

Triploid hybrid progeny derived from 'Fina' clementine × ' Nadorcott' tangor crosses were shown to originate from unreduced diploid megagametophytes produced by SDR. Under this model, multilocus HTA was used to map genetically the position of the centromere in all nine *Citrus* linkage groups. Identification of the centromere positions paves the way for developing simple methods to determine the mechanism of unreduced gamete formation in a large range of genotypes. Understanding how the distance between genes and centromeres varies across an entire genome, and the effect of sexual polyploidisation, will enable better modelling of the inheritance of favourable traits and thus enhance breeding schemes.

The inference of the physical location of the centromeres from the whole genome sequence of *Citrus* anchored on the genetic map revealed one acrocentric, four metacentric and four submetacentric chromosomes. Partial crossover interference was observed in at least one arm of each chromosome, with variation between arms. Total interference was found in four chromosome arms, although no interference was observed between the different arms of the same chromosome. For haploid clementine gametes, a decrease in the rate of recombination was observed in centromeric and pericentromeric regions, which contained a low density of genic sequences. On chromosomes VIII and IX, these regions of limited recombination extended beyond the pericentromeric regions and affected areas richer in genic sequences. This should produce strong linkage disequilibrium for a large number of genes, resulting in stable epistatic interactions and retention of multilocus-controlled traits over successive generations, but will also maintain multi-trait associations. The genomic region corresponding to a genetic distance less than 5 cM from a centromere ranged from 5.0 Mb in chromosome VII to 30 Mb in chromosome III. In total, this represented 47% of the *Citrus* genome and 23% of the genic sequences. These findings will be helpful for optimising breeding strategies based on sexual hybridisation and also for better understanding the genetics of triploid *Citrus*.

Additional files

Additional file 1: Characteristics and genetic/physical positions of the SSR, InDel and SNP markers used in mapping of triploid progenies to determine the positions of the centromeres.

Additional file 2: χ^2-test for allelic segregation and heterozygosity restitution percentages (%HR) at each locus, LOD SDR/FDR for markers with HR < 50% and χ^2-test for partial interference for markers with HR > 66.67%.

Additional file 3: Analysis at individual level of the origin (SDR/FDR) of clementine 2n gametes. For each of the nine LGs, we selected the marker closest to the estimated position of the centromere and computed the LOD of the probability of the individual multilocus allelic configuration under SDR/FDR hypotheses (Cuenca *et al.*, [24]). No more than one marker in heterozygosity was observed for each individual and the LOD value supported SDR as the origin of the 2n gamete.

Additional file 4: Observed heterozygosity restitution (HR frequency) values for markers (squares) and theoretical functions for SDR with partial interference (x-axis indicates Kosambi's distance) and no interference models (x-axis indicates Haldane's distance).

Additional file 5: Heterozygous and homozygous profiles for 2n gametes analysed by HTA using four markers flanking the position of the centromere in each LG. N: Number of triploid hybrids for each profile; M: Name of the molecular marker; HR%: Heterozygosity restitution value; HE: Heterozygous; HO: Homozygous.

Additional file 6: Distribution of the number of crossovers observed in 2n gametes on each arm of each LGs.

Additional file 7: Analysis of crossover interference in clementine. Heterozygous (HE, ab) and homozygous (HO, aa and bb) profiles of 2n gametes analysed using 11 markers. The location of the centromere on LG 3 was estimated using HTA. Numbers in grey are the number of 2n gametes with different crossover events on one arm and between arms (0, 1, 2, 3 and 4). Numbers in bold correspond to percentages of crossovers observed in each chromosome arm. Potential interference between arms was tested for each chromosome using χ^2-analysis based on the contingency table of the number of crossovers (grey numbers) in each chromosome arm.

Additional file 8: Variation in recombination rate along the different clementine chromosomes (Chr). The x-axis shows the physical position in megabases along each chromosome and the y-axis represents the ratio of genetic distance to physical distance (cM/Mb). The bars beneath the x-axis indicate the approximate locations of the centromeres (CI). * These data have been calculated from up and down intervals (no marker on the genetic map on the considered genomic segment).

Additional file 9: Relationship between the physical location (x-axis), the genetic distance between centromere and markers on the clementine genetic map (black dots) and the proportion of gene sequences (blue bars) along all chromosomes. Bars beneath the x-axis indicate the approximate locations of the centromeres location (CI).

Abbreviations

Chr: Chromosome; CCO: Complementary crossovers; FDR: First-division restitution; HTA: Half-tetrad analysis; HR: Heterozygosity restitution; InDel: Insertion-deletion; IC: Interference coefficient; LG: Linkage group; LOD: Logarithm of the odds ratio; MCO: Multiple crossovers; SDR: Second-division restitution; SSR: Simple sequence repeat; SNP: Single nucleotide polymorphism.

Competing interests

The authors declare that they have no competing interests.

Authors' contributions

PO and LN conceived and designed the experiments. JJ and PA performed the experiments and provided plant material. MH and PA analysed the data. JC and PO provided a statistical method for centromere mapping. PA and PO wrote the manuscript. All authors read and approved the final manuscript.

Acknowledgements

This work was supported by a grant [AGL2011-26490] from the Ministry of 'Economía y Competividad' – Fondo Europeo de Desarrollo Regional (FEDER) and a grant [Prometeo 2008/121] from the Generalitat Valenciana, Spain.

References

1. Hall AE, Keith KC, Hall SE, Copenhaver GP, Preuss D. The rapidly evolving field of plant centromeres. Curr Opin Plant Biol. 2004;7:108–14.
2. Copenhaver GP, Nickel K, Kuromori T, Benito MI, Kaul S, Lin X, et al. Genetic definition and sequence analysis of Arabidopsis centromeres. Science. 1999;286(5449):2468–74.
3. Ma J, Wing RA, Bennetzen JL, Jackson SA. Plant Centromere organization: a dynamic structure with conserved functions. Trends Genet. 2007;23(3):134–9.
4. Jiang J, Birchler JA, Parrot WA, Dawe RK. A molecular view of plant centromeres. Trends Plant Sci. 2003;8(12):570–5.
5. Houben A, Schubert I. DNA and proteins of plant centromeres. Curr Opin Plant Biol. 2003;6:554–60.
6. Nie H, Li Q, Zhao X, Kong L. Genetic positioning of centromeres through half-tetrad analysis in gynogenetic diploid families of the Zhikong Scallop (*Chlamysfarreri*). Mar Biotechnol. 2013;15:1–15.
7. Singh K, Ishii T, Parco A, Huang N, Brar DS, Khush GS. Centromere mapping and orientation of the molecular linkage map of rice (Oryzasativa L). PNAS. 1996;93:6163–8.
8. Guyomard R, Mauger S, Tabet-Canale K, Martineau S, Genet C, Krieg F, et al. A type I and type II microsatellite linkage map of rainbow trout (Oncorhynchusmykiss) with presumptive coverage of all chromosome arms. BMC Genomics. 2006;7:302.
9. Park TH, Kim JB, Hutten RCB, van Eck HJ, Jacobsen E, Visser RGE. Genetic positioning of centromeres using half-tetrad analysis in a 4x-2x cross population of potato. Genetics. 2007;176:85–94.
10. Copenhaver GP, Keith KC, Preuss D. Tetrad analysis in higher plants. a budding technology. Plant Physiol. 2000;124:7–15.
11. Harlan JR, De Wet JMJ. On O. Winge and a prayer: the origins of polyploidy. Botan Rev. 1975;41:361–90.
12. Bretagnolle F, Thompson JD. Gametes with the somatic chromosome number: mechanisms of their formation and role in the evolution of autopolyploid plants. New Phytol. 1995;129:1–22.
13. Ramsey J, Schemske DW. Pathways, mechanisms, and rates of polyploidy formation in flowering plants. Annu Rev Ecol Syst. 1998;29:467–501.
14. Ramsey J, Schemske DW. Neopolyploidy in flowering plants. Annu Rev Ecol Syst. 2002;33:589–639.
15. Tavoletti S, Bingham ET, Yandell BS, Veronesi F, Osborn TC. Half tetrad analysis in alfalfa using multiple restriction fragment length polymorphism markers. Proc Natl Acad Sci U S A. 1996;93:10918–22.
16. Rhoades MM, Dempsey E. Induction of chromosome doubling at meiosis by the elongate gene in maize. Genetics. 1966;54(2):505–22.
17. Mendiburu AO, Peloquin SJ. Gene-centromere mapping by 4x-2x matings in potatoes. Theor Appl Genet. 1979;54:177–80.
18. Veilleux R. Diploid and polyploid gametes in crop plants: mechanisms of formation and utilization in plant breeding. Plant Breed Rev. 1985;3:253–88.
19. Carputo D, Barone A, Frusciante L. 2n gemetes in the potato: essential ingredients for breeding and germplasm transfer. Theor Appl Genet. 2000;101:805–13.
20. Ramanna MS, Jacobsen E. Relevance of sexual polyploidization for crop improvement—a review. Euphytica. 2003;133:3–18.
21. Esen A, Soost RK. Unexpected triploids in citrus: their origin, identification and possible use. J Hered. 1971;62:329–33.
22. Geraci G, De Pasquale F, Tusa N. Percentages of spontaneous triploids in progenies of diploid lemons and mandarins. In: Grierson W, editor. Proceedings of Second International Citrus Congress.International Society of Citriculture, vol. 2. 1977. p. 596–7.
23. Bastiaanssen HJM, Van Den Berg PMMM, Lindhout P, Jacobsen E, Ramanna MS. Postmeiotic restitution in 2n-egg formation of diploid potato. Heredity. 1998;81:20–7.
24. Cuenca J, Froelicher Y, Aleza P, Juarez J, Navarro L, Ollitrault P. Multi*locus* half-tetrad analysis and centromere mapping in citrus: evidence of SDR mechanism for 2n megagametophyte production and partial chiasma interference in mandarin cv 'Fortune'. Heredity. 2011;107:462–70.
25. Douches DS, Quiros CF. Genetic strategies to determine the mode of 2n egg formation in diploid potatoes. Euphytica. 1988;38(3):247–60.
26. Luro F, Maddy F, Ollitrault P, Rist D. Identification of 2n gamete parental origin and mode of nuclear restitution of spontaneous triploid citrus hybrids. In: Proceedings of 9th International Citrus Congress. Orlando: International Society of Citriculture; 2000. p. 168–9. 3–7 December.
27. Crespel L, Gudin S. Evidence for the production of unreduced gametes by tetraploid Rosa hybrida L. Euphytica. 2003;133:65–9.
28. Kauffman EJ, Gestl EE, Kim DJ, Walker C, Hite JM, Yan G, et al. Microsatellite–centromere mapping in the Zebrafish (Daniorerio). Genomics. 1995;30:337–41.
29. Bastiaanssen HJM, Ramanna MS, Sawor Z, Mincione A, Steen A, Jacobsen E. Pollen markers for gene–centromere mapping in diploid potato. Theor Appl Genet. 1996;93:1040–7.
30. FAOSTAT [http://faostat.fao.org/site/567/DesktopDefault.aspx?PageID=567#ancor]
31. Ollitrault P, Terol J, Chen C, Federici CT, Lotfy S, Hippolyte I, et al. A reference genetic map of C. clementina hort. ex Tan.; citrus evolution inferences from comparative mapping. BMC Genomics. 2012;13:593.
32. Wu AG, Prochnik S, Jenkins J, Salse J, Hellsten U, Murat F, et al. Sequencing of diverse mandarin, pummelo and orange genomes reveals complex history of admixture during citrus domestication. Nat Biotechnol. 2014;32:656–62.
33. Aleza P, Juárez J, Hernández M, Pina JA, Ollitrault P, Navarro L. Recovery and characterization of a Citrus clementina Hort ex Tan Clemenules haploid plant selected to establish the reference whole Citrus genome sequence. BMC Plant Biol. 2009;9:110.
34. Ollitrault P, Dambier D, Luro F, Froelicher Y. Ploidy manipulation for breeding seedless triploid citrus. Plant Breed Rev. 2008;30:323–52.
35. Aleza P, Juárez J, Cuenca J, Ollitrault P, Navarro L. Recovery of citrus triploid hybrids by embryo rescue and flow cytometry from 2x X 2x sexual hybridisation and its application to extensive breeding programs. Plant Cell Rep. 2010;29:1023–34.
36. Chen C, Lyon MT, O'Malley D, Federici CT, Gmitter J, Grosser JW, et al. Origin and frequency of 2n gametes in Citrus sinensis x Poncirus trifoliata and their reciprocal crosses. Plant Sci. 2008;174:1–8.
37. Ollitrault P, Terol J, Garcia-Lor A, Bérard A, Chauveau A, Froelicher Y, et al. SNP mining in C. clementina BAC end sequences; transferability in the Citrus genus (Rutaceae), phylogenetic inferences and perspectives for genetic mapping. BMC Genomics. 2012;10:13–3.
38. Esselink GD, Nybom H, Vosman B. Assignment of allelic configuration in polyploids using the MAC-PR (microsatellite DNA allele counting-peak ratios) method. Theor Appl Genet. 2004;109:402–8.
39. Cuppen E. Genotyping by allele-specific amplification (KASPar). Cold Spring Harb Protoc 2007; 172–173. doi: 10.1101/pdb.prot4841.
40. Cuenca J, Aleza P, Vicent A, Brunel D, Ollitrault P, Navarro L. Genetically based location from triploid populations and gene ontology of a 3.3-Mb genome region linked to alternaria brown spot resistance in citrus reveal clusters of resistance genes. PLoS One. 2013;8(10):e76755.
41. Zhao H, Speed TP. Statistical analysis of ordered tetrads. Genetics. 1998;150:459–72.
42. Griffiths A, Miller J, Suzuki D, Lewontin R, Gelbart W. An Introduction to Genetic Analysis. 6th ed. New York, USA: W. H. Freeman and Company; 1996.
43. Zhao H, Speed TP. Statistical analysis of half-tetrads. Genetics. 1998;150:473–85.
44. Lander ES, Botstein D. Mapping Mendelian factors underlying quantitative traits using RFLP linkage maps. Genetics. 1989;121:185–99.

45. Altschul SF, Gish W, Miller W, Myers EW, Lipman DJ. Basic local alignment search tool. J Mol Biol. 1990;3:403–10.
46. Esen A, Soost RK. Precocious development and germination of spontaneous triploid seeds in citrus. J Hered. 1973;64:147–54.
47. Geraci G, Esen A, Soost RK. Triploid progenies from 2x–2x crosses of citrus cultivars. J Hered. 1975;66:177–8.
48. Esen A, Soost RK, Geraci G. Genetic evidence for the origin of diploid megagametophytes in Citrus. J Hered. 1979;70:5–8.
49. Wakana A, Iwamasa M, Uemoto S. Seed development in relation to ploidy of zygotic embryo and endosperm in polyembryonic Citrus. Proc Int Soc Citricult. 1981;1:135–9.
50. Soost R. Breeding citrus-genetics and nucellarembryony. In: Abbott AJ, Atkin R, editors. Improving Vegetatively Propagated Crops. London: Academic Press; 1987. p. 83–110.
51. Ferrante SP, Lucretti S, Reale S, De Patrizio A, Abbate L, Tusa N, et al. Assessment of the origin of new citrus tetraploid hybrids (2n = 4x) by means of SSR markers and PCR based dosage effects. Euphytica. 2010;173:223–33.
52. Errico A, Cammareri M, Conicella C. Meiotic nuclear restitution mechanisms in a triploid lily. Caryologia. 2005;58(4):367–73.
53. Brownfield L, Kohler C. Unreduced gamete formation in plants: mechanisms and prospects. J Exp Bot. 2011;62:1659–68.
54. Barone A, Gebhardt C, Frusciante L. Heterozygosity in 2n gametes of potato evaluated by RFLP markers. Theor Appl Genet. 1995;91(1):98–104.
55. Vorsa N, Rowland LJ. Estimation of 2n megagametophyteheterozygosity in a diploid blueberry (Vacciniumdarrowi Camp) clone using RAPDs. J Hered. 1997;88(5):423–6.
56. Dewitte A, Van Laere K, Van Huylenbroeck J. Use of 2n fametes in plant breeding. In: Ibrokhim Y, editor. Plant Breeding. Abdurakhmonov: InTech; 2011. p. 59–88.
57. Douches DS, Maas DL. Comparison of FDR and SDR derived tetraploid progeny from 2x x 4x crosses using haploids of Solanumtuberosum L. that produce mixed modes of 2n eggs. Theor Appl Genet. 1998;97:1307–13.
58. Mok DWS, Peloquin SJ. Three mechanisms of 2n pollen formation in diploid potatoes. Can J Genet Cytol. 1975;17:217–25.
59. Kidane-Mariam HM, Arndt GC, Macaso-Khwaja AC, Peloquin SJ. Comparisons between 4x × 2x hybrid and open-pollinated true potato seed families. Potato Res. 1985;28:35–42.
60. David JL, Boudec P, Gallais A. Quantitative genetics of 4x-2x hybrid populations with First-Division Restitution and 2nd-Division Restitution 2n gametes produced by diploid parents. Genetics. 1995;139(4):1797–803.
61. Buso JA, Boiteux LS, Tai GCC, Peloquin SJ. Chromosome regions between centromeres and proximal crossovers are the physical sites of major effect loci for yield in potato: genetic analysis employing meiotic mutants. PNAS. 1999;96(4):1773–8.
62. Zhu C, Sun Y, Yu X, Tong J. Centromere localization for bighead carp (Aristichthysnobilis) through Half-Tetrad analysis in diploid gynogenetic families. PLoS One. 2013;8(12):e82950.
63. Zhao H, Speed TP, McPeek MS. Statistical analysis of crossover interference using the chi-square model. Genetics. 1995;139:1045–56.
64. Zhao H, Speed TP. On genetic map functions. Genetics. 1996;142:1369–77.
65. Werner JE, Douches DS, Freyre R. Use of half-tetrad analysis to discriminate between 2 types of 2n egg formation in a potato haploid. Genome. 1992;35:741–5.
66. Teuscher F, Brockmann GA, Rudolph PE, Swalve HH, Guiard V. Models for chromatid interference with applications to recombination data. Genetics. 2000;156:1449–60.
67. Giraut L, Falque M, Drouaud J, Pereira L, Martin OC, Mézard C. Genome-Wide crossover distribution in arabidopsis thaliana meiosis reveals sex-specific patterns along chromosomes. Plos Genetics 2011, doi:10.1371/journal.pgen.1002354.
68. Drouaud J, Mercier R, Chelysheva L, Berard A, Falque M, Martin O, et al. Sex-specific crossover distributions and variations in interference level along Arabidopsis thaliana chromosome 4. PLoS Genet. 2007;3:e106.
69. Lian J, Yin Y, Oliver-Bonet M, Liehr T, Ko E, Turek P, et al. Variation in crossover interference levels on individual chromosomes from human males. Hum Mol Genet. 2008;17(17):2583–94.
70. Broman KW, Rowe LB, Churchill GA, Paigen K. Crossover interference in the mouse. Genetics. 2002;160:1123–31.
71. Levan A, Fredga K, Sandberg AA. Nomenclature for centromeric position on chromosomes. Hereditas. 1964;52:201–20.
72. Dobzhansky T. Translocations involving the third and fourth chromosomes of Drosophila melanogaster. Genetics. 1930;15:347–99.
73. Sherman JD, Stack SM. Two-dimensional spreads of synaptonemal complexes from solanaceous plants VI high-resolution recombination nodule map for tomato (Lycopersiconesculentum). Genetics. 1995;141:683–708.
74. Gill SK, Gill BS, Endo TR, Taylor T. Identification and high-density mapping of gene-rich regions in chromosome group I of wheat. Genetics. 1996;144:1883–91.
75. Haupt W, Fischer TC, Winderl S, Fransz P, Torres-Ruiz RA. The centromere1 (CEN1) region of Arabidopsis thaliana: architecture and functional impact of chromatin. Plant J. 2001;27:285–96.
76. Chen M, Prestin G, Barbazuk WB, Goicoechea JL, Blackmon B, Fang G, et al. An integrated physical and genetic map of the rice genome. Plant Cell. 2002;14(3):537–45.
77. Anderson LK, Doyle GG, Brigham B, Carter J, Hooker KD, Lai A, et al. High resolution crossover maps for each bivalent of Zea mays using recombination nodules. Genetics. 2003;165:849–65.
78. Talbert PB, Henikoff S. Centromeres convert but don't cross. PLoS Biol. 2010;8:e1000326. doi:1000310.1001371/journal.pbio.1000326.
79. Wang G, Zhang X, Jin W. An overview of plant centromeres. J Genet Genomics. 2009;36(9):529–37. doi: 10.1016/S1673-8527(08)60144-7.
80. Sandhu D, Gill K. Gene-containing regions of wheat and the other grass genomes. Plant Physiol. 2002;128:803–11.
81. Erayman M, Sandhu D, Sidhu D, Dilbirligi M, Baenziger PS, Gill KS. Demarcating the gene-rich regions of the wheat genome. Nucleic Acids Res. 2004;32(12):3546–65.
82. Tian Z, Rizzon C, Du J, Zhu L, Bennetzen JL, Jackson SA, et al. Do genetic recombination and gene density shape the pattern of DNA elimination in rice long terminal repeat retrotransposons? Genome Res. 2009;19(12):2221–30.
83. Liu Z, Yue W, Li D, Wang R, Kong X, Lu K, et al. Structure and dynamics of retrotransposons at wheat centromeres and pericentromeres. Chromosoma. 2008;117:445–56.

Permissions

All chapters in this book were first published in BMCPB, by BioMed Central; hereby published with permission under the Creative Commons Attribution License or equivalent. Every chapter published in this book has been scrutinized by our experts. Their significance has been extensively debated. The topics covered herein carry significant findings which will fuel the growth of the discipline. They may even be implemented as practical applications or may be referred to as a beginning point for another development.

The contributors of this book come from diverse backgrounds, making this book a truly international effort. This book will bring forth new frontiers with its revolutionizing research information and detailed analysis of the nascent developments around the world.

We would like to thank all the contributing authors for lending their expertise to make the book truly unique. They have played a crucial role in the development of this book. Without their invaluable contributions this book wouldn't have been possible. They have made vital efforts to compile up to date information on the varied aspects of this subject to make this book a valuable addition to the collection of many professionals and students.

This book was conceptualized with the vision of imparting up-to-date information and advanced data in this field. To ensure the same, a matchless editorial board was set up. Every individual on the board went through rigorous rounds of assessment to prove their worth. After which they invested a large part of their time researching and compiling the most relevant data for our readers.

The editorial board has been involved in producing this book since its inception. They have spent rigorous hours researching and exploring the diverse topics which have resulted in the successful publishing of this book. They have passed on their knowledge of decades through this book. To expedite this challenging task, the publisher supported the team at every step. A small team of assistant editors was also appointed to further simplify the editing procedure and attain best results for the readers.

Apart from the editorial board, the designing team has also invested a significant amount of their time in understanding the subject and creating the most relevant covers. They scrutinized every image to scout for the most suitable representation of the subject and create an appropriate cover for the book.

The publishing team has been an ardent support to the editorial, designing and production team. Their endless efforts to recruit the best for this project, has resulted in the accomplishment of this book. They are a veteran in the field of academics and their pool of knowledge is as vast as their experience in printing. Their expertise and guidance has proved useful at every step. Their uncompromising quality standards have made this book an exceptional effort. Their encouragement from time to time has been an inspiration for everyone.

The publisher and the editorial board hope that this book will prove to be a valuable piece of knowledge for researchers, students, practitioners and scholars across the globe.

List of Contributors

Xin Chen
The Institute of Tropical Bioscience and Biotechnology (ITBB), Chinese Academy of Tropical Agricultural Sciences (CATAS), Haikou 571101, PR China
Key Laboratory of Biology and Genetic Resources of Tropical Crops, Ministry of Agriculture, Haikou 571101, PR China

Jing Xia
Institute for Systems Biology, Jianghan University, Wuhan 430056, China
Department of Computer Science and Engineering, Washington University in St. Louis, St. Louis, Missouri, MO 63130, USA

Zhiqiang Xia
The Institute of Tropical Bioscience and Biotechnology (ITBB), Chinese Academy of Tropical Agricultural Sciences (CATAS), Haikou 571101, PR China
Key Laboratory of Biology and Genetic Resources of Tropical Crops, Ministry of Agriculture, Haikou 571101, PR China

Hefang Zhang
The Institute of Tropical Bioscience and Biotechnology (ITBB), Chinese Academy of Tropical Agricultural Sciences (CATAS), Haikou 571101, PR China
Key Laboratory of Biology and Genetic Resources of Tropical Crops, Ministry of Agriculture, Haikou 571101, PR China

Changying Zeng
The Institute of Tropical Bioscience and Biotechnology (ITBB), Chinese Academy of Tropical Agricultural Sciences (CATAS), Haikou 571101, PR China
Key Laboratory of Biology and Genetic Resources of Tropical Crops, Ministry of Agriculture, Haikou 571101, PR China

Cheng Lu
The Institute of Tropical Bioscience and Biotechnology (ITBB), Chinese Academy of Tropical Agricultural Sciences (CATAS), Haikou 571101, PR China
Key Laboratory of Biology and Genetic Resources of Tropical Crops, Ministry of Agriculture, Haikou 571101, PR China

Weixiong Zhang
Institute for Systems Biology, Jianghan University, Wuhan 430056, China
Department of Computer Science and Engineering, Washington University in St. Louis, St. Louis, Missouri, MO 63130, USA

Wenquan Wang
The Institute of Tropical Bioscience and Biotechnology (ITBB), Chinese Academy of Tropical Agricultural Sciences (CATAS), Haikou 571101, PR China
Key Laboratory of Biology and Genetic Resources of Tropical Crops, Ministry of Agriculture, Haikou 571101, PR China

Luis Fernando Samayoa
Misión Biológica de Galicia, Spanish National Research Council (CSIC), P.O. Box 2836080 Pontevedra, Spain

Rosa Ana Malvar
Misión Biológica de Galicia, Spanish National Research Council (CSIC), P.O. Box 2836080 Pontevedra, Spain

Bode A Olukolu
Department of Crop Science, North Carolina State University, Raleigh, North Carolina 27695, USA

James B Holland
Department of Crop Science, North Carolina State University, Raleigh, North Carolina 27695, USA
U.S. Department of Agriculture-Agricultural Research Service, Plant Science Research Unit, Raleigh, North Carolina 27695, USA

Ana Butrón
Misión Biológica de Galicia, Spanish National Research Council (CSIC), P.O. Box 2836080 Pontevedra, Spain

Nasheeman Ashraf
Plant Biotechnology Division, CSIR- Indian Institute of Integrative Medicine, Sanat Nagar, Srinagar J&K-190005, India

Deepti Jain
National Institute of Plant Genome Research, Aruna Asaf Ali Marg, New Delhi 110067, India

Ram A Vishwakarma
Medicinal Chemisrty Division, CSIR- Indian Institute of Integrative Medicine, Canal Road, Jammu J&K 180001, India

Ai-Hua Wang
Key Laboratory of Plant Resources Conservation and Sustainable Utilization, South China Botanical Garden, Chinese Academy of Sciences, Guangzhou 510650, China
University of Chinese Academy of Sciences, Beijing 100049, China

Ye Sun
Key Laboratory of Plant Resources Conservation and Sustainable Utilization, South China Botanical Garden, Chinese Academy of Sciences, Guangzhou 510650, China

Harald Schneider
Department of Life Sciences, Natural History Museum, London SW75BD, UK

Jun-Wen Zhai
College of Landscape Architecture, Fujian Agriculture and Forestry University, Fuzhou 350002, China

Dong-Ming Liu
Key Laboratory of Plant Resources Conservation and Sustainable Utilization, South China Botanical Garden, Chinese Academy of Sciences, Guangzhou 510650, China

Jin-Song Zhou
College of Chinese Traditional Medicine, Guangzhou University of Chinese Medicine, Guangzhou 510006, China

Fu-Wu Xing
Key Laboratory of Plant Resources Conservation and Sustainable Utilization, South China Botanical Garden, Chinese Academy of Sciences, Guangzhou 510650, China

Hong-Feng Chen
Key Laboratory of Plant Resources Conservation and Sustainable Utilization, South China Botanical Garden, Chinese Academy of Sciences, Guangzhou 510650, China

Fa-Guo Wang
Key Laboratory of Plant Resources Conservation and Sustainable Utilization, South China Botanical Garden, Chinese Academy of Sciences, Guangzhou 510650, China

Vicent Arbona
Laboratori d'Ecofisiologia i Biotecnologia, Departament de Ciències Agràries i del Medi Natural, Universitat Jaume I, E-12071 Castelló de la Plana, Spain

Domingo J Iglesias
Institut Valencià d'Investigacions Agràries (IVIA), Moncada, Spain

Aurelio Gómez-Cadenas
Laboratori d'Ecofisiologia i Biotecnologia, Departament de Ciències Agràries i del Medi Natural, Universitat Jaume I, E-12071 Castelló de la Plana, Spain

Rugang Yu
National Key Laboratory of Crop Genetics and Germplasm Enhancement; Engineering Research Center of Horticultural Crop Germplasm Enhancement and Utilization, Ministry of Education of P.R.China; College of Horticulture, Nanjing Agricultural University, Nanjing 210095, P.R. China
School of Life Sciences, Huaibei Normal University, Huaibei, Anhui 235000, P.R. China

Yan Wang
National Key Laboratory of Crop Genetics and Germplasm Enhancement; Engineering Research Center of Horticultural Crop Germplasm Enhancement and Utilization, Ministry of Education of P.R.China; College of Horticulture, Nanjing Agricultural University, Nanjing 210095, P.R. China

Liang Xu
National Key Laboratory of Crop Genetics and Germplasm Enhancement; Engineering Research Center of Horticultural Crop Germplasm Enhancement and Utilization, Ministry of Education of P.R.China; College of Horticulture, Nanjing Agricultural University, Nanjing 210095, P.R. China

Xianwen Zhu
Department of Plant Sciences, North Dakota State University, Fargo, ND 58108, USA

Wei Zhang
National Key Laboratory of Crop Genetics and Germplasm Enhancement; Engineering Research Center of Horticultural Crop Germplasm Enhancement and Utilization, Ministry of Education of P.R.China; College of Horticulture, Nanjing Agricultural University, Nanjing 210095, P.R. China

Ronghua Wang
National Key Laboratory of Crop Genetics and Germplasm Enhancement; Engineering Research Center of Horticultural Crop Germplasm Enhancement and Utilization, Ministry of Education of P.R.China; College of Horticulture, Nanjing Agricultural University, Nanjing 210095, P.R. China

Yiqin Gong
National Key Laboratory of Crop Genetics and Germplasm Enhancement; Engineering Research Center of Horticultural Crop Germplasm Enhancement and Utilization, Ministry of Education of P.R.China; College of Horticulture, Nanjing Agricultural University, Nanjing 210095, P.R. China

Cecilia Limera
National Key Laboratory of Crop Genetics and Germplasm Enhancement; Engineering Research Center of Horticultural Crop Germplasm Enhancement and Utilization, Ministry of Education of P.R.China; College of Horticulture, Nanjing Agricultural University, Nanjing 210095, P.R. China

Liwang Liu
National Key Laboratory of Crop Genetics and Germplasm Enhancement; Engineering Research Center of Horticultural Crop Germplasm Enhancement and Utilization, Ministry of Education of P.R.China; College of Horticulture, Nanjing Agricultural University, Nanjing 210095, P.R. China

Wenzhu Yang
Department of Plant Molecular Systems Biotechnology and Crop Biotech Institute, Kyung Hee University, Yongin 446-701, Korea
Department of Crop Genomics and Genetic Improvement, Biotechnology Research Institute, Chinese Academy of Agricultural Sciences, Beijing 100081, China
National Key Facility for Crop Gene Resources and Genetic Improvement (NFCRI), Biotechnology Research Institute, Chinese Academy of Agricultural Sciences, Beijing 100081, China

Jinmi Yoon
Department of Plant Molecular Systems Biotechnology and Crop Biotech Institute, Kyung Hee University, Yongin 446-701, Korea

Heebak Choi
Department of Plant Molecular Systems Biotechnology and Crop Biotech Institute, Kyung Hee University, Yongin 446-701, Korea

Yunliu Fan
Department of Crop Genomics and Genetic Improvement, Biotechnology Research Institute, Chinese Academy of Agricultural Sciences, Beijing 100081, China
National Key Facility for Crop Gene Resources and Genetic Improvement (NFCRI), Biotechnology Research Institute, Chinese Academy of Agricultural Sciences, Beijing 100081, China

Rumei Chen
Department of Crop Genomics and Genetic Improvement, Biotechnology Research Institute, Chinese Academy of Agricultural Sciences, Beijing 100081, China
National Key Facility for Crop Gene Resources and Genetic Improvement (NFCRI), Biotechnology Research Institute, Chinese Academy of Agricultural Sciences, Beijing 100081, China

Gynheung An
Department of Plant Molecular Systems Biotechnology and Crop Biotech Institute, Kyung Hee University, Yongin 446-701, Korea

Jonghoon Lee
Department of Plant Science, Plant Genomics and Breeding Institute, and Research Institute of Agriculture and Life Sciences, College of Agriculture and Life Sciences, Seoul National University, Seoul 151-921, Republic of Korea

Nur Kholilatul Izzah
Department of Plant Science, Plant Genomics and Breeding Institute, and Research Institute of Agriculture and Life Sciences, College of Agriculture and Life Sciences, Seoul National University, Seoul 151-921, Republic of Korea
Indonesian Research Institute for Industrial and Beverage Crops (IRIIBC), Pakuwon, Sukabumi, Indonesia

Murukarthick Jayakodi
Department of Plant Science, Plant Genomics and Breeding Institute, and Research Institute of Agriculture and Life Sciences, College of Agriculture and Life Sciences, Seoul National University, Seoul 151-921, Republic of Korea

Sampath Perumal
Department of Plant Science, Plant Genomics and Breeding Institute, and Research Institute of Agriculture and Life Sciences, College of Agriculture and Life Sciences, Seoul National University, Seoul 151-921, Republic of Korea

Ho Jun Joh
Department of Plant Science, Plant Genomics and Breeding Institute, and Research Institute of Agriculture and Life Sciences, College of Agriculture and Life Sciences, Seoul National University, Seoul 151-921, Republic of Korea

Hyeon Ju Lee
Department of Plant Science, Plant Genomics and Breeding Institute, and Research Institute of Agriculture and Life Sciences, College of Agriculture and Life Sciences, Seoul National University, Seoul 151-921, Republic of Korea

Sang-Choon Lee
Department of Plant Science, Plant Genomics and Breeding Institute, and Research Institute of Agriculture and Life Sciences, College of Agriculture and Life Sciences, Seoul National University, Seoul 151-921, Republic of Korea

Jee Young Park
Department of Plant Science, Plant Genomics and Breeding Institute, and Research Institute of Agriculture and Life Sciences, College of Agriculture and Life Sciences, Seoul National University, Seoul 151-921, Republic of Korea

Ki-Woung Yang
Department of Plant Science, Plant Genomics and Breeding Institute, and Research Institute of Agriculture and Life Sciences, College of Agriculture and Life Sciences, Seoul National University, Seoul 151-921, Republic of Korea
Department of Horticulture, Sunchon National University, Suncheon 540-950, Republic of Korea

Il-Sup Nou
Department of Horticulture, Sunchon National University, Suncheon 540-950, Republic of Korea

Joodeok Seo
Joeun Seed, #174, Munbang-Ri, Cheonhan-Myun, 367-833 Goesan-Gu, Chungcheongbuk-Do, Korea

Jaeheung Yoo
Joeun Seed, #174, Munbang-Ri, Cheonhan-Myun, 367-833 Goesan-Gu, Chungcheongbuk-Do, Korea

Youngdeok Suh
Joeun Seed, #174, Munbang-Ri, Cheonhan-Myun, 367-833 Goesan-Gu, Chungcheongbuk-Do, Korea

Kyounggu Ahn
Joeun Seed, #174, Munbang-Ri, Cheonhan-Myun, 367-833 Goesan-Gu, Chungcheongbuk-Do, Korea

Ji Hyun Lee
Research Center for Biobased Chemistry, Korea Research Institute of Chemical Technology, Daejeon 305-600Yusong-Gu, Republic of Korea

Gyung Ja Choi
Research Center for Biobased Chemistry, Korea Research Institute of Chemical Technology, Daejeon 305-600Yusong-Gu, Republic of Korea

Yeisoo Yu
Arizona Genomics Institute, School of Plant Sciences, University of Arizona, Tucson, Arizona 85721, USA

Heebal Kim
Department of Agricultural Biotechnology, Seoul National University, Seoul 151-921, Republic of Korea
CHO & KIM genomics, Seoul National University Mt.4-2, Main Bldg. #514, SNU Research Park, NakSeoungDae, Seoul 151-919Gwanakgu, Republic of Korea

Tae-Jin Yang
Department of Plant Science, Plant Genomics and Breeding Institute, and Research Institute of Agriculture and Life Sciences, College of Agriculture and Life Sciences, Seoul National University, Seoul 151-921, Republic of Korea

Trees-Juen Chuang
Genomics Research Center, Academia Sinica, Taipei 11529, Taiwan

Min-Yu Yang
Genomics Research Center, Academia Sinica, Taipei 11529, Taiwan

Chuang-Chieh Lin
Genomics Research Center, Academia Sinica, Taipei 11529, Taiwan

Ping-Hung Hsieh
Genomics Research Center, Academia Sinica, Taipei 11529, Taiwan

Li-Yuan Hung
Genomics Research Center, Academia Sinica, Taipei 11529, Taiwan

Jie Chen
Beijing Key Laboratory of Grape Science and Enology, and CAS Key Laboratory of Plant Resources, Institute of Botany, Chinese Academy of Sciences, Beijing 100093, P. R. China
University of Chinese Academy of Sciences, Beijing 100049, P. R. China

Nian Wang
Key Laboratory of Plant Germplasm Enhancement and Speciality Agriculture, Wuhan Botanical Garden, Chinese Academy of Sciences, Wuhan 430074, China

Lin-Chuan Fang
Key Laboratory of Plant Germplasm Enhancement and Speciality Agriculture, Wuhan Botanical Garden, Chinese Academy of Sciences, Wuhan 430074, China

Zhen-Chang Liang
Beijing Key Laboratory of Grape Science and Enology, and CAS Key Laboratory of Plant Resources, Institute of Botany, Chinese Academy of Sciences, Beijing 100093, P. R. China

Shao-Hua Li
Beijing Key Laboratory of Grape Science and Enology, and CAS Key Laboratory of Plant Resources, Institute of Botany, Chinese Academy of Sciences, Beijing 100093, P. R. China

Ben-Hong Wu
Beijing Key Laboratory of Grape Science and Enology, and CAS Key Laboratory of Plant Resources, Institute of Botany, Chinese Academy of Sciences, Beijing 100093, P. R. China

Frederike Horn
Max Planck Institute for Plant Breeding Research, Carl-von-Linné Weg, 50829 Cologne, Germany

Antje Habekuß
Julius Kühn Institute, Erwin-Baur-Straße 27, 06484 Quedlinburg, Germany

Benjamin Stich
Max Planck Institute for Plant Breeding Research, Carl-von-Linné Weg, 50829 Cologne, Germany

Jinquan Chao
Ministry of Agriculture Key Laboratory of Biology and Genetic Resources of Rubber Tree/ State Key Laboratory Breeding Base of Cultivation and Physiology for Tropical Crops, Rubber Research Institute, Chinese Academy of Tropical Agricultural Sciences, Danzhou, Hainan 571737, PR China

Yueyi Chen
Ministry of Agriculture Key Laboratory of Biology and Genetic Resources of Rubber Tree/ State Key Laboratory Breeding Base of Cultivation and Physiology for Tropical Crops, Rubber Research Institute, Chinese Academy of Tropical Agricultural Sciences, Danzhou, Hainan 571737, PR China

Shaohua Wu
Ministry of Agriculture Key Laboratory of Biology and Genetic Resources of Rubber Tree/ State Key Laboratory Breeding Base of Cultivation and Physiology for Tropical Crops, Rubber Research Institute, Chinese Academy of Tropical Agricultural Sciences, Danzhou, Hainan 571737, PR China

Wei-Min Tian
Ministry of Agriculture Key Laboratory of Biology and Genetic Resources of Rubber Tree/ State Key Laboratory Breeding Base of Cultivation and Physiology for Tropical Crops, Rubber Research Institute, Chinese Academy of Tropical Agricultural Sciences, Danzhou, Hainan 571737, PR China

Andrey B Shcherban
Institute of Cytology and Genetics, Lavrentiev ave. 10, Novosibirsk 630090, Russia

Kseniya V Strygina
Institute of Cytology and Genetics, Lavrentiev ave. 10, Novosibirsk 630090, Russia

Elena A Salina
Institute of Cytology and Genetics, Lavrentiev ave. 10, Novosibirsk 630090, Russia

Jaana Vuosku
Genetics and Physiology Unit, University of Oulu, P.O. Box 3000, FI-90014 Oulu, Finland
Current address: Natural Resources Institute Finland (Luke), Rovaniemi Unit, FI-96301 Rovaniemi, Finland

Suvi Sutela
Genetics and Physiology Unit, University of Oulu, P.O. Box 3000, FI-90014 Oulu, Finland

Johanna Kestilä
Genetics and Physiology Unit, University of Oulu, P.O. Box 3000, FI-90014 Oulu, Finland

Anne Jokela
Genetics and Physiology Unit, University of Oulu, P.O. Box 3000, FI-90014 Oulu, Finland

Tytti Sarjala
Natural Resources Institute Finland (Luke), Parkano Unit, Kaironiementie 15, FI-39700 Parkano, Finland

Hely Häggman
Genetics and Physiology Unit, University of Oulu, P.O. Box 3000, FI-90014 Oulu, Finland

Xiaodong Wang
College of Life Science and Technology, Huazhong University of Science and Technology, Wuhan 430074, China
Key Laboratory of Cotton and Rapeseed, Ministry of Agriculture, Institute of Industrial Crops, Jiangsu Academy of Agricultural Sciences, Nanjing 210014, China

Yan Long
National Key Lab of Crop Genetic Improvement, Huazhong Agricultural University, Wuhan 430070, China
Institute of Biotechnology, Chinese Academy of Agricultural Sciences, Beijing 100081, China

Yongtai Yin
College of Life Science and Technology, Huazhong University of Science and Technology, Wuhan 430074, China

Chunyu Zhang
National Key Lab of Crop Genetic Improvement, Huazhong Agricultural University, Wuhan 430070, China

Lu Gan
College of Life Science and Technology, Huazhong University of Science and Technology, Wuhan 430074, China

Liezhao Liu
College of Agronomy and Biotechnology, Southwest University, Chongqing 400716, China

Longjiang Yu
College of Life Science and Technology, Huazhong University of Science and Technology, Wuhan 430074, China

Jinling Meng
National Key Lab of Crop Genetic Improvement, Huazhong Agricultural University, Wuhan 430070, China

Maoteng Li
College of Life Science and Technology, Huazhong University of Science and Technology, Wuhan 430074, China

Romina Petersen
Department of Molecular Genetics, Johannes Gutenberg-University, Mainz D-55128, Germany

Haris Djozgic
Department of Molecular Genetics, Johannes Gutenberg-University, Mainz D-55128, Germany

Benjamin Rieger
Department of Molecular Genetics, Johannes Gutenberg-University, Mainz D-55128, Germany

Steffen Rapp
Department of Molecular Genetics, Johannes Gutenberg-University, Mainz D-55128, Germany

Erwin Robert Schmidt
Department of Molecular Genetics, Johannes Gutenberg-University, Mainz D-55128, Germany

Xianjun Peng
Key Laboratory of Plant Resources, Institute of Botany, the Chinese Academy of Sciences, Beijing 100093, People's Republic of China

Qingqing Wu
Key Laboratory of Plant Resources, Institute of Botany, the Chinese Academy of Sciences, Beijing 100093, People's Republic of China
University of the Chinese Academy of Sciences, Beijing, People's Republic of China

Linhong Teng
Key Laboratory of Plant Resources, Institute of Botany, the Chinese Academy of Sciences, Beijing 100093, People's Republic of China

Feng Tang
Key Laboratory of Plant Resources, Institute of Botany, the Chinese Academy of Sciences, Beijing 100093, People's Republic of China

Zhi Pi
Key Laboratory of Plant Resources, Institute of Botany, the Chinese Academy of Sciences, Beijing 100093, People's Republic of China

Shihua Shen
Key Laboratory of Plant Resources, Institute of Botany, the Chinese Academy of Sciences, Beijing 100093, People's Republic of China

Dong Zhang
Plant Genome Mapping Laboratory, University of Georgia, Athens, GA 30602, USA
Institute of Bioinformatics, University of Georgia, Athens, GA 30602, USA

Wenqian Kong
Plant Genome Mapping Laboratory, University of Georgia, Athens, GA 30602, USA
Department of Crop and Soil Sciences, University of Georgia, Athens, GA 30602, USA

Jon Robertson
Plant Genome Mapping Laboratory, University of Georgia, Athens, GA 30602, USA

Valorie H Goff
Plant Genome Mapping Laboratory, University of Georgia, Athens, GA 30602, USA

Ethan Epps
Plant Genome Mapping Laboratory, University of Georgia, Athens, GA 30602, USA

Alexandra Kerr
Plant Genome Mapping Laboratory, University of Georgia, Athens, GA 30602, USA

Gabriel Mills
Plant Genome Mapping Laboratory, University of Georgia, Athens, GA 30602, USA

Jay Cromwell
Plant Genome Mapping Laboratory, University of Georgia, Athens, GA 30602, USA

Yelena Lugin
Plant Genome Mapping Laboratory, University of Georgia, Athens, GA 30602, USA

Christine Phillips
Plant Genome Mapping Laboratory, University of Georgia, Athens, GA 30602, USA

Andrew H Paterson
Plant Genome Mapping Laboratory, University of Georgia, Athens, GA 30602, USA
Institute of Bioinformatics, University of Georgia, Athens, GA 30602, USA
Department of Crop and Soil Sciences, University of Georgia, Athens, GA 30602, USA
Department of Plant Biology, University of Georgia, Athens, GA 30602, USA
Department of Genetics, University of Georgia, Athens, GA 30602, USA

Pablo Aleza
Centro de Protección Vegetal y Biotecnología, Instituto Valenciano de Investigaciones Agrarias (IVIA), Moncada, Valencia, Spain

José Cuenca
Centro de Protección Vegetal y Biotecnología, Instituto Valenciano de Investigaciones Agrarias (IVIA), Moncada, Valencia, Spain

María Hernández
Centro de Protección Vegetal y Biotecnología, Instituto Valenciano de Investigaciones Agrarias (IVIA), Moncada, Valencia, Spain

José Juárez
Centro de Protección Vegetal y Biotecnología, Instituto Valenciano de Investigaciones Agrarias (IVIA), Moncada, Valencia, Spain

Luis Navarro
Centro de Protección Vegetal y Biotecnología, Instituto Valenciano de Investigaciones Agrarias (IVIA), Moncada, Valencia, Spain

Patrick Ollitrault
Centro de Protección Vegetal y Biotecnología, Instituto Valenciano de Investigaciones Agrarias (IVIA), Moncada, Valencia, Spain
CIRAD, UMR AGAP, Avenue Agropolis - TA A-75/02 F-34398, Montpellier, France